Recent Progress in Slow Sand and Alternative Biofiltration Processes

Recent Progress in Slow Sand and Alternative Biofiltration Processes

Edited by

Prof. Dr.-Ing. habil. Rolf Gimbel
IWW Rheinisch-Westfälisches Institut für Wasser, and Universität Duisburg-Essen, Germany

Prof. Dr. Nigel J.D. Graham
Imperial College London, UK

and

Prof. Dr. M. Robin Collins
University of New Hampshire, USA

IWA
Publishing

Published by IWA Publishing, Alliance House, 12 Caxton Street, London SW1H 0QS, UK
Telephone: +44 (0) 20 7654 5500; Fax: +44 (0) 20 7654 5555; Email: publications@iwap.co.uk
Web: www.iwapublishing.com

First published 2006
Reprinted 2007

Printed by Lightning Source

British Library Cataloguing in Publication Data
A CIP catalogue record for this book is available from the British Library

Library of Congress Cataloging- in-Publication Data
A catalog record for this book is available from the Library of Congress

ISBN: 1843391201
ISBN13: 9781843391203

Contents

Preface

Slow sand filtration is typically cited as being the first "engineered" process in drinking water treatment. Proven modifications to the conventional slow sand filtration process, the awareness of induced biological activity in riverbank filtration systems, and the growth of oxidant-induced biological removals in more rapid-rate filters, e.g. biological activated carbon, demonstrate the renaissance of biofiltration as a treatment process that remains viable for both small, rural communities and major cities. Biofiltration is expected to become even more common in the future as efforts intensify to decrease the presence of disease-causing microorganisms and disinfection by-products in drinking water, to minimize microbial regrowth potential in distribution systems, and where operator skill levels are emphasized.

As a contribution to this growing interest in slow sand and alternative biofiltration systems, the editors with the assistance of others, have held three previous international conferences on this theme beginning in London (November 1988), New Hampshire (USA-October 1991), and London and Amsterdam (April 1996). A fourth conference, held in Mülheim, Germany (May 2006), aimed to build on the success and momentum of the previous meetings by providing an updated perspective on the physical, chemical, biological, and operational factors affecting the performance of slow sand filtration (SSF), riverbank filtration (RBF), soil-aquifer treatment (SAT), and biological activated carbon (BAC) processes. The main themes of the conference encompassed: comparable overviews of biofiltration systems; slow sand filtration process behavior, treatment performance and process developments; and alternative biofiltration process behaviors, treatment performances, and process developments.

Compiled from the contributors to the *4th International Slow Sand and Alternative Biological Conference*, this book provides a state-of-the-art assessment on a variety of biofiltration systems from studies conducted around the world. The authors collectively represent a perspective from 23 countries and include academicians, biofiltration system users, designers, and manufacturers.

The editors would like to give special recognition to the conference sponsors and supporters including the IWA (UK), Federal Ministry of Education and Research (Germany), IWW Water Centre (Germany), RWE Aqua (Germany), Thames Water Utilities (UK), RWW (Germany), Amsterdam Water Supply (The Netherlands), Zürich Water Supply (Switzerland), AWWA Research Foundation (USA), University of Duisburg/Essen (Germany), Imperial College London (UK), and the Water Treatment Technology Assistance Center at the University of New Hampshire (USA). The editors extend appreciation to the distinguished members of the Programme Advisory Panel (listed separately) for their assistance in the selection of the conference papers. Finally, the editors wish to gratefully acknowledge Stefan Panglisch, Vaso Partinoudi, and especially Hans-Joachim Mälzer for their administrative assistance. The conference was truly an international enterprise and such collaborative efforts are just one reason why biological filtration will continue its worldwide evolution.

R. Gimbel (University of Duisburg-Essen, Germany)
N.J.D. Graham (Imperial College London, UK)
M.R. Collins (University of New Hampshire, USA)
May 2006

Conference programme advisory panel

Mr René van der Aa: Amsterdam Water, The Netherlands

Dr Josefin Abrahamsson: Stockholm Water Company, Sweden

Prof. Dr Gary Amy: UNESCO-IHE Delft, The Netherlands

Prof. Dr Luiz Di Bernardo: University of Sao Paulo, Brazil

Prof. Dr Hans-Curt Flemming: IWW Water Centre, Germany

Prof. Dr Peter Huck: University of Waterloo, Canada

Prof. Dr Ma Jun: Harbin Institute of Technology, China

Prof. Dr Barry Lloyd: CEHE University of Surrey, UK

Dr Hans-Joachim Mälzer: IWW Water Centre, Germany

Prof. Dr Nobutada Nakamoto: Shinshu University, Japan

Prof. Dr Hallvard Ødegaard: Norwegian University of Science/Technology, Norway

Prof. Dr C. Ojha: Indian Institute of Technology/University of Roorkee, India

Mr Tony Rachwal: Thames Water Utilities/RWE, UK

Prof. Dr Scott Summers: University of Colorado, USA

Mrs Bénédicte Welte: Eau de Paris, France

Mrs Dipl.-Chem.-Ing. Ninette Zullei-Seibert: Westfälische Wasser- und Umweltanalytik GmbH, Germany

Part I

GENERAL OVERVIEW

1 Integrated comparison of biofiltration in engineered versus natural systems

Gary Amy*, Kenneth Carlson, M. Robin Collins***, Jörg Drewes****, Steffen Gruenheid*****, and Martin Jekel*******

*UNESCO-IHE Institute for Water Education, Delft, The Netherlands
**Colorado State University, Fort Collins, CO USA
***University of New Hampshire, Durham, NH USA
****Colorado School of Mines, Golden, CO USA, and
*****Technical University of Berlin, Berlin, Germany

Abstract An integrated comparison is made between four biofiltration processes/systems encompassing both engineered/mechanical processes; slow sand filtration (SSF) and biological (rapid) sand filtration (BSF); and natural systems; river bank filtration (RBF) and soil aquifer treatment (SAT) in terms of performance, reliability, and robustness. This comparative assessment takes into account differences in design/operations (particularly required operator intervention) as well as projected water quality performance. The robustness of each process/system is also considered from the perspective of its role as a single barrier within a multi-barrier system as well as its efficacy within the context of multi-objective (-contaminant) treatment, and its potential to serve as a complete stand-alone treatment system.

Keywords Biofiltration, slow sand filtration, bank filtration, soil aquifer treatment

Introduction

Filtration through porous geological media and biodegradation in soil systems are natural processes that influence water quality during natural groundwater recharge processes. Analogs of these natural processes have been used as drinking water and wastewater treatments processes under the classification of natural treatment systems. However, these same phenomena, granular media filtration and biodegradation, have also been exploited within engineered/mechanical unit processes or systems. The purpose of this paper is to compare these natural and engineered/mechanical systems in terms of their objectives, performances, and constraints.

The focus will be on river bank filtration (RBF) and soil aquifer treatment (SAT) as natural drinking water and (advanced) wastewater treatment systems; and on slow sand filtration (SSF) and biological rapid sand filration (BSF) as engineered/mechanical drinking water treatment systems. It is noteworthy that SAT practice in North America and Europe has been in the advanced treatment of secondary effluents with the objective of promoting

edited by Rolf Gimbel, Nigel J.D. Graham and M. Robin Collins. ISBN: 9781843391203. Published by IWA Publishing, London, UK.

indirect potable reuse and hence there is a linkage to RBF. Thus, with an indirect potable reuse perspective of SAT, all can be considered as drinking water treatment (DWT) processes/systems.

All of these processes/systems; SSF, BSF, RBF, and SAT; can be classified as physico-biological processes; all are biologically- and physically-driven. As physico-biological processes, they are biologically-driven in terms of organic transformations and/or mineralization and possible inorganic (e.g., nitrogen) transformations, and physically-driven in terms of filtration theory (straining/depth collection of turbidity (particle/colloids) and microorganisms). Only BSF can be considered as potentially chemically-driven in terms of possible chemical coagulant addition. A unifying perception of all of these processes is envisioning each as a *soil column*, given that each can be simulated by laboratory- or pilot-scale soil column experiments with appropriate scale-up considerations.

Engineered/mechanical (bio)filtration systems

Slow sand filtration (SSF)

Historically, slow sand filtration (SSF) was the first engineered/mechanical filtration process used in drinking water treatment. In SSF, untreated (surface) water slowly percolates through a bed of porous sand. Below the sand bed is a layer of gravel for support and an underdrain system that collects the filtered water. The influent raw water is introduced over the surface of the filter, and then drained from the bottom. While SSF involves a depth of granular media (sand), most contaminant removal occurs near the top/surface in association with a clogging layer known as a Schmutzdecke. Important aspects of a SSF include:

- Height of Supernatant Water: 0.7–2.0 meters
- Media Depth: 0.75–1.25 meters
- Schmutzdecke Thickness: $\leq$ 25 mm
- Hydraulic Loading (Filtration) Rate (HLR): 0.1–0.3 m/hour
- Empty Bed Contact Time (EBCT): 3–10 hours
- Filter Media: 0.15–0.45 mm (ES)
- Typical Operational Cycle (Schmutzdecke removal): > 30 days (30–300 + days)
- Possible Amendments: GAC *Sandwich* Layer

SSF is effective in the removal of turbidity and microorganisms, with more limited removal of natural organic matter (NOM), measured as dissolved organic carbon (DOC), and biodegradable organic matter (BOM), measured as biodegradable DOC (BDOC) or assimilable organic carbon (AOC). SSF performance is influenced by source water characteristics, temperature, and the extent of surface ripening. As a consequence of the biologically-active Schmutzdecke, the SSF can be considered as both a biological as well as physical process. Attributes of the SSF process are its low cost, operational simplicity, applicability to slow/decentralized systems, and the possibility of it functioning as a stand alone treatment system (except for the option of post disinfection). Limitations of the SSF process are high land requirements (constraining its large system/centralized system applicability), need for the occasional removal/restoration of the Schmutzdecke (a required operational intervention making SSF initially less effective until reestablishment of the Schmutzdecke), and the perception that SSF is not a very robust process because of its shallow media depth

Table 1 SSF Performance Summary

Water Quality Parameter	Removal Capacity
Turbidity	< 1 NTU
Coliform Bacteria	1 – 3 log units
Enteric Viruses	2 – 4 log units
Giardia Cysts	2 – 4+ log units
Cryptosporidium Oocysts	> 4 log units
Dissolved Organic Carbon (DOC)	< 15 – 30 %
Biodegradable DOC (BDOC)	< 80 %
Assimilable Organic Carbon (AOC)	< 65 %
THM Precursors	< 20 – 35 %
Iron/Manganese (Fe/Mn)	> 67 %

and (over)reliance on the Schmutzdecke for surface straining. Process performance can be improved through modifications such as a roughing filter or addition of a GAC sandwich layer; moreover, pilot studies have demonstrated the feasibility of preozonation. From a different perspective, it is noteworthy that there has been a renewed interest in implementation of SSF in developing countries as a tool for meeting the millennium development goals (MDGs). A performance summary of SSF appears in Table 1. As will be shown in the next section, DOC, BDOC, AOC removals are comparable to BSF as a consequence of the Schmutzdecke.

Biological (rapid) sand filtration (BSF)

The SSF process chronologically evolved into rapid sand filtration (RSF), a process that involves chemical pretreatment (coagulation/flocculation) and relies on depth filtration for removal of turbidity and microorganisms, with only limited bioactivity (abiotic if prechlorination is practiced). A well-designed and -operated RSF can achieve a turbidity approaching 0.1 NTU and is conservatively capable of at least 2-log removal of Giardia cysts and 1-log removal of (enteric) viruses (based on North American regulations). With the emergence of ozone as a primary disinfectant/oxidant, the conversion of NOM into BOM (i.e., DOC into biodegradable DOC (BDOC)), can transform the process of RSF into biological (rapid) sand filtration (BSF) or simply biofiltration where, besides turbidity and microorganisms, some DOC removal in the form of BDOC is realized.

Attributes of BSF include all of the features of RSF plus partial biostabilization of ozonated water. Among its limitations, BSF is not a stand-alone process but one that is coupled with an oxidation (ozonation) step as well as part of a process train that possibly includes coagulation/flocculation and sedimentation; it requires periodic backwashing (regular operator intervention); and it approaches but does not achieve true biostability in terms of assimilable organic carbon (AOC). (In the Netherlands, a biostable water is defined as a water with AOC $\leq$ 10 ug/L). Because it is preceded by ozonation, with oxygen as an ozone decomposition product, BSF is practiced in drinking water treatment as an aerobic process, although nitrate removal can be achieved under anoxic conditions in special groundwater treatment applications of BSF. A major process constraint is bromate (BrO_3^-) formation, a carcinogenic by-product formed as a consequence of ozone oxidation/transformation of bromide ion (Br^-). Important aspects of a SSF include:

- Media Depth: 1–2 meters
- Hydraulic Loading (Filtration) Rate (HLR): 5–25 m/hour
- Empty Bed Contact Time (EBCT): <5 to >10 minutes
- Typical Operational Cycle (Backwashing): 12–96 Hours

Design/operation considerations of SSF and BSF

The fundamental design/operational parameters describing SSF and RSF are media depth (D), hydraulic loading rate (HLR), and empty bed contact time (EBCT) with these three parameters related according to: HLR = D/EBCT. Both of these engineered/mechanical filtration systems require some operational intervention in the form of (less frequent) periodic removal of the Schmutzdecke for SSF, or (more frequent) periodic backwashing for BSF. Based on a performance comparison, SSF and BSF are generally comparable in terms of achievable water quality; the main differences being the much lower HLR and much higher EBCT for SSF attributable to the Schmutzdecke, the key equalizing factor in terms of generally comparable water quality performance. While SSF and BSF may not provide true biostability, they can significantly reduce chlorine demand, permitting maintenance of a chlorine residual with a lower chlorine dose.

Natural treatment systems

River bank filtration (RBF)

River bank filtration (RBF) and, to a lesser extent, lake bank filtration (LBF) have been practiced in Europe, particular Germany, for decades as a natural drinking water treatment process that is capable of removals of microorganisms, DOC, nitrogen (N), and trace organic compounds. In lakes and slow flowing (low turbulence) rivers, a clogging layer (Schmutzdecke) develops which is periodically removed during scouring. While straining and/or die-off (inactivation) of microorganisms occur during saturated flow of bank filtrate through an aquifer, biodegradation is the primary removal mechanism for organic contaminants. Thus, RBF/LBF is a *sustainable* treatment system that does not require operational intervention to restore its capacity. In RBF/LBF, important design/operational factors include travel distance and travel time (or hydraulic residence time (HRT)); these factors are analogs of D and EBCT for the SSF and BSF processes, respectively. EBCT is related to HRT through porosity (ϕ): HRT = EBCT/ϕ. In RBF/LBF, the pumping rate at the recovery well(s) together with the resultant hydraulic gradient establish a Darcy flow velocity, which is an analog of HLR for the SSF and BSF processes. The Darcy velocity or HLR are related to the interstitial velocity (v) through porosity (ϕ): v = HLR/ϕ.

Other similar natural systems include aquifer recharge and recovery (ARR), aquifer storage and recovery (ASR), and aquifer storage, treatment, and recovery (ASTR). ARR represents a variation on the bank filtration theme whereby surface water is first diverted into a recharge basin and then infiltrates into the groundwater for subsequent recovery by a pumping well(s). While ARR is generally an aerobic process because of oxygenation during infiltration cycles, RBF (or LBF) can transition to anoxic conditions if oxygen is depleted and nitrate is present, or even to Fe/Mn reduction; ultimately, oxic and/or anoxic microbial activities are limited by the available BDOC. In contrast to RBF, ARR involves both unsaturated and saturated zone passage. ASR is similar to ARR except that an injection well is used instead of an infiltration basin. If different wells are used for injection and recovery, separated by some travel distance/time, then the ASR technology becomes ASTR. Thus, rigorously, it is ASTR that represents an alternative form of RBF (or LBF). A major difference between all of these natural systems versus engineered/mechanical systems is that dilution (with native groundwater) plays a role in the former, with the influence represented by percent (%) bank filtrate or infiltrate or injectate (as defined by an appropriate conservative tracer). Another major difference is the wider range of oxidation-reduction potential (ORP) conditions possible; if Fe/Mn reducing conditions do evolve, there may be a need for post treatment (oxidation) of recovered bank filtrate. RBF (or LBF) has been successfully applied to sites with permeabilities (conductivities) ranging from 10^{-4} to 10^{-2} m/s, and has been reported to achieve removals of a wide range multiple contaminants, suggesting its potential role as a total treatment system targeting:

- Turbidity
- DOC
- Bacteria, Protozoa, and Viruses
- Organohalogens (AOX)
- Trace Organics Compounds (pesticides, pharmaceutically active compounds (PhACs), endocrine disrupting compounds (EDCs))
- Nitrogen (ammonia and nitrate)
- Algal Toxins

A summary of DOC and trace organic compound removals appears in Table 2. These results are derived from a field study of a lake bank filtration (LBF) site in Germany

Table 2 Biodegradability of Organic Contaminants based on Field Data (Lake Tegel, Berlin, Germany) and Corresponding Soil Column Data

Parameter or Compound	Field Sites: LBF (HRT $=\leq$ 4 mos); ASTR (HRT $=\leq$ 50 days)		Soil Column Experiments (HRT = 30 days, 15 °C)	
	Oxic (ASTR)	**Anoxic (LBF)**	**Oxic**	**Anoxic**
DOC	30 % (3 days) 40 % (30 days)	<25% (30 days) 35 % (4 months)	20 % (3 days) 45% (30 days)	20 % (3 days) 30 % (30 days)
Sulfamethoxazole	Partial, slow removal (50 days ⇒ 55%)	Partial, slow removal (4 mo ⇒ 75%)	95%*	65%*
1,5-Naphthalene-sulfonic acid	No removal, very stable under aerobic and anoxic conditions; 10 – 20 % removal by dilution		10%*	5%*
1,7-Naphthalene-sulfonic acid	60% removal (50 days)	55% removal (4 months)	98%*	55%*
2,7-Naphthalene-sulfonic acid	85% removal (50 days)	50% removal (4 months)	99%*	75%*

*10× Lake concentration

and associated soil column experiments. The field results demonstrate that significant DOC removals are achievable, corresponding to a field definitions of BDOC because of the long residence times, and that soil column experiments provide a good simulation of LBF and ASTR in terms of DOC. The trace organic results show that some compounds are biodegradable under oxic and/or anoxic conditions while some are persistent under both conditions; it is noteworthy that soil column simulations are poor although a difference in experimental conditions (initial concentration) is noted. Good microbial removals were also observed at the LBF site in Germany with bacteriophages reduced from 1 to 44 pfu/100 mL in the lake water to <1 pfu/100 mL (MDL) in all samples after 3 months of travel time.

Soil aquifer treatment (SAT)

Soil aquifer treatment (SAT) is essentially the same as the ASTR system described for RBF (or LBF) except that it is defined within the context of advanced wastewater treatment of secondary effluent. Treatment occurs in both the unsaturated zone and saturated zone, with storage in the saturated zone (aquifer). SAT recharge/infiltration basin operation requires alternating wet/dry cycles in order to restore permeability and maintain infiltration rates. Based on two Arizona field sites, the following are representative operational conditions of SAT recharge/infiltration:

- Wetting/Drying Cycle: 1-week/1-week
- Periodic Surface Disking (to restore permeability)
- Water Depth: 0.3–0.6 m
- Infiltration Rates: Initial: 30–60 cm/day; Final: 3–6 cm/day

Removal trends of DOC are presented in Table 3 for one Arizona field site. After a travel time of about two weeks, the DOC approaches that of the native groundwater. The specific UV absorbance (SUVA) at 254 nm increases with time/distance, generally suggesting the preferential removal of non-humic over humic NOM. At this particular site, treatment occurs almost exclusively through the unsaturated zone with initial oxic (aerobic) conditions succeeded by anoxic conditions.

Trace organic compound removal by SAT was investigated in soil column experiments, with results presented in Table 4. Chlorinated flame retardants persisted under both oxic (aerobic) and anoxic conditions; propyphenazone persisted under oxic conditions but was degraded under anoxic conditions; and the remainder of the compounds could be broken down under both oxic and anoxic condition but more effectively under oxic conditions.

Table 3 DOC Removal at Arizona Field Site

Sample	Travel Distance (m)	Travel Time (days)	Reclaimed Water (%)	DOC (mg/L)	SUVA (L/m-mg)
Recharge Basin	0	0	100	14.1	1.26
Shallow Monitoring Well	5	7	100	4.84	2.38
Deep Recovery Well	30–40	14	100	0.98	2.46
Native G.W.	—	—	0	0.50	—

Table 4 Soil Column Experiments with Spiked Trace Organic Compounds (derived from 1 m column studies assuming linear removal)

Compound	Aerobic (ng $L^{-1}d^{-1}$)	Anoxic (ng $L^{-1}d^{-1}$)
TCEP	0	0
TCIPP	0	0
Propyphenazone	0	32.4
Diclofenac	93–130	28.2
Naproxen	65.7	19.8
Ketoprofen	71.6	31.0
Ibuprofen	39.4	33.3
Gemfibrozil	33.0	21.2

Microbial removals by SAT were assessed through a bacteriophage tracer study at one of the Arizona field sites. The attenuation rate observed over the first 3 meters was 0.85 log for MS2 and 1.5 log for PRD1. Extrapolation of these results suggests a 7 log removal over about 30 meters. No positives were observed in the recovery well (two weeks of travel time; 30-40 meters of travel distance). There were detectable enteric viruses in the secondary effluent introduced into the pond. Corresponding soil column experiments indicated a ~1-log reduction of bacteriophage over 1 meter and up to a 2-log reduction of poliovirus.

Comparative assessment

One can compare the engineered/mechanical systems versus natural systems by contrasting their performance in terms of: (i) media depth (D) versus travel distance, (ii) EBCT versus travel (residence) time, and (iii) HLR versus Darcy (or interstitial) flow velocity.

A comparison of the four different processes/systems appears in the following tables. Table 5 contrasts design/operational parameters/conditions for BSF, SSF, RBF (or LBF),

Table 5 Design/Operational Parameters/Conditions

Parameter or Condition/Process	BSF	SSF	RBF (and LBF)	SAT
Depth/Travel Distance	$\leq$ 2 m	$\approx$ 1 m	$\geq$ 10 m	$\geq$ 30 m
Residence/Travel Time	$\leq$ 10 min ($\leq\approx$ 0.01 day)	$\geq$ 2 hours ($\geq\approx$ 0.1 day)	$\geq$ 1 day	$\geq$ 2 weeks
Hydraulic Loading/Infiltration Rate	$\geq$ 5 m/hour ($\geq$ 120 m/day)	$\leq$ 0.5 m/hour ($\leq$ 12 m/day)	$\leq$ 1 m/day	$\leq$ 1 m/day
Schmutzdecke/ Clogging Layer	no	yes	yes	yes
Flow Regime	saturated	saturated	unsaturated and saturated	unsaturated and saturated
Redox	oxic (aerobic)	oxic (aerobic)	oxic (aerobic) and anoxic	oxic (aerobic) and anoxic

Table 6 Qualitative Performance Comparison

Parameter/Process	BSF	SSF	RBF (or LBF)	SAT
Effective Turbidity Removal	✓✓	✓✓	✓✓	✓✓
Effective DOC Removal	✓	✓	✓✓	✓✓
Biostability	✓	✓	✓✓	✓✓
Effective Trace Organics Removal	×	×	✓✓	✓✓
Effective Nitrogen Removal	×	×	✓✓	✓✓
Effective Microbial Removal	✓	✓	✓✓	✓✓

and SAT. The natural systems clearly embody longer travel distances (depths), greater travel (residence) times, lower infiltration/hydraulic loading rates, variable (unsaturated and saturated) flow regimes, and variable redox (oxic and anoxic) conditions. It is noteworthy that all except BSF embody a Schmutzdecke (clogging layer), an attribute in terms of water quality but a constraint in terms of loading rate. Comparing BSF with SSF, the former is operationally more complex in terms of chemical (coagulant) addition and relatively frequent backwashing requirements. While the Schmutzdecke is influential, it is not a necessary component of RBF (or LBF) if there is a long enough travel time/distance; conversely, the Schmutzdecke is an essential component of SSF given its small media depth. In SAT, a Schmutzdecke is virtually unavoidable in terms of feed-water solids.

Table 6 provides a qualitative summary of the relative performance of the different processes/systems in terms of water quality. It is clear that RBF (or LBF) and SAT are more rigorous in terms of multi-contaminant objectives. In particular, under normal design and operation, BSF and SSF are generally not capable of removing trace organics or nitrogen, although some ammonia removal by SSF is possible. Moreover, BSF and SSF only approach biostability while RBF and SAT, with adequate travel times and distances, can achieve biostability (i.e., AOC < 10 ug/L). Because of greater travel times and distances, RBF and SAT can realize greater DOC and microbial removals.

Table 7 provides a more quantitative comparison of process/system performance. While the same trends shown in Table 6 are evident, these quantitative projections clearly show the robustness of the natural systems, RBF (or LBF) and SAT, as a barrier.

Table 7 Quantitative Performance Comparison

Parameter/Process	BSF	SSF	RBF (or LBF)	SAT
Turbidity (NTU)	$\leq$ 1NTU	$\leq$ 1NTU	$\leq$ 1NTU	$\leq$ 1NTU
DOC Removal	$\geq$ 15%	$\geq$ 15%	$\geq$ 50%	$\geq$ 50%
Biostability: BDOC Removal	$\geq$ 50%	$\geq$ 50%	< MDL	< MDL
Trace Organics Removal	—	—	$\geq$ 50%(except for few persistent PhACs)	$\geq$ 50%(except for few persistent PhACs)
(Total) Nitrogen Achievable	—	—	$\leq$ 2mg/L	$\leq$ 2mg/L
Microbial Removal (Viruses)	$\geq$ 2-log	$\geq$ 2-log	$\geq$ 4-log	$\geq$ 4-log

BSF, SSF, RBF, and SAT are all physico-biological processes. RBF (and SAT) are more robust than BSF (and SSF) in terms of embodying lower loading rates, greater travel (residence) times, and greater travel distances (depths). RBF (and SAT) embody features of BSF (depth filtration) as well as features of SSF (Schmutzdecke/clogging layer). RBF has relevance to small/decentralized systems as an alternative to SSF. RBF is an alternative to above ground BSF (or RSF) while SAT provides an opportunity for indirect potable reuse. RBF and SAT versus BSF and SSF constitute a more robust barrier, are more multi-objective (-contaminant), and embody more sustainability as a consequence of being more biologically- than physically-driven. Because RBF and SAT lower BDOC and AOC levels to < MDL, there is an opportunity to eliminate chemical disinfection; this is the case in terms of the LBF in Berlin, Germany as well as other German RBF systems. Other attributes of the natural systems include the range of ORP conditions for removal of trace organics, and the ability to manage dilution with native groundwater for water quality purpose. While there has been much emphasis on multi-barrier approaches to drinking water treatment, there should also be equal emphasis on the robustness of a given barrier. A biofiltration process/system with a very long/great travel time/distance can, in fact, be considered as a multi-barrier, conceptually with similar processes in series (e.g., 10 meters of travel distance in RBF versus 1 m of depth in BSF or SSF).

The results presented herein conceptually promote the concept of an engineered biofilter (BSF and SSF) as a soil column analog of RBF (or LBF) and SAT and, as such, demonstrate that soil columns can be used as pilot simulations of RBF (or LBF) and SAT. Such an approach provides a basis for defining appropriate design and operational parameters for implementation of these natural systems for drinking water and advanced wastewater treatment. Natural biofiltration systems, if properly designed and operated, are as robust, reliable, and high-performance as their engineered/mechanical system counterparts; there is a need for a change in perception to envision them as robust systems that can be engineered; i.e., *engineered* natural systems; for high performance and reliability. Thus, the water industry should consider a *paradigm shift* to further embrace natural, sustainable processes such as RBF & SAT.

Future directions

There has been a recent trend toward replacement of granular media filtration by (low pressure) membrane filtration in both drinking water and wastewater treatment applications. Low-pressure membranes have been coupled with biological wastewater treatment through the concept of membrane bioreactors (MBRs). There has been recent interest in extending MBRs into drinking water treatment applications, preceded by ozonation to create ample BDOC. Moreover, there has been recent work in implementing ozonation as a pretreatment step prior to RBF or SAT, and even as an intermediate step in an RBF/ARR sequence.

2 Removal of microorganisms by slow sand filtration

Y. J. Dullemont[1], J. F. Schijven[2], W.A.M. Hijnen[3], M. Colin[1], A. Magic-Knezev[4], W.A. Oorthuizen[5]

[1]Amsterdam Water Supply, Amsterdam, The Netherlands
[2]National Institute of Public Health and the Environment, Bitlhoven, The Netherlands
[3]Kiwa Research, Nieuwegein, The Netherlands
[4]Het Waterlaboratorium, Haarlem, The Netherlands
[5]Dune Water Company South Holland, Voorburg, The Netherlands.

Abstract Slow sand filtration experiments at pilot scale were conducted to determine removal of microorganisms. Using bacteriophage MS2 as a model virus for enteroviruses, it was found that viruses were the most critical microorganisms for the removal efficiency of slow sand filtration because they were removed the least (1.7–2.2 $\log_{10}$) and were affected the least by the presence of the Schmutzdecke at low temperatures ($< 10\,^{\circ}$C). Removal of *E. coli* was about 2 $\log_{10}$ higher, probably due to straining in the Schmutzdecke. Removal of viruses and bacteria by slow sand filtration did not depend on the seeding concentration. *E. coli* was found to be a useful model bacterium for campylobacters that were removed at least as well or about one $\log_{10}$ more. At higher temperatures (15–20 °C), removal of microorganisms by slow sand filtration, including viruses, increased by about two $\log_{10}$. It was shown that a Schmutzdecke that was scraped off was restored within 53 days. Removal of oocysts of *Cryptosporidium parvum* that were seeded for 100 days was highly efficient (5.3 $\log_{10}$). Only a few percent of the retained oocysts could be recovered, implying degradation and/or predation of the oocysts. Removal of clostridium spores and centric diatoms was much less efficient than of the oocysts. Furthermore, it was shown that clostridium spores persisted very well in the slow sand filter. Given this difference in removal and persistence, it was concluded that spores and diatoms are not useful model organisms for oocysts.

Keywords Slow sand filtration, Schmutzdecke, removal microorganisms, *Campylobacter lari*, *E. coli*, *Cryptosporidium* oocysts, MS2 bacteriophage, *Clostridium perfringens*, *Stephanodiscus hantzchii*.

Introduction

The Dutch Drinking Water Act (2001) requires that concentrations of (entero)viruses, oocysts of *Cryptosporidium* and cysts of *Giardia* in drinking water are so low that an annual infection risk of 10^{-4} per person is not exceeded. To demonstrate compliance, drinking water companies need to conduct a quantitative microbiological risk assessment. As part of this

 Recent Progress in Slow Sand and Alternative Biofiltration Processes edited by Rolf Gimbel, Nigel J.D. Graham and M. Robin Collins. ISBN: 9781843391203. Published by IWA Publishing, London, UK.

risk assessment, removal of these pathogens or appropriate indicator or model organisms by drinking water treatment needs to be known.

Slow sand filtration is a drinking water treatment step applied by the drinking water companies of Amsterdam and the Hague. Sofar, monitoring of the removal of faecal indicator bacteria and clostridium spores did not provide sufficient information on the removal of these pathogens.

The current study aimed to determine removal of viruses, bacterial spores, bacteria, oocysts of *Cryptosporidium* and centric diatoms by slow sand filtration. This encompasses a wide spectrum of microorganism sizes (26 nm – 7 µm). Bacteriophage MS2 was studied as a model virus for enteroviruses. *E. coli* was evaluated as model bacterium for campylobacter. Clostridium spores and small centric diatoms (*Stephanodiscus hantzchii*), were studied as model organisms for oocysts of *Cryptosporidium* because of their high persistence in the aqueous environment. Because of their persistence, oocysts, spores and diatoms may accumulate in a sand filter over time, which may confound the interpretation of their actual removal.

Special emphasis was given to the role of the so-called Schmutzdecke or filter cake on top of slow sand filters. Periodically, the Schmutzdecke is scraped off to prevent clogging of the filter, but this may adversely affect the performance of the slow sand filter. Therefore, the contribution of the Schmutzdecke to the removal of microorganisms and the time to restore after scraping were investigated at a temperature range from 8 to 19°C.

In addition, an effect of concentrations of microorganisms in the influent was studied because peak concentrations can be of overriding importance for risk analysis.

When passing a slow sand filter, microorganisms are removed from the water by inactivation (degradation and/or predation), physical straining and attachment to the sand grains. Attached microorganisms may also detach and then pass the sand filter. Breakthrough curves were constructed in order to provide insight into these removal processes. Modelling of these breakthrough curves will be conducted for quantifying these processes, but is not the scope of the current paper (Schijven *et al.*, in preparation).

Materials and methods

Microorganisms

- Bacteriophage MS2, as a conservative model virus. Due to its negative charge it attaches to sand as much as or less than other viruses.
- *Escherichia coli* WR1, as model bacterium. Due to its negative charge it attaches poorly like MS2.
- *Campylobacter lari*, as model for pathogenic Campylobacter.
- Spores of *Clostrium perfringens* D10 as model for *Cryptosporidium parvum*
- Centric diatoms *Stephanodiscus hantzschii* as model for *Cryptosporidium parvum*
- Oocysts of *Cryptosporidium parvum*

Microorganisms were enumerated as described in Hijnen *et al.* (2004), Hijnen and Schijven (2003) and Schijven (2000).

Table 1 Slow sand filtration experiments

Exp.	Filter	Start date of experiment	Age of Schmutzdecke (days)	Average grain size (mm)	Temp. (°C)	Microorganisms
A	F1	Sep 10, 2001	553	0.27	14	MS2
B	F1	Feb 04, 2002	12	0.27	10	MS2, WR1
C	F2	Feb 04, 2002	81	0.27	10	MS2, WR1
D	F2	Feb 18, 2002	4	0.27	10	MS2, WR1
E	F3	Jan 17, 2005	56	0.53	7̃.0	MS2, WR1, *C. lari*
F	F1	Jan 31, 2005	1105	0.27	9̃.9	WR1, *C. lari*
G	F4	Mar 21, 2005	137	0.53	13	MS2, WR1
H	F4	Apr 4, 2005	4	0.53	14	MS2, WR1
I	F4	May 23, 2005	53	0.53	16	MS2, WR1
J	F1	Feb 15,2005	1121	0.27	8̃.2–19	Oocysts , D10, diatoms

MS2 = bacteriophage MS2; WR1 = *Escherichia coli* WR1; *C.lari* = *Campylobacter lari;* D10 = spores of *Clostridium perfringens*; occysts = oocysts of *Cryptosporidium parvum; diatoms = Stephanodiscus hantzschii*
F1/F3: slow sand filters after rapid sand filtration, ozonation and active carbon filtration
F2/F4: slow sand filters after rapid sand filtration.

Slow sand filters (pilot plant)

Series of experiments were conducted in 2002 and in 2005 at pilot plant scale. Slow sand filters with a filter depth of approximately 1.5 m and a surface area of 2.56 m^2 were operated in parallel and under the same conditions as the plant scale slow sand filters. The flow rate was 0.3 m^3/hour. Table 1 lists the experiments. In experiments A-D, MS2 and WR1 were seeded for 24 hours onto the slow sand filters at a low concentration (10^4–10^5 /l), followed by 24-hours seeding at a 1000 times higher concentration (10^7–10^8/l). In experiments E-I, microorganisms were seeded for 24 hours only at the high concentration.

The effect of the presence of the Schmutzdecke was studied by comparing experiment A and B, and C and D, respectively. Filter 1 and 2 differed in the influent that was applied to these filters. The influent of Filter 1 was ozonated/active carbon filtrated, but that of Filter 2 not. Therefore, the Schmutzdecke of Filter 2 needed to be scraped off more frequently than that of Filter 1. Experiments G, H and I were conducted in order to investigate the time that is needed for a Schmutzdecke to restore. Breakthrough of the microorganisms was monitored for about one week in each experiment.

In experiment J, oocysts of *C. parvum* (260 /l),spores of *C*.perfringens D10 (10^3/l) and centric diatoms, *Stephanodiscus hantzschii*, (260 /l) were seeded simultaneously onto Filter 1 during a period of approximately 100 days. Breakthrough of these microorganisms was monitored for a period 250 days.

Results and discussion

Figure 1a-b shows the breakthrough curves of MS2 and WR1 from experiments B, C and D. Maximum breakthrough concentrations are determined by inactivation, attachment to the sand grains and straining in the pores in the sand. After seeding of the microorganisms was

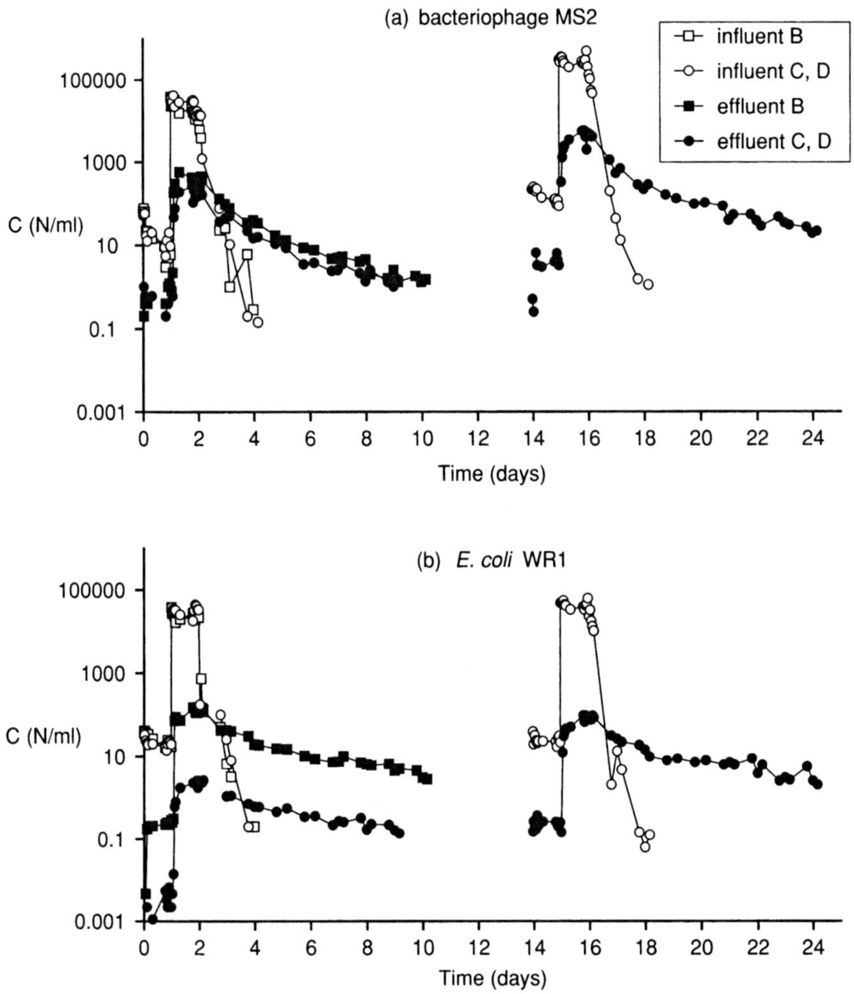

Figure 1 Breakthrough curves of experiments B, C (left) and D (right).

stopped, tailing of the breakthrough curves was observed, due to slow detachment of previously attached microorganisms. Removal of the microorganisms was calculated from the seeding concentrations C_0 and the maximum breakthrough concentrations C_{max} (Table 2).

Removal of bacteriophage MS2 in experiments B, C and D was very similar, indicating little, if any, effect of the presence of a Schmutzdecke, whereas removal of *E. coli* WR1 was 2 $\log_{10}$ higher when the Schmutzdecke was present. Because *E. coli* is much larger (about 1.2 μm) than bacteriophage MS2 (26 nm), this indicates straining of *E. coli* by the Schmutzdecke but not of MS2.

In experiment A, removal of MS2 was higher. In that case, temperature was higher, and, possibly, MS2 may have been removed more efficiently due to a more biologically active sand filter (predation).

Table 2 Removal $\log_{10}(C_0/C_{max})$

	MS2		WR1	
Experiment	**Low C_0**	**High C_0**	**Low C_0**	**High C_0**
A	>2.2	3.5	-	-
B	1.7	1.8	2.1	2.3
C	1.8	2.2	3.9	4.2
D	1.7	1.9	2.0	2.8
	MS2	WR1	*C. lari*	
E	0.57	2.5	3.9	
F	-	4.8	4.8	
G	3.4	4.9	-	
H	2.8	3.1	-	
I	3.9	5.6	-	
	Oocysts	D10	Diatoms	
J	5.3	3.8	1.6	

— = not done.

Regression analysis of removal of bacteriophage MS2 in experiments B, C and D, and of *E. coli* WR1 removal in experiments B and D as a function of seeding concentrations showed that removal did not change significantly with the seeding concentration.

The very low removal of bacteriophage MS2 in experiment E possibly reflects an operational failure of the slow sand filtration (Table 2). In the coagulation/sedimentation step prior to the sand filtration, accidentally too much caustic soda was added that temporarily may have increased pH of the sand filter, whereby MS2 was removed less, because it attaches less at high pH. Removal of *E. coli* WR1 in experiment E was similar to that of filters where the Schmutzdecke was scraped off (experiments B, D and H). This suggests that in experiment E, the Schmutzdecke was not effective.

Note that removal of *E. coli* WR1 in experiment F was much more efficient, even higher than in experiment B (same filter). The Schmutzdecke of this filter was not scraped off since then. Experiments E and F showed the same or more removal of *Campylobacter lari* compared to the removal of *E. coli*. This is supported by the study of Hijnen *et al.* (2004), where about one $\log_{10}$ more removal was reported of environmental campylobacters than of environmental *E. coli* at full plant scale. Note that removal of MS2 in experiment G was as high as in experiment A, demonstrating that an effectively working Schmutzdecke appeared to affect removal of MS2 too. After scraping of the Schmutzdecke (experiment H), removal of WR1 is again about 2 $\log_{10}$ less efficient, as was demonstrated before in experiments B, C and D. This time also the removal of MS2 is reduced by 0.6 $\log_{10}$. Dizer *et al.* (2004) also found a relatively small enhanced removal of bacteriophages in the presence of a Schmutzdecke. After 53 days (experiment I), the removal efficiency of Filter 4 was fully restored, *i.e.* removal was even higher than in experiment G, probably because operational temperature was higher as well.

Figure 2 shows the breakthrough curves of MS2 and WR1 from experiments G, H and I. The shape of these curves differs from those of experiments B, C and D (Figure 1). Because

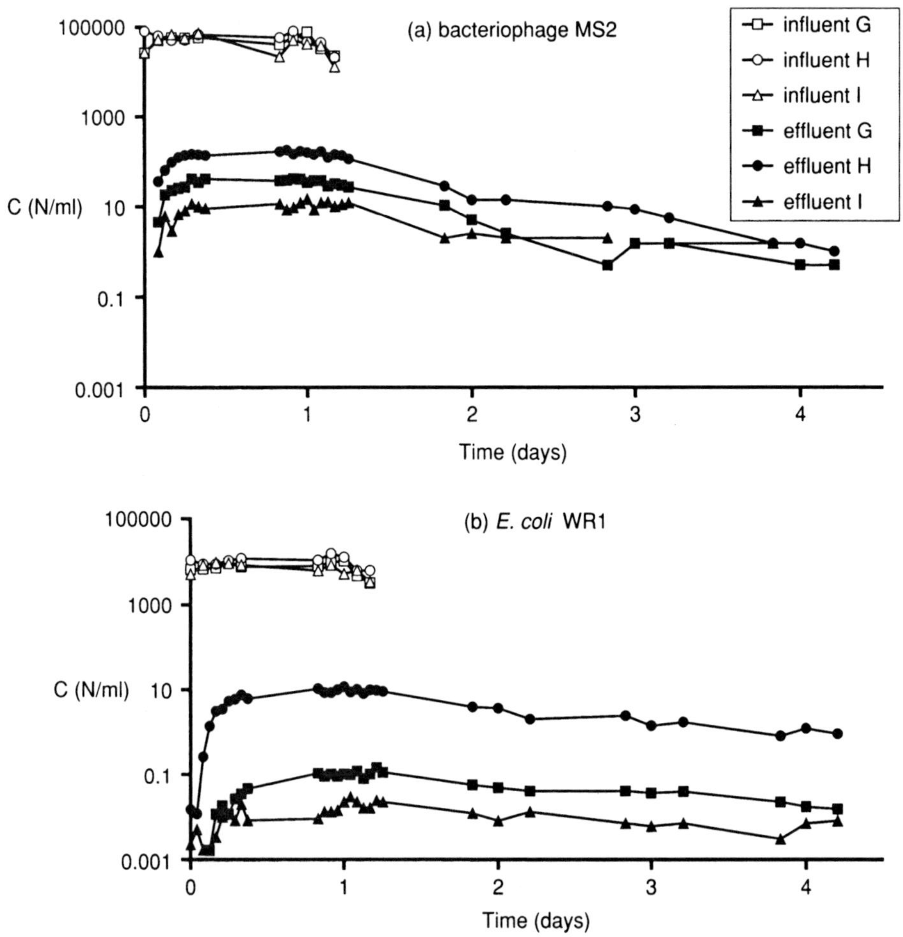

Figure 2 Breakthrough curves of experiments G, H and I.

more microorganisms were retained in experiments G, H and I than in B, C and D, the maximum breakthrough concentrations are relatively lower and approximately at the same level as the tails of the breakthrough curves.

In experiment J, high removal of oocysts (5.3 $\log_{10}$) and spores (3.8 $\log_{10}$) but relatively low removal of centric diatoms (1.6 $\log_{10}$) was found (Table 2 and Figure 3). In total 20 oocysts were observed in a total sample volume of 36000 litre of effluent. In total 336 spores were observed in a total sample volume of 2200 litre and 218 diatoms were counted in a total sample volume of 29 litre. The low removal of centric diatoms was most likely due to environmental centric diatoms of the same cell size as the seeded diatoms (4–7 μm) that were present in the influent (median concentration 76 per litre; $n = 22$). Concentrations of environmental oocysts and spores of sulphite-reducing clostridia in the influent were below the detection limit.

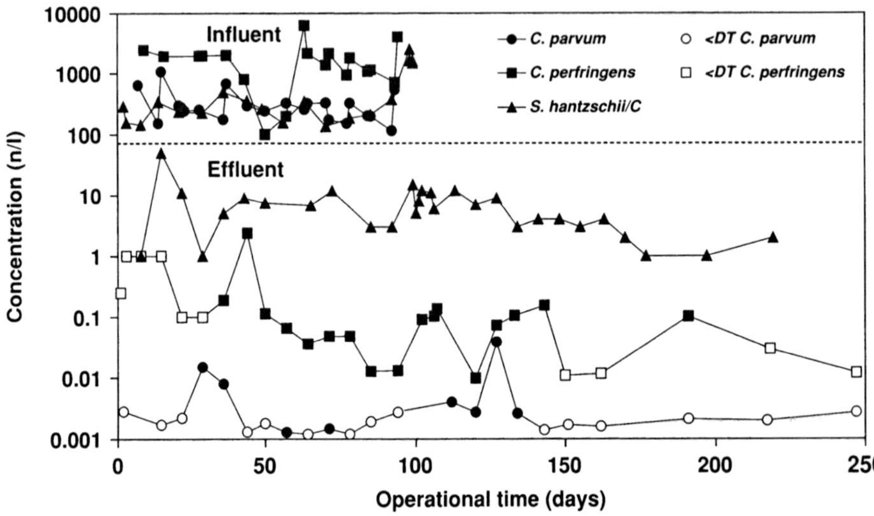

Figure 3 The concentrations of C. *parvum, C. perfringens* (SSRC) and *S. hantzschii* (CD) in influent and effluent of the tested filter (open symbols are results below the indicated detection limit, DT)

A small fraction of the oocysts, spores and diatoms was able to pass the sand filters as is demonstrated by the long tails of the breakthrough curves at approximately the same level as the maximum breakthrough concentrations (Figure 3).

After 182 and 252 days of filtration the numbers of *C. parvum* oocysts and spores of sulphite-reducing clostridia (SSRC) that were retained in the filter bed were determined by extraction from sand samples at different depths up to 123 cm (Figure 4). The highest numbers of these organisms were observed near the surface of the filter bed.

After 182 days, approximately 1.6% of the total number of seeded oocysts could be recovered in the first 30 cm of the sand filter and after 252 days only 0.2%. Of the seeded spores, about 40% could be recovered in the first 30 cm of the sand filter after 182 days and were still at the same level after 252 days. Apparently, almost all of the seeded oocysts that were retained in the sand filter had disappeared, whereas most of the seeded spores were still present. This disappearance of retained oocysts can be explained by disintegration of the accumulated oocysts but also by predation by zooplankton. Both King *et al.* (2005) and Hutchison *et al.* (2005) showed reduction of accumulated oocysts in water and soil, respectively. By comparing reduction in autoclaved and non-autoclaved surface water, King et al [2005] demonstrated the role of micro-organisms in reducing oocyst numbers in the water. The role of bacteria in this process has not been documented to our knowledge but predation of oocysts by free living protozoa and rotifers have been presented in literature (Fayer *et al.*, 2000; Scott *et al.*, 2001 and 2003).

Conclusions

- Because bacteriophage MS2 may be considered as a conservative model virus and because they were removed the least and affected the least by the presence of the

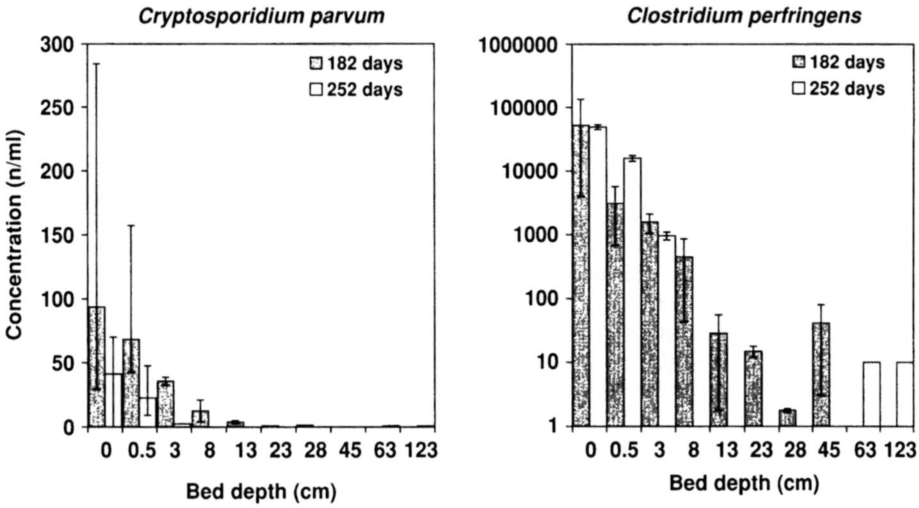

Figure 4 Concentrations of oocysts and bacterial spores (identified as *C. perfringens* by molecular typing).

Schmutzdecke at low temperatures (<10 °C), viruses are the most critical pathogens for the removal efficiency of slow sand filters,

- Removal of *E. coli* was about 2 $\log_{10}$ more, probably due to straining in the Schmutzdecke.
- At higher temperatures (15–20 °C), removal of microorganisms by slow sand filtration, including viruses, was found to be about 2 $\log_{10}$ higher.
- Removal of viruses and bacteria by slow sand filtration did not depend on the seeding concentration.
- The Schmutzdecke could restore within 53 days.
- *E. coli* is a useful model bacterium for campylobacters that were removed at least as well of about one $\log_{10}$ more.
- Mature slow sand filters with a Schmutzdecke are very effective in removing oocysts of *Cryptosporidium parvum*. Recovery of retained oocysts in the sand filter indicated that the retained oocysts were removed by disintegration and/or predation.
- Highly persistent clostridium spores and centric diatoms are not good model organisms for removal of oocysts by slow sand filtration. Because of their persistence spores and diatoms may accumulate in a sand filter, thereby confounding interpretation of their removal.

References

Anonymous (2001). Besluit van 9 januari 2001 tot wijziging van het waterleidingbesluit in verband met de richtlijn betreffende de kwaliteit van voor menselijke consumptie bestemd water. Staatsblad van het Koninkrijk der Nederlanden, 31:1–53.

Dizer, H., G. Grützmacher, H. Bartel, H.B. Wiese, R. Szewzyk and J.M. López-Pila (2004). Contribution of the colmation layer to the elimination of coliphages by slow sand filtration. Wat. Sci. and Technol. **50**(2), 211–214.

Hijnen, W.A.M. and Schijven, J.F. (2003) The elimination capacity of slow sand filtration for micro-organisms – Pilot plant and laboratory research. Kiwa report, BTO 2002.133, in Dutch.
Hijnen, W.A.M., Schijven, J.F., Bonné, P., Visser, A. and Medema, G.J. (2004). Elimination of viruses, bacteria and protozoan oocysts by slow sand filtration. Water Sci. Technol. **50**:147–154.
Fayer, R., Trout, J.M., Walsh, E. and Cole, R. (2000). Rotifers ingest oocysts of Cryptosporidium parvum. J. Eukaryot. Microbiol. **47**(2): 161–163.
Hutchison, M.L., Walters, L.D., Moore, T., Thomas, D.J.I. and Avery, S.M. (2005). Fate of pathogens present in livestock wastes spread onto fescue plots. Appl. Environ. Microbiol. **71**(2): 691–696.
King, B.J., Keegan, A.R., Monis, P.T. and Saint, C.P. (2005). Environmental temperature controls Cryptosporidium oocyst metabolic rate and associated retention of infectivity. Appl. Environ. Microbiol. **71**(7): 3848–3857.
Schijven, J.F. (2000). Virus removal from groundwater by soil passage. Thesis Technical University Delft, ISBN 90-646-4046-7.
Scott, R., Matsushita, E. and Warren, A. (2001). Protozoan predation as a mechanism for the removal of *Cryptosporidium* oocysts from wastewaters in constructed wetlands. Wat. Sci. Tech. **44**(11–12); 194–198.
Scott, R., May, E., Ramirez, E., and Warren, A. (2003). Predation of Cryptosporidium oocysts by protozoa and rotifers: implications for water quality and public health. Wat. Sci. Techn. **47**(3): 77–83.

3 Assessing the role of the schmutzdecke in pathogen removal in riverbank and slow sand filtration

M. Unger, M.R. Collins

Water Treatment Technology Assistance Center, Department of Civil Engineering, University of New Hampshire, Durham, NH, USA

Abstract Results from a series of studies on laboratory-scale sand columns confirmed that *E. coli* removals in slow-rate biological filters occur primarily at the interface. These interfacial removals were impacted by schmutzdecke ripening state, empty bed contact time, biological activity, temperature, and protistan population.

Keywords Schmutzdecke, pathogen removal, riverbank filtration, slow sand filtration

Introduction

The microbial removal role of the interface between the overlaying water and the media bed whether in a slow sand filter (SSF) or a riverbank filtration (RBF) system has been a subject of much debate over the years. In slow sand filtration, this interface is defined as the schmutzdecke, which in German is translated as "dirty skin". It has been characterized visually as a gelatinous mat in which microorganisms thrive and where the major portion of removals occur (Hendricks et al., 1991).

The SSF schmutzdecke can vary widely and sometimes seasonally. In general, the schmutzdecke consists of numerous forms of life including algae, plankton, diatoms, protozoa, rotifers, and bacteria and is where inert suspended particles can be mechanically strained, organic matter and nitrogenous compounds broken down, and microorganisms entrapped (Huisman and Wood, 1974). Although not specifically characterized, the microbial treatment performance of the SSF schmutzdecke and RBF bed-river interface layers are considered similar (Partinoudi, 2004; Wang et al., 2002). Any pathogen removal mechanism at work in SSF will also influence a riverbank filter. The major differences between a riverbank filter and a slow sand filter are the increased length and heterogeneity of the riverbank, but these differences do not apply to the interface.

Unfortunately, the exact role this interface layer or schmutzdecke exerts in microbial removals in both SSF and RBF systems has not been well-defined. The need to establish or re-establish a schmutzdecke layer after filter cleaning or riverbed scouring events in order to once again provide water of acceptable quality needs to be ascertained so that appropriate operational methods can be implemented if required. For example, after scouring events,

 Recent Progress in Slow Sand and Alternative Biofiltration Processes edited by Rolf Gimbel, Nigel J.D. Graham and M. Robin Collins. ISBN: 9781843391203. Published by IWA Publishing, London, UK.

RBF extraction well pumping rates could be decreased or stopped, or laterals further from the river could be selected, until the riverbed schmutzdecke is re-established. Such operating techniques are needed to ensure public health risks are minimized for all biological filtration systems.

The specific objectives of this research include:

(1) Ranking the relative importance of various media characteristics and operational conditions on biofiltration removals of *E. coli* bacteria.
(2) Determining the extent to which biological activity contributes to pathogen removal in a biological filter. Specifically, do the extracellular polymeric excretions of the biofilm present in drinking water treatment enhance the "stickiness" of filter media?
(3) Assessing the effect of a sudden removal of the schmutzdecke on pathogen removal in the event of scouring or cleaning and a filter's ability to recover from such an event.
(4) Estimating the potential influence of protistan predators in bacterial removal at the interface of a biological filter.

Methods

In order to satisfy these objectives, five studies were conducted on laboratory scale sand columns under varying operation and design conditions as described below with spiked concentrations of *E. coli* to assess removal efficiency.

Experimental setup

Each sand filter column was constructed in 4.8 cm inner diameter glass chromatography columns (Kontes, Vineyard, NJ). Norprene tubing connected a constant flow raw water tank to the top of each column, with energy provided by peristaltic pumps. Raw water from the Durham, NH water treatment plant, which is drawn from the Oyster River, was used to ripen the filters. The sand filter columns were ripened and/or challenged in series or parallel depending on the individual study's experimental design.

Experimental microorganism

A strain of *E. coli* resistant to amoxicillin and streptomycin was cultured in agar broth in the Water Treatment Technology Assistance Center (WTTAC) lab at the University of New Hampshire. During an experimental challenge, 20 L raw water would be spiked with stock *E. coli* to a goal concentration of 10^4 MPN/100mL. *E. coli* were enumerated with the IDEXX Quanti-Tray 2000.

Biofilm characteristics

Four analyses were used to characterize the biofilm colonizing filter media: a general biomass measurement, which measured lipid phosphate, total carbohydrate and total protein to estimate the mass of extrapolymeric substances (EPS) excreted, and CO_2 respiration to monitor biological activity. Biomass in the schmutzdecke (and occasionally at various depths within the filter) was monitored using the lipid phosphate extraction introduced by White (1979) and later refined by Findlay et al. (1989). Phospholipids are common to all cells and act as an indicator of viable biomass because they are turned over rapidly during metabolism (Vestal and White, 1989). Lipids were extracted using a mixture of chloroform, methanol, and water, dried under a nitrogen stream, and oxidized with potassium persulfate to liberate phosphate, which is quantified colorimetrically with a spectrophotometer.

Total carbohydrates were quantified using the phenol-sulfuric acid method of Dubois et al. (1956), which produces a golden-brown color that is read colorimetrically.

Total proteins were measured with a Modified Lowry Protein Assay Kit (Pierce, Rockford, IL). The kit simplifies Lowry's (1951) method and includes bovine serum albumin (BSA) as a standard for development of a calibration curve.

Biological activity in the schmutzdecke was determined by measuring CO_2 production by microorganisms on a known mass of sand incubated 24 hours at 25°C. For the temperature dependence study (Section 3.3), samples were incubated at the influent water temperature. The total CO_2 released was measured on a Licor 6252 CO_2 analyzer. Respiration could then be normalized to CO_2 released by gram dry weight of sample per hour.

Counting protists

Samples were taken from random locations within the schmutzdecke of a sand filter column with a 3 mm diameter plastic straw to a depth of 5 mm. The mass of the samples was determined before they were diluted with Poland Spring water. Forty microliters of diluted sample were placed on a slide under a $25mm^2$ coverslip. Slides were observed at 400x magnification, and protists were counted and grouped into categories: flagellates, amoebae, attached ciliates, and unattached ciliates.

Results and discussion

Five studies were designed to isolate the influence of various media characteristics and operational parameters on bacterial removal.

Four factor orthogonal array

In the first study, an orthogonal array of sand filter columns was used to evaluate the effect of selected media characteristics and operational parameters on pathogen removal. Factors were assigned as displayed in Table 1.

After packing the columns with sand according to the orthogonal array, the filters were challenged with raw water spiked with *E. coli*. Filters then operated for three weeks filtering Oyster River (Durham, NH) water before being challenged again. An amount of sand equivalent to 7.5 min EBCT was scraped from the top of each column, and the filter operated for 24 additional hours before challenging a third time. Log removal results for each of the three challenges are presented in Figure 1.

Table 1 Assignment of factor levels to sand columns in orthogonal array.

Column #	Size[a]	HLR - hydraulic loading rate (m/h)	Biomass	EBCT -empty bed contact time (min)
1	large	0.2	Virgin	15
2	large	0.2	Acclimated	60
3	large	0.6	Virgin	60
4	large	0.6	Acclimated	15
5	small	0.2	Virgin	60
6	small	0.2	Acclimated	15
7	small	0.6	Virgin	15
8	small	0.6	Acclimated	60

[a] The entire stock was sieved through a #16 sieve (1.18 mm). All sand retained was considered "large" and all sand passing was considered "small".

Analysis of variance (ANOVA) of the four factors mentioned above was conducted on log removal data from the challenges of ripened and scraped filters. EBCT was indicated to be the most significant factor of the four tested, contributing 41% and 57% of the total variation in the data for the ripened and scraped filters, respectively. Higher removals were observed with higher EBCTs, suggesting schmutzdecke depth, while dependant on HLR for the same EBCT, may not be a good indicator of overall removals. A schmutzdecke may be shallow at lower HLR but still may be equivalent as a removal barrier to a deeper

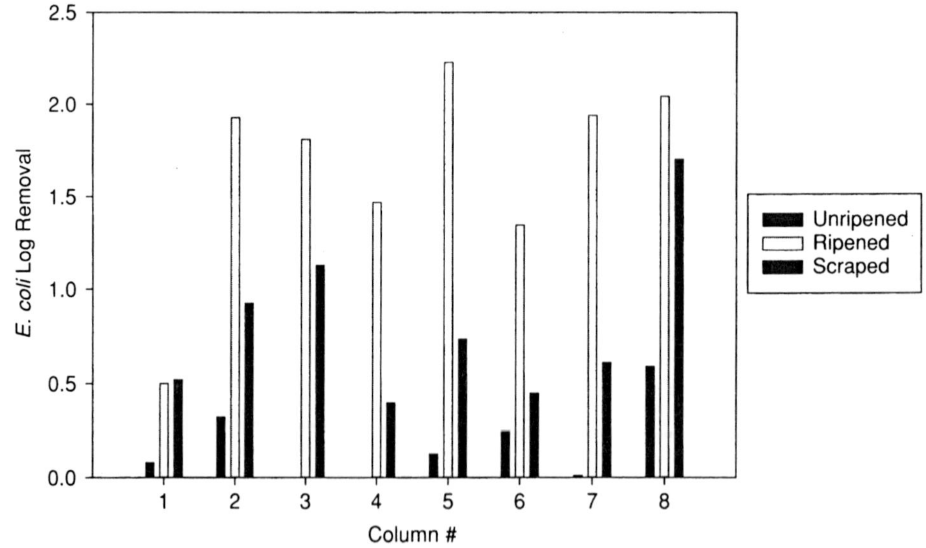

Figure 1 Log removal of *E. coli* in designed sand columns under varying conditions before ripening, after ripening with an established schmutzdecke, and after scraping the schumtzdecke.

schmutzdecke developed at higher HLR, provided they both have the same EBCT. Media size was also statistically significant at the 90% confidence level for the ripened filters, thereby confirming higher removals with smaller media. After scraping, however, HLR was more significant than media size, suggesting that recovery from a cleaning or scouring event could be enhanced with higher flow rates to accelerate schmutzdecke development.

Differing levels of biomass in the filters before schmutzdecke development had no statistically significant effect on removal, which implies that after a certain period of filter run time (less than three weeks in this case), ripening of the schmutzdecke is more dependant on the contribution of the influent water than on the previously established biological community. In other words, whether or not a filter started with a pre-established biofilm made no difference in its ability to remove *E. coli* after three weeks of ripening under the conditions of this study.

Scouring simulation

The goals of the second study included quantifying pathogen removal with depth in a biological filter and also reevaluating the ability of a filter to recover from a sudden loss of the schmutzdecke, for example riverbed scouring or slow sand filter scraping. Two trains of four pilot columns were filled with 22.5 cm of sand, connected in series and allowed to ripen for two weeks on Oyster River raw water, thus simulating a filter 90 cm deep. Both trains were then challenged with *E. coli*, one in series to assess the overall removal efficiency and one in parallel to assess the efficiency of each individual section. Following the challenge, the parallel columns were allowed to continue filtering Oyster River water for two more weeks in order to enable simultaneous development of a schmutzdecke in each column.

Another, identical series of columns operated simultaneously and was used to evaluate the status of the biofilm using the methods described in Section 2.3. Selected results for the series- and parallel-ripened columns are displayed in Table 2.

The results of the columns ripening in series indicate that the majority of total coliform removals and biomass content are associated with the first column, which contained the

Table 2 Relationship between total coliform removal and biomass (as phospholipid) for various filter ripening conditions, July 2005.

	Challenge in Parallel After 2 weeks Ripening in Series		**Challenge in Parallel After 2 weeks Ripening in Parallel**	
	Log Removal of Total Coliforms	**Biomass, Top 2.5 cm (nmol PO4 / gdw)**	**Log Removal of Total Coliforms**	**Biomass, Top 2.5 cm (nmol PO4 / gdw)**
Col A (0 – 22.5 cm)	1.3	36 ± 2	1.6	42 ± 4
Col B (22.5 – 45 cm)	0.8	14 ± 7	1.6	29 ± 2
Col C (45 – 67.5 cm)	0.2	9 ± 1	1.4	31 ± 2
Col D (67.5 – 90 cm)	0.3	11 ± 4	1.6	39 ± 2
Full Train of 4 Challenged in Series (90 cm total)	2.1			

schmutzdecke. Although not statistically significant, there appeared to be a correlation between coliform removals and biomass levels. As was the case in the first study, the original amount of biomass on the sand of a filter appeared to have no bearing on the filter's ability to remove coliforms after two weeks of additional ripening in parallel. Total coliform removal in Column A, the first column of the series, (which, in the first two weeks, had ripened significantly compared to the following three columns) was equivalent to removal in the later three columns after all four had operated for two additional weeks in parallel. This result suggests that both a slow sand filter can recover from a scraping and a riverbank can recover from scouring in two weeks or less under warm temperatures.

Influence of temperature on Schmutzdecke redevelopment

A third study examined the influence of water temperature on schmutzdecke redevelopment after scraping or scouring removed a significant portion of the top of the filter. Initial ripening in series was identical to that in study 2, but only 3 columns of 22.5 cm length were used.

After two weeks ripening, the columns were again connected in parallel, but for this study, one series was taken to a cold temperature room held at a constant 8° C. The previous day, Durham raw water had been transported to the cold temperature room and mixed continuously with an impeller. Water was replenished with water from the Oyster River every 48 hours. Meanwhile, the second series of columns remained at the WTP. After four days of ripening, each column segment was challenged with an identical challenge solution consisting of raw water spiked with *E. coli*.

The study generated the following results:

(1) Warm biological columns outperformed cold biological columns with respect to *E. coli* removal.

(2) Biomass levels as quantified by phospholipids and EPS (carbohydrates and proteins) were not significantly different between respective cold and warm biological columns under the conditions of this study. The activity of the biological columns, however, as quantified by CO_2 respiration, was higher in the warmer columns. (Figure 2)

(3) Biological removals between the first column (Column A) and the remaining columns (Columns B and C) were not significantly different even though the first column received three additional weeks of ripening time.

These results suggest:

- biomass and EPS levels (carbohydrates and proteins) do not correlate to *E. coli* removals; however, removals appear to be more closely related to biological activity as defined by CO_2 respiration, which in turn is affected by temperature, and
- bacteria removals again are more dependant on interface ripening or schmutzdecke development than starting biomass levels.

Influence of L/d (filter depth to media diameter) ratio and empty bed contact time (EBCT)

Results from the first study (Section 3.1) identified EBCT as having a significant influence on removal. A fourth study aimed to narrow the scope of the first study to focus on the

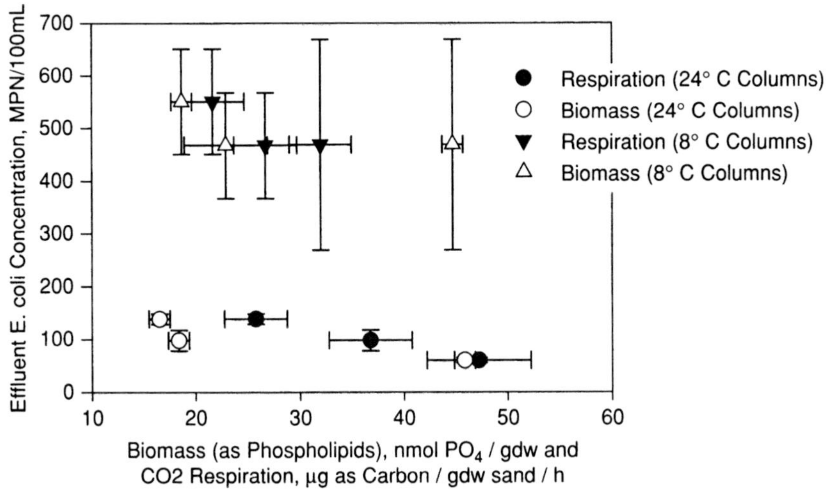

Figure 2 Correlation between schmutzdecke biofilm and *E. coli* removal in sand columns at different temperatures. Influent concentration was 21000 ± 6000 MPN/100 mL to cold columns (80°C) and 19000 ± 3000 MPN/100 mL to warm columns (240°C).

impact of EBCT and L/d, the ratio of the length of the filter to the effective diameter of the media. Filter columns contained one of two sand media configurations ("small" with effective size (ES) = 0.25 mm and uniformity coefficient (UC) = 1.5, or "large" with ES = 0.5 mm and UC = 1.5) and one of three HLR ("fast" 0.6 m/h, "medium" 0.4 m/h, and "slow" 0.2 m/h). Column design factors and their corresponding log *E. coli* removals are presented in Figure 3. These data indicate that the L/d ratio did not influence *E. coli* removal to the same extent as EBCT. In addition, above a certain EBCT (20 min for this study), there

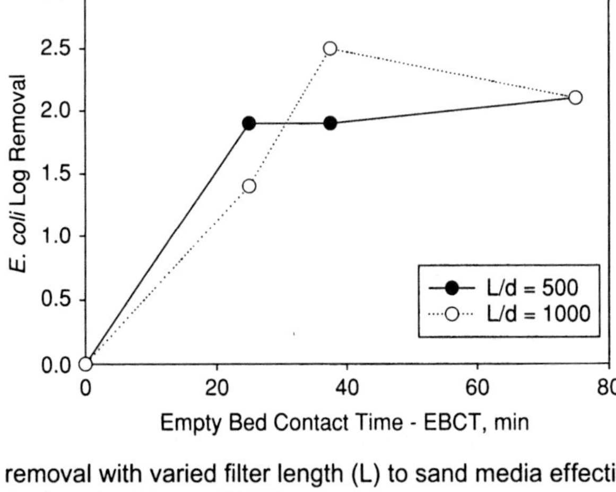

Figure 3 *E. coli* removal with varied filter length (L) to sand media effective size (d), *L*/*d*, ratios and empty bed contact times, EBCT.

was no significantly greater *E. coli* removal, which suggests the effect of depth filtration on bacteria removal in biological filters is minimal compared to removal mechanisms at the interface between the water and filter medium.

Potential role of protistan predation on *E. coli* removal

The potential ability of naturally occurring protists to remove challenge *E. coli* was estimated according to the procedure presented in Section 2.4. Flagellates were at least one order of magnitude more abundant than amoebae and ciliates. The population of flagellates in the top 5 mm of the schmutzdecke was calculated to be $6 \pm 2 \times 10^6$ per cm^2. A first estimate of consumption rate was calculated using the size specific uptake rate reported by Kinner and coworkers (1998) of 0.77 bacteria / flagellate / hour, which was lower than other uptake rates reported in the literature and specific to a size range of bacteria, which includes *E. coli* (0.8–1.5 μm).

This uptake rate implies a potential bacterial consumption rate of $5 \pm 2 \times 10^6$ per cm^2/hour. *E. coli* were spiked into the influent water at a concentration of 2100 ± 200 MPN/100 mL, and 61 ± 1 MPN/100 mL were recovered in the effluent. At a flow rate of 18 mL/min onto a column with 4.8 cm diameter, this corresponds to an actual removal rate of 1.3×10^2 bacteria/cm^2/hour, four orders of magnitude less than the potential bacterial consumption rate of $5 \pm 2 \times 10^6$ per cm^2/hour.

This calculation is based on many assumptions:

(1) The *E. coli* used to spike the influent water are the only bacteria being consumed by flagellates.
(2) All flagellates are feeding continuously.
(3) Flagellates occupy the schmutzdecke uniformly in space.

Despite these assumptions, the calculated estimated consumption rate significantly exceeds the actual removal rate observed in the filter, and the contribution of other protists besides flagellates would even further increase the potential removal. This premliniary result suggests predation by protists is potentially a very significant bacterial removal mechanism in biological filters and is worthy of additional study.

Conclusions

(1) EBCT appeared to be the most important operational parameter influencing bacterial removal in sand filters, but above a certain EBCT, there was no significantly greater *E. coli* removal.
(2) Biological activity in the schmutzdecke as measured by CO_2 production correlates well to *E. coli* log removal (r^2 value of 0.9241; data not shown), but extrapolymeric substances (EPS) as measured by carbohydrates and proteins, did not show a significant correlation.
(3) A sudden removal of the first 22.5 cm of a filter dramatically reduced bacterial removal efficiency, but efficiency was recovered in less than four days. Water temperature affected the extent of recovery, with cold filters (8° C) achieving a significantly lower ultimate removal efficiency than warm filters (24° C).

(4) The overall uptake rate of the number of protists observed in the schmutzdecke of a sand filter column was calculated to be roughly four orders of magnitude greater than the actual *E. coli* removal rate measured. The calculated uptake potential of flagellates alone was $5 \pm 2 \times 10^6$ per cm^2/hour.

Acknowledgements

The authors gratefully acknowledge:

- USEPA for funding this research through the New England Water Treatment Technology Assistance Center (WTTAC) at the University of New Hampshire.
- Dr. Nancy Kinner and Dr. Gabriela Martinez, University of New Hampshire for assistance with the protist predation study.

References

Dubois, M., Gilles, K.A., Hamilton, J.K., Rebers, P.A., Smith, F. (1956) Colorimetric method for determination of sugars and related substances. *Anal. Chem.* **28**, 350–356.

Findlay, R.H. (1989) Efficacy of phospholipid analysis in determining microbial biomass in sediments. Appl. and Env. Microbiol. **55**(11), 2888.

Hendricks, D. et al. (1991) *Manual of Design for Slow Sand Filtration.* AWWARF, Denver, CO.

Huisman, L. and Wood, W.E. (1974) *Slow Sand Filtration.* Geneva, World Health Organization.

Kinner, N.E., Harvey, R.W., Blakesless, K., Novarino, G., and Meeker, L.D. (1998) Size-Selective Predation on Groundwater Bacteria by Nanoflagellates in an Organic-Contaminated Aquifer. *Applied and Environmental Microbiology*, **64**(2), 618–625.

Lowry, O.H., Rosebrough, N.J., Farr, A.L., Randal, R.J. (1951) Protein measurement with the folin phenol reagent. *J. Biol. Chem.* **193**, 265–275.

Partinoudi, V. (2004) Riverbank Filtration as a Viable Pretreatment and Treatment Method. MS Thesis, University of New Hampshire, Durham, NH.

Vestal, J.R. and White, D.C. (1989). Lipid analysis in microbial ecology: quantitative approaches to the study of microbial communities. *Bioscience* **39**(8), 535.

Wang, J.Z. (2002) Evaluation of Riverbank Filtration as a Drinking Water Treatment Process. AWWARF, Denver, CO.

White, D.C. (1979) Determination of the Sedimentary Microbial Biomass by Extractible Lipid Phosphate. *Oceologia* **40**, 51–62.

4 30 years of RWW's practical experience with an advanced microbiological water treatment system for Ruhr river water - "the muelheim treatment process 1976–2006"

Gerhard Bundermann*

*RWW Rheinisch-Westfälische Wasserwerksgesellschaft mbH, Am Schloß Broich 1 - 3, D-45479 Mülheim an der Ruhr

Abstract The "Muelheim Process" for producing drinking water from surface water of the Ruhr river was developed in 1976 by the RWW Rheinisch-Westfälische Wasserwerksgesellschaft, Mülheim an der Ruhr, Germany. Although this treatment process is in operation for 30 years now it is still a modern and most efficient multy-barrier-system against polluted raw water with waste water input up to 30 %. Several bio-filtration steps like slow sand filtration, artificial ground water recharge, biological double layer filtration and permanently operated BAC filtration are combined with traditional water purification systems such as flocculation and sedimentation steps, mechanical filtration and oxidation by ozone. Particular substance as well as different dissolved organic matter, especially DOC, pesticides, pharmaceuticals and micro-organisms, pathogenes and parasites are reliably removed from the raw water input. Since 2003 RWW is able to produce drinking water by this process at the Styrum-East water work without using any chlorine or chlorine product either for oxidation during the treatment system or for final water disinfection. Because of the intentional bio-degradation process due to the reduction of DOC and the removal of pathogenes the regrowth-potential for micro-organisms in the drinking water is very low, so that it can be pumped into the distribution pipe network after a final UV-light irradiation.

Keywords "Muelheim Process" for treating Ruhr river water, RWW-Muelheim, slow sand filtration, artificial ground water recharge, ozone technology, double layer filtration, BAC filtration, bio-degradation, DOC-reduction, removal of parasites / pathogenes, pharmaceuticals, pesticides, regrowth potential for micro-organisms, UV-light irradiation, disinfection of drinking water.

Introduction

In the Federal Republic of Germany the production of drinking water from natural ground water resources has a long tradition. It is more than 120 years ago that the construction of central water supply units started just in the big cities and most crowded regions of the country. Depending on the rising demand of drinking water it became also necessary to use as well surface water from rivers. Especially in the North-Rhine-Westphalian industrial zone of the Ruhr river basin the use of surface water is of high importance.

For the realization of the water catchment process from rivers normally treatment steps are used which are created by the nature itself and developed by mankind such as bank filtration and ground water recharge. Main topic of those "indirect" catchment systems is the optimization of raw water quality and water purification process often managed by microbiological support. In case of strong raw water pollution it might be necessary to complete those "indirect" systems by additional treatment steps for getting acceptable drinking water quality, expecially microbiological filtration steps and different processes.

"The Muelheim Treatment Process" which was developed exactly 30 years ago and which is in the focus of this report is based on such an advanced microbiological water treatment combination. The process and the experience are making sure for this long period now that advanced biofiltration process is a most effective step with excellent progress due to the water quality requirements of the future.

Presentation of RWW

The RWW Company (Rheinisch-Westfälische Wasserwerksgesellschaft mbH) with its head-quarter in Muelheim-Ruhr is one of Germany's biggest water service companies.

It was founded in 1912 as an union of three already existing water works in this region. Initiators of this foundation were several municipalities and the private RWE Company. The roots were located in Muelheim in the lower Ruhr river valley where the three operating water units were producing drinking water.

From the year of foundation the service area of the RWW Company has been growing up constantly until today. The number of operating water plants has been increasing from three to actually nine units. **Figure 1** is giving a short documentation of RWW today.

Development of the water catchment system and the water treatment process

In the northern part of its water service area RWW is only using natural ground water resources from the "Haltern Sands", an important upper cretaceous marine layer and aquifer system of the Muensterland region.

In the southern area of RWW the surface water treatment from the river Ruhr is in the focus.

For the first existing three water works in the lower Ruhr river valley it was quite easy to operate the catchment systems. It was managed just by the bank filtration process. Infiltrated river water mixed up with natural ground water of the porous aquifer system of the Ruhr river valley was gathered by siphon well systems, then the water was disinfected by chlorine and finally pumped into the distribution pipe network. This very simple process could only be operated in a successful way under conditions of low turbidity ranges of the raw water and of nearly constant flow rates of the river which was not regulated at that time.

At the beginning of the 20th century however the situation changed basicly. Because of the upcoming industrialization and the increasing population rate the water works got rising

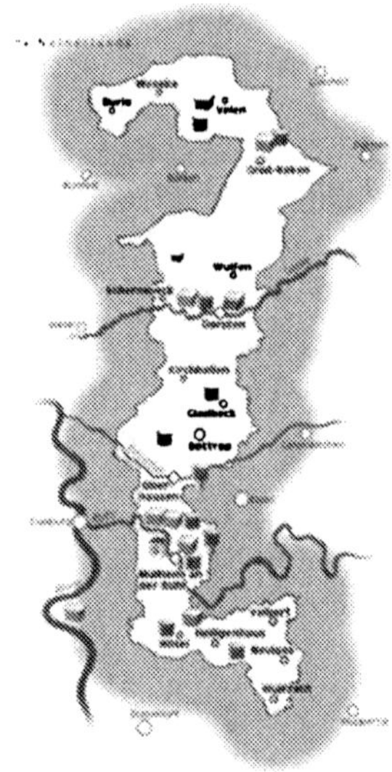

Supply district:	850 sq. km.
Year of foundation:	1912
Shareholders:	RWE Comp. 80 %, local municipalities 20 %
Total number of staff:	500
Number of water production units: 9	3 natural ground water works, 6 surface water works
Length of pipe network:	3000 km.
Annual water production:	80 million cu. m.

Figure 1 Main features of RWW

problems. Large amounts of untreated waste water caused heavy pollution of the Ruhr river water with clogging effects and colmation of the river banks. Decreasing flow rates because of a massive water exportation from the Ruhr river valley and the upcoming construction of dams for water power plants soon made the bank filtration process impossible.

It was the slow sand filtration process and the artificial ground water recharge to solve the problem since 1902 along the Ruhr river valley (**figure 2**).

In the summer time the typical run off of the Ruhr river has its lowest flow rate at 15 cu. m per second. Under such condition the Ruhr river water is consisting of a 30 percentage of treated waste water input. The raw water is infiltrated into the natural aquifer system in the Ruhr river valley by slow sand filter basins. After a 1–2 days lasting soil passage in the natural quaternary aquifer system the water is gathered by siphon well systems.

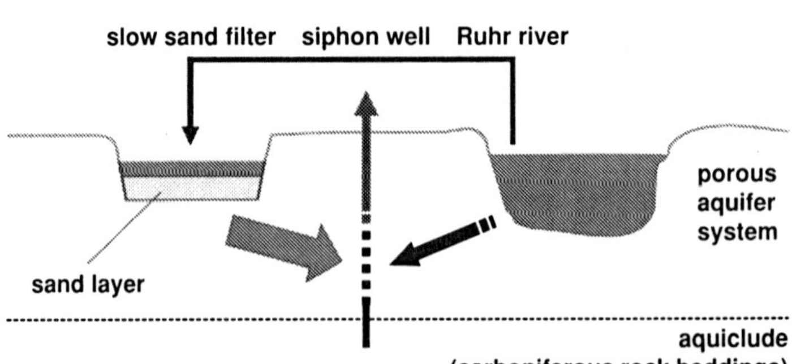

Figure 2 Slow sand and Infiltration process in the Ruhr River Valley since 1900

In the sand filters a mechanical filtration process with the removal of particular substance and turbidity from the raw water can be watched as well as a dominating microbiological treatment. The concentration of many raw water pollutants and especially the DOC is being decreased by a complex process dominated by hydromechanical dispersion, diffusion, adsorption, coagulation, sedimentation and even bio-degradation... Extremely high concentrations of ammonia (>5 ppm) coming from the waste water input caused a lot of problems to the drinking water production process in previous times. This problem was extremely present in the winter time at low water temperatures. RWW then often had a lot of problems to complete the nitrification process only by sand filtration step and the short time lasting soil passage. When finally chlorine for disinfection was added the water was of very bad odor and smell and the hygienical risk of regrowth of micro-organisms in the network system was at high level. The only way of escape from this situation up to the 60ies of the 20th century was to remove the ammonia by chemical oxidation with chlorine ("break-point-chlorination-process").

But new difficulties soon appeared: By adding high concentrations of free chlorine (max. 50 ppm) to the raw water AOX was produced by reaction with DOC as well as high concentrations of haloforms (max. 100 ppb).

The pollution of the Ruhr river and the presence of nutritive substance were producing mass growth of algae on top of the slow sand filters. Even in the summer time new hygienical problems by metabolisms like volatile oil production occurred. The reaction of the components: chlorine, ammonia, DOC and metabolics of algae were a daily present problem for RWW at that time.

A new German drinking water act in 1975 and the restriction by law of using chlorine for any drinking water treatment process stopped RWW's activities very abruptly. The company had to look for a new water treatment technology for producing hygienical blameless drinking water quality.

Development and characteristics of the "Muelheim Process" of the Dohne, Kettwig and Styrum-West water works

It was in 1976 that RWW was very successful to develop a new water treatment combination of advanced microbiologically dominated steps – the so called Muelheim Treatment Process – for producing drinking water from surface water.

This project was scientifically supported by the Technical University of Karlsruhe / Germany – Prof. Dr. H. Sontheimer – and financially sponsored by the German Bundesminister für Forschung und Technologie. RWW's water plant to test the new technology was the Muelheim-Dohne unit where up to this time the process of "break-point-chlorination" was in operation (**figure 3**):

- raw water catchment directly from the run off of the Ruhr river
- flocculation and sedimentation (system pulsator) by adding large amounts of chlorine, powder of limestone, aluminium-sulfate, jelly of silicate...
- rapid sand filtration
- GAC filtration for the removal of residual chlorine
- infiltration into the quaternary aquifer system of the Ruhr river valley by negative wells for artificial ground water recharge and soil passage

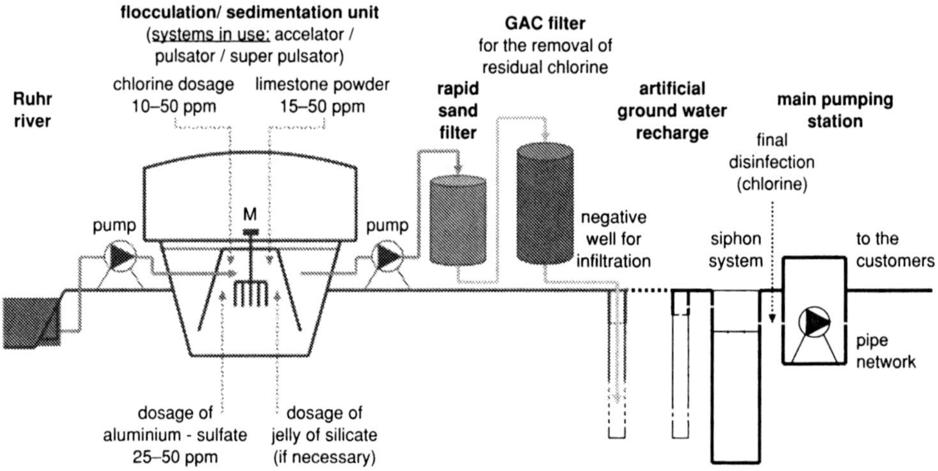

Figure 3 RWW's water treatment process ("Break-Point-Chlorination") before 1976

- re-catchment of the soil filtrated water, disinfection by chlorine
- pumping station and distribution pipe network

After a several months lasting research program at the Dohne water work the new technology was completed: The "Muelheim Treatment Process" as a multy-barrier-system against water pollutants was initially producing good and successful data. Instead of the former use of chlorine the oxidation of the water by ozone was realized combined with an adsorption process in biological activated carbon filters (BAC).

The "Muelheim Process" of the Dohne unit is operating since that time in the following way (**figure 4**):

- raw water catchment directly from the run off of the Ruhr river with pre-ozonation
- flocculation and sedimentation by adding Poly-aluminium-chloride (PAC)
- main-ozonation
- biological double layer filtration
- BAC filtration
- infiltration into the natural quaternary aquifer system for ground water recharge
- re-catchment by siphon well systems after a soil passage of 1–2 days
- final disinfection by chlorine (0,1 – 0,15 ppm), deacidification by sodium-hydroxide
- pumping station and distribution pipe network

Main target of the "Muelheim Process" is the degradation process of DOC for reducing the need of disinfectants in the drinking water and for avoiding the re-growth of micro-organisms (particularly pathogens) in the pipe network.

Concerning different problematical and disturbing substances the "Muelheim Process" is also producing excellent results: **Figure 5** is showing its efficiency e. g. by the data of turbidity, ammonia, DOC, HPC 20°C / 36°C, coliforms and E.-coli as well as of parasites and pesticides.

The positive experience with the "Muelheim Process" running at the Dohne water work enabled RWW to publish this new technology [1]. The company decided at the end of

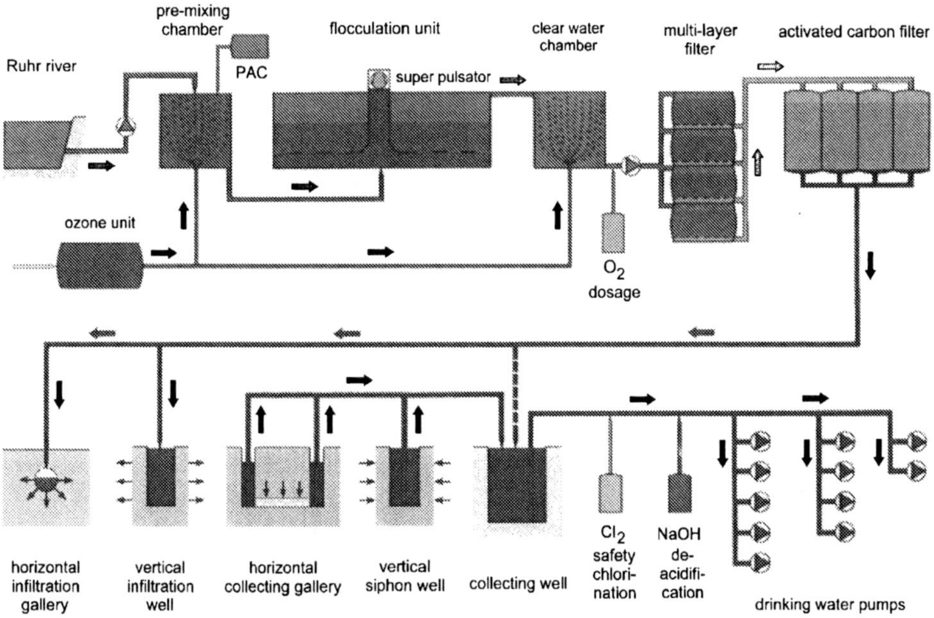

Figure 4 Water work Muelheim-Dohne, actual treatment process since 1976

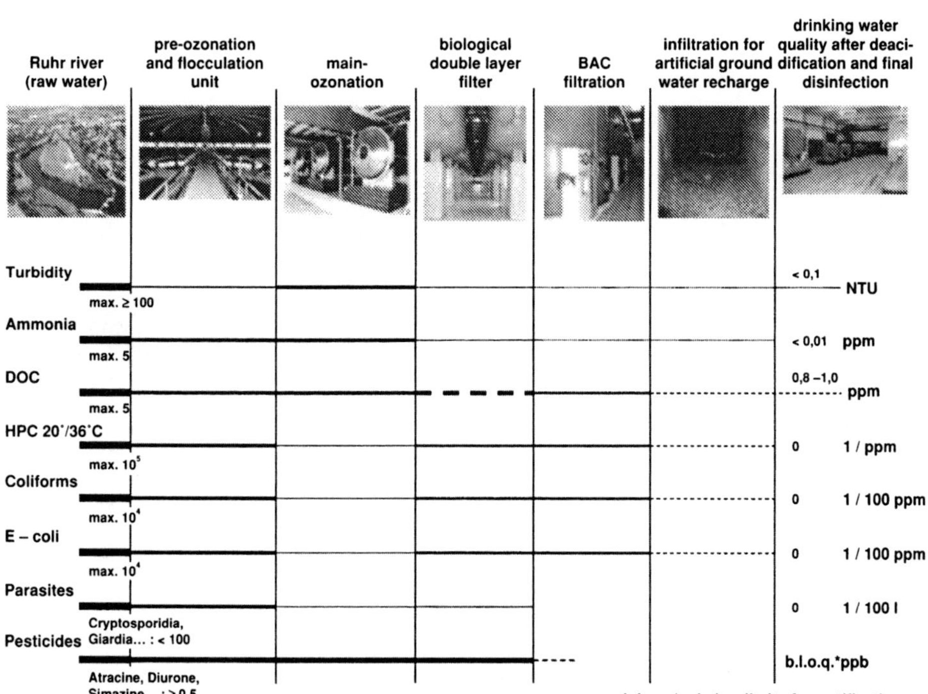

Figure 5 Removal of raw water compounds at the Styrum-West, Dohne and Kettwig Units

the 1970 decade to verify the "Muelheim Process" from the Dohne plant in a similar way at the Essen-Kettwig, Muelheim-Styrum/East and Muelheim-Styrum/West water works as well.

The "Muelheim Process" of the Styrum-East water work

A little bit different was the situation at the Muelheim-Styrum/East water work, one of the biggest units of RWW.

At the end of the 1970 decade the water work was producing drinking water in the "classical" way only by operating a lot of slow sand filters in its water catchment area including artificial ground water recharge process and soil filtration. This kind of initial water treatment step was to maintain further on.

That's why the typical steps of the "Muelheim Process" like ozonation and BAC-filtration were added to this process so that in 1981 the restored Muelheim-Styrum/East water work with its new treatment process started the drinking water production in the following way (**figure 6**):

- raw water catchment directly from the run off of the Ruhr river
- mechanical and microbiological pre-treatment of the water by slow sand filtration process, artificial ground water recharge and a 1–2 days lasting soil passage
- re-catchment of the soil filtrated water and additional ozonation step
- biological double layer filtration and BAC filtration
- UV-light irradiation for final disinfection (since 2003), deacidification by sodium-hydroxide
- pumping station and distribution pipe network.

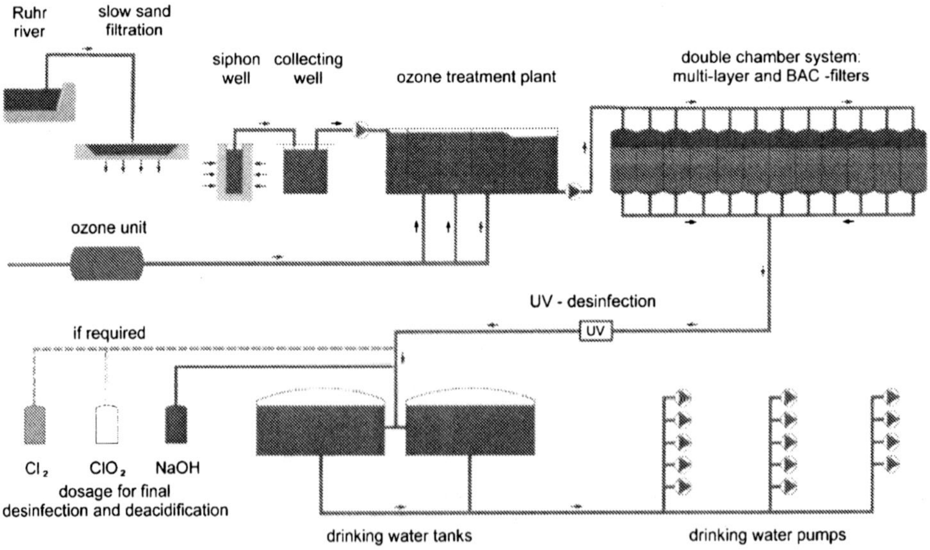

Figure 6 Water work Muelheim-Styrum/East, actual treatment process since 1981

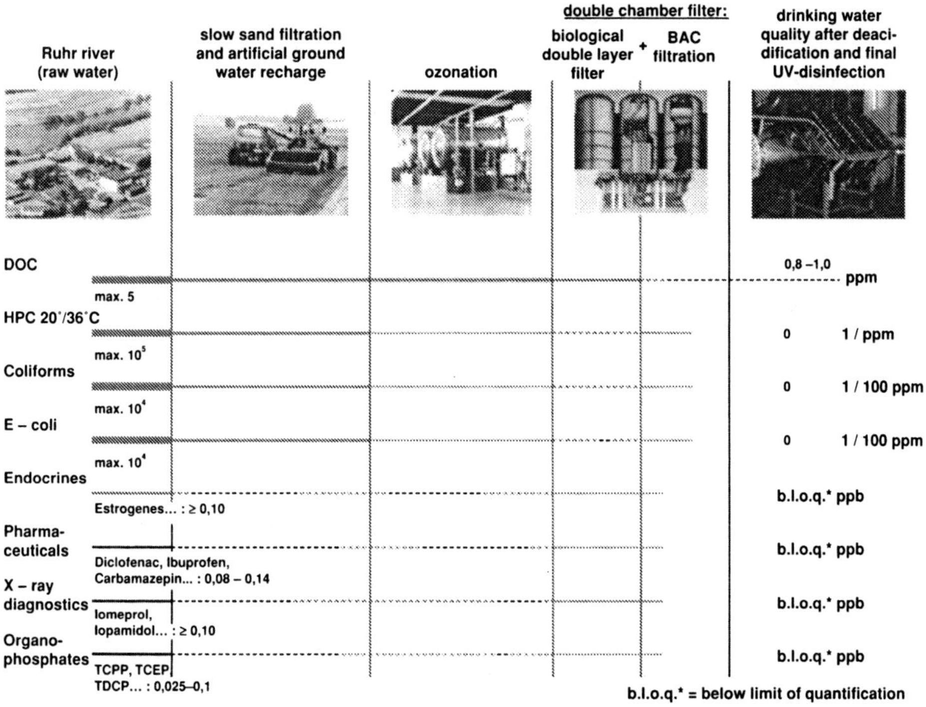

Figure 7 Removal of raw water compounds e. g. at the Styrum-East Water Unit

Concerning this Styrum-East variation of the "Muelheim Process" RWW was also very successful in getting good results of high standards and good drinking water quality: **Figure 7** is showing the DOC, HPC 20°C / 36°C, coliforms and E.-coli data and even organic substances like endocrines, pharmaceuticals, X-ray diagnostics and organophosphates.

Conclusions

Biofiltration and biological water treatment steps for the drinking water production are very well known from literature and international publications [2].

Starting from bank filtration and the "classical" slow sand filtration process RWW developed the "Muelheim Process" in 1976 in two variations.

Main target of this water treatment combination is to decrease the concentration of DOC by bio-degradation process and to minimize different water pollutants – even organic traces – by oxidation and adsorption technology. The removal of particles, micro-organisms and pathogens from the raw water is also of interest.

The chemical oxidation by ozone and the biological treatment step in activated carbon filters in combination with additional water treatment systems like flocculation, sedimentation, adsorption, slow sand filtration and artificial ground water recharge are producing excellent drinking water quality from polluted surface water.

Since 2003 RWW is very successful to run the complex process completely free from chlorine only by using UV-light irradiation for final water disinfection. A special research program in cooperation with the Board of Public Health Administration could establish the positive results of a very low re-growth potential for micro-organisms in the distribution pipe network.

30 years of good experience with such a multy-barrier-system make sure that there is no need for any use of chlorine in the treatment process either for chemical oxidation or for final disinfection.

This report is dedicated to the experts who had the vision of the "Muelheim Process": Prof. Dr. Heinrich Sontheimer†, Ewald Heilker, Heinrich Nolte, Prof. Dr. Martin Jekel et al.

References

[1] Sontheimer, H.; Heilker, E., Jekel, M.; Nolte, H.; Vollmer, H. (July 1978) The Muelheim Process. *Journ. AWWA*, 393

Heilker, E. (November 1979) The Muelheim Process for Treating Ruhr River Water *Journ. AWWA*, 623

Jekel, M. & Sontheimer, H. (1978) Biologisch-adsorptive Trinkwasseraufbereitung in Aktivkohle-Filtern, Statusseminar des BMFT/DVGW *Schriftenreihe Wasser*, **101**

[2] Hubele, C. (1985) Adsorption und biologischer Abbau von Huminstoffen in Aktivkohlefiltern – Dissertation, Universität Karlsruhe (TH)

Werner, P., Die Bedeutung biologischer Prozesse bei der Wasseraufbereitung durch Filtration – in: Wagner, R. (Hrsg.): Wasser-Kalender 1987, Jahrbuch für das gesamte Wasserfach, Berlin: Erich Schmidt Verlag (1986), 92–116

Sontheimer, H. & Haist-Gulde, B. (1992) Sicherheit der Trinkwasserversorgung nach Stoßbelastungen unter besonderer Beachtung der Entfernung von Mikroverunreinigungen mit Aktivkohlefiltern *gwf, Wasser Abwasser*, **133**, **5**, 239–246

5 Advances in hybrid membrane filtration system for drinking water production

Yoshimasa Watanabe*, Tatsuhiko Suzuki, Tooru Morita***, Genzou Ozawa****

*Department of Environmental Engineering, Hokkaido University, Kita 13, Nishi 8, Sapporo 060-8628, Japan
**Maezawa Industries, Inc., 5-11 Naka-cho, Kawaguchi, Saitama Prefecture 322-8556, Japan
***Sumitomo Electric Fine Polymer, Inc., 4-5-33 Kitahama, Chuo-ku, Osaka, 541-0041, Japan

Abstract This paper contains the results obtained in a pilot plant study about the hybrid MF membrane filtration system combining the other processes with submerged membrane. A newly developed PTFE (polytetrafluoroethylene) MF membrane was submerged in a tank where powdered activated carbon adsorption and biological/chemical oxidation were simultaneously achieved. Under the membrane flux of 0.8–1.0 m/day, trans-membrane-pressure was almost constant for the pilot plant operation time of 50 days and much longer operation can be possible without chemical washing of the membrane when an asymmetrical PTFE membrane with the skin layer thickness of 30 μm is used. The filtrate quality was excellent such as 0.017 1/cm of E260, 1.17 mg/L of TOC, 0.01 mg/L of total manganese and ammonia nitrogen, while the raw water quality was as follows: E260, concentration of TOC, total manganese and ammonia were 0.099 1/cm, 2.43 mg/L, 0.105 mg/L and 0.23 mg/L, respectively.

Keywords Hybrid membrane system, PTFE membrane, powdered activated carbon, oxidation

Introduction

There are several papers[1,2,3,4,5,6)] describing the combination of coagulation and carbon adsorption with the membrane filtration. The authors[7,8)] have developed a hybrid MF membrane system which combines the submerged MF membrane and the other processes such as powdered activated carbon adsorption and chemical/biological oxidation. In the early research where the polyethylene MF membrane was used, possible membrane flux was 0.4 m/day even though the membrane backwashing with hypo-chloride solution every few hours was applied. The authors[9)] have recently developed a PTFE (*polytetrafluoroethylene*) hollow fiber MF membrane. The PTFE is mechanically and chemically robust material. It has particularly strong resistance to chemical agents compared with other organic membrane materials. In a long-term operation of the membrane filtration plant, periodical chemical washing is necessary, and it is important that membrane permeability recovers to the initial level by the chemical washing. Strong chemicals can be used for washing, such as 2% sodium chloride solution or 3,000mg Cl_2/l hypochlorite solution, in the PTFE membrane. So the recovery rate in chemical washing is expected to be very high and the lifetime of the membrane will be estimated to be long.

In order to evaluate the permeability of submerged PTFE membrane when it is used for the hybrid system, the pilot plant study has been conducted. Based on the experimental results, the PTFE membrane has been improved over several years. The structure of PTFE membrane was changed from symmetrical to asymmetrical, and the thickness of the skin layer of the membrane has become thinner. With these improvements, membrane flux has been increasing gradually in the pilot plant experiments.

This paper deals with the operation performance of the pilot plant and the filtrate quality of a hybrid MF membrane system using the four types of PTFE membranes.

Materials and methods

Hybrid membrane system

The hybrid MF membrane system is a combination of submerged membrane and the other processes such as the powdered activated carbon adsorption and chemical / biological oxidation. The membrane system has been developed to purify raw waters with low quality containing a lot of soluble matter such as biodegradable organics, humic substances, manganese and ammonia nitrogen. In the hybrid system, soluble less-biodegradable organics are adsorbed to the powdered activated carbon, and suspended particles including powdered activated carbon are separated by the membrane filtration. The soluble biodegradable organics, manganese and ammonia nitrogen are biologically or chemically oxidized. In the case of chemical oxidation (with pre-chlorination), soluble manganese is oxidized with chlorine and the catalytic reaction of powdered activated carbon, and the oxidized manganese is removed by membrane separation. Ammonia nitrogen is also oxidized by chlorine in a pre-chlorination tank. In the case of biological oxidation (without pre-chlorination), the iron oxidizing bacteria and ammonia oxidizing bacteria, which are condensed in submerged membrane tank, oxidize the soluble manganese and ammonia, respectively.

A schematic diagram of the pilot plant is shown in Fig. 1. The volume of membrane submerged tank and the surface area of submerged membrane were 4 m^3 and 86–120 m^2, respectively. Detention time in the mixing tank was 10–15 min. The raw water was fed into the mixing tanks. As described later, four types of PTFE membrane were used. When the first, second and third type of membrane were used, the raw water to the pilot plant was taken from the existing water purification plant, which had already contained the powdered activated carbon in the concentration of 5–30 mg/L. In the first tank, hypoc-hlorite was added when the chemical oxidation was applied, and the sludge containing biomass and activated carbon were returned from the membrane submerged tank and mixed with the raw water in the second tank. The same powdered activated carbon (average diameter of 10 μm) was dosed into the second tank at the constant concentration of 13 mg/L when the fourth PTFE membrane was used. Poly-Aluminium Chloride was added to coagulate the small

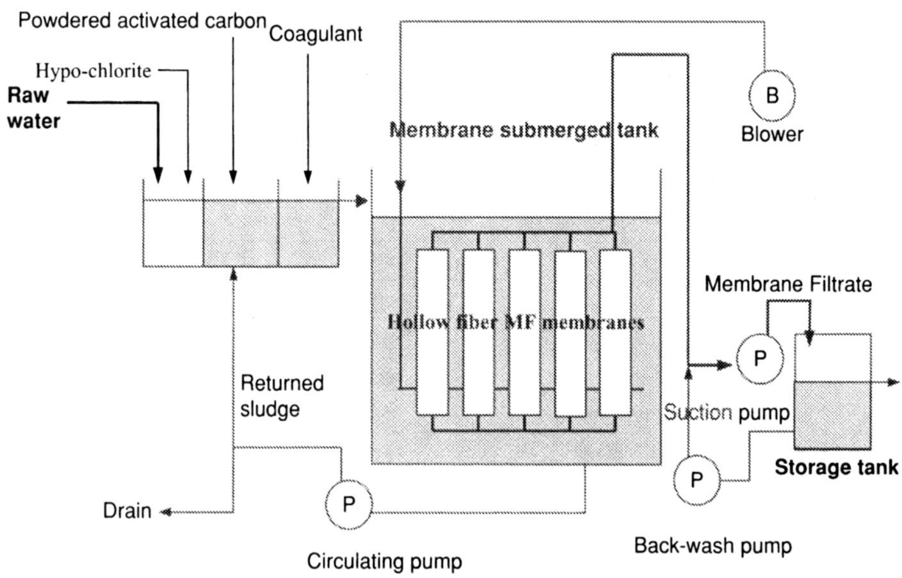

Figure 1 Schematic diagram of hybrid MF membrane system combined with powdered activated carbon adsorption and chemical/biological oxidation

suspended particles in the third tank. The pre-treated water was fed into the submerged tank where the hollow fiber PTFE membranes were submerged and intermittent aeration was performed to supply the oxygen to the microorganisms. Air was supplied for 1 min. with the intensity of 0.64 Nm^3/min. every 4 min.

Experimental conditions

The raw water in the pilot plant study was surface water from Chitose river. It contained a lot of humic substances, soluble manganese and ammonia nitrogen. The average raw water quality is given in Table 1. The dosage of hypo-chlorite in the mixing tank was 4–6 mg Cl_2/l. Coagulant dosage was 2 to 3 mg Al/L.

Experimental results

Permeability of PTFE membrane

Pilot plant with a symmetric PTFE membrane was operated from June 2002 to October 2002. Figure 2 shows the change of the permeate flux, trans-membrane-pressure (TMP) and raw water temperature during the experiment. In this experiment, the hybrid MF membrane system was operated without pre-chlorination. The average flux was relatively low at less than 0.3 m/day and the TMP increased to 70 kPa after 5 months of operation.

To improve the permeability of the PTFE membrane, the structure of membrane was changed from symmetrical to asymmetrical. The asymmetrical hollow fiber MF membrane consists of the skin layer and supporting layer as shown in Photo 1. Nominal pore size of

Table 1 Average raw water quality in pilot plant study

Temperature	°C	11.5
pH		7.11
Turbidity	NTU	16.54
UV absorbance at 220nm	1/cm	0.411
UV absorbance at 260nm	1/cm	0.099
TOC	mg/1	2.43
DOC	mg/1	2.29
THMFP	mg/1	0.086
Manganese	mg/1	0.100
Soluble manganese	mg/1	0.074
Ammonia Nitrogen	mg/1	0.22

the skin layer is 0.1 μm. The permeability of three types of asymmetrical membranes was compared in the pilot plant experiment. The first and second type of membrane have the skin layer thickness of 60 and 30 μm, respectively. The third one has the skin layer thickness of 30 μm and its pore density is larger compared with the first and second asymmetrical membrane. From November 2002, the first asymmetrical membrane was used in the pilot plant. Figure 3 shows the change of the membrane flux, TMP and raw water temperature with increasing operation time. The pilot plant was operated without pre-chlorination from June 25 to November 13. The average membrane flux in this period was 0.53 m/day. It is

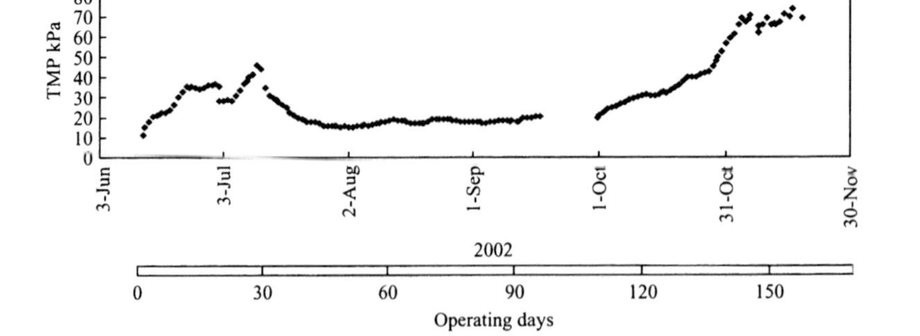

Figure 2 Change of membrane flux, TMP and raw water temperature in the experiment with symmetric PTFE membrane from January 2002 to November 2002

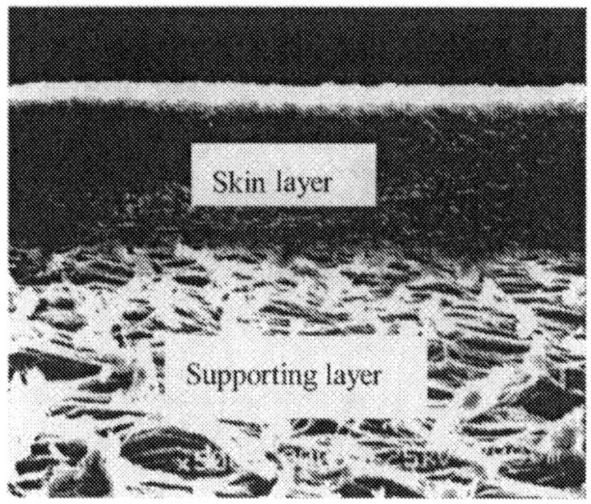

Photo 1 Cross-section of asymmetric PTFE membrane

about 1.7 times higher than that in the symmetrical membrane. Therefore, it is clear that the asymmetrical structure improved the permeability of the PTFE membrane. Chemical membrane washing was carried out twice during this operation period.

From January to October, 2004, the second asymmetrical membrane was used. In this period, the plant was operated with pre-chlorination. The change of the membrane flux, TMP and raw water temperature is shown in Fig. 4. The membrane flux was set at 0.5 m/day.

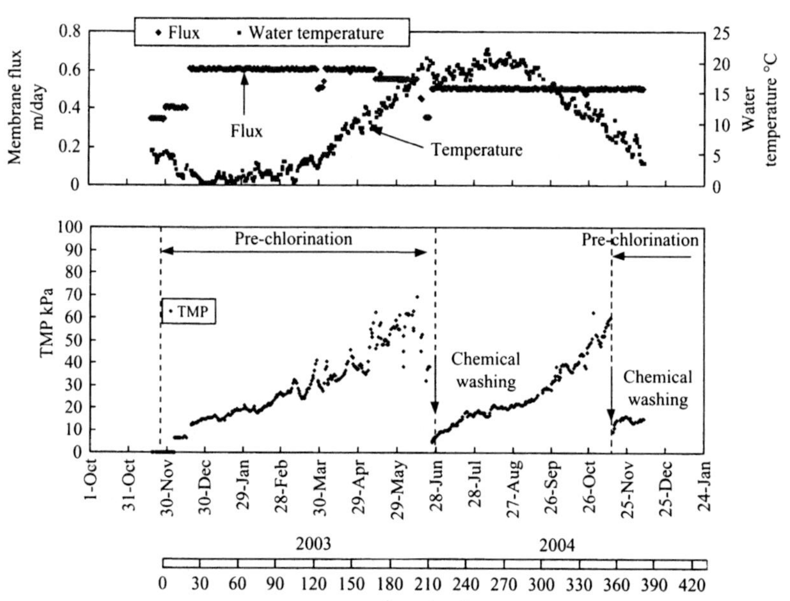

Figure 3 Change of membrane flux, TMP and raw water temperature in the experiment with the first asymmetrical membrane from November 2002 to January 2003

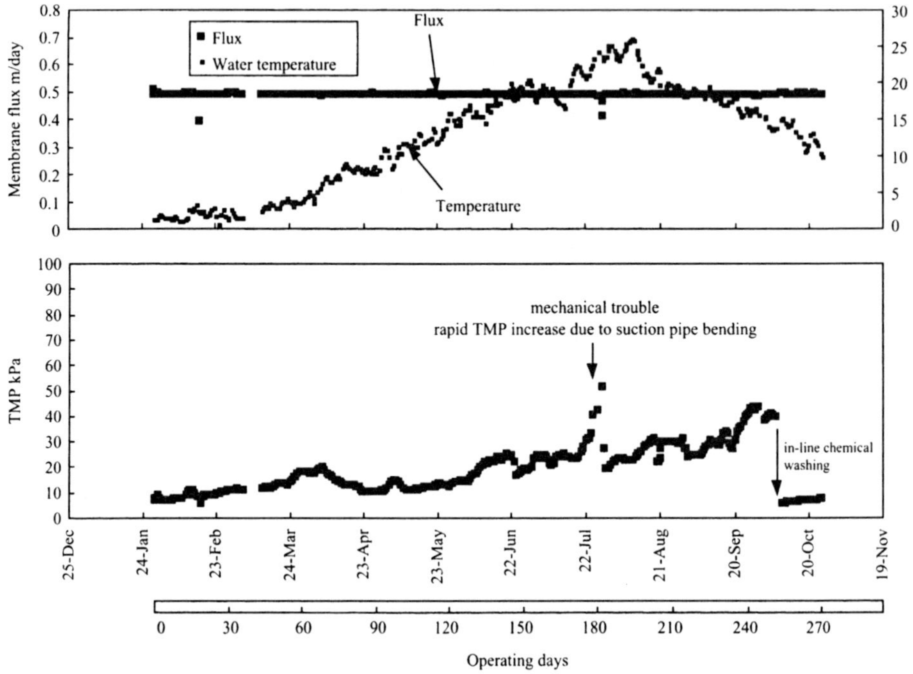

Figure 4 Change of membrane flux, TMP and raw water temperature in the experiment with the second asymmetrical membrane from January to October 2004

After 8 months of operation, the TMP was less than 50 kPa. The increase rate of TMP was slower than that in the first asymmetrical membrane.

The third type of asymmetrical membrane was used from September 2004. The changes of the membrane flux, TMP and raw water temperature with increasing operation time is shown in Fig. 5. It demonstrates that TMP was very stable and it was less than 10 kPa under a high flux of 1.0 and 0.8 m/day. It is about 2.5 times higher than that in the symmetrical membrane.

Filterate quality in hybrid membrane system

The hybrid membrane system is able to efficiently remove the soluble matter such as organics, manganese, and ammonia nitrogen. The soluble manganese and ammonia nitrogen were oxidized biologically or chemically and small humic substances were adsorbed to the powdered activated carbon. Figure 6 shows the removal efficiency of TOC, E260 and THMFP when the third asymmetrical PTFE membrane was used. In the experiment, dosage of powdered activated carbon was fixed at 13 mg/L. Figure 7 shows the comparison in the removal efficiency of the soluble manganese and ammonia nitrogen between chemical oxidation with pre-chlorination and biological oxidation without pre-chlorination. As previously reported by the authors[9], chemical oxidation is necessary to oxidize soluble manganese when the

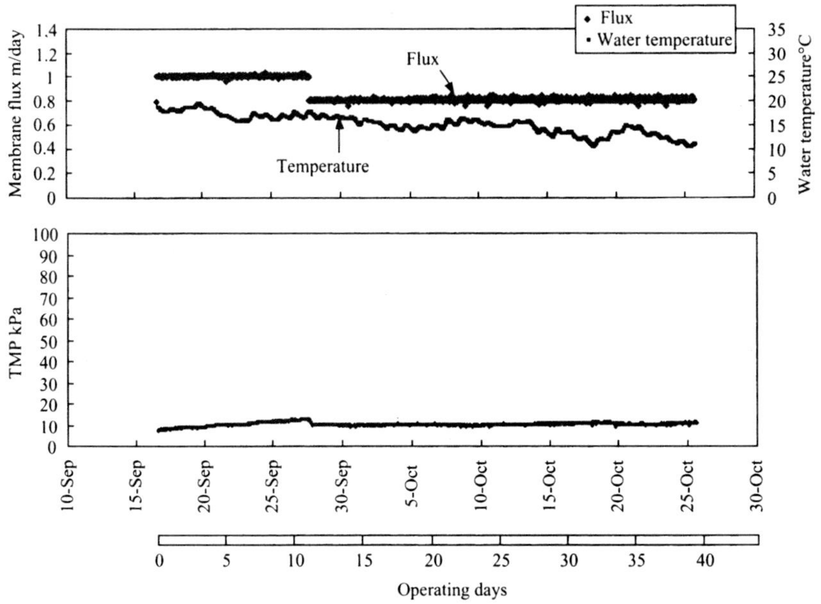

Figure 5 Change of membrane flux, TMP and raw water temperature in the experiment with the third asymmetrical membrane

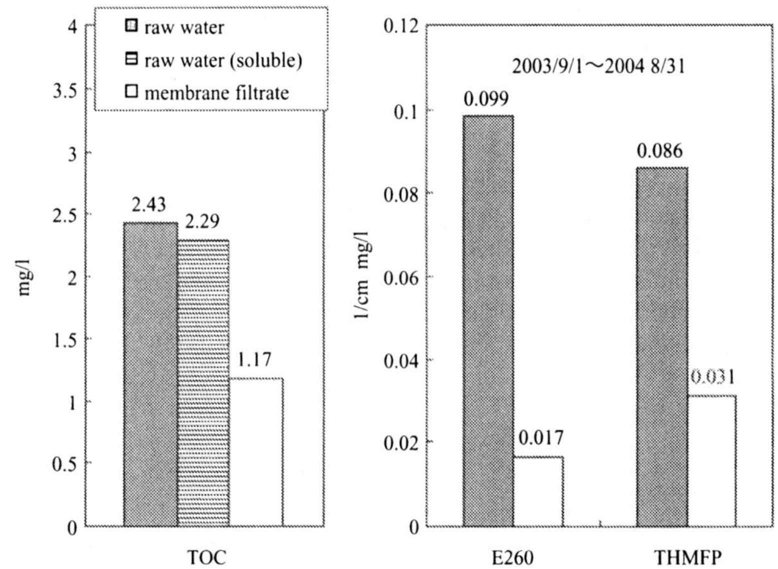

Figure 6 Removal efficiency of TOC, E260 and THMFP

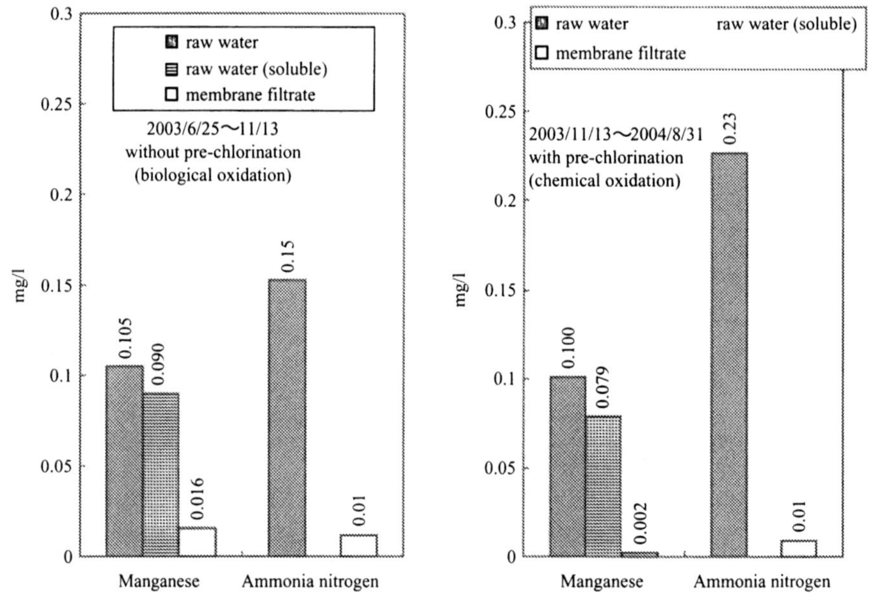

Figure 7 Effect of pre-chlorination on removal efficiency of manganese and ammonia nitrogen

raw water temperature become less than 10 °C. The authors also reported that improved filtrate quality can contribute to keep a higher flux.

Conclusions

The authors developed the hybrid MF membrane system which combines the membrane filtration and the other processes such as powdered activated carbon adsorption, coagulation and biological /chemical oxidation. This paper dealt with the experimental results obtained in the pilot plant of the hybrid membrane system. Symmetrical and asymmetrical PTFE membranes with different thickness and pore density of the skin layer were used. With asymmetrical membrane, thinner the thickness and higher pore density, higher the membrane flux. Very high removal efficiency was obtained in TOC, E260, soluble manganese and ammonia nitrogen.

References

[1] Lee, D.J. and Lee, S.H. (2000) Effect of coagulation conditions on membrane filtration characteristics in coagulation – microfiltration process for water treatment. *Environmental Science & Technology*, **34**, 3780–3788.

[2] Guigui, C., Rouch, J.C. and Durand-Bourlier (2002) Impact of coagulation condition on the in-line coagulation/UF process for drinking water, *Desalination*, **147**, 95–100.

[3] Choksuchart, P. Heran, and Grasmik, A. (2002) Ultrafiltration enhanced by coagulation in an immerged membrane system. *Desalination*, **145**, 265–272.

[4] Jang, N.Y., Watanabe, Y. and Minegishi, S. (2005) Performance of ultrafiltration membrane process combined with coagulation/sedimentation. *Wat. Sci. Technol.* **51** (6), No.6, 209–219.
[5] Snoeyink, V.L., Campos, C. and Martnas, B, J (2000) Design and performance of powdered activated carbon/ultrafiltration systems. *Wat. Sci. Technol*, **42**(12), 1–10.
[6] Matsui, Y. Fukuda, Y., Inoue, T., Matsushita, T. Aiki, N. and Mima, S. (2005) Enhancing adsorption-membrane hybrid system with microground activated carbon. *Wat. Sci. Technol.: Water Supply*, **5/6**(4), 189–197.
[7] Suzuki, T., Watanabe, Y., Ozawa, G. and Ikeda, S. (1998) Removal of soluble organics and manganese in a hybrid MF hollow fiber membrane system. *Desalination*, **117**, 119–130.
[8] Watanabe, Y., Kimura, K.and Suzuki, T. (2000) Membrane application to water purification process in Japan – development of hybrid membrane system. *Wat. Sci. Technol.*, **41**(10/11), 9–16.
[9] Suzuki, T., Morita, T., Watanabe, Y. and Ozawa, G. (2004) Hybrid MF membrane system with PTFE hollow fiber in water water purification. In *Proc. of 10th Korea/Japan Symposium on Water Environment*, Gifu-city, 85–93 October

Part II

SLOW SAND FILTRATION – PROCESS BEHAVIOUR

6 Chironomid midges: the forgotten water industry engineers?

S.P. Hurley* and R.S. Wotton**

*Research & Development, Thames Water Utilities Ltd, Kempton Park AWT, Feltham Hill Road, Hanworth, TW13 6XH, U.K.
**Department of Biology, UCL, Gower Street, London WC1E 6BT, U.K.

Abstract Slow sand filters appear simple, yet their efficient functioning is dependent on extensive microbial and invertebrate communities, with chironomid midges often the most conspicuous organisms. Known to package material, they may occur in huge numbers and can be regarded as biological engineers. A series of trials demonstrated that midge larvae have a consistent effect on headloss development in pilot scale filters. It is suggested that the larvae reduce blockage at the filter surface through their burrowing action and by the packaging of materials, both the compacting of small food particles into larger faecal pellets and the production of silk dwelling tubes which become covered in adsorbed detritus and dissolved organic matter.

Keywords Chironomid midges, larvae, headloss, slow sand filters, covers, schmutzdecke.

Introduction

Water companies are under increasing pressure to control midge flies (Diptera: Chironomidae) on their property, as the public perceives treatment works as a source of the 'nuisance swarms' common near every water body throughout the year. The covering of filters is one possible solution and previous work has shown the filtrate quality of a slow sand filter to be uncompromised by a cover at a treatment works with enhanced pre-treatment (Hurley 2004). That study showed covering to have the added advantage of increasing filter run length dramatically. However, it has been suggested that covered filters may be less effective when subjected to the higher surface loading found where pre-treatment is minimal. In these circumstances the role of midge larvae may be of increased importance in maintaining the porosity of the sand surface layer, and thus the performance of slow sand filters.

Previous workers have looked at the engineering role of invertebrates in both natural habitats and artificial systems such as slow sand filters (Berkenbusch and Rowden 2003, Jones *et al* 1994, Wotton *et al* 1996, Wotton and Hirabayashi 1999), although there has been some controversy over the use of the term engineering as applied to invertebrates (Jones *et al* 1997, Power 1997). The action of chironomids burrowing through the schmutzdecke and sand surface layers, and packaging organic matter into tubes and faecal pellets might have an effect on the filter's hydraulic performance, just as animals in other systems have been shown to increase the porosity of sediments (Aller and Aller 1992, Hargrave 1976).

To conduct repeatable trials, small-scale pilot filters were developed. These could be sampled and manipulated with ease yet still closely modelled a full-size operational slow sand filter. Headloss is the key physical parameter in slow sand filtration, since it controls the interval between the periodic cleans required as the filter surface clogs. The burrowing and packaging carried out by midge larvae is likely to reduce this clogging so filters where larvae were present should show a lower rate of headloss development than filters where larvae were absent (for a given feed water quality).

Methods

The pilot filters were constructed from 25 litre containers with a horizontal cross sectional area of 700 cm^2 and a height of 45 cm. Filter media consisted of 15 cm of sand supported by 5 cm of gravel. This allowed 20 cm of supernatant water, which was allowed to overflow ensuring a constant driving head. The filters had an outlet tube at the base which could be raised or lowered to control filtrate flow rate. Following difficulties measuring headloss, manometers were added for experiment 3. Previous headloss measurements were calibrated against this new design. Headloss figures have been normalised for temperature and flow rate as appropriate. Data were analysed using GLM ANOVA, with *post-hoc* Tukey tests.

Organic matter was added to the filters in the form of 5 litre suspensions of schmutzdecke from an operational filter and samples taken to compare their concentrations. To test the effects of midge larvae, some filters were supplied with a raw suspension containing larvae and others with a suspension passed through a 212 μm mesh to remove larvae. During Experiment 3, those larvae removed by sieving schmutzdecke suspension were added to the unsieved portion. In this way the difference between the two treatments was exaggerated.

Experiment 1 used ten pilot filters, covered to exclude light and other aerial inputs. The experiment ran for 12 days, with a single schmutzdecke suspension addition occurring on day 6. Five of the filters received sieved suspension, five received unsieved suspension. Flow rate and head loss was measured every 3 days, with flows subsequently readjusted to give a 20 cmh^{-1} vertical flow rate. Sand cores were taken every 6 days for the enumeration of midge larvae. Three cores were taken from each filter using a 2.25 mm^2 cuvette pushed into the sand to a depth of 2 cm.

Experiment 2 was identical to the first, except that the filters were open to the atmosphere.

Sampling protocols were the same during Experiment 3 which lasted 40 days. This experiment used fifteen pilot filters, five control filters which received no additions having been added. The filters were again covered but this time multiple additions were made (Table 1). Treatments were switched during the experiment in order to investigate whether chironomid establishment time had an effect.

Table 1 Patterns of schmutzdecke suspension additions in covered filters. O = no suspension added, U = unsieved suspension added, F = sieved suspension added.

Day Number	Control Filters	Initially Sieved Filters	Initially Unsieved Filters
12	O	F	U
19	O	F	U
26	O	U	F
33	O	U	F

Results and discussion

Experiment 1

There were a significantly higher number of larvae in samples from filters which received unfiltered suspension than in filters which received filtered suspension (Table 2). A significant difference in headloss development was also shown (Table 3, Figure 1). This suggests that the presence of midge larvae was reducing the rate of headloss development.

Experiment 2

Again there were a significantly higher number of larvae in those filters which received unsieved suspension (Table 2), but there was no corresponding difference in headloss development on this occasion (Table 3, Figure 1). The lack of headloss development in open filters might suggest some other biological activity occurs, such as the growth of a straining algal layer (Nakamoto *et al* 2000, Sakai and Nakamoto 1994).

Experiment 3

The number of midge larvae in filters receiving the different treatments were again significantly different (Table 2). However we can see that there are changes in numbers over the course of the experiment (Figure 2). *Post-hoc* Tukey tests show that there were initially higher numbers of midge larvae in those filters which received unfiltered suspension (Figure 2). However after the switch in treatments was made, there was no significant

Table 2 ANOVA table of comparisons of total numbers of midge larvae in core samples from pilot filters. DF = degrees of freedom, F = variance ratio, P = probability.

	Experiment 1 (Covered)			Experiment 2 (Open)			Experiment 3 (Covered)		
Factor	DF	F	P	DF	F	P	DF	F	P
Date	1	22.70	0.000	1	23.65	0.000	4	56.58	0.000
Treatment	1	78.83	0.000	1	5.56	0.032	2	145.95	0.000
Interaction	1	22.70	0.000	1	5.56	0.032	8	16.93	0.000

Table 3 ANOVA table of comparisons of headloss figures for pilot filters. DF = degrees of freedom, F = variance ratio, P = probability.

	Experiment 1 (Covered)			Experiment 2 (Open)			Experiment 3 (Covered)		
Factor	DF	F	P	DF	F	P	DF	F	P
Date	4	12.67	0.000	4	2.53	0.056	14	48.36	0.000
Treatment	1	19.88	0.000	1	0.25	0.620	2	175.22	0.000
Interaction	4	12.98	0.000	4	0.67	0.619	28	4.85	0.000

difference between the filters receiving the suspensions. Larvae numbers in the control filters were significantly lower than in the other filters after the additions commenced.

There was a clear difference in headloss development between the treatments (Figure 3, Table 3). However *post-hoc* Tukey tests show that this was due to differences between the control filters and the filters receiving suspensions. At no point was there a significant difference between the two suspension treatments. On the final day there was no significant

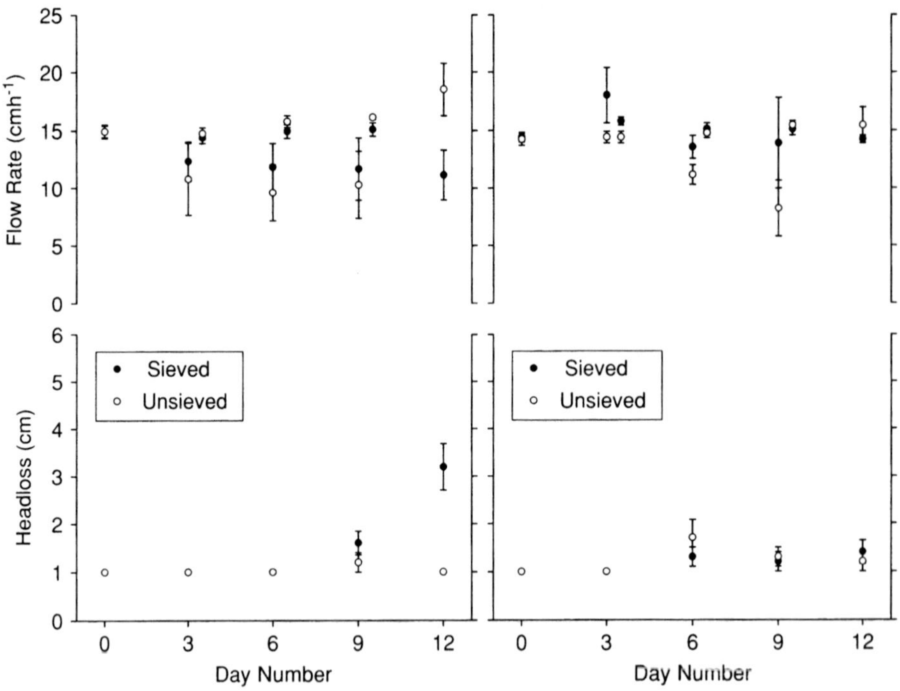

Figure 1 Flow rate and headloss comparisons for (a) covered and (b) open pilot slow sand filters. Filters received an addition of 5 litres of schmutzdecke suspension after sampling on day 6. Five filters received suspension that was sieved to remove midge larvae five received unsieved suspension. Numbers are means of five samples with standard errors. Flow rates were adjusted on each visit, both pre and post adjustment values are shown.

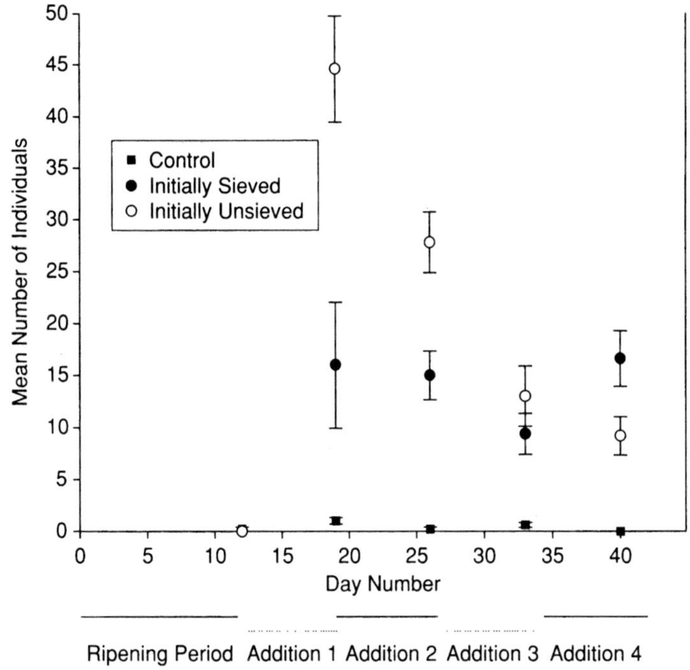

Figure 2 Total number of midge larvae found in core samples taken from covered pilot slow sand filters. The raw water source was common to all filters throughout. Those filters labelled initially sieved received 5 litres of schmutzdecke suspension, which had been sieved to remove larvae on the 12th and 19th days. They then received unsieved suspension on the 26th and 33rd days. Those labelled initially unsieved received the opposite regime of additions. Control filters received no additions of suspension. Core samples were taken before the scheduled additions were made. Numbers are means of five samples with standard errors, each sample being the combined result of three 2.25 cm^3 cores from a filter.

difference between the control filters and those which initially received unfiltered suspension, whereas there continued to be a difference between the control filters and those which initially received sieved suspension. This goes some way to support the apparent trend of increased headloss development in those filters which initially received sieved suspension. A longer experiment may have confirmed this speculation, but the increasing blockage of the surface layer prevented maintenance of the required flow through those filters which initially received sieved suspension, preventing any extension to the planned trial.

In Experiment 3, 12 days were allowed to pass before any additions of suspension were made rather than the 6 days of the previous experiments. This was because samples of the filtrate in Experiments 1 and 2 showed that water quality was improving throughout the length of the trials, suggesting continued ripening of the filters. This supports operational observations of increased ripening periods for covered filters. Previous work has shown un-ripened filters to be unable to cope with sudden increases in raw water turbidity, even in the short term (Cullen and Letterman 1985). The same study found that such turbidity

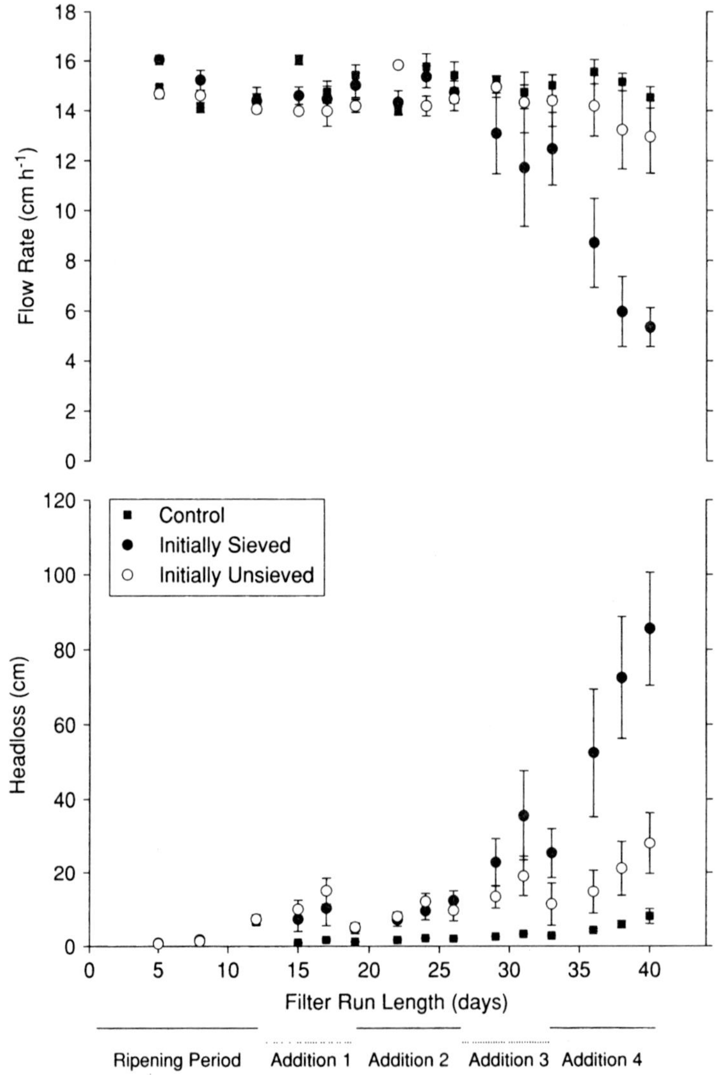

Figure 3 Flow rate and normalised headloss data for small-scale covered pilot slow sand filters. The raw water source was common to all filters throughout. Those filters labelled 'initially sieved' received 5 litres of schmutzdecke suspension, which had been sieved to remove midge larvae on the 12th and 19th days. They then received unsieved suspension on the 26th and 33rd days. Those labelled 'initially unsieved' received the opposite regime of additions. Control filters received no additions of suspension. Samples were taken before the scheduled additions were made. Numbers are means of five samples with standard errors. Flow rates were adjusted on each visit, only post adjustment values are shown for clarity.

'spikes' are, however, easily absorbed by a ripened filter, further indicating the importance of the biological community within the bed. The extension in ripening time for all the filters allows greater confidence in the conclusions on the possible effect of midge larvae on headloss development in Experiment 3.

The exact reasons for the slower ripening of a covered filter are unclear. The lack of obvious algal growth at the surface is likely to have an effect, perhaps through a reduction in the straining effect of the surface layer (Sakai and Nakamoto 1994), or the reduction in biological surfaces for development of colonies of micro-organisms. It is possible that the presence of algae and associated micro-organisms accelerates the ripening of open filters. In particular the release of dissolved organic matter (DOM) such as polysaccharides, by both living and decaying algae (Mann 1988), might influence the development of the biology of open filters, it is certainly important to the growth of bacteria and associated grazers (Salonen and Hammar 1986). DOM is important for its physical effects as well as its role in the food chain. Wotton (2002) has stressed the role of exopolymers in promoting flocculation, and the artificial addition of polymers to enhance ripening of filters has been investigated, although with limited success (Jellison *et al* 2000).

The first two, shorter, experiments showed only slight headloss increases, but illustrate differences between covered and open filters. The absence of midge larvae in filters which received a pulse of organic matter appeared to cause an increase in the headloss of covered filters, but not open filters (Table 3, Figure 1). However these were very short trials, a fact which must be borne in mind when interpreting the results.

The final experiment was more reminiscent of an operational filter run, the longer time period resulting in a more pronounced headloss development. The control filters, which received only the feed water, showed a minimal headloss development over the course of the experiment (Figure 3). This suggests that covered filters, if protected from periodic deteriorations in water quality, might be able to function well with feed water of a lower quality than that found previously (Hurley 2004). The other two treatment groups show similarly low headloss development until the time of the switch in treatments after 28 days. After this, the headloss of those filters that had not established a midge larval community started to rise rapidly, approaching terminal headloss at the end of the trial. This occurred despite the addition of midge larvae with the organic pulses. However the filters which already had an established midge larval community continued to operate well.

This suggests that it is important for any midge larval community to have time to establish itself before being faced with high organic loadings. Those filters that received larvae early in the experiment would have benefited from the packaging of material, through the compacting of small food particles into larger faecal pellets and by the production of silk dwelling tubes which become covered in adsorbed detritus (Wotton and Hirabayashi 1999), before the filter surface had had time to be blocked by the organic additions. In turn, such a substratum might be easier for midge larvae to colonise than one having already received additions of organic matter, and which was already showing signs of blocking. In such circumstances larvae may have been less likely to settle (Chaloner and Wotton 1996) and may have remained in the water column whilst searching for a more suitable substratum (Lancaster 1990, Winterbottom *et al* 1997, Zimmer *et al* 2001).

Conclusion

Midge larvae have a significant effect on the development of headloss in a slow sand filter and attempts to remove these 'engineers' from the system must take into account the changes in filter operation that result.

At the last slow sand filter conference, Wotton *et al* (1996) suggested that midge larvae have an effect on filter operation and asked whether this role was being overlooked. This study provides experimental evidence on the role midge larvae play in controlling filter headloss development and suggests that water engineers and biologists have started a mutually beneficial dialogue.

The views expressed in this paper are those of the authors and not necessarily those of Thames Water Utilities Ltd. The authors would like to thank TWUL and BBSRC for their funding and Dr Helen Evans, Dr Michael Chipps and Robin Bayley of TWUL R&D for practical support.

References

Aller RC and Aller JY (1992) Meiofauna and solute transport in marine muds. *Limnology and Oceanography* **37**(5), 1018–1033.

Berkenbusch K and Rowden AA (2003) Ecosystem engineering - moving away from 'just-so' stories. *New Zealand Journal of Ecology* **27**(1), 67–73.

Chaloner DT and Wotton RS (1996) Substratum preferences by larvae of three species of midge (Diptera: Chironomidae). *Hydrobiologia* **339**, 93–99.

Cullen TR and Letterman RD (1985) The Effect of Slow Sand Filter Maintenance On Water-Quality. *Journal American Water Works Association* **77**(12), 48–55.

Hargrave BT (1976) The central role of invertebrate faeces in sediment decomposition. In *The role of terrestrial and aquatic organisms in decomposition processes* Anderson, J M and Macfadyen, A. Blackwell Science, Oxford pp322.

Hurley SP (2004) The role of macroinvertebrates in slow sand filtration. Unpublished PhD Thesis, University College London.

Jellison KL Dick RI and Weber-Shirk ML (2000) Enhanced ripening of slow sand filters. *Journal of Environmental Engineering* **126**(12), 1153–1157.

Jones CG Lawton JH and Shachak M (1997) Ecosystem engineering by organisms: Why semantics matters. *Trends in Ecology and Evolution* **12**(7), 275–275.

Jones CG Lawton JH and Shachak M (1994) Organisms as ecosystem engineers. *Oikos* **69**(3), 373–386.

Lancaster J (1990) Predation and drift of lotic macroinvertebrates during colonization. *Oecologia* **85**(1), 48–56.

Mann KH (1988) Production and use of detritus in various fresh-water, estuarine, and coastal marine ecosystems. *Limnology and Oceanography* **33**(4), 910–930.

Nakamoto N Yasuda M Sakai M and Nozaki K (2000) Oxygen production and bubble formation by algal mat on the bed in a slow sand filter pond. *Verh.Internat.Verein.Limnol.* **27**, 3092–3096.

Power M (1997) Ecosystem engineering by organisms: Why semantics matters - Reply. *Trends in Ecology and Evolution* **12**(7), 275–276.

Sakai M and Nakamoto N (1994) Role Of Filamentous Algae In A Slow Sand Filter. *Proceedings Of The International Association For Theoretical & Applied Limnology, Congress In Barcelona,1992* **25**, 1993–1995.

Salonen K and Hammar T (1986) On the importance of dissolved organic-matter in the nutrition of zooplankton in some lake waters. *Oecologia* **68**(2), 246–253.

Winterbottom JH Orton SE and Hildrew AG (1997) Field experiments on the mobility of benthic invertebrates in a southern English stream. *Freshwater Biology* **38**(1), 37–47.
Wotton RS (2002) Water purification using sand. *Hydrobiologia* **469**(1-3), 193–201.
Wotton RS Chaloner DT Armitage PD (1996) The Colonisation, Role In Filtration & Potential Nuisance Value Of Midges In Slow Sand Filter Beds. In *Advances In Slow Sand & Alternative Biological Filtration* Graham, N. J. D. and Collins, R. Wiley pp 157.
Wotton RS and Hirabayashi K. (1999) Midge larvae (Diptera, Chironomidae) as engineers in slow sand filter beds. *Water Research* **33**(6), 1509–1515.
Zimmer KD Hanson MA Butler MG and Duffy WG (2001) Size distribution of aquatic invertebrates in two prairie wetlands, with and without fish, with implications for community production. *Freshwater Biology* **46**(10), 1373–1386.

7 Role of algal growth and photosynthesis in slow sand filters as an advanced wastewater treatment

Noriyasu IWASE*, Satoshi KINOSHITA, Makoto KOJIMA*** and Nobutada NAKAMOTO******

*Fukuoka Branch Office, Nippon Jogesuido Sekkei Co., Ltd., 2-2 Tsunaba-cho, Hakata-ku, Fukuoka-city, Fukuoka, 812-0024, Japan
**Main Office, Nippon Jogesuido Sekkei Co., Ltd., 1-9-15 Kaigan, Minato-ku, Tokyo, 105-0022, Japan
***Hiroshima Branch Office, Nippon Jogesuido Sekkei Co., Ltd., 2-1-23 Kokutaiji-cho, Chuou-ku, Hiroshima-city, Hiroshima, 730-0042, Japan
****Department of Applied Biology, Faculty of Textile Science and Technology, Shinshu University, 3-15-1 Tokida, Ueda-city, Nagano, 386-8567, Japan

Abstract Bench-scale experiments were conducted to assess the role of algal growth and photosynthesis in slow sand filters (SSFs) as an advanced wastewater treatment. Algal growth and photosynthesis enhanced by sunlight caused larger amount of biomass mud in the sand bed without any additional increase of head loss; oxygen supplied by algal photosynthesis activated biodegradation, and the system could be operated for 38 days without any maintenance. Roughing filtration as a preliminary treatment was observed to be effective to reduce the load of suspended matter to SSFs and minimize the head loss increase. Based on a life-cycle cost analysis, about 30% of total cost might be saved on full-scale SSFs compared with rapid sand filters (RSFs).

Keywords Slow sand filter; rapid sand filter; algal growth; photosynthesis; oxygen; advanced wastewater treatment; Life-cycle cost (LCC)

Introduction

Low operational cost and small energy consumption are distinctive characteristics of slow sand filters (SSFs) compared with rapid sand filters (RSFs). In Japan SSFs are used in water purification plants where the quality of raw water is relatively pure, but some reports suggest that they are applicable even where the raw water is more polluted, including the advanced wastewater treatment (Eighmy *et al.*, 1988; Ellis, 1987; Farooq and Al-Yousef, 1993). Moreover, it has been demonstrated that the sunlight enhances the growth of filamentous algae in the filtration tank and oxygen produced by photosynthesis promotes the treatment capacity (Nakamoto *et al.*, 1996).

 Recent Progress in Slow Sand and Alternative Biofiltration Processes edited by Rolf Gimbel, Nigel J.D. Graham and M. Robin Collins. ISBN: 9781843391203. Published by IWA Publishing, London, UK.

Bench-scale experiments were conducted to assess the applicability of SSFs as an advanced wastewater treatment; the effect of physical straining, biodegradation, algal growth and photosynthesis in the sand bed was estimated, as well as the effect of roughing filtration as a preliminary treatment.

Methods

Bench scale experiment plant

Bench scale experiment plants were set up in an existing wastewater plant. Raw water for this experiment was taken from the final sedimentation tank before chlorination.

Slow sand filtration tank was made of polyethylene, of which size was 1,390mm (inside diameter) and 1,545mm (depth). Concrete blocks of 100mm's height were laid at the bottom of the tank for underdrainage. Gravel was laid with depth of 300mm between the concrete blocks and sand bed, with effective size of gravel varied from 2mm to 40mm. Sand bed was of 550mm's depth, using sand commonly adopted in rapid sand filters (i.e. effective size 0.608mm and uniformity coefficient 1.39). The surface of sand bed was kept 500mm's depth from the top of each tank in order to supply enough sunlight for algal photosynthesis. The device for pre-chlorination was also attached.

Upflow roughing filter tank of the same size and material was situated at higher level of 200mm than the SSF tank considering the head loss. Roughing filter was of 850mm's depth, using plastic hollow column as filter medium, of which size was about 35mm (inside diameter), 50mm (length) and 1mm (thickness).

Experimental setups

Table 1 shows the experimental conditions. **Figure 1** and **Figure 2** show the system flow diagram of each Run.

Run1 and Run2 were intended to assess the physical straining process in SSFs without microbial activity and hence the raw water was pre-chlorinated to inhibit the microbial activity. In Run1 roughing filter was also added as a pretreatment in order to retain larger solids.

Run3 and Run4 were intended to assess the biological removal process of SSFs. In Run3 the filtration tank was covered with shading to avoid sunlight and hence the algal growth and photosynthesis.

Table 1 Experimental conditions

					Average raw water quality			
Run	**Start date**	**Roughing Filter**	**Pre-chlorination**	**Shading Cover**	**Water temp. °C**	**DO mg/L**	**EC μs/cm**	**pH -**
1	Dec.	o	o	x	18.0	3.1	2,484	7.0
2	1999	x	o	x				
3	Dec.	x	x	o	16.0	2.7	1,562	7.3
4	2001	x	x	x				

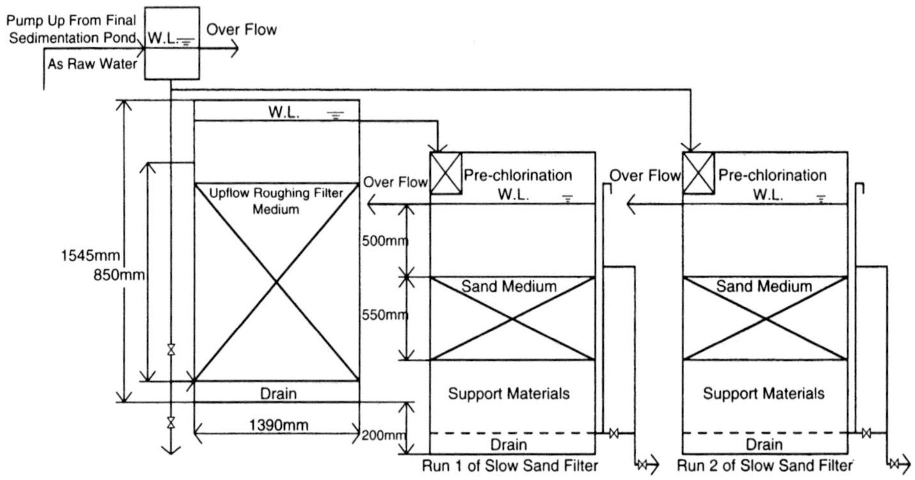

Figure 1 System flow diagram in Run1 and Run2

Each run was done until total head loss reached the range of about 500mm and 1000mm.

Life-cycle cost analysis

The cost of construction and operation of full-scale SSFs was estimated by draft designing for some given treatment capacities; functions were formulated using treatment capacity as a parameter for required area, construction cost and operation cost. The cost of RSFs was estimated using functions prepared for the advanced wastewater treatment plant (JSWA, 1999).

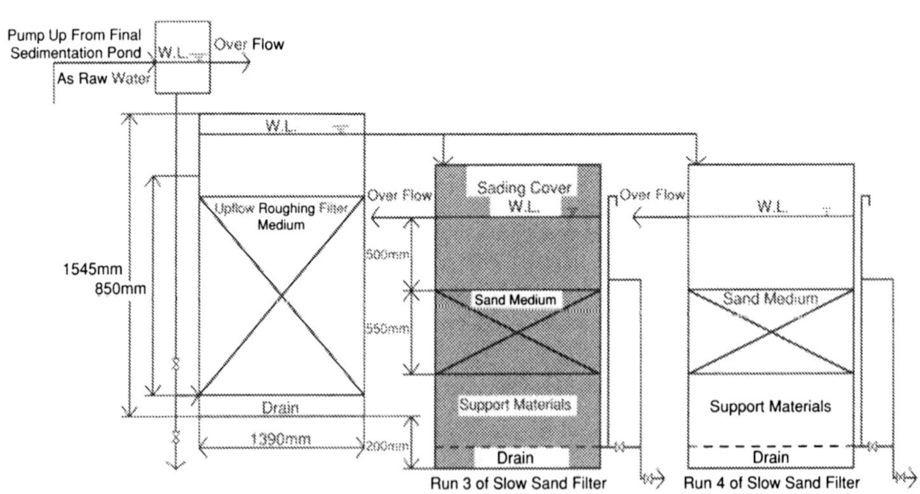

Figure 2 System flow diagram in Run3 and Run4

Table 2 Functions for required area, construction and Operation

		Unit	Function	Exchange to 10^3 * euro
For RSFs	Required Area	m^2	$= 0.0165 * Q + 1.852$	
	Construction	10^6 yen	$= 353.14 * Q^{0.4646}$	$7.178 * 353.14 * Q^{0.4646}$
	Operation	10^6 yen	$= 0.9075 * Q^{0.9220}$	$7.178 * 0.9075 * Q^{0.9220}$
For SSF	Required Area	m^2	$= 0.2365 * Q + 0.9938$	
	Construction	10^6 yen	$= 154.42 * Q^{0.723}$	$7.178 * 154.42 * Q^{0.723}$
	Operation	10^6 yen	$= 0.7500 * Q^{0.8}$	$7.178 * 0.7500 * Q^{0.8}$

Q = Treatment Capacity; $10^3 * m^3$/day
1000 yen = 7.178 euro

Functions obtained for SSFs and used for RSFs are shown in **Table 2**. Annual cost for chemicals and electricity is counted for operation cost, whereas sludge disposal and operating personnel expenses are not included.

Life-cycle cost was calculated for 50 years under the conditions shown in **Table 3.**

Results and discussion

Roughing filter and pre-chlorination

Figure 3 shows the daily increase of normalized head loss and **Table 4** shows the average water quality of inlet and outlet in Run1 and Run2. In Run1 and Run2 the pre-chlorinated filter was under operation for 14 days before clogging (i.e. stop because of a rising of head loss rapid). Total removal rate of SS in both Runs was almost the same level. Roughing filtration in Run1 was observed to be effective to reduce the load of suspended matter to SSF and prevent the increase of the head loss.

Table 3 Conditions of Life-cycle cost analysis

	Conditions	Notes
Life cycle	50 years	
Useful years of SSFs	75 years	There is an existing plant in Japan under operation for more than 100 years. Calculated Construction cost is multiplied by 50/75.
Useful years of RSFs	33 years	Weighted average value of useful years of a sewage treatment plant (JSWA, 2002). Calculated Construction cost is multiplied by 50/33.
Discount rate for net present value	0.0%	The official discount rate in Japan is 0.10%.
Land price	30,000 yen/m^2 (=215€/m^2)	

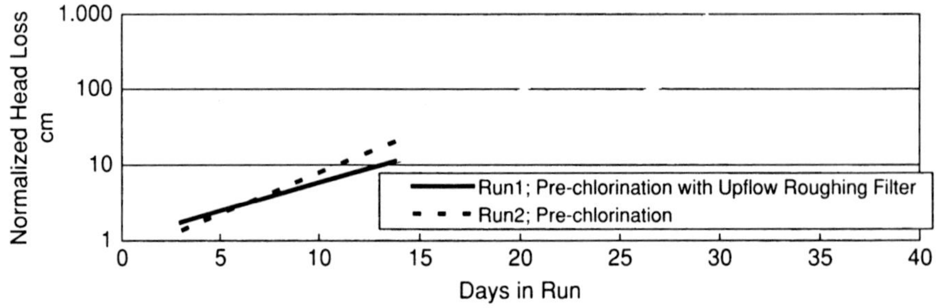

Figure 3 Daily Increase of Normalized Head Loss in SSF at Run1 and Run2

Table 4 Average water Quality of Inlet and Outlet at Run1 and Run2

			Formula	Water temp. °C	pH -	Dissolved Oxygen mg/L	Dissolved Oxygen %	SS mg/L
Inlet			A	18	7.0	3.1	34%	9.7
Outlet		Upflow Roughing Filter	B	17.7	7.0	2.1	23%	5.9
		Removal Coefficient	C=(1-B/A)* 100	-	-	-	-	39%
	Run 1	Filtered	D	16.8	7.0	2.0	22%	2.1
		Removal Coefficient	E=(1-D/B)* 100	-	-	-	-	64%
		Total removal Coefficient	F=(1-D/A)* 100					78%
	Run 2	Filtered	G	16.6	7.0	2.2	24%	2.2
		Removal Coefficient	H = (1-G/A)* 100	-	-	-	-	77%

Shading cover and sunlight

Figure 4 shows the daily increase of normalized head loss and **Table 5** shows the average water quality of inlet and outlet in Run3 and Run4. In both Runs as long as 38 days were required before clogging, longer than Run1 and Run2 (i.e. because head loss had been increasing in both Runs , filter rates had been decreased from 26th days). The removal rate of SS was improved compared with in Run1 and Run2. These indicate the importance of biodegradation process.

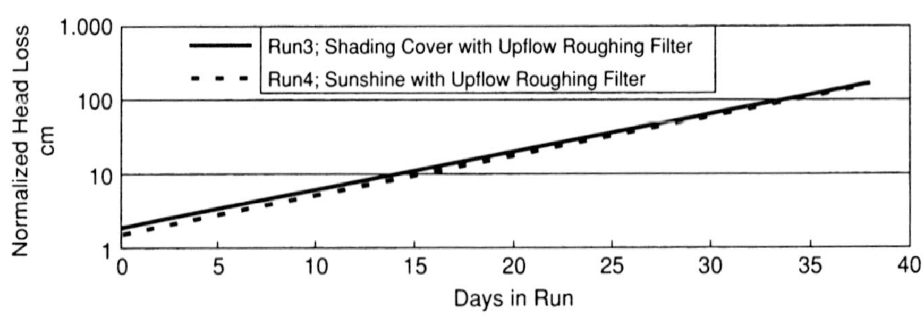

Figure 4 Daily increase of Normalized Head Loss in SSF at Run3 and Run4

Table 5 Average water Quality of Inlet and Outlet at Run3 and Run4

			Formula	Water temp. °C	pH -	Dissolved Oxygen mg/L	Dissolved Oxygen %	SS mg/L
Inlet			A	16	7.4	5.1	53%	7.8
Outlet	Run 3	Shading cover	B	13	7.3	3.7	36%	0.9
		Removal Coefficient	C=(1-B/A)*100	-	-	-	-	89%
	Run 4	Sunshine	D	12	7.1	4.5	43%	0.7
		Removal Coefficient	E=(1-D/A)*100	-	-	-	-	91%

Figure 5 and **Figure 6** show the daily increase and vertical density of Chlorophyll a (Chl-a) plus Pheophytin a (Pheo-a) in the sand bed. Concentration of suspended solids retained in the sand bed is also shown in **Figure 6**. Suspended solids retained in the sand bed are almost the same in both Runs, while in Run4 larger amount of biomass was observed than in Run3, which suggests that sunlight in Run4 enhanced algal growth and photosynthesis.

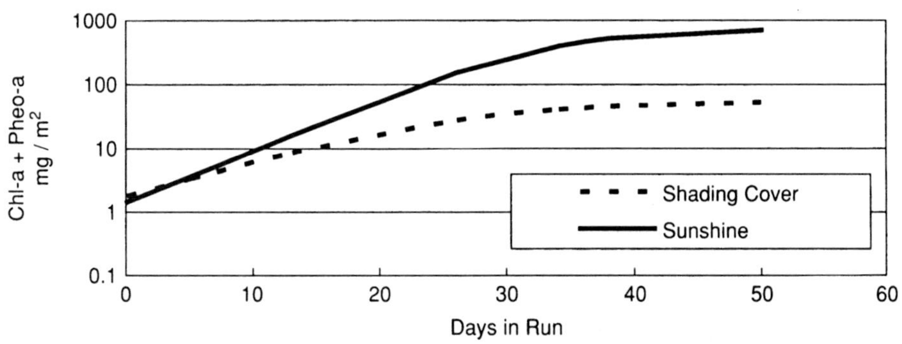

Figure 5 Daily increase of mass of algae in the sand bed at Run3 and Run4

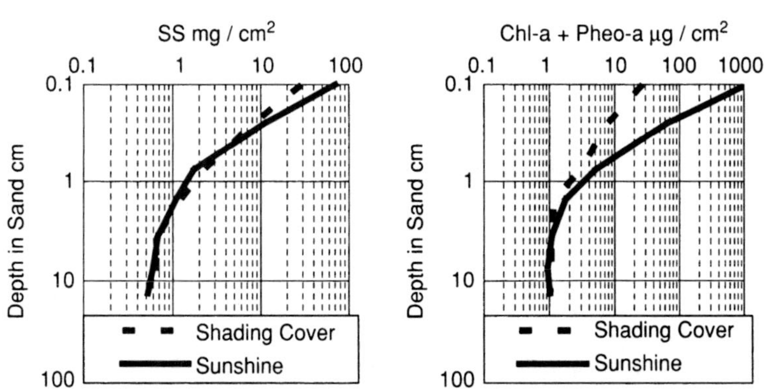

Figure 6 Vertical profiles of SS and Chl-a + Pheo-a in the sand bed

Table 6 Mass balance in Run3 and Run4 for 38 days (unit: g/m^2)

		formula	Run3 Shading Cover	Run4 Sunshine
DO	Load of Inlet	A	817.3	813.9
	Load of Outlet	B	669.8	811.6
	Decreaced	C = A-B	147.5	2.3
	Production	C = K * I-C	0.0	150.6
	Consumption	E = C + D	147.5	152.9
VS	Load of Inlet	F	758.3	752.0
	Load of Outlet	G	306.7	286.6
	Removed	H = F-G	451.6	465.4
	Decomposition	I = H-J	375.6	389.4
	Anaerobic Decomposed	L = I-E/1.2549*	258.1	267.6
	Traped	J	76.0	110.0
Ratio of Oxygen / VS		$K = C_{Run3}/I_{Run3}$		

Almost the same level of head loss and suspended solids in both Runs may be explained that the oxygen-supply by algal photosynthesis activated biodegradation process in Run4. In fact, dissolved oxygen concentration in the filter effluent in Run4 was higher than Run3.

Mass balance

Table 6 shows the mass balance in Run3 and Run4 for operational 38 days. Anaerobically decomposed VS is rather too much compared with aerobically decomposed VS. It might be helpful to study the case of other wastewater treatment plant or otherwise water purification plant.

Life-cycle cost analysis

Life-cycle cost calculated is shown in **Figure 7**. About 30% of total cost might be saved by SSFs instead of RSFs. The rate becomes relatively advantageous for SSFs so that scale becomes small.

Conclusions

Following results are obtained through bench-scale SSF experiments using secondary treated wastewater as raw water: a) algal growth and photosynthesis enhanced by sunlight cause larger amount of biomass mud in the sand bed without any additional increase of head loss; oxygen supplied by algal photosynthesis activates biodegradation; b) SSF can be operated for about 40 days without any maintenance; c) upflow roughing filtration as a preliminary treatment is effective to minimize the head loss increase.

Based on a life-cycle cost analysis, about 30% of total cost can be saved on full-scale SSFs compared with RSFs.

It is concluded that SSFs might be applicable to advanced wastewater treatment.

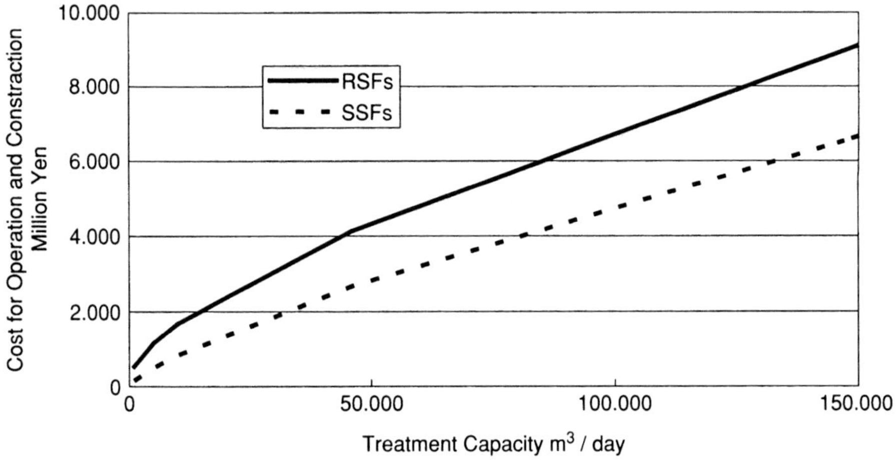

Figure 7 Life-cycle cost of SSFs and RSFs

Acknowledgements

Support of Mr. Hideo Takita is greatly appreciated. We are grateful to Mr. Chowi Kan and staff of the Technical Services Department of NJS for their discussion. We thank staff of the Hiroshima branch office of NJS for management of experiment plant. Shinko-Tec Co., Ltd. did great service to us about operating of experiment plant. We also thank Mr. Toshihiko Tamama for his discussion and carefully reading manuscript very much.

References

Eighmy, T.T., Collins, M.R., Spanos, S.K. and Fenstermacher, J. (1988) Microbial Population Distributions and Benzoate Mineralization Kinetics in a Municipal Slow Sand Filter. *Wat. Sci. Tech.*, **20** ,11–12.

Ellis, K.V. (1987) Slow Sand Filtration as a Technique for the Tertiary Treatment of Municipal Sewages. *Wat. Res.*, **21** (4), 403–410.

Farooq, S., Al-Yousef, A.K. (1993) Slow Sand Filtration of Secondary Effluent. *Environmental Engineering*, **119** (4-6) , 615–630.

Nakamoto, N., Yamamoto, M., Sakai, M., Nozaki, K., Iwase, N., Yasuda, M., Kitada, T., and Hirabayashi, K. (1996) Role of Filamentous Diatom as an Automatic Purifier in a Slow Sand Filter. In *Proc. 3rd International Slow Sand/Advanced Biological Filtration Conference*, 139–148.

Japan Sewage Works Association (1999). A guideline for planning and designing sewer system in river basin (in Japanese).

Japan Sewage Works Association (2002). Design manual for effective sewage treatment facilities in metropolitan areas (in Japanese).

8 Development pattern of filamentous diatom and its condition related with midge larvae in slow sand filter

N. NAKAMOTO* and H. KATO**

*Department of Applied Biology, Shinshu University, Ueda, Nagano 386-8567, Japan
**Present address: Aqua Environmental Business Division, Development Group, R&D Centre, Yamaha Motor Co. Ltd., Iwata, Shizuoka 438-8501, Japan

Abstract Growth pattern of filamentous diatom on slow sand filter bed was checked. During the cold season in February, cell condition of diatom became active along the filter run. However, in August, development of algal mat was rapid until one week and cell condition became active. Then cell condition became bad and small number of *Melosira* colony was abundant on the sand bed. It was caused by grazing activity of midge larvae. However, filter efficiency of resistance and clearness of filtrate was not influenced.

Keywords *Melosira*, cell condition, filamentous diatom, midge, grazing activity, head loss

Introduction

Slow sand filtration has been recognized as a biological purification system passing through the sand filter under slow water current. It has been considered the main purification process was physical process of trapping small particle by small sand layer. We emphasized the role of filamentous diatom as an automatic purifier in slow sand filter bed (Nakamoto et al. 1996). Filamentous diatom is usually dominant in the open slow sand filter bed which water source is a surface water of a river in Japan. It is necessary to remove the accumulated silt material on the surface of sand bed to keep better filter condition. Abundant casting shell of midge larvae on the surface of a filter pond was sometimes remarkable in warm period. We checked developmental pattern of algal mat with special reference to the algal cell condition. Grazing effect of midge larvae on filamentous diatom was remarkable in warm period. In this report, filter condition did not influenced when filamentous diatom was grazed in warm period. Animal activity is one of the key factors to keep the better filter condition.

Study plant and method

The study was done at the Wakata water works, Takasaki city, Japan which had 10 slow sand filters. Area of one filter bed is 1,250 m^2. Surface water from a river flows into a sedimentation basin, and the supernatant water flows into the slow sand filters. Residence

 Recent Progress in Slow Sand and Alternative Biofiltration Processes edited by Rolf Gimbel, Nigel J.D. Graham and M. Robin Collins. ISBN: 9781843391203. Published by IWA Publishing, London, UK.

time of water is about 8 hours in a sedimentation basin. And the water depth of slow sand filter was about one meter and the speed of filtering rate was 3 to 6 m per day during study period. The maximum water temperature of monthly average was 22.9C in July and the minimum was 4.1C in January. The normal turbidity level of intake water was about 0.4 to 2.2 mg per litre. In summer, the raw water into the water works sometimes became turbid water of 7 to 23 mg per litre due to storm event. The surface skin of sand bed is usually scraped after about every 25 days of filter run at this water works.

Algal mat on the sand bed was sampled quantitatively during a filter run using an algal mat sampler (Nakamoto, 1988, 1993). Three spots (10 $cm^2 \times 3$) were taken in each filter bed. Sampling was done 3 times in every other month from December 2003 to October 2004. Chlorophyll *a* content per area in algal mat was analyzed by composite samples by Lorenzen method (1967). Filamentous diatom of *Melosira varians* dominated during filter run in all season. Cell condition of *Melosira* colony was evaluated by colony length of cell number and the presence of protoplasm in the cell using microscopic observation. Number of midge larvae and body size were also counted in the samples.

Results and discussion

Developmental patterns of algal mat as in chlorophyll *a* on the sand filter are shown in Figure 1 and 2. In February, the initial stock was about 2 chl. *a* mg m^{-2} at the beginning.

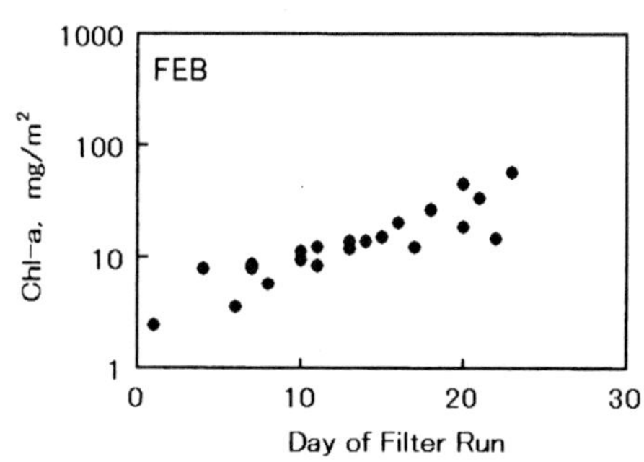

Figure 1 Growth of algal mat as chl-a during the filter run in February.

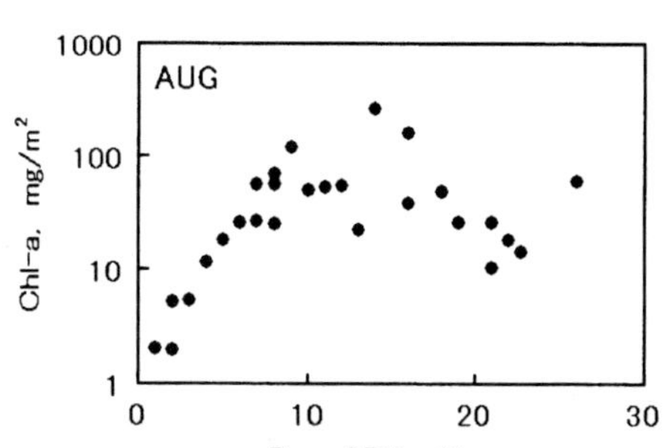

Figure 2 Growth of algal mat as chl-a during the filter run in August.

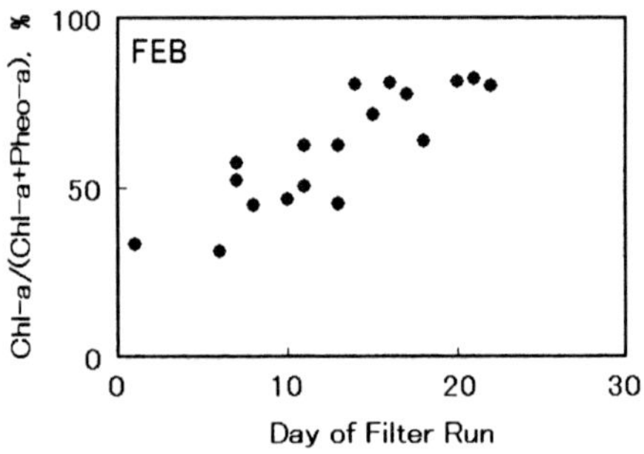

Figure 3 Change of chlorophyll to total pigments during the filter run in February.

The biomass increased logarithmically and reached about 50 chl. *a* mg m^{-2} on the 20th day. In August, the remarkable logarithmic growth of algal mat kept only 10 days. The chlorophyll amount exceeded over 100 chl. *a* mg m^{-2} then the amount decreased. Algal activity was evaluated by the ratio of chlorophyll *a* to total pigments (chlorophyll *a* and pheophytin *a*) in Figure 3 and 4. In February, algal activity was bad condition at the beginning and its value became well condition during the filter run. In August, algal activity was bad at the beginning. And it became well condition until 10 days then its activity decreased. These patterns during the filter run were similar to biomass change. This indicates that the period for full algal mat with active condition of algal colony in the filter bed takes many days in cold season. On the other hand, the time to development of well algal mat needs only 10 days in warm period. However, it seems that the long filter run is not good for algal condition in warm period.

The chain length of *Melosira* colony was checked in August. At the initial day, several cell number of chain was only observed. Then, chain length expanded rapidly until one week. After that, the chain length became again small. Small colony which was composed with small number of cell had almost none of protoplasm at the initial days and at the long filter run (Figure 5 and Figure 7). The cell of the long chain with full of protoplasm (full) appeared during the filter run. This cell was under healthy condition on the filter bed. However, after 10 days, the empty cell of protoplasm (emp) and the cell with the partial full of protoplasm (sep) became remarkable. This cell condition was caused by the grazing

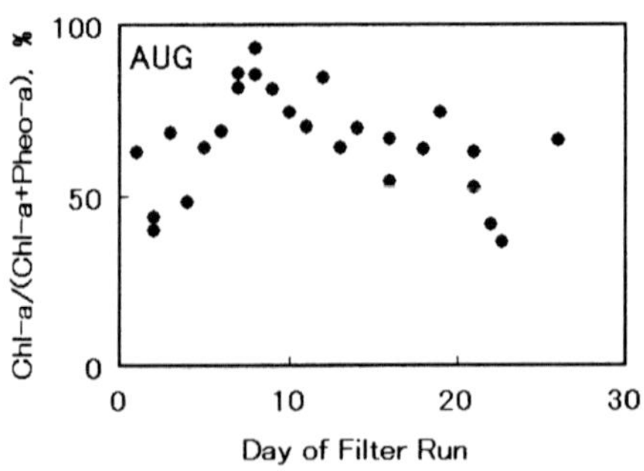

Figure 4 Change of chlorophyll to total pigments during the filter run in August.

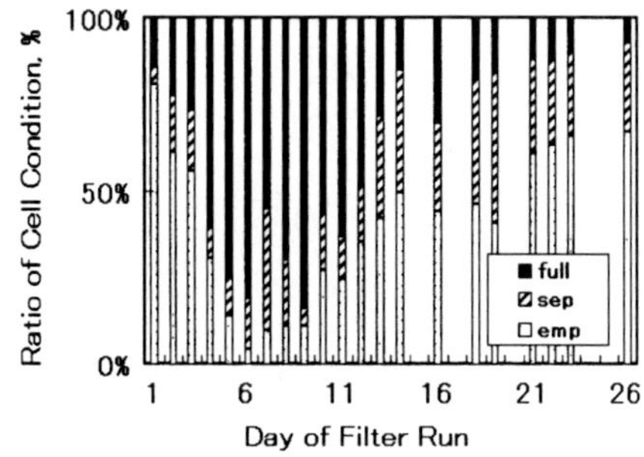

Figure 5 Cell condition change of *Melosira* during the filter run in August.

activity of midge larvae. There was abundant faecal pellet of midge (Chironomidae) larvae on the filter bed. We checked the biomass of midge on the filter bed (Figure 6). Number of midge larvae increased with filter run. After one week, the larger size of midge dominated. Grazing activity of midge larvae on *Melosira* became larger after one week. As a result, there were so many empty cells of filamentous diatom, small chain of colony, and unhealthy diatom on the bed in case of long filter run. However, it was confirmed that there were healthy cells of diatom and long chains of filamentous diatom in long filter run in February.

We checked the filter resistance from the Water works data. In warm period, there was no evidence of increase of filter resistance with filter run (Figure 8) in spite of the in load increase of suspended matter to the filter bed (Figure 9). It means that small animals including midge larvae are active on and beneath the sand surface in warm period. Growth of algae provides food for animals. Animals on the sand bed and beneath the sand surface trap suspended matter in water. Nakamoto et al. 1997 reported the purity of filtrate water was evaluated with suspended matter of chlorophyll during filter run. After the one week, the filtrate water kept the clearness for long filter run. In case of longer filter run in warm period, a remarkable growth of filamentous green algae and snails are observed successively after filamentous diatom bloom. It may indicate that animal activity is important to eliminate suspended matter and to keep small filter resistance. We emphasized the role of filamentous diatom to keep better condition of slow sand filter (Sakai and Nakamoto 1994). It was one side of role of algal activity.

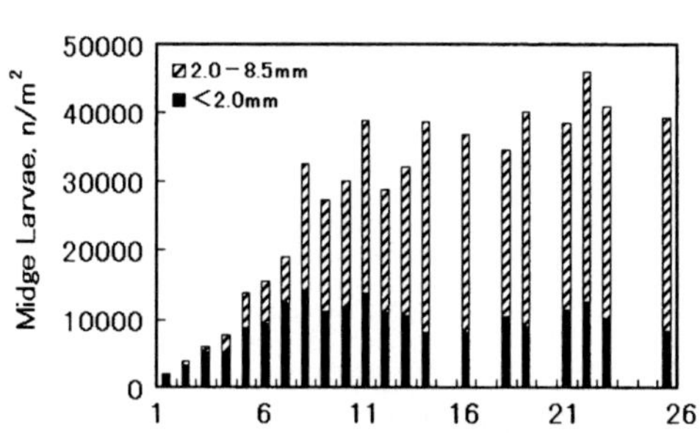

Figure 6 Change of number and body size of midge larvae during the filter run in August.

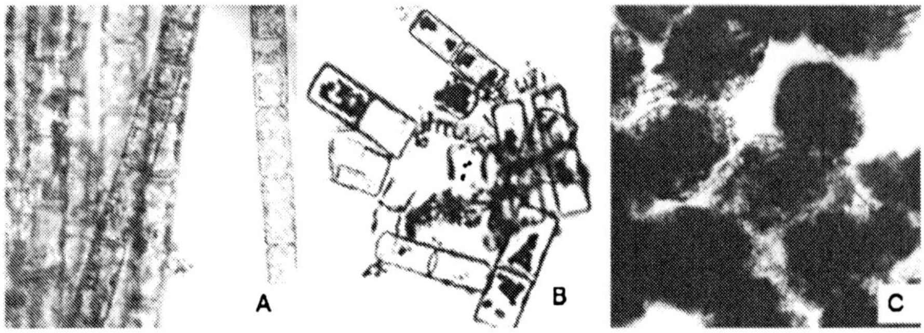

Figure 7 Healthy and active cell of *Melosira varians* with full of protoplasm (A), unhealthy cells of broken colony and cells without protoplasm (B) and faecal pellets by midge larvae (C).

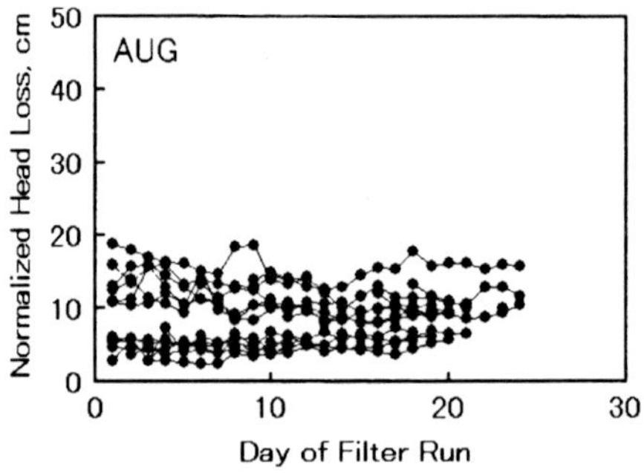

Figure 8 Filter resistance (Normalized Head Loss) during the filter run in August.

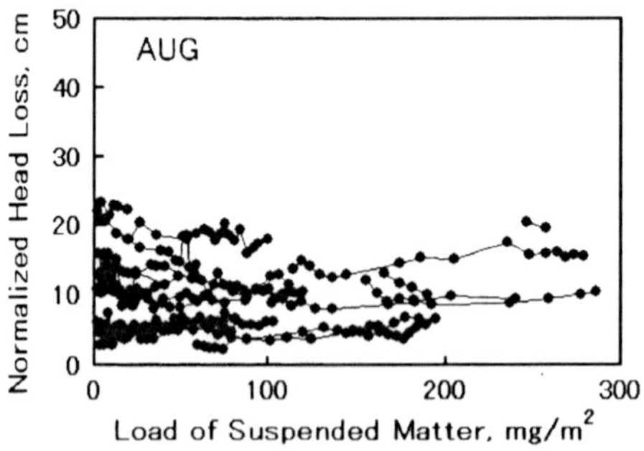

Figure 9 NHL change with the load increase of suspended matter on the filter bed in August.

Conclusions

Growth of algae provides the better food for midge larvae. The activity of small animal community including midge in the filter pond eliminates suspended matter and keeps better condition of slow sand filter. Growth of algae gives better condition for heterotrophic organisms. Slow sand filter is not simple physical and biological process. It is real ecological purification system by use of the activity of all the organisms. It is better to rename Ecological Purification System instead of slow sand filtration.

Acknowledgements

We are grateful to Takasaki City for our field survey. Thanks are also due to Dr. K. Hirabayashi and Mr. M. Takeda for their kind discussion.

References

Lorenzen, C. J. (1967) Determination of chlorophyll and pheo-pigments: spectrophotometoric equations. *Limnol. Oceanogr.* 12:343–346.

Nakamoto, N. (1988) An algal mat sampler for a slow sand filtration basin. *J. Jap. Wat. Wks. Ass.* 57(10):17–19. (in Japanese with English summary).

Nakamoto, N. (1993) Schmutzdecke sampler reduces filter damage. *Opflow* (American Water Works Association) 19(7):1–4.

Sakai, M. and Nakamoto, N. (1994) Role of filamentous algae in slow sand filter. *Verh. Internat. Verein. Limnol.* 25:1993–1995.

Nakamoto, N., Yamamoto, M., Sakai, M., Nozaki, K., Iwase, N., Yasuda, M., Kitada, T. and Hirabayashi, K. (1996) Role of filamentous diatom as an automatic purifier in a slow sand filter. p138–149. in Graham, N. and Collins, R. (ed.) *Advances in slow sand and alternative biological filtration.* John Wiley & Sons Ltd.

Nakamoto, N., Iwase, N., Nozaki, K. and Sakai, M. (1997) Inconsiderable amount of phytoplankton leakage by algal mat peeling from slow sand filter. *Jap. J. Wat. Treat. Biol.* 33(4):179–186.

Wotton, R. S. (2002) Water purification using sand. *Hydrobiologia* 469:193–201.

9 The effect of water temperature on the slow sand filter process

H.S. Jabur*

*SWECO VIAK, Jönköping, Sweden

Abstract Temperature plays an essential and comprehensive role in relation to the slow sand filter process. It considerably affects both the capacity of the filter and the filtrate water quality. The effects can be demonstrated by comparing the results of pilot and full-scale studies on slow sand filters conducted in Hungary, Algeria and Sweden with different water temperatures and climate conditions.

Keywords Temperature, slow sand filter, algae.

Introduction

The water temperature is a particularly important parameter in relation to the operation of slow sand filters. It affects the biological and chemical oxidation processes and the algae growth significantly.

The increase of the water temperature accelerates the bacteriological degradation of organic matters, increases the consumption of oxygen and the production of carbon dioxide. In cold climates, low temperatures reduce the efficiency of purification in slow sand filters. The falling temperature during autumn is causing a massive death of algae, which causes a release of organic matter (Huisman and Wood 1974). It has also been showed that the barrier effect against bacteria and viruses in slow sand filters is reduced at low temperatures (Cary S. Logsdon 1994).

Furthermore, the temperature has an effect on the adsorption efficiency of the filter media, which in general deteriorates with increased temperature, and it is also an important factor for the filter velocity and the hydraulic operation. Water viscosity and hydraulic conductivity of the porous media are also temperature dependent.

Gas formation in fluids occurs when the dissolved amount of gas exceeds the saturation point at a certain pressure and at certain temperature. It causes the formation of gas bubbles and air binding in the pores of the media, which increase the filter resistance (Jabur, Mårtensson, Öllös 2005).

 Recent Progress in Slow Sand and Alternative Biofiltration Processes edited by Rolf Gimbel, Nigel J.D. Graham and M. Robin Collins. ISBN: 9781843391203. Published by IWA Publishing, London, UK.

Experiments

We have had the opportunity to carry out pilot and full scale studies on slow sand filters in Hungary, Algeria and Sweden, where different temperatures and climate conditions could be compared.

From 1970 to 1974, the Technical University of Budapest conducted a series of pilot plant experiments at the Budapest Water Works with different pre-treatment of Danube water, using parallel open and covered slow sand filters. The water temperature varied between 7 to 20 °C (Jabur 1975).

From 1989 to 1995, several pilot slow sand filter plant studies were conducted at the University of Chlef, Algeria, to design and test an optimum model for dam-water treatment in the middle region of the country, by using open and covered slow sand filters as the main purification step after proper pre-treatment. The water temperature was 14–27 °C (Jabur and Maamer 1999).

From 1997 to 1999, pilot- and full-scale tests have been performed in order to optimise the design and operation of slow sand filters at Häggeberg Water Works in Jönköping, Sweden. The tests have included different filter sand, hydraulic conditions, open and covered filter columns etc. The water temperature was 2–14 °C (Jabur and Mårtensson 1999).

Results and discussion

The different results from the three countries gave a broad spectrum of operation conditions from cold clear raw water with low content of NOM and nutrients treated at sub-arctic conditions, to polluted river water with high content of nutrients in central Europe to, finally, warm water from rain water reservoirs treated at sub-tropical condition.

The correlation between the temperature and changes in dissolved oxygen before and after a slow sand filter for an 18 months period at Häggeberg Water Works in Jönköping, Sweden, is shown in figure 1. Figure 2 represents the correlation between temperature and changes in pH, COD_{Mn}, TOC, and BOD_7 (Jabur and Mårtensson 1999). Both figures can clearly show that the filtrated water quality is strongly dependent on the temperature.

The above-mentioned studies show that the oxidation of the organic matters in Algeria is faster and more intensive than in Hungary and Sweden. The consumption of dissolved oxygen is much higher and, in some cases, may determine the filter run length, to the extent that the filter must be cleaned before reaching the design pressure drop (figure 3). This could be observed even with pre-treated raw water with relatively good water quality. COD_{Mn} removal in Algeria was in average 35–45 %, while in Hungary and in Sweden about 25 %.

The establishment of the micro-flora in new sand needs several days in Algeria, several weeks in Hungary and several months in Sweden.

The effect of climate conditions, water temperature, sunlight conditions etc., can also be seen when comparing the capacity of the open and covered slow sand filters.

In Algeria the operation of slow sand filters for pre-treated dam water, in covered filters at a filtration rate of 0,1–0,2 m/h, last more than 6 months, while open filters results in 2–5 months operation period. This is due to the fast and dense growths of algae. In Hungary

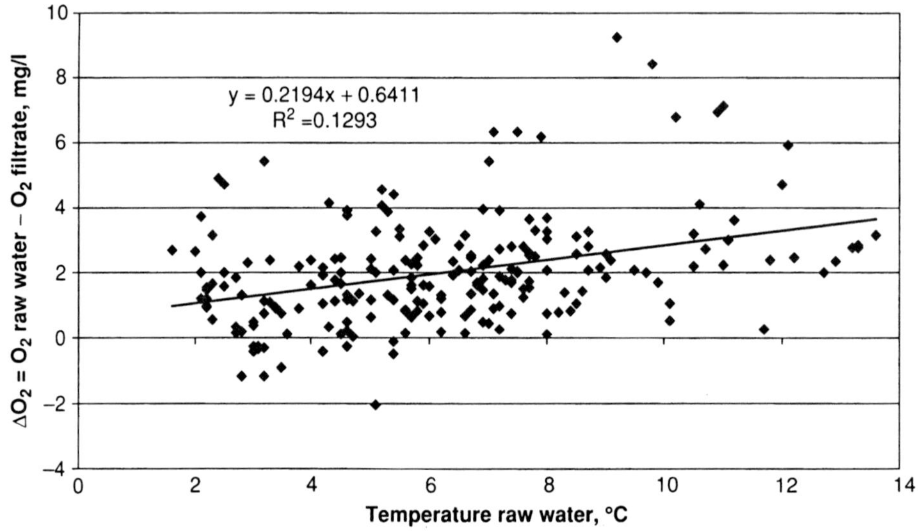

Figure 1 Oxygen consumption at Häggeberg WW vs. temperature.

covered filters take only 20–35 % longer operation time than open filters, depending on water quality. In Sweden, the difference between open and covered filters was small and mainly reliant on weather conditions. Up to 20–25 % longer operation time was achieved in covered filters.

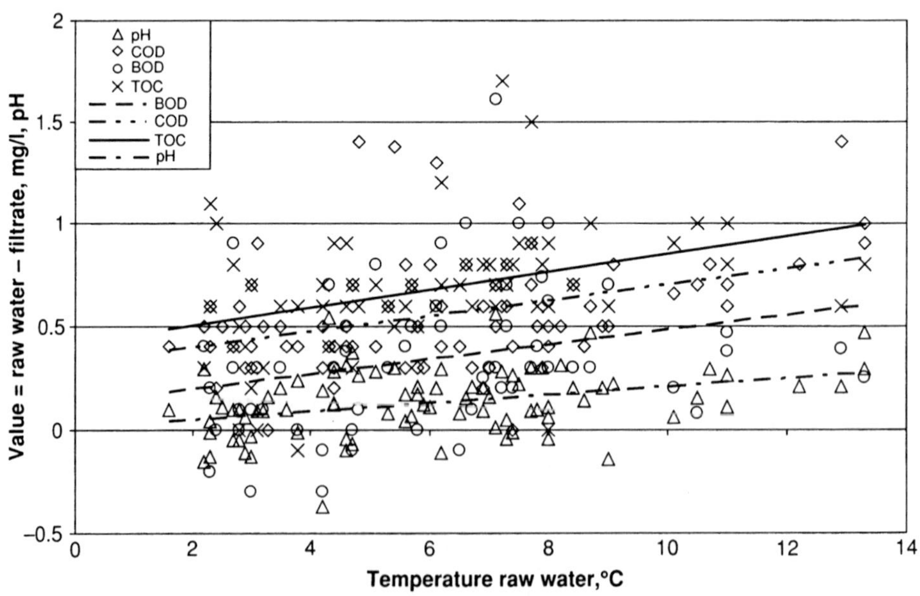

Figure 2 Changes in pH, BOD, COD and TOC at Häggeberg WW vs. temperature.

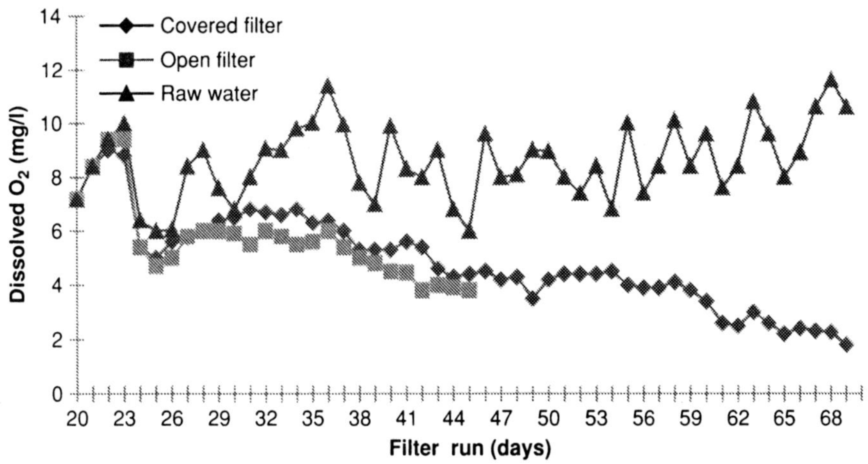

Figure 3 Slow sand filtration of reservoir water (without pre-treatment) in Algeria.

With regards to the dissolved organic substances, BOD_7, COD, TOC, bacteriological tests and other main water parameters, no significant differences can be observed between open and covered filters in all the three studies.

Conclusion

Water temperature is an important parameter for the design and operation of slow sand filters. It affects both the capacity and the filtrate water quality. The filter efficiency to remove organic matter and the oxygen demand are temperature dependent, why it is an essential factor in choosing proper pre-treatment and open or covered filters.

References

Huisman L. and Wood W. E. (1974) Slow Sand Filtration. *WHO Geneva.*

Jabur, Husam S. (1975) Slow Sand Filters and its Rule in Water Treatment. *Report. Technical University of Budapest.*

Jabur, Husam S. and Mårtensson, J. (1999) Optimising of slow sand filters. *Svenskt vatten AB, VA-Forsk 1999-17*, Sweden

Jabur Husam S. and Maamer M. (1999) Surface Water Treatment in Algeria. *Hidrologiai Közlöny, Vol 79, No 1, Budapest*

Jabur, Husam S., Mårtensson, J., Öllös G. (2005) Some Notes on Hydraulics and a Mathematical Description of Slow Sand Filtration. *Vatten 61, Lund.*

Task Committee on Slow Sand Filtration (1994) Slow Sand Filtration. *Edited by Cary S. Logsdon, N.Y.*

10 Mishaps linked to incorrect use of slow sand filters

J. Mårtensson*, H. S. Jabur*

*SWECO VIAK, Jönköping, Sweden

Abstract Attempts have been made to adopt the slow sand filter technology to modern computerised operation, speed up the filters and use slow sand filters in other applications then the original surface water treatment. By studying the results of such modern "development" it is clear though, that some of the old experiences have been put aside and in the same time the safe operation of slow sand filters have been put at risk. When slow sand filters are used in new applications or when the operation conditions are changed the result must be evaluated by studying proper hydraulic, chemical and microbiological parameters.

Keywords Breakthrough, coliform bacteria, algae, under-pressure, hydraulic design.

Common factors causing mishaps in slow sand filtration

The design and operation of slow sand filters, SSF, have mainly been based on experiences during almost two centuries. During the past decades attempts have been made to adopt the technology to modern PLC operation, speed up the filters and use SSF in other applications then the original surface water treatment. By studying the results of such modern "development" it is clear though, that some of the old experiences have been put aside and in the same time the safe operation of SSF have been put at risk.

By using SSF in new applications or by changing the operation conditions there are obvious risks for deterioration in the water quality. Too many examples can be shown where basic operation thumb rules have been overruled, sometimes deliberately, sometimes as a result of bad knowledge. The problem can become serious when the SSF are supposed to be a barrier against microorganisms but only separates suspended solids.

Often the result of bad design and operation is breakthrough of bacteria, algae etc. The problems are mainly related to factors as high filter velocity, intermittent operation, and bad hydraulic design causing under-pressure or local "vacuum". Hydraulic errors can easily be made by e.g. excluding an outlet weir above the sand level or, as some of the largest SSF plants in Sweden have practised, overfilling sand or using the filters as reservoirs, which gives about the same effect; under-pressure dragging pollutions deep down in the sand and a release of gases, mainly oxygen, into the sand causing rapid deterioration of filter capacity. The result can be craters on the surface with very coarse sand. The velocities in these craters are much higher than the intended why microorganisms present in the inlet water can slip through. Since SSF usually are the final treatment step before disinfection, and water samples are taken after chlorination or UV, the malfunction of the microbiological barrier often is undetected. Control parameters as turbidity, COD and E-coli are in this respect inefficient, since E-coli is one of the first organisms to perish in the competition, and the COD removal is about the same no matter how the filters are operated. Usually coliform bacteria and especially algae are much better indicators, but there are situations where the coliforms also can play games with you. One example is a major outbreak of coliform bacteria in May 2005, traced back to SSF used in groundwater treatment. In this example there are indications that the coliforms actually grows in the filters.

Case study Skillingaryd

The water supply is based on groundwater from two wells in a huge sand formation. The process includes two SSF to remove low amounts of iron and manganese, dosing of bi-carbonate to increase the alkalinity and sodium hydroxide to adjust the pH from about 7,8 to 8,3. Disinfection is made by UV-irradiation. The waterworks had been in operation for seven years and a manual cleaning of the sand surface is made each one or two years. The staff had noticed that the pH increases during summer, at some occasions up to pH 9,0. By only operating one of the filters the pH could be kept somewhat lower.

After analyses 2005-05-25, it was found that the distributed drinking water contained high amounts of coliform bacteria. New samples showed that the groundwater was free from coliforms while the filtrate from the SSF contained high amounts. The filter in operation was taken out of production and temporary chlorination was introduced. The manganese problem (0,03–0,08 mg/l) was considered less important for the time being. The filter was in operation to waste until cleaning 2005-06-13 when the water level was reduced slowly and samples were taken of the filter sand at about 1, 5, 10, 20, 50 and 70 cm under the schmutzdecke. At the lab drained but moist sand was mixed with sterile water and shaken. The water phase was analysed with respect to heterotrophic bacteria 20°C, coliform bacteria 35°C, E-coli 44°C, streptococcus faecalis, sulphide reducing clostridium, Pseudomonas aeruginosa and iron bacteria (the upper sample). The sand was also analysed for glowing losses as mg/g DS. A special sample was taken of the schmutzdecke and analysed regarding algae and other microorganisms by a limnologist. The used methods and parameters were the same that had been used in earlier pilot- and full-scale tests at Jönköping and Nässjö surface waterworks.

The analyses showed that the filters contained an unusually high amount of heterotrophic bacteria at all levels in the sand (Fig. 1). This type of bacteria should be present and dominates the biological decomposition of organic matter, but the amount was about 5–10 times higher than at earlier studied waterworks. The amount of E-coli was, as usual, very low. A few bacteria were present at the sand surface but none deeper into the sand. Pseudomonas aeruginosa, a relatively resistant indicator organism of faecal origin, was not found. The intended biological removal of small amounts of iron and manganese can't be confirmed since no iron bacteria were found. The mechanisms might just as well be chemical oxidation and physical separation in the sand. A brown layer was found about 20–30 cm down in the filter.

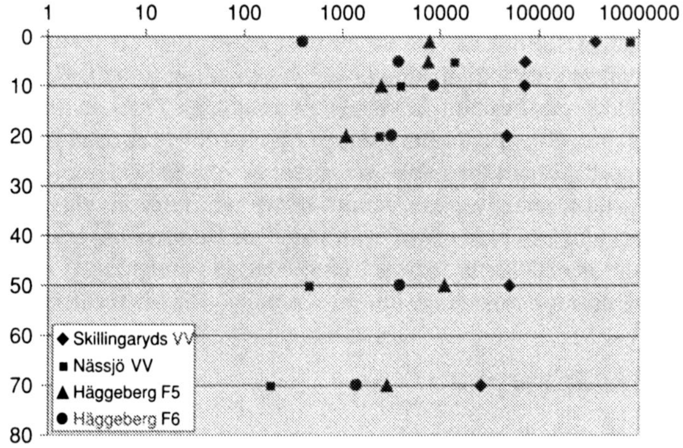

Figure 1 Heterotrophic 7d bacteria (No/g dry sand) at different sand depth (cm)

The SSF was full of coliform bacteria (Figure 2). Coliforms can be present in soil, water and faeces and since other faecal indicators are absent it is likely that we in this case are facing an unusual but reported exception; the coliforms have found a niche and can thrive in the filter. The limnological survey showed a huge amount of living and dead algae in the schmutzdecke. A few diatoma and green algae species dominated the algae flora completely and a large number of nematodes and mosquito larvae were also noticed. Diatoma shells could be seen as kieselguhr. The schmutzdecke looked similar to the sediments in a lake. The design of the SSF is correct hydraulically and no evidence for under-pressure or local vacuum can be traced. The amount of organic matter deep down in the sand is normal.

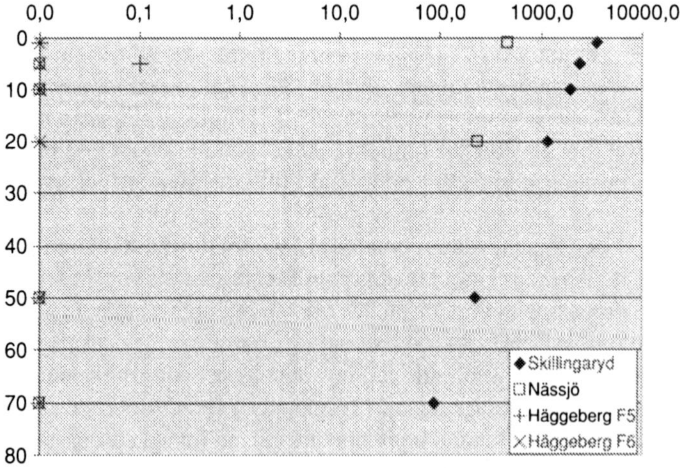

Figure 2 Coliform bacteria (No/g dry sand) at different depth (cm)

The fact that diatoma are present but no capacity problems have been reported indicates penetration of the schmutzdecke, most likely made by the nematodes etc.

Possible explanation

Surface water usually contains high O_2 and low CO_2 levels where the organic matters are not easily biodegradable. The growth of algae is usually limited by the supply of nutrients, why it is the bacteria in the sand that make the major biodegradation of organic matter. In this process carbon dioxide is released and the pH decreases. Gradually the available food is reduced and the competition gets tougher with the sand depth. Coliform bacteria, and especially E-coli, can't usually survive under these circumstances. This seems to be valid also for pathogen bacteria etc and results in the barrier effect. Studies indicate that some coliform species can survive a few days, especially in cold water and then gradually perish. If suitable nutrients are present and the conditions are favourable a certain growth can occur.

The main difference between the SSF in Skillingaryd vs. the application in e.g. Jönköping and Nässjö, which are used in this study as reference, is that the source water is groundwater. In groundwater all kind of trace minerals can be expected and is not limiting the growths of microorganisms. The content of COD is low, < 1 mg/l, why the microbiological activity also should be low. The reality shows the opposite and is likely caused by available aggressive CO_2, 3–4 mg/l, which is easily available for algae, why the pH increases. Especially diatoma forms in this case a primary production of biomass that later is food to higher organisms. Other organisms as the nematodes and mosquito larvae reuse the organic matter and since they are mobile they keep the schmutzdecke open for the water to enter. When the organisms in the schmutzdecke die, easy available organic matter is released once again and somewhere in the food chain the coliforms has found their niche. Since analyses normally are made on drinking water it is likely that coliform bacteria have been present from time to time in the filtrate before. What happened in May 2005 was that a heavy rain disturbed the flow through the filter causing extra many bacteria to enter the filtrate, more than the UV could handle.

A couple of other waterworks have been built in Sweden with this process. In two of them there are problems with massive algae growth. At one waterworks the inlet was changed to let the water enter the filter in a cascade above the water level, this reduced the problems somewhat. A number of waterworks are built where the groundwater first is aerated or passing a gravel- or plastic filter before entering an infiltration bed. At these waterworks, without a free water volume, no bacteria problems have been reported.

During October all filter sand was renewed and one filters taken into operation to waste. All kinds of bacteria could be found in the first filtrate samples, but gradually the number of different species became less and less. After 2–3 weeks the E-coli was gone and after a month only a few coliforms was detected. In December the number of bacteria was less than the limits in the Water Standards. The next step was to add NaOH in the distribution well before the SSF to limit the source of the primary production of biomass and evaluate the results. If this doesn't help aeration is the next step. We have to wait until spring to see.

Conclusions

Massive breakthrough of coliform bacteria can be a result of an unexpected process in the SSF, caused by a massive primary production of algae, when using SSF on a groundwater. Generally, if the function of a SSF should be evaluated the samples must be taken on the filtrate and cannot be taken as granted, COD and E-coli are inefficient; check out algae and more resistant indicator organisms. If craters in the sand are found, check the hydraulic conditions and filter design, errors can result in breakthrough of bacteria and low capacity.

References

Jabur, Husam S. and Mårtensson, J. (1999) Optimising of slow sand filters. *Svenskt vatten AB, VA-Forsk 1999-17.*

Jabur, Husam S. and Mårtensson, J. (2003) Rinsing methods of slow sand filters. *Svenskt vatten AB, VA-Forsk 2003-39.*

11 Dissolved oxygen issues with granular activated carbon sandwich™ slow sand filtration

M.E.J. Steele*, H.L. Evans*, J. Stephens*, A.J. Rachwal*, B.A. Clarke**

*Thames Water Utilities, Reading, UK
**Centre for Environmental Health Engineering, University of Surrey, Guildford, UK

Abstract The requirement to comply with the revised pesticide limits in the 1980 EC Drinking Water Directive lead to the development and successful implementation of the GAC Sandwich™ filter by Thames Water in the 1990s. Recent research provides further information into the impact of the GAC layer on slow sand filter (SSF) treatment. Experimental work comprised trials undertaken on operational and pilot scale SSFs. Results suggested that particular care is required when operating GAC Sandwich™ filters in order to ensure that an adequate dissolved oxygen (DO) supply is provided to the filter's microbiological community.

Keywords Dissolved oxygen, GAC Sandwich™, aerobic respiration, anaerobic zones.

Introduction

Research was undertaken by Thames Water (TW) to determine the optimum means of complying with EC pesticide standards (80/778/EEC, 1980), focusing on ozonation and granular activated carbon (GAC) adsorption. The GAC Sandwich™ filter (GAC-SSF) was developed, in which 100mm GAC was sandwiched between SSF sand (Bauer *et al*, 1996). Pre-ozonation was also introduced to extend the bed life of the GAC and to enhance treatment (Graveland 1996). Pre-ozonation increases the biodegradability of organic compounds, thus encouraging the biological colonisation of GAC. Evidence in support of microbial colonisation of GAC filters is abundant (AWWA, 1981, Servais *et al*, 1996). Whether this is more likely on GAC media compared to other environments (e.g. sand) is unclear (Den Blanken 1982, Mallevialle & Duguet, 1988). The comparatively large external surface area provided by irregular grain shape and the presence of macro-pores which micro-organisms can enter (AWWA, 1981), provide GAC with a potential advantage over sand.

The microbial colonisation of SSF media is fundamental to effective treatment. The colonisation rate is dependant on temperature, substrate availability, microbial species and the population density of organisms in the influent water. In 1993, Foster *et al* suggested that the GAC-SSF was exerting a greater demand for dissolved oxygen (DO) compared to the sand-only SSFs, although this additional demand was not seen to compromise performance.

DO is the amount of oxygen (O_2) dissolved in a given volume of water (Chapman, 1996). Atmospheric oxygen will dissolve in water at the air-water interface until saturation is reached for a given set of environmental conditions. Water may be 'super-saturated' via oxygen supplementation, for example by photosynthesis or ozonation. DO levels determine whether biological changes in a water body are brought about by aerobic or anaerobic processes, the latter being associated with undesirable end-products. Opinions as to what constitutes a minimum DO value for sustaining an aquatic biological community vary. The World Health Organisation (1996) states values should be kept as near to saturation as possible (e.g. 9mg/l at 20°C), whilst Chapman (1996) reported that values below 5mg/l might adversely affect the survival of biological communities. Huisman and Wood (1974) stated that SSF filtrate DO levels should be maintained above 3mg/l to avoid anaerobic conditions.

Factors Affecting Dissolved Oxygen Levels in GAC Sandwich™ Filters

Atmospheric Pressure: Atmospheric oxygen dissolves into a water body at the air-water interface, this being governed by atmospheric pressure, mass flow in the top water reservoir and concentration of DO in the water.

Microbial Respiration and Photosynthesis: The microbial respiratory demands placed on the DO supply in a SSF 24hr/day are partially offset by photosynthetic oxygen production during daylight hours. This results in diurnal DO trends for uncovered SSFs (MWB 1961-2).

Filter Run Length (time between skimming): As a filter run progresses, the microbiological community grows (and with it, its respiratory demands for DO). This is demonstrated by a declining trend in minimum DO values with time elapsed since skimming.

Filtration Rate: Influent water provides a supply of oxygen, the magnitude of which will be governed by the SSF filtration rate. High filtration rates (which maximise DO supply) also accelerate head-loss development, thus shortening runs and reducing the accumulation of algae through autochthonous growth. This, in turn, reduces algal respiratory demands for DO (Toms & Bayley, 1988). Hence, higher filtration rates increase the DO made available to the microbiological community, and lessen the respiratory demands due to shortened bed runs.

Cleaning Period: A TW SSF is drained prior to skimming. With absence of influent DO supply, the organisms surviving within the biofilm (surrounding the sand grains) will be starved of DO. Extended 'down-time' heightens the likelihood that obligatory aerobic micro-organisms are adversely affected. Once flow is restored, the initial filtrate is likely to be stripped of DO as reduced chemical species are oxidised.

Season: The solubility of oxygen in water declines as temperatures rise. Temperature also impacts on microbiological metabolic rates. Hence, the effects of microbiological respiratory DO depletion and photosynthetic DO supplementation become more pronounced in the warmer seasons. Photoperiod will also have an effect (Toms & Bayley, 1988).

Pre-treatment Processes: e.g. pre-ozonation enhances DO levels (Malley *et al*, 1993).

Impact of Dissolved Oxygen Levels on GAC Slow Sand Filtration

A SSF's predominantly aerobically respiring microbial community requires adequate DO supplies to treat water. Should DO levels become depleted and anaerobic respiration ensue, the micro-organisms involved use electron acceptors other than oxygen to oxidise a compound, for example nitrate (NO_3^-), ferric iron (Fe^{3+}), sulphate (SO_4^{2-}) or carbon dioxide (CO_2^-, Brock *et al*, 2000). Obligatory aerobic micro-organisms may die and/or detach when the environment is unsuitable. The products of anaerobic respiration are highly undesirable both operationally and in terms of filtrate quality. Recent in-house research indicated that periods of DO depletion may occur in GAC Sandwich™ filters, particularly during the onset of a filter run and when operational problems occur. In order to comply with TW Best Operating Practice (BOP) procedures, TW has focused research to investigate these issues.

Materials and methods

Nine trials were undertaken on TW's full-sized SSFs and two trials using pilot SSFs. Trials varied in length from 2–60 days and were undertaken at three TW advanced water treatment works (AWTWs). GAC-SSFs at two of these works was preceded by river abstraction, reservoir storage, pre-ozonation, rapid gravity filtration (RGF) and post ozonation (Run 1–7). At the third site, river derived water was treated directly by pre-ozonation, coagulation, rapid upflow filtration and post ozonation, without a reservoir storage period (Runs 8 & 9).

The pilot plant trials comprised four 150mm columns, each of which operated independently, using feed water from the inlet of an operational SSF:

- Run 10 investigated the impact of post-skim ramp-up rate (staged flow increases) on DO levels in a GAC-SSF. Column 1 was ramped up (to nominal flow rates) over 3 days as per TW's current operational practice. Columns 2–4 were ramped up within shorter time periods.
- Run 11 investigated the impact of media composition on the DO levels in a SSF. Column 1 comprised a sand-only medium whilst Columns 2–4 operated as GAC-SSFs where the GAC layer varied in its age (virgin GAC, and GAC operated to 21000 bed volumes (bv) and 67000bvs respectively, where 10000bvs equates to approximately 6 months operation).

Trial design details are shown in Table 2. For every trial, intensive (hourly) sampling took place during the initial (12–96) hours after start-up (following skimming). Water quality testing is detailed in Table 1.

Results and discussion

Low filtrate DO levels appear typical for GAC Sandwich™ filters following routine skimming (Figure 1). These trends were not linked to supernatant DO variation. It is postulated that, although the filter is drained and hence the supply of oxygen from filtering water is removed, the majority of the microbiological community within the bed remains viable. However, the rate of transfer of oxygen from the air in the media pore spaces is predicted to be lower than that from saturated water in a flowing filter. The presence of biofilm probably prolongs the survival of micro-organisms under these undesirable conditions. SSF media is rich in organic material and will be microbiologically active immediately prior to skimming. Thus carbon and nitrogen cycle nutrients will continue to be available, however, obligatory aerobic micro-organisms will require an oxygen supply in order to process these substrates.

Table 1 Water Quality Testing

Parameter	Test Kit	Method	Reference
Ammonia as Nitrogen	LCK 204, HACH Lange Ltd.	Phenate Method	L1500NH_3 F, APHA *et al*, 1995, pp. 4–80
Nitrite as Nitrogen	LCK 341, HACH Lange Ltd.	Colorimetric Method	4500 NO_2^-, APHA *et al*, 1995, pp. 4–83
Nitrate as Nitrogen	Various Lab. Apparatus, SEAL Discrete Analyser (SEAL Analystical Ltd.)	Nitrite Determination as above. Total Oxidised Nitrogen (TON) determination using Hydrazine Reduction Method, (TON-NO_2^- = NO_3^-)	4500 NO_3^-, APHA *et al*, 1995, pp. 4–87
Turbidity	2100P Turbidity Kit, HACH Lange Ltd	Nephelometric Measurement	2130B; APHA et al, 1995, pp. 2–9
Total Coliforms	Colilert-18, DST, IDEXX Laboratories Inc.	Defined Substrate Technology	APHA *et al*, 1998
Escherichia coli	Colilert-18, DST, IDEXX Laboratories Inc.	Defined Substrate Technology	APHA *et al*, 1998
Dissolved Oxygen, DO	6562 YSI Probe, 6920–40 Sonde, YSI Incorporated	Membrane Electrode Method	4500 OG, APHA *et al*, 1995 pp. 4–102
Temperature	6560 YSI Probe, 6920–40 Sonde, YSI Incorporated	Thermistor Method	2550 B, APHA *et al*, 1995, pp. 2–59
pH	6920 YSI Probe, 6920–40 Sonde, YSI Incorporated	Electrometric Methods	4500 H^+ B, APHA *et al*, 1995, pp. 4–65
Oxidation-Reduction Potential, ORP	6920 YSI Probe, 6920–40 Sonde, YSI Incorporated	Electrometric Methods	2580B, APHA *et al*, 1995, pp. 2–74
Particulate Organic Carbon, POC	Various Lab Apparatus, O. I. Analytical Carbon Analyser (Model 1010, O. I. Corporation)	Wet Oxidation Method	HMSO 1979, ISBN 0117514586
Chlorophyll-*a*	Various Lab. Apparatus / For online monitoring: Minitracka MkII Chelsea Tech. Group	Spectrophotometic Method / Fluorimetric Measurement	HMSO 1980, ISBN 0117516740

Available DO is consequently stripped from remaining water (e.g. within the biofilm). If demand exceeds this supply then anaerobic respiration may begin, forming anaerobic zones. If DO levels are low for a prolonged period then strictly aerobic micro-organisms may begin to die and/or detach. Once the oxygen supply is resumed to the filter then facultative anaerobes

Table 2 Trial Details

	Run 1	**Run 2**	**Run 3**	**Run 4**	**Run 5**	**Run 6**	**Run 7**	**Run 8**	**Run 9**	**Run 10**	**Run 11**
AWTW	Site 1, Bed A	Site 1, Bed A	Site 1, Bed B	Site 2, Bed C	Site 2, Bed C	Site 1, Bed D	Site 1, Bed D	Site 3, Bed E	Site 3, Bed E	Pilot Plant	Pilot Plant
Time	October	June	June	Feburary	May	June	July	January	August	May	June
Online Monitoring	DO, ORP, Temperature, pH	DO, ORP, Temperature, pH	DO, ORP, Temperature, pH	DO, Temperature, pH	DO, Temperature, pH	DO, Temperature, pH	DO, Temperature, pH	DO, ORP, Temperature, pH	DO, ORP, Temperature, pH	DO, ORP, pH	DO, ORP, pH
Sampling Points	Influent, Supernatant, Filtrate	Influent, Supernatant, Filtrate	Influent, Supernatant, Filtrate	Influent, Supernatant, Filtrate	Influent, Supernatant, Filtrate	Influent, Supernatant, Filtrate	Influent, Supernatant, Filtrate	Supernatant, At Schmutzdecke, Just above GAC Just Below BAC Filtrate	Supernatent, 30mm below Sand Just above GAC Just Below GAC Filtrate	Supernatant, Filtrate	Supernatant, Filtrate
Operational Mangement (Surface Loading Rates in $m^3/m^2/h$)	Routine Skim, Ramped up to Nominal Flow Rates of 0.35 over first 3 days	Routine Skim Ramped up to Nominal Flow Rates of 0.35 over first 3 days	Routine Skim Ramped up to Nominal Flow Rates of 0.35 over first 3 days	Routine Skim Ramped up to Nominal Flow Rates of 0.45 over first 3 days	Re-sanded Bed. Operated at 0.2 for first 10 days. Subsequently ramped up to Nominal Flow rates of 0.45	Problematic Skim, Operated at reduced flow rates for first 4 days. Subsequently ramped up to nominal flow rates of 0.35	Routine Skim. Operated at reduced flow rates for first 4 days. Subsequently ramped up to nominal flow rates of 0.35	Routine Skim. Re-started at Nominal flow rates of 0.2	Routine Skim. Re-started at Nominal flow rates of 0.2	Routine Skim. Varied Ramp up Rates between Columns	Routine Skim. Ramped up to Nominal flow rates of 0.35 over first 3 days

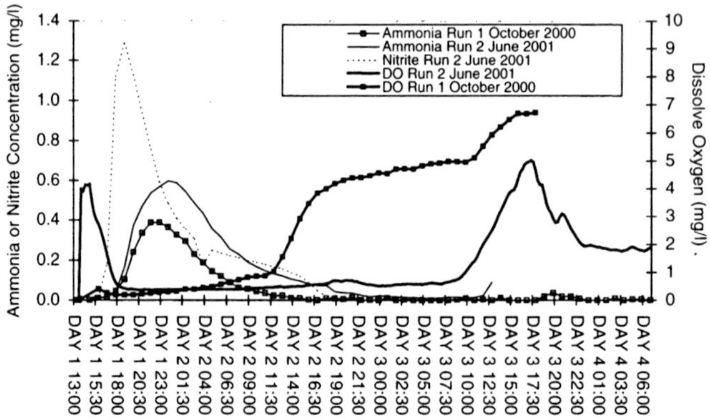

Figure 1 SSF Filtrate DO, Ammonia and Nitrite Concentrations following Return to Service after Routine Skim - comparing Run 1 (October 2000) and Run 2 (June 2001).

will switch back to aerobic respiration, as this is metabolically favoured (Lehninger, 1975). Thus, when the oxygen supply to the bed is resumed during the onset of the subsequent run, the initial water which passes through the media will be depleted of its DO content. Hence, it is not surprising to find that filtrate DO levels typically recover as flow (bringing in an oxygen supply) is ramped up after skimming.

The extent and duration of the post-skim DO depletion period is influenced by operational and environmental factors. Low DO values were sustained for longer during the summer months (Figure 1), and particularly if a SSF had been drained for an extended time period due to operational factors (Figure 2). In addition, DO levels took longer to recover from the drain-down period when a SSF was ramped up to its nominal flow rates over a longer period of time (Figure 7). The current ramp-up procedure (over 3 days) was based on an historical requirement for a SSF to pass a bacteriological test (which took 24–48 hours), prior to any flow rate changes. Lower flow rates during the 3-day ramp-up period also restrict the

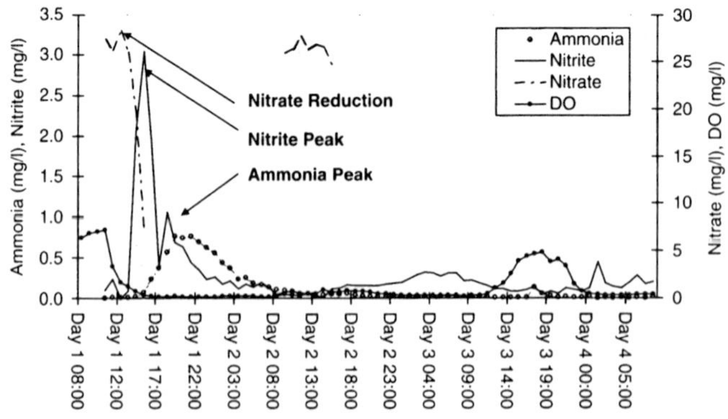

Figure 2 Ammonia, Nitrate, Nitrite and DO Trends for Run 6 Subsequent to an Operational Problem during Skimming, Run 6, June 2002.

water volume that the newly skimmed filter contributes to supply. Another consideration is the requirement for a so-called, 'ripening period', for a newly skimmed SSF before it is subjected to the same operating conditions as a mature SSF (Ellis, 1980). However, results from Run 10 indicated that detrimental effects, in terms of bacteriological filtrate quality, might not necessarily result from ramping a SSF up in a shortened time period. For this trial, all four columns performed similarly in terms of bacteriological filtrate quality (results not shown here). This suggests that rather than the microbiological re-colonisation of the surface sand layers (e.g. schmutzdecke development), the recovery of the remaining microbial community as a whole is required in order for a newly skimmed SSF to achieve acceptable treatment. This is more likely to occur within a shorter time period if the oxygen supply to the DO-deprived microbial community is maximised, which is achieved by increasing the flow rate through the SSF more quickly than is current practice. Hence, it is suggested that the microbial community recovers more quickly as a result of shortened ramp-up period and that contaminant penetration (and/or detachment) is less likely, despite higher loading rates.

Water quality testing suggested that anaerobic respiration takes place within a GAC-SSF during the onset of a run. Oxidation-reduction potential (ORP) data (not reported here) suggested that a reduced environment prevailed. Elevated ammonia and nitrite concentrations recorded in filtrate immediately after skimming indicated that ammonification and denitrification (respectively) occurred under the prevailing depleted oxygen environment (Figure 1 & Figure 2). Production of ammonia via ammonification is thought more likely than its production from the reduction of nitrite because this is inefficient in metabolic terms (Lehninger, 1975) and achievable by few bacteria (Brock *et al*, 2000). Furthermore, results from Run 9 demonstrated that ammonia concentrations were initially elevated throughout the filter media (with depth), whilst nitrite concentrations were only recorded from the GAC layer and below (Figure 3 & Figure 4). This pointed to the anaerobic reduction of nitrate by micro-organisms inhabiting the GAC media, presumably as a result of low DO levels within this layer. During the summer months GAC was seen to reduce DO levels by approximately 6–12mg/l, compared to 2–4mg/l in the winter months (Figure 5).

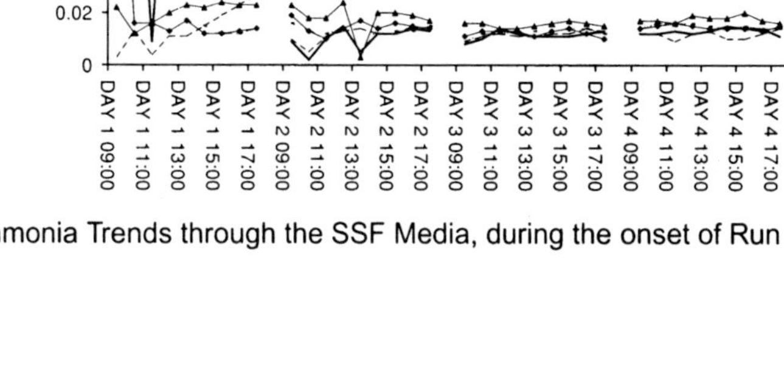

Figure 3 Ammonia Trends through the SSF Media, during the onset of Run 9, July/August 2003

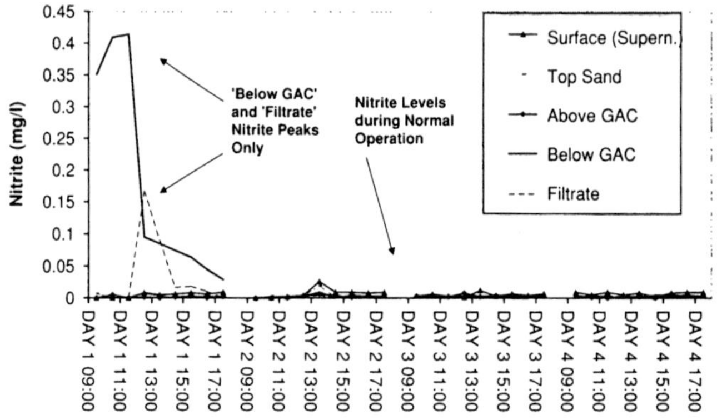

Figure 4 Nitrite Trends through the SSF Media, during Run 9, July/August 2003

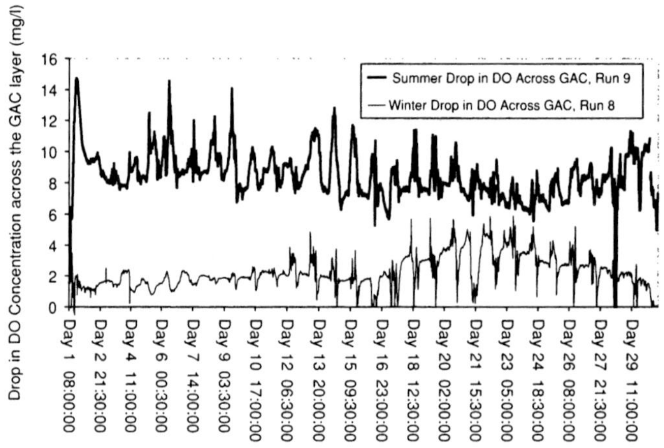

Figure 5 Comparing the Dissolved Oxygen concentrations across the GAC layer in a GAC-SSF during a winter and summer run.

In pilot scale trials (Run 11) the GAC-SSFs exerted a greater DO demand compared to a sand-only SSF (Figure 6). However, after approximately a year of its operation (21000bvs), the additional DO demand exerted by the GAC was not seen to increase with further operation. The GAC-SSFs with GAC of different ages (21000bvs & 67000bvs) exhibited similar characteristics to one another[1]. This implies that the GAC was colonised relatively quickly, but thereafter its DO requirements were relatively constant, suggesting a stabilisation in microbial community size. Bauer *et al* (1996) previously observed that removal of total organic carbon (TOC) by a GAC-SSF fell from 90% with virgin GAC, stabilising at 30–40% after 24000bvs, compared to 15–20% TOC removal by sand-only SSFs. It is proposed that

[1] The overnight drop in DO levels recorded for the 21000bvs GAC Sandwich^{TM} column resulted from a decline in flow rate when operational control was not possible.

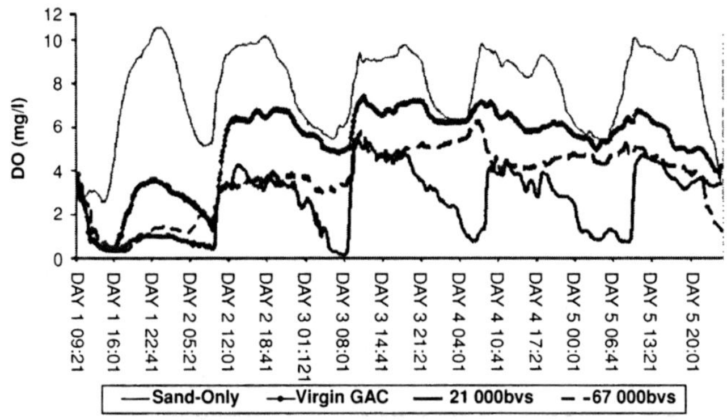

Figure 6 Filtrate DO Trends for Columns 1–4 during the Initial 5 Days of Run 11, June 2003

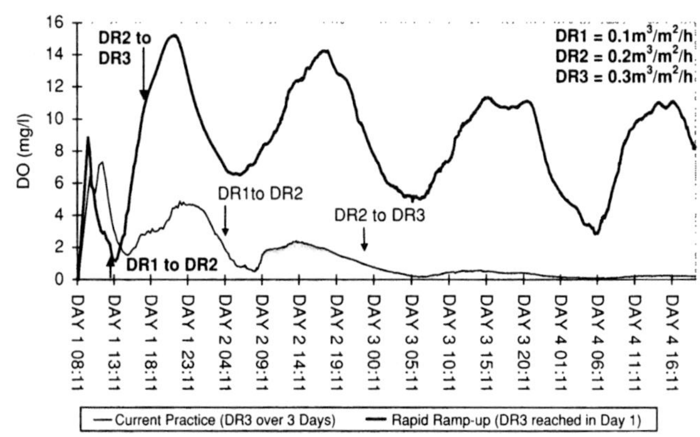

Figure 7 Post-skim ramp-up rate (i.e. staged increase in surface loading rate) and its impact on DO for Columns 1 & 4, Run 10, May 2003

there are links between microbial growth rate and biologically available TOC in the GAC layer.

After recovering from the initial depletion period, DO levels within the SSFs were seen to fluctuate diurnally, which was broadly consistent with the discussed patterns of microbiological photosynthetic and respiratory activity. Occasionally, during the summer months, DO levels dropped during the later stages of a run (night-time troughs <3mg/l). Thus 2 stages in the GAC-SSF's run (onset and end) may represent periods when it is vulnerable to developing anaerobic zones. Appropriate operational control measures can minimise this risk (e.g. SSFs prioritised for skimming based on individual filtrate DO monitoring).

It is possible that DO depletion across the GAC layer is a physico-chemical effect (Foster *et al*, 1993). According to Prober *et al* (1975), up to 19mgO_2/gGAC can be adsorbed onto GAC (cited by AWWA, 1981), however, additional data (Runs 8 & 9, not reported here) linked microbial colonisation of the GAC to the reduction in DO levels recorded across it (e.g. pH dropped across the GAC layer). Furthermore, results (Figure 6) indicated that virgin GAC (with more available adsorption sites) exerted a lower DO demand compared to the older GAC, implying that this was due to microbial activity, not a physico-chemical effect.

In summary, given that GAC is reportedly a favourable media for microbial activity and also that evidence exists to suggest that its presence in a GAC-SSF increases a SSF's DO consumption, it is suggested that a GAC Sandwich™ filter may be more vulnerable to low DO conditions compared to sand-only SSFs. This effect is most likely caused by the additional microbial activity made possible by the activated carbon media. In light of these findings, all GAC-SSFs at 2 TW AWTWs were fitted with individual DO monitors. If there is a perceived problem regarding an individual GAC-SSF then the resultant enhanced monitoring programme will usually include DO measurement. Where it is not currently possible to use dedicated DO monitors, a number of mobile monitors can be installed. This research has also lead to a heightened awareness of DO issues with GAC-SSFs (that treat eutrophic waters), which in turn has facilitated improved management by TW.

Conclusions

- Low filtrate DO levels are observed for GAC Sandwich™ filters (12–48 hours) following routine skimming (summer and autumn operating conditions). DO levels typically recover as flow is ramped up after skimming. The extent and duration of DO depletion is influenced by operational and environmental factors.
- Water quality testing during the onset of a GAC Sandwich™ filter run suggested that anaerobic respiration may occur within the bed.
- Shortening the time period used to return a SSF to service following routine skimming may not adversely impact on filtrate quality and may improve typically low DO levels within the SSF, thereby quickening its recovery.
- DO levels fluctuate diurnally, as a result of microbiological photosynthetic and respiratory activity. Low DO levels at night present periods during which the SSF is vulnerable to the development of anaerobic zones, particularly during the summer months.
- DO levels were seen to drop at the later stages of a SSF run during the summer months. Ideally DO monitoring should enable the identification of problematic filters, such that they can be prioritised for skimming.
- DO levels drop with depth through the SSF. The presence of the GAC layer in the GAC Sandwich™ filter appeared to exert an additional DO demand, which is thought to result from its capacity for microbial colonisation. This implies that a GAC Sandwich™ filter is more vulnerable to low DO conditions compared to a sand-only SSF.
- Results indicated that virgin GAC located in a GAC Sandwich™ filter becomes microbiologically active within a relatively short time period. Thereafter the DO demand exerted by the microbiological population inhabiting this layer is relatively stable from one year to the next. On a shorter time scale, the DO demand associated with the GAC layer is influenced by season, with DO consumption rates increasing markedly during the summer.

The impact of a GAC layer upon SSF DO concentration, immediately after cleaning and during low flow conditions with warm feed water temperatures, should be understood and managed by SSF monitoring and operational procedures to maintain overall treatment plant final water quality.

The views expressed are those of the authors and not necessarily those of Thames Water.

Acknowledgments

Thames Water Utilities are acknowledged for their funding. Particular thanks are due to TW Operational staff for their assistance and to Robin Bayley for his advice and support.

References

APHA, AWWA & WEF (1995) "*Standard Methods for the Examination of Water and Wastewater*", 19th Edition, APHA, AWWA & WEF, Franson, M.A.H., (Ed.).
AWWA, (1981), "An assessment of Microbial Activity on GAC", *J. AWWA*, **73**(8), 447–454.
Bauer, M.J., Colbourne, J.S., Foster, D.M., Goodman, N.V., & Rachwal, A.J., (1996), "GAC Enhanced Slow Sand Filtration (GAC Sandwich)^{TM}", *Advances in Slow Sand and Alternative Biological Filtration*, Collins, R. & Graham, N. (Ed.s), John Wiley & Sons, Inc., pp. 223–232.
Brock, T.D., Madigan, M.T., Martinko, J.M., & Parker, J., (2000), "*Biology of Microorganisms*", Ninth Edition, Prentice Hall International, Inc.
Chapman, D., (1996), *Water Quality Assessments*", 2nd Edn, UNESCO, WHO & UNEP E&F Spon.
den Blanken, J.G., (1982), "Microbial Activity in Activated Carbon Filters", *Journal of the Environmental Engineering Division*, Proceedings of ASCE, **108**(EE2), April 1982, 405–425.
EEC (1980), Council of European Communities: Council Directive relating to the Quality of Water Intended for Human Consumption, 80/778/EEC. *Journal of the European Communities*, August 1980.
Ellis, K.V., (1980), "Slow Sand Filtration", *CRC Critical Reviews in Env. Control*, **15**(4), 315–354.
Foster, D.M., Chipps, M., Holmes, M., Renton, P., Finan, J., Bayley, R., Kay, A., & Mackinson, S., (1993), "*Effect of GAC Sandwich on DO*", TW Briefing Note: Ref KAWT/BN01, Unpublished Text.
Graveland, A., (1996), "Application of Biological Activated Carbon Filtration at Amsterdam Water Supply", *Water Supply*, **14**(2), 233–241.
HMSO, (1979) "The Determination of TOC, TOD and Related Determinands", & (1980) "Determination of Chlorophyll-*a* in Aquatic Environments" *Her Majesty's Stationary Office.*
Huisman, L., & Wood, W.E., (1974), "*Slow Sand Filtration*", World Health Organisation, Geneva.
Lehninger, A.L., (1975), "*Biochemistry – the Molecular Bias of Cell Structure and Function*", Second Edition, The John Hopkins University School of Medicine, Worth Publishers, Inc.
Malley, J.P., Eighmy, T.T., Collins, M.R., Royce, J.A., & Morgan, D.F., (1993), The Performance and Microbiology of Ozone-Enhanced Biological Filtration", R&T, *J AWWA*, Dec 1993, 47–57.
Mallevialle, J., & Duguet, J.P., (1988), "Comparison Between Activated Carbon and Slow Sand Filtration in the Treatment of Surface Waters", *Slow Sand Filtration: Recent Developments in Water Treatment Technology*, Graham, N.J.D. (Ed.), Ellis Horwood Ltd., 265–280.
MWB (1961–62), "40th Report: 1961–1962", E.W. Taylor, *Metropolitan Water Board*, 58–59.

Servais, P., Cauchi, B., & Billen, G., (1996), "Experimental Study and Modelling Bacterial Activity in Biological Activated Carbon Filters", *Water Supply*, **14**(2), 223–231.

Toms, I.P., & Bayley, R.G., (1988), "Slow Sand Filtration: an approach to practical issues", *Slow Sand Filtration: Recent Developments in Water Treatment Technology*, Graham, N.J.D., Ellis Horwood

WHO, (1996), "*Guidelines for Drinking-water Quality*" 2nd Ed., WHO, Geneva.

12 Numerical simulation of slow sand filtration and parameter estimation of relevant processes

M. Rödelsperger*

*DVGW-Technologiezentrum Wasser (TZW), Karlsruher Str. 84, 76139 Karlsruhe, Germany

Abstract A numerical model system was developed to simulate the operational behaviour and the performance of slow sand filters. The program permits a very flexible implementation of model concepts. Properties of the filter design and operating characteristics may be freely combined with algorithms which mathematically describe the processes occuring in the filter. This concerns removal of particles, microbial activities, development of the schmutzdecke, growth of algae in the supernatant as well as headloss in the schmutzdecke and the filter bed. The user can choose from a library of algorithms, which were developed on the basis of literature data and results from laboratory experiments. By means of appropriate software interfaces the program is able to exchange data with the automatic calibration module PEST (Doherty 2004). This allows estimation of process parameters by comparing the results of simulation with data from real filter runs via an iterative fitting procedure. The program was successfully applied on interpretation of results from laboratory experiments.

Keywords Slow sand filtration, simulation, computer program, parameter estimation

Introduction

Design and operation of slow sand filters frequently is based on experiences out of the long history of application of this process. Usually however it is difficult to transfer experiences from one filter to another since differing environmental and operating conditions as well as varying raw water qualities and requirements to the treatment goal may occur. To overcome this problem, modeling concepts are required by which the relevant processes may be quantified satisfactory.

Precise computation of the operational behaviour is usually limited to individual characteristics, as far as their mathematical description is accessible through an analytic solution. In reality however a multiplicity of mutually overlaying and affecting processes takes place within a filter, resulting in a system which in its complexity can only be solved using numeric methods.

Process modelling mainly serves two different purposes:

- Predictive simulation of filter behaviour
- Evaluation of experiments to verify model concepts and to estimate model parameters

A numerical simulation program that is able to achieve both objectives and may be parameterised individually to realise different model concepts was developed and is described in this paper.

Methods

Concept

To simulate slow sand filtration a semi-deterministic numerical model was developed and realised as a computer program named SSF.EXE. The intention was to provide high flexibility and to allow the realisation of different model approaches inside the same software environment.

To fit model parameters on the basis of experimental results SSF.EXE was designed to co-operate with the program PEST.EXE (version 9.11, S.S. Papadopulos & Associates Inc., 2005), which can be downloaded as freeware from http://www.sspa.com/pest.

Programming and program properties

SSF.EXE was developed using the integrated development environment and compiler of PowerBASIC (version 3.50, © Robert Zale, PowerBASIC Inc., 1997) and the PowerTOOLS utilities (version 2.0, © Dirk Hilger, Kirschbaum Software GmbH, 1993). The program may be run from the DOS-command line on IBM compatible PCs (numeric co-processor recommended). It doesn't provide any graphical user interface and hence is merely to be seen as a module, which later will possibly be integrated into a pre- and post-processing environment. Nevertheless it may be used on its one, though this requires some skills from the user. Input and output data are exchanged between SSF.EXE and the user and/or the program PEST.EXE via ASCII-text files and CSV-data files. The latter may be easily edited using Microsoft® Excel. SSF.EXE can be run in two different modes which are described below.

Program structure

Figure 1 shows a simplified flowchart of the main routine, which computes the actual filter simulation. After its initialisation, during which configuration data are read from files SSF.CFG and SSF.FIT (only in parameter estimation mode, see below), the program is structured in nested loops, ordered on five levels.

1st level loop

Calculation is done by processing the filter runtime t_{end} in constant output time steps Δt. After each output time step results are stored to output files SSF.CSV and SSF.OUT (only in parameter estimation mode, see below).

2nd level loop

To enhance accuracy, output time steps are subdivided into increments Δt_c of variable length in the given range $\Delta t_{c,min} \ldots \Delta t_{c,max}$ ($\Delta t_{c,max} \leq \Delta t$). At start time, Δt_c is set to $\Delta t_{c,min}$. Later on, Δt_c is automatically adjusted during the process simulation subroutine (see below).

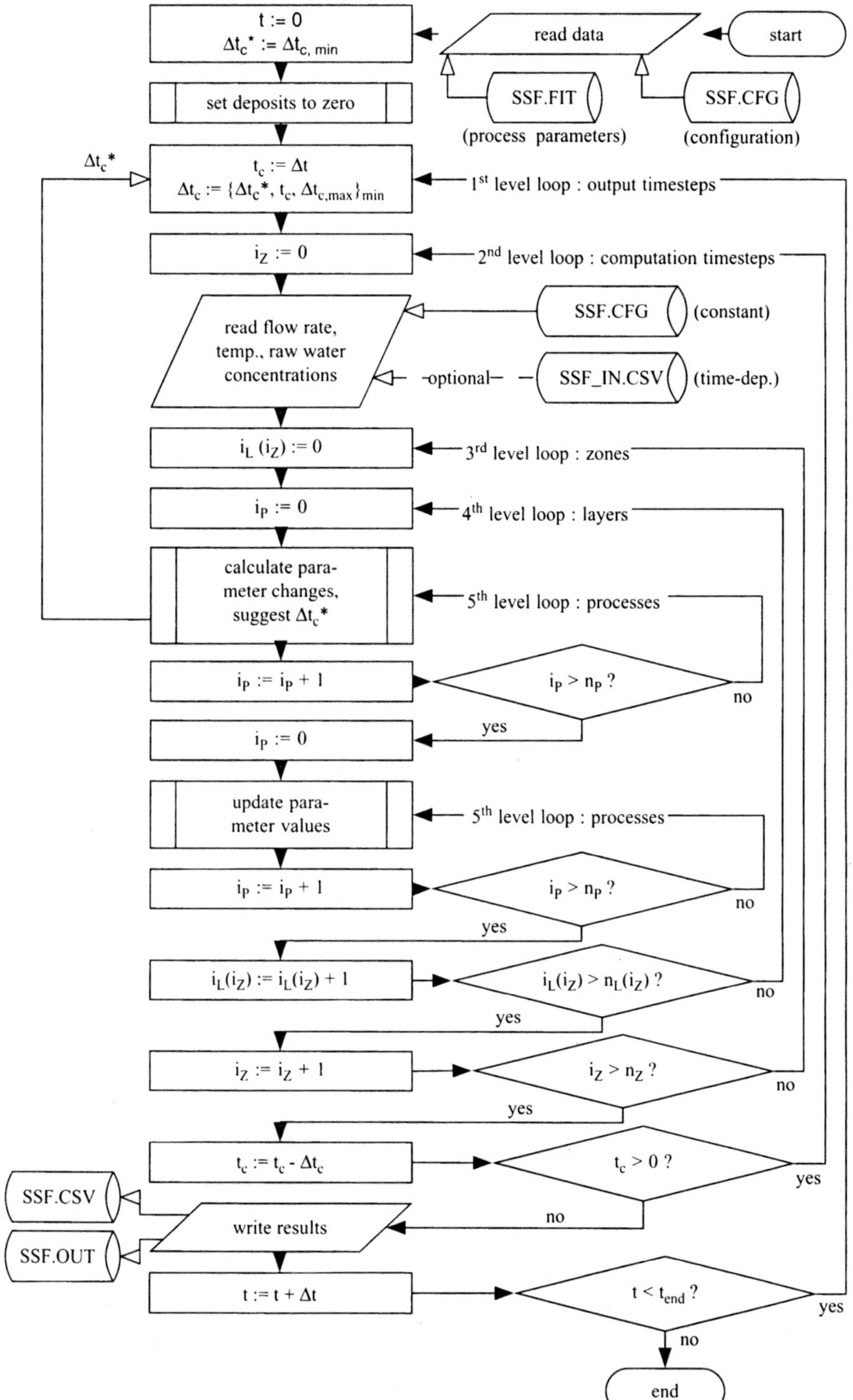

Figure 1 Simplified flowchart of the main routine of SSF.EXE

3rd and 4th level loop

The filter is divided into zones (supernatant, additional layer if so, sand bed), which on their part may be subdivided into layers. The number of zones n_Z as well as the number of layers per zone $n_L(i_Z)$ may be defined individually. It is possible to define layer heights as constant or as progressively increasing throughout one zone (see below). Layers are considered as mixed cells, so dispersion or anisotropy effects are not taken into account.

During computation, the concentrations of dissolved and suspended substances are changed gradually along the filterlength according to the modelled processes (influent to one layer = effluent from the above layer). Concentrations of the influent to the upmost layer of the upmost zone (= raw water) as well as operating data (flow rate, temperature) are assumed constant during an output time step. They may be provided as constant values throughout the runtime in file SSF.CFG or optionally as time dependent values in file SSF_IN.CSV. In the latter case the values are interpolated linearly to output timesteps Δt.

5th level loops

A library of freely configurable algorithms to describe the modelled processes mathematically (e.g. particle retention, microbial metabolism, headloss buildup) is integrated into the program. In a first 5th level loop an arbitrary number n_P of processes to be modelled are computed. Each of them is defined by the algorithm to be used, the zone, in which it occurs, and the parameters, which are changed in the course of the process (concentrations of dissolved or suspended substances, components of deposit) or which are calculated by the algorithm (headloss). To each process the appropriate process parameters are read from files SSF.CFG and/or SSF.FIT (only in parameter estimation mode, see below). In a second 5th level loop the changes of parameters, which were stored in the first 5th level loop, are added to the current parameter values. By this means it is made sure, that all processes are based on the same parameter values, so the sequence of their execution is irrelevant.

Automatic adaption of spatiotemporal discretisation

Temporal discretisation

In principle the use of a numerical model reduces the accuracy of the simulation compared to an analytical solution, because time steps of finite size Δt_c are applied. This deviation may be minimised by decreasing their size depending on the rate of parameter changes. Instead of using time steps of constant size throughout the simulation, which would slow down program execution unnecessarily, an automatic adaption of Δt_c is applied.

After computing an algorithm in the first 5th level loop, all relative parameter changes $r_x = | \Delta x/x |$ are evaluated. In this context, x denotes concentrations or deposits, that are affected by the according process and whose change Δx is proportional to the size of the current computation time step Δt_c. To suggest the length of the next computation time step, a maximum value $r_{x,max}$ is applied, that has to be given by the user. A computation time step, that would cause a relative change of $r_x = r_{x,max}$ should be $\Delta t_c^* = \Delta t_c \cdot (r_{x,max}/r_x) = \Delta t_c \cdot r_{x,max} \cdot | x/\Delta x |$. If for any reasons $\Delta x = 0$, Δt_c^* is set to $\Delta t_{c,max}$. If Δt_c^* results to be outside the range $\Delta t_{c,min} \ldots \Delta t_{c,max}$, it is set to the lower or upper limit, respectively. After Δt_c^*

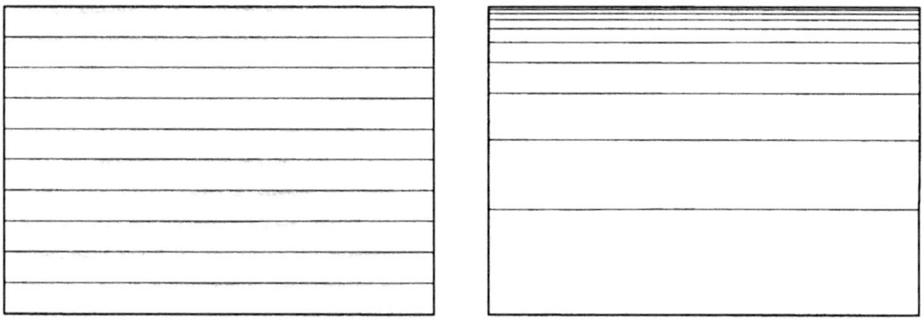

Figure 2 Layering of zones (left: linear, $h(i_L) = h(i_L - 1)$; right: progressive, $h(i_L) = h(i_L - 1) \cdot 1.5$)

was calculated for all relevant parameters, all processes, all layers, and all zones during the current time step Δt_c, the smallest value of all the Δt_c^* is applied to the next cycle of the 2nd level loop as new Δt_c. To meet the end of the current output timestep Δt, the last Δt_c has to be reduced accordingly. By choosing appropriate values for $r_{x,max}$, $\Delta t_{c,min}$ and $\Delta t_{c,max}$ the user keeps control over the adaption process.

Spatial discretisation

Since usually the extent of parameter changes is greatest at the top of the filter, in order to enhance spacial resolution it is possible to subdivide zones into layers of increasing height along the filter depth automatically, giving a ratio $h(i_L) / h(i_L - 1) > 1$ (see figure 2, right part).

When changes of concentrations by microbial degradation processes are to be calculated, layers are further split up into a number of sublayers dynamically in a similar manner like described above for adaption of Δt_c. This is necessary, because the differential equations describing microbial degradation cannot be integrated over filter depth, as this is possible in the case of the algorithm for the calculation of particle retention.

Algorithms

Particle retention

Particle retention is calculated using the well known equation by Iwasaki (1937). Deposition is calculated from a mass balance. Different filter models may be applied to describe the dependency of the filter coefficient on specific deposit.

Headloss

The headloss of the clean filter bed Δp_0 is calculated using equations by Carman-Kozeny and Darcy. Herein the dependency of viscosity on temperature is regarded. The buildup of headloss Δp with increasing specific deposit σ is calculated by the equation $\Delta p / \Delta p_0 = 1 + k_1 \cdot \sigma + k_2 \cdot \sigma^2$. By choosing appropriate values for k_1 and k_2 this allows the application of models proposed by Boucher (1947), Mays and Hunt (2005) and Woodward and Ta (1988).

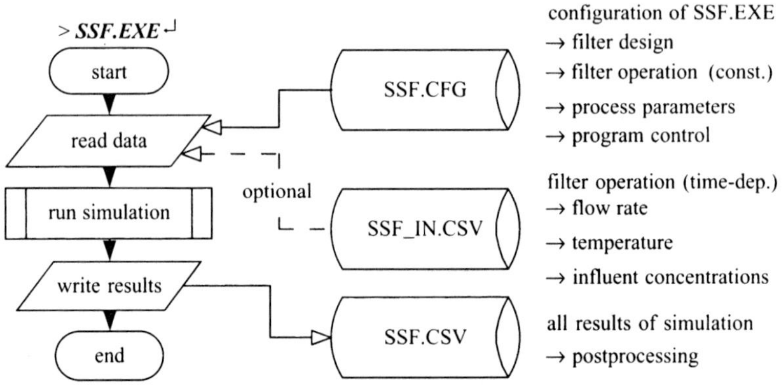

Figure 3 Simplified flowchart of program execution in stand alone-mode

Microbial degradation

Microbial degradation and buildup of biomass is calculated using mathematics according to Monod kinetics and stoichiometry of the relevant reactions. Dependency of the growth rate on two limiting substrates as well as inhibition by a substance, for example oxygen in the case of denitrification, may be regarded. Temperature dependency is calculated following the Arrhenius equation.

Program execution

Stand alone mode

If SSF.EXE is started from the command line with no further parameters (>SSF.EXE ↵, see figure 3), the program performs the simulation of one filter run using data that are to be provided in an ASCII file named SSF.CFG. These data include program control parameters, the description of filter design and operation conditions, as far as they are constant, as well as the parameter values of the simulated processes. Optionally, values of flow rate, temperature and/or raw water concentrations may be handed over in an additional file SSF_IN.CSV if they shall be considered to be time-dependent.

Parameter estimation mode

In this mode additional data are required. Starting values as well as upper and lower variation limits of the process parameters to be estimated are passed in a file named SSF.FIT. Measured values of characteristics that are influenced by the corresponding processes (e.g. headloss or effluent concentrations) are given in a file named SSF_OBS.CSV.

In a first step SSF.EXE is started from the command line with the parameter PEST (>SSF.EXE PEST ↵, see figure 4). Subsequently input data are processed similarly to stand alone mode, but no simulation is run. Instead some files are created automatically that are required by the program PEST.EXE. Subsequently these files may be modified by the user to optimise the fitting procedure (Doherty, 2004).

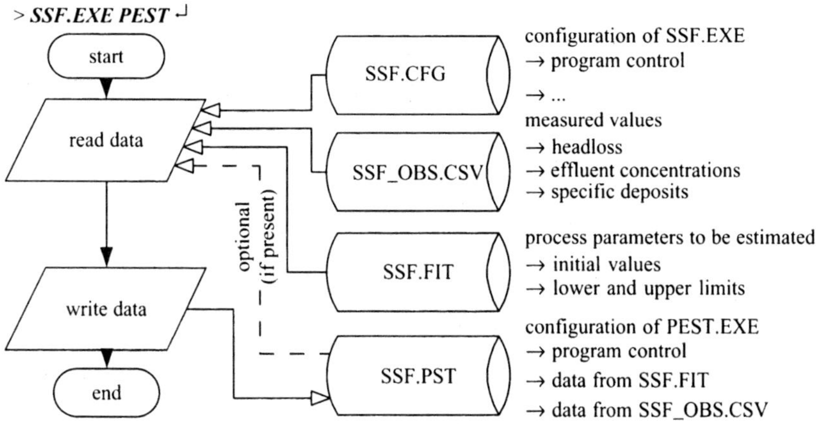

Figure 4 Simplified flowchart of initialisation in parameter estimation mode

In the following step PEST.EXE is started with command line parameter SSF (>PEST.EXE SSF ↵, see figure 5) which invokes the iterative parameter estimation procedure. During this procedure PEST.EXE starts SSF.EXE with command line parameter FIT (>SSF.EXE FIT ↵; SSF.EXE runs one simulation reading current values of process parameters from SSF.FIT and passes the results to PEST.EXE in file SSF.OUT; PEST.EXE compares the results with observation data and decides if the matching is satisfactory; if not, PEST.EXE modifies the process parameters within the given variation limits, stores them into SSF.FIT and starts SSF.EXE again with command line parameter PEST and so on, until the deviation between measured and simulated values conforms to the termination criteria.

To ensure a 1 : 1 relation between measured and simulated values, results are interpolated linearly to observation times before stored in file SSF.OUT.

Results

As an example, figure 6 shows measured concentrations during a laboratory experiment compared to simulated values after automatic estimation of process parameters using SSF.EXE and PEST.EXE. The agreement seems to be satisfactory.

Conclusions

The program SSF.EXE presented in this paper may serve as an useful tool in development of modelling concepts for slow sand filtration but still further improvements are to be implemented. This concerns especially the modelling of Schmutzdecke, for which appropriate algorithms are already integrated but not yet tested.

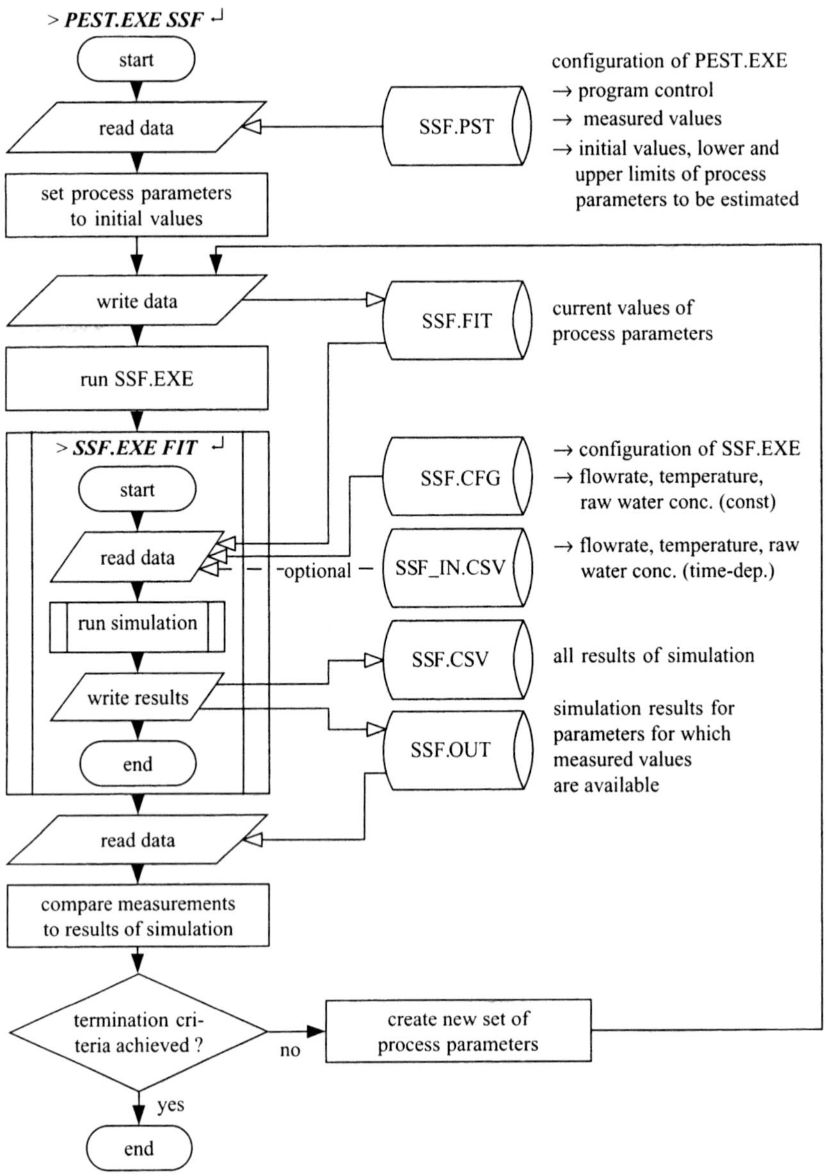

Figure 5 Simplified flowchart of program execution in parameter estimation mode

Acknowledgements

This study was conducted with subsidy of the German Ministry of Education and Research (BMBF) and is part of the joint research project "Adjusted water treatment technologies and water distribution under regional conditions" (02 WT 280/A4).

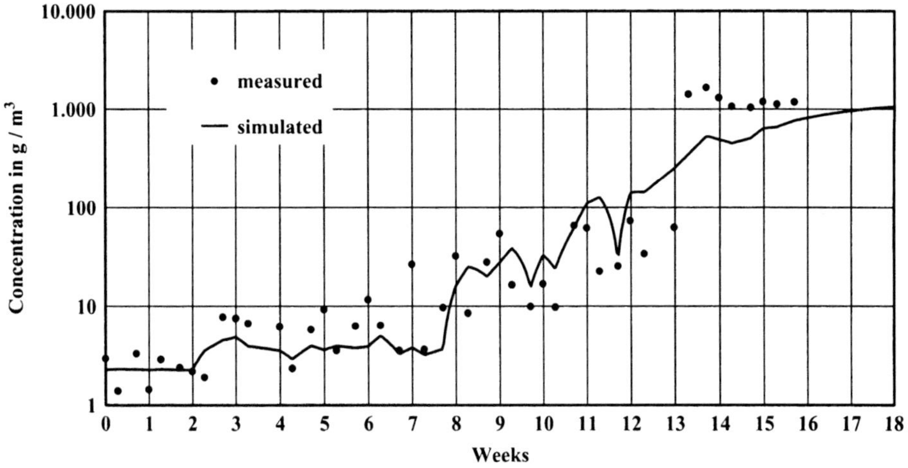

Figure 6 Comparision between measured concentrations during a laboratory experiment and results of simulation after automatic estimation of process parameters.

References

Boucher, P.J. (1947): A new measure of the filterability of fluids with application to water engineering. J. Inst. Civ. Engrs. **27**, 415–446 (ISSN 0368–2455).

Doherty, J. (2004): PEST-model-independent parameter estimation. Software manual, 5th edn, http://www.sspa.com/pest.

Iwasaki, T. (1937): Some notes on sand filtration. J. AWWA **29**(10), 1591–1602 (ISSN 0003-150X).

Mays, D.C. and Hunt, J.R. (2005): Hydrodynamic aspects of particle clogging in porous media. Environ. Sci. Technol. **39**(2), 577–584 (ISSN 0013-936X).

Woodward, C.A. and Ta, C.T. (1988): Developments in modelling slow sand filtration. In: Graham, N.J.D. (ed.): Slow Sand Filtration, 349–367 (ISBN 0-7458-0585-X).

13 Use of a novel simulation model to define the behaviour of covered and uncovered slow sand filters

L. C. Campos*, S. R. Smith, N. J. D. Graham****

*School of Civil Engineering, Federal University of Goiás Praça Universitária S/N Setor Universitário CEP 74605-220 Goiânia-Goiás-Brazil
**Department of Civil & Environmental Engineering, Imperial College London, South Kensington, London SW7 2AZ, UK

Abstract A new deterministic simulation model of the SSF process has been developed which incorporates fundamental physico-chemical and biological dynamics within a classical filtration framework. The model is based on a temporal and spatial finite difference method and is being used to investigate and assess some of the fundamental aspects and mechanisms of the SSF process. Simulations of the model for uncovered and covered SSF have provided information about the spatial and temporal development of interstitial biomass within the sand and the headloss profile, which is consistent with physical observations. The importance and rôle of the schmutzdecke layer in the SSF process has been demonstrated.

Keywords Slow sand filtration; deterministic model; process simulation; covering SSF

Introduction

In potable water treatment slow sand filters (SSF) are normally designed and operated without a cover over the filter bed. In recent years there has been a growing interest in covering filter beds to minimise winter freezing problems and to reduce algal growth (eg. in Yorkshire Water and Stockholm Water). In climates where seasonal algal blooms are prevalent, covering improves filter run lengths under these conditions (Haarhoff and Cleasby, 1991). However, uncertainty remains as to whether this effect is due solely to the absence of algae in the surface deposits, or in conjunction with the absence of algae in the sand media (Graham, 1991). In contrast, some authors (Huisman and Wood, 1974) argue that the decreased biological activity of the schmutzdecke observed in covered filters may potentially compromise the effectiveness of the sand bed for water filtration.

Thus, whilst the comparative benefits of covered and uncovered filters have been demonstrated in general terms, there is still a need to understand, define and quantify the filtration process mechanisms and interactions operating in these filters. A mathematical simulation model can offer a valuable and effective way to understand and evaluate the various dynamic

 Recent Progress in Slow Sand and Alternative Biofiltration Processes edited by Rolf Gimbel, Nigel J.D. Graham and M. Robin Collins. ISBN: 9781843391203. Published by IWA Publishing, London, UK.

Table 1 Details of the pilot SSF filtration runs used in the modelling.

Filter	Run	Date of operation	Duration of the run (day)	(hour)	Media depth (m)
Uncovered SSF B	1	16/07/1991 to 23/10/1991	99	2376	0.70
	2	25/10/1991 to 25/03/1992	152	3648	0.65
	3	26/03/1992 to 15/08/1992	142	3408	0.59
Covered SSF C	1	16/07/1991 to 14/09/1991	60	1440	0.70
	1	15/10/1991 to 25/03/1992	162	3888	0.63
	3	26/03/1992 to 21/12/1992	270	6480	0.59

processes operating within covered and uncovered slow sand filters. A deterministic-mechanistic model has been developed (Campos, 2002) to obtain a fundamental understanding of individual processes believed to contribute to the purification mechanisms in SSF and which are responsible for gradually blocking the filter. The mathematical model has been applied in this paper to explore the behaviour of covered and uncovered slow sand filters using a reasonably comprehensive data set provided by Thames Water Utilities Ltd.

Methodology

For the evaluation described in this paper, experimental data relating to consecutive filtration runs of one uncovered filter (SSF B) and another covered filter (SSF C) were used; information concerning these runs is given in Table 1. These data are part of long timescale measurements of physico-chemical and biological water quality parameters that were monitored routinely for the influent and filtrate of pilot scale slow sand filter units at the Thames Water Kempton Park Advanced Water Treatment (AWT) facility from July 1991 to July 1993.

The Kempton Park AWT pilot plant has two parallel process streams with a total capacity of approximately 5 Ml/d. Both process streams include two stages of ozonation in combination with rapid gravity filtration, SSF and granular activated carbon (GAC). Each stream has two rapid filters, two GAC filters and three slow sand filters, with the total flow being split between the individual filters. The pre-ozone contactor is a single stage tank, while each main ozone contactor is made up of four tanks in series. In stream 1 (SSF 1), the GAC filters follow the slow sand filters, while in stream 2 (SSF 2) they come before them. SSF B and SSF C is part of SSF 1 and the mean values for water quality are shown in Table 2.

The model developed by Campos (2002) considered two different forms of SSF, where the filter bed is covered or not. The uncovered filter was conceptualised as a reactor receiving full light interception. Three different zones are identified: the supernatant water, the schmutzdecke layer, and the sand bed. By contrast, the covered filter was conceptualised as a closed reactor without light interception. Therefore, supernatant water quality is unchanged and no schmutzdecke development occurs in the covered system. Nevertheless, particles are retained and microbial biomass develops within the sand bed, contributing to progressive headloss. The schmutzdecke is defined as a uniform layer on the top of the sand formed by filamentous algae. The SSF model assumes a linear function to simulate the growth of the schmutzdecke with time, and that a new layer is produced each day.

Table 2 Mean values for influent and effluent concentrations for SSF 1.

	SSF 1		
Parameter	**Influent**	**Effluent**	**t-Test**
Turbidity (NTU)	0.79	0.09	$P < 0.05$
Chlorophyll-a (mg l^{-1})	3.9×10^{-3}	1.21×10^{-4}	$P < 0.05$
Bacteria counts (cells per ml)	13.2	3.9	$P = 0.05$
POC (mg l^{-1})	0.29	0.18	$P < 0.05$
TOC (mg l^{-1})	4.9	3.9	$P < 0.05$
Phosphate (mg l^{-1})	0.50	0.56	$P = 0.13$
Ammonia (mg l^{-1})	3.0×10^{-3}	3.5×10^{-3}	$P = 0.43$
Nitrate(*)(mg l^{-1})	6.49	6.66	$P = 0.19$

The model is based on classical filtration theory and a temporal and spatial finite difference method, in which changes in particle removal and headloss are related to the bulk specific deposit in the schmutzdecke and filter media. The total bulk specific deposit is assumed to consist of biomass, inert material and non-living particulate organic carbon (POC). The biomass was formed by the removal of algae and removal/growth of bacteria and protozoa within the pores of the schmutzdecke and sand beds. In the covered filter, algae were ignored (Campos, 2002). The dynamic interactions between the principal groups of microorganisms including: algae, bacteria and protozoa, are modelled using Monod-type kinetic equations. The filtration performance of the filter media is defined in the model by the removal of particulate material from water and is represented by a combination of headloss and filtration coefficient functions.

The computer program was developed in Visual Basic with Application to Excel for the model simulation and the data required for simulation consist of influent water quality concentrations (inert material, chlorophyll-a, bacteria, protozoa, non-living POC, dissolved organic carbon (DOC), phosphorus, nitrate-nitrogen, ammonium-nitrogen), temperature, flow rate, filter surface area, supernatant water level, filter bed depth, duration of filtration runs, and several other input parameters (Campos, 2002). A full description of the model developed by Campos (2002) has been given by Campos *et al.* (2006).

Results and discussion

The results discussed in this paper were obtained mainly from the simulation of the uncovered SSF B and covered SSF C using the best set of the values of parameters calibrated for filtration Run 1 (Campos, 2002).

The results show that headloss decreases with depth in the sand bed of both filters (Fig. 1) and the largest headloss occurred in the top 2 cm of sand. These observations, including the spatial dimensions and magnitude of the predicted headloss response are in close agreement with practical measurements of headloss in experimental and operational slow sand filters (Huisman and Wood, 1974; Barret and Silverstein, 1988). Headloss in the schmutzdecke increases with time and decreases with depth (Fig 1a). Layer zero is immediately above the surface of the sand and layer −90 is the uppermost layer of the schmutzdecke. Accordingly,

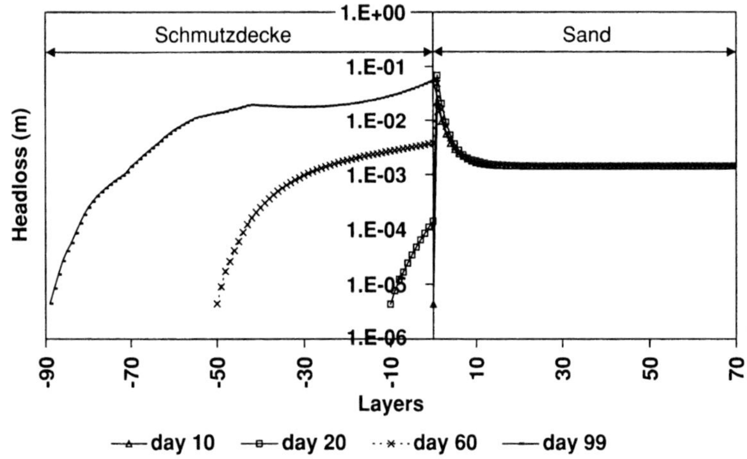

(a)

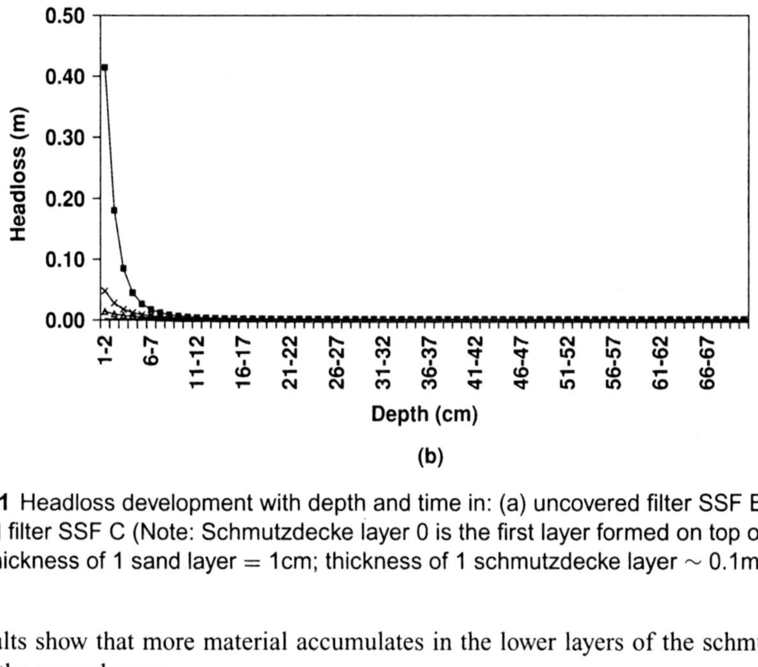

(b)

Figure 1 Headloss development with depth and time in: (a) uncovered filter SSF B; (b) covered filter SSF C (Note: Schmutzdecke layer 0 is the first layer formed on top of the sand; thickness of 1 sand layer = 1cm; thickness of 1 schmutzdecke layer ~ 0.1mm).

the results show that more material accumulates in the lower layers of the schmutzdecke than in the upper layers.

The profile of headloss (per unit depth) with filter depth was expected to correspond with the bulk specific deposit, and this was confirmed by the simulation (Fig. 1). Fig 1a shows that for the uncovered filter the schmutzdecke was responsible for a substantial contribution to the overall headloss and the largest proportion of the headloss in the schmutzdecke (and by implication the bulk specific deposit) was located within the lower layers.

The total bulk specific deposit in the sand decreased sharply with depth in both filters (Fig. 2). Similar trends in the vertical distribution of deposit have been observed in

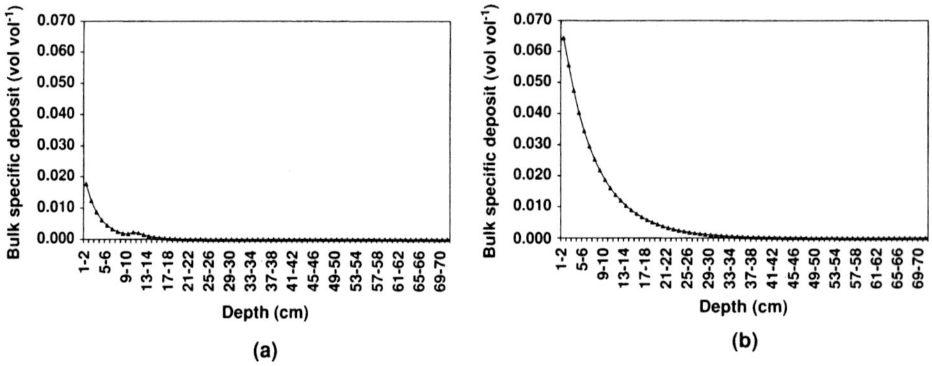

Figure 2 Bulk specific deposit in the sand at the end of Run 1 of (a) uncovered SSF B and (b) covered SSF C.

practical SSF studies (eg. Duncan, 1988; Eighmy *et al.*, 1994; Yordanov *et al.*, 1996). In the schmutzdecke, the largest proportion of bulk specific deposit was located within the lower layers, as shown by the headloss profile (Fig. 2a).

It is observed that the amount of biomass in both filters increases rapidly during the initial 10 days of operation (Fig. 3), although the absolute quantity of biomass was very different in the two cases since, in the covered filter, there is no algal development in the supernatant water. After the initial increase the biomass decreases in the uncovered filter, while it continues to increase in the covered filter. The decline in biomass concentration in the uncovered bed appears to correlate with and was possibly the cause of the decreasing headloss in the sand bed with time (Fig. 4).

The biomass predicted in the uncovered SSF B (Fig. 3) is larger than that in the covered SSF C, which is in agreement with Campos *et al.* (2002). However, in terms of magnitude, the prediction of biomass in these filters is larger than that measured by Campos *et al.*

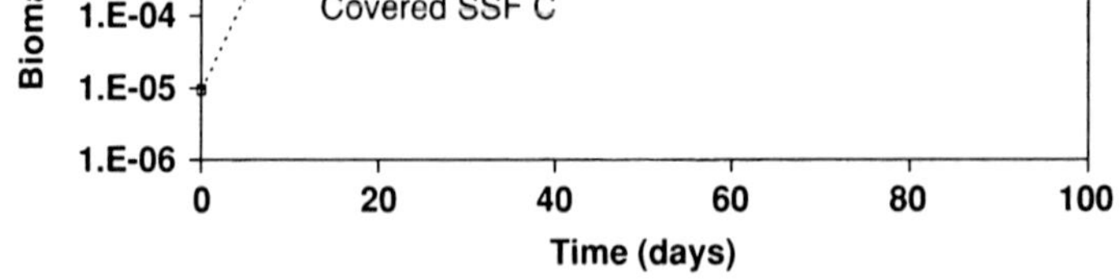

Figure 3 Development of biomass in (a) uncovered SSF B, and (b) covered SSF C (Note: Biomass concentration plotted on logarithmic scale).

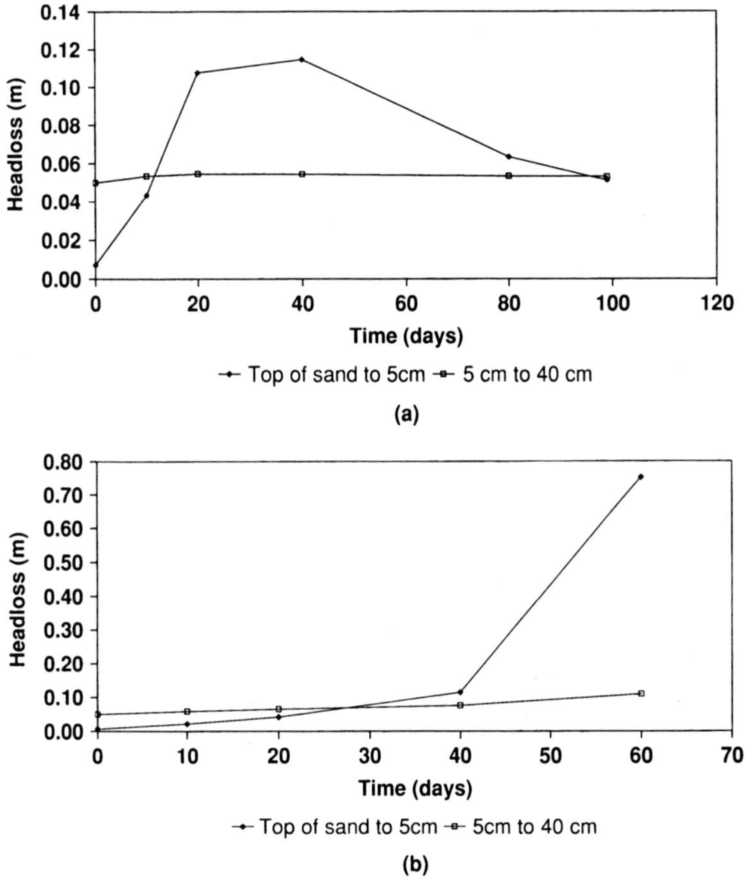

Figure 4 Headloss in the top 40 cm of the sand of: (a) uncovered filter SSF B; (b) covered filter SSF C.

(2002). This could be explained because SSF B and SSF C are located after ozonation, while Campos *et al.* (2002) measured the microbial biomass in filters located after GAC. The biomass predicted in SSF B is consistent with the maximum biomass of approximately 1 mg C cm^{-3} measured by Duncan (1988) in uncovered filters.

The importance and role of the schmutzdecke are demonstrated based on changes in the 'lag phase', and in the variation of the thickness and biomass of the filamentous algae. The 'lag phase' represents the time period that elapsed from the start of the filtration run for the growth of the first schmutzdecke layer, and consequently, when filtration across the schmutzdecke was simulated. The total absolute specific deposit in the top 10 cm of sand increased as the 'lag phase' was extended (Fig. 5). This is because more impurities reach the sand bed when schmutzdecke is not developed, emphasising the role of the schmutzdecke in removing particles in uncovered filters. Hirschi and Sims (1991) suggested that increasing in schmutzdecke growth improves the collector efficiency of the filter. Huisman and Wood

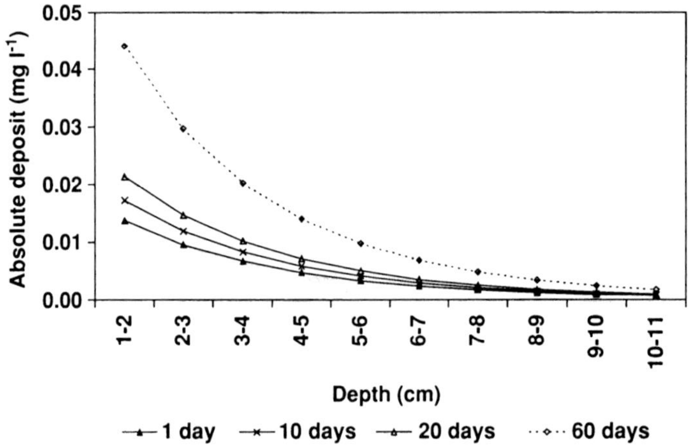

Figure 5 Influence of 'lag phase' of the schmutzdecke on total absolute specific deposit in the top 10 cm of sand at the end of Run 1 of uncovered SSF B (Note: Layer 0 is the first layer to be formed on top of the sand).

(1974) also pointed out that less suspended matter reaches the filter medium when the schmutzdecke is well established, and this may lengthen the interval between successive filter cleaning of uncovered filters. In the model, the schmutzdecke thickness is determined by the assumed growth rate. Running the model with different growth rates showed that the resulting total headloss was inversely related to the schmutzdecke thickness (Fig. 6).

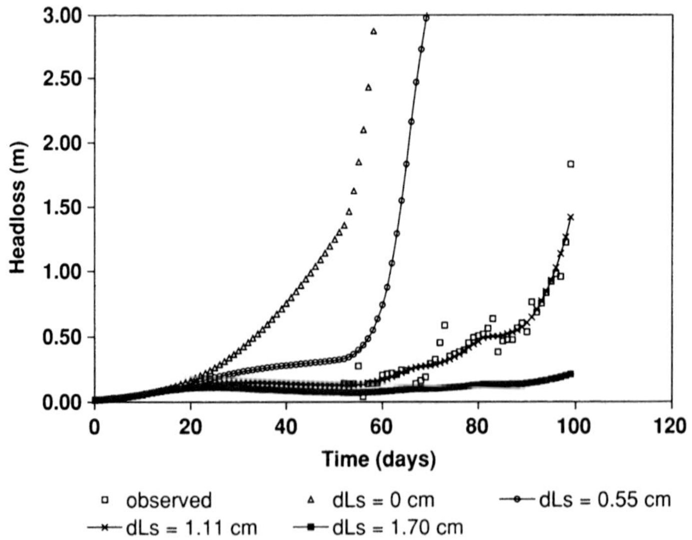

Figure 6 Influence of total thickness of schmutzdecke on total headloss during Run 1 of uncovered SSF B.

Without schmutzdecke (dLs = 0 cm) or with only a thin schmutzdecke layer (dLs = 0.55 cm), total headloss markedly increased, shortening the filtration run length. However, total headloss significantly reduced with increasing thickness of schmutzdecke (dLs = 1.7 cm).

Conclusions

The main conclusions of this work are:

1. The simulation results presented in this work show that the SSF model provides a potential tool for investigating the behaviour of covered and uncovered slow sand filters;
2. The highest contribution to headloss in SSF is located in the top few centimetres of the sand and bottom layers of the schmutzdecke. The schmutzdecke is found to contribute substantially to the total headloss compared to the sand bed;
3. The total bulk specific deposit in the sand decreases sharply with depth in both filters. Similar trends in the vertical distribution of deposit have been observed in practical SSF studies;
4. The amount of biomass in both filters increases rapidly during the initial 10 days of operation and it is larger in the uncovered SSF B than that in the covered SSF C, which is in agreement with Campos *et al.* (2002);
5. The schmutzdecke acts as a porous filter medium in uncovered filters, preventing the fast clogging of the sand bed. The necessary period for establishment of the schmutzdecke (lag phase) and its thickness are important factors affecting headloss development.

Acknowledgements

The authors wish to acknowledge the important contribution of Thames Water Utilities Ltd for providing extensive pilot plant data, and the assistance of Prof S. Chapra and Dr M. Schütze with certain aspects of this work.

References

Campos, L.C. (2002) *Modelling and simulation of the biological and physical processes of slow sand filtration.* Ph.D. Thesis, University of London, UK.

Campos, L.C., Su, M.F.J., Graham, N.J.D., and Smith, S.R. (2002) Biomass development in slow sand filters. *Water Res.*, **36**(18), 4543–4551.

Campos, L.C., Smith, S.R., and Graham, N.J.D. (2006) A deterministic model of slow sand filtration - I. Model development. *J. Environ. Eng.*, ASCE, in press.

Barret, J.M. and Silverstein, J. (1988) The effects of high-carbon and high-coliform feedwaters on the performance of slow sand filters under tropical conditions. In *Slow Sand Filtration: Recent Developments in Water Treatment Technology*, N.J.D.Graham (Ed.), Ellis Horwood Ltd., UK, 231–251.

Duncan, A. (1988) The ecology of slow sand filters. In *Slow Sand Filtration: Recent Developments in Water Treatment Technology*, N.J.D.Graham (Ed.), Ellis Horwood Ltd., UK, 281–304.

Eighmy, T.T., Spanos, S.K., Royce, J., Collins, M.R., and Malley Jr, J.P. (1994) Microbial activity in slow sand filters. In *Slow Sand Filtration*, Collins, M.R. and N.D.J. Graham, (Eds.), AWWA, 218–240.

Graham, N.J.D. (1991) Research needs in slow sand filtration. *International Slow Sand Filtration Workshop*, 27–30 October, Durham, USA.

Haarhoff, J. and Cleasby, J.L. (1991) Biological and physical mechanisms in slow sand filtration. In *Slow Sand Filtration*, Logsdon, G.S. (Ed.), Task Committee on Slow Sand Filtration, ASCE, 19–68.

Hirschi, S.D. and Sims, R.C. (1991) Particles and microorganisms in slow rate sand filtration. *International Slow Sand Filtration Workshop*, 27–30, October, USA.

Huisman, L. and Wood, W.E. (1974) *Slow Sand Filtration*, World Health Organisation, Geneva.

Yordanov, R.V., Lamb, A.J., and Melvin, M.A.L. (1996) Biomass characteristics of slow sand filters receiving ozonated water. In *Advances in Slow Sand and Alternative Biological Filtration.* N.J.D. Graham and R. Collins (Eds.), John Wiley & Sons, UK, 107–118.

14 Characteristics of straining of *Escherichia coli* in saturated porous media

J. W. Foppen*, M. van Herwerden* and J.F. Schijven**

*UNESCO-IHE Institute for Water Education, Delft, The Netherlands (j.foppen@unesco-ihe.org),
**National Institute of Public Health and the Environment, Microbiological Laboratory for Health Protection, Bilthoven, The Netherlands (j.schijven@rivm.nl)

Abstract Existing studies on the straining of bacteria in saturated porous media arrive at strained amounts of bacteria via modelling, but never have measured the amount of strained bacteria directly. We carried out a series of column experiments with ultra-pure quartz sand and *Escherichia coli* ATCC 25922. At the end of each column experiment, the flow was reversed and the reversed flow effluent at the top of the column was collected and analysed. We expected that reversing the flow would dislodge *E. coli* strained in dead end pores, because strained particles are considered to be trapped within a pore and not attached to the surface of the quartz grains. The results indicated that straining was not dependent on the grain diameter, but -most likely- on the pore size distribution of the sediments. In addition, the amount of strained bacteria was linearly dependent on the *E. coli* concentration of the suspension and when *E. coli* concentrations exceeded 2×10^8 cells/mL, the amount of strained bacteria remained constant. This indicated the presence of a certain maximum pore volume available for straining. Finally, for the 0–5 cm column length range we used, straining was almost linearly dependent on column length. This finding was not in correspondence with existing straining models that attribute non log-linear retention in the sediment as a function of transport distance to straining. Non log-linear behaviour was not present in our experiments. These observations will assist in developing a transient mass balance model that considers straining and kinetic attachment.

Keywords *Escherichia coli*, straining, colloid filtration theory, porous media.

Introduction

Straining, which is defined as the trapping of (bio)colloids in pores that are too small to allow passage, has not received much attention in literature. The process is not included in colloid filtration theory (Bradford et al., 2003) and existing studies that do consider straining (Bradford et al., 2003; Foppen et al., 2005) arrive at values for strained colloid masses based on modelling breakthrough curves and retention profiles, but do not measure the amount of strained (bio)colloids directly. We carried out a series of column experiments

 Recent Progress in Slow Sand and Alternative Biofiltration Processes edited by Rolf Gimbel, Nigel J.D. Graham and M. Robin Collins. ISBN: 9781843391203. Published by IWA Publishing, London, UK.

with ultra-pure quartz sand and *Escherichia coli* ATCC 25922. At the end of each column experiment, the flow was reversed and the reversed flow effluent at the top of the column was collected and analysed. We expected that reversing the flow would dislodge *E. coli* strained in dead end pores, because strained particles are considered to be trapped within a pore and not attached to the surface of the quartz grains. Therefore, the aim of this paper is to experimentally assess the advantages and disadvantages of the reverse flow method and to qualitatively describe the characteristics of the straining process.

Methods

Bacterial suspensions

Bacterial suspensions were prepared in a similar way as described by Foppen and Schijven (2005). Bacteria were suspended in demineralised water to concentrations ranging between 2×10^7 and 1.2×10^9 cells/mL. The EC- value of the suspension was 1–3 μS/cm.

Porus media

Ultrapure quartz sand (J.T. Baker) with a diameter of 1 mm was ground with a mortar and sieved into 4 size fractions (180–212, 75–90, 45–53, 38–45 μm). To remove impurities at the surface of the grains that might affect bacteria attachment, the sand was combusted at 800 °C for 4 h. Then, the sand was acid-washed by soaking in 12 N HCl for 24 h, and rinsed with deionised water until the electrical conductivity of the rinse water was less than 1 μS/cm. Two experiments were carried out with soda-lime glass microspheres (Whitehouse Scientific Ltd., Chester, England) of various sizes (38–45 and 180–212 μm). Prior to the experiments, the spheres were rinsed sequentially with acetone, hexane and concentrated HCl, followed by repeated rinsing with demineralised water until the electrical conductivity was close to zero.

Column experiments

Column experiments were conducted in borosilicate glass columns with an inner diameter of 2.5 cm (Omnifit, Cambridge, U.K.) with polyethylene frits (25 μm pore diameter) and one adjustable endpiece. The columns were packed wet with one of the quartz or glass bead size fractions with vibration to minimize any layering or air entrapment. Column sediment length varied between 0.25 and 5 cm. Prior to each experiment, the packed column was equilibrated by pumping (Watson-Marlow 101U/R) 10 pore volumes of demineralised water through the column at a constant discharge rate of 7.0 mL/min. Usually, a suspension of *E. coli* was pumped for 10 minutes (loading phase) at the same constant discharge rate followed by 8 minutes of *E. coli* free demineralised water (deloading phase). Then, the flow direction was reversed for 10 minutes and the effuent at the top end of the column was collected. In order to assess the effect of the duration of loading and deloading on the amount of strained bacteria mass, in a number of cases, durations were varied. To assess the effect of concentration on strained mass, the concentration of the influent bacteria suspension was varied between 2×10^7 cells/mL and 10^9 cells/mL (Table 1).

Table 1 Overview of experimental conditions, calculated parameters and mass balances. For the calculation of η_0, the following parameter values were assumed: temperature $T = 293$ K, bacteria diameter $a_p = 1.49$ μm, porosity $\varepsilon = 0.44$–0.50, Hamaker constant $H = 6.5 \times 10^{-21}$ J, particle density $\rho_p = 1055$ kgm^{-3}, fluid density $\rho_{fl} = 1000$ kgm^{-3}, and fluid voscosity $\varpi = 1.3 \times 10^{-3}$ kgsm^{-1}

Experimental parameters					**Calculated**		**Mass balance (fraction of applied cells)**		
grain size (μm)	type of sediment	Duration (min)	C_0 ($\times 10^7$) (cells/mL)	col. length (cm)	η_0 (-)	α (-)	Effluent	Reversed flow effluent	Non-recovered
180–212	quartz	18	4.14	5			0.92	0.05	0.03
180–212	quartz	18	4.67	2			0.90	0.02	0.08
180–212	quartz	18	3.75	1			0.98	0.01	0.01
180–212	quartz	18	17.80	1	4.226E-03	9.91E-03	0.90	0.03	0.07
180–212	quartz	18	120.24	1			0.95	0.01	0.04
180–212	glass	18	4.68	5			0.95	0.01	0.04
75–90	quartz	18	4.32	5			0.84	0.07	0.09
75–90	quartz	18	4.54	2			0.88	0.05	0.08
75–90	quartz	18	4.48	1	1.112E-02	1.26E-03	0.97	0.01	0.02
75–90	quartz	18	13.87	1			0.93	0.04	0.03
75–90	quartz	18	67.07	1			0.96	0.01	0.03
45–53	quartz	18	4.22	5			0.71	0.10	0.19
45–53	quartz	18	4.63	2			0.78	0.05	0.16
45–53	quartz	18	4.59	1			0.82	0.05	0.13
45–53	quartz	18	12.15	1	2.026E-02	1.30E-03	0.86	0.03	0.11
45–53	quartz	18	101.36	1			0.81	0.02	0.17
38–45	quartz	18	4.41	5			0.44	0.17	0.39
38–45	quartz	18	3.76	2			0.75	0.09	0.16
38–45	quartz	18	4.27	1			0.79	0.06	0.16
38–45	quartz	18	17.98	1	2.489E-02	1.55E-03	0.70	0.05	0.25
38–45	quartz	18	90.12	1			0.83	0.01	0.16
38–45	quartz	6	4.38	1			0.69	0.12	0.31
38–45	quartz	6	4.69	0.75			0.78	0.09	0.22
38–45	quartz	6	4.61	0.5			0.85	0.06	0.15
38–45	quartz	6	4.53	0.25			0.96	0.03	0.04
38–45	glass	18	4.53	5			0.88	0.02	0.10

In order to assess the effect of transport distance, column lengths were varied between 0.25 and 5.0 cm. The *E. coli* concentration was determined using optical density measurements (at 220 nm) with a UV-visible spectrophotometer (Cecil 1021, Cecil Instruments Inc.) and a 1-cm cuvette. Absolute cell numbers were deduced after calibration with plate counts.

Results and discussions

Breakthrough curves

Breakthrough was rapid and without retardation in all cases (Fig. 1a-c). For the 5 cm columns (Fig. 1a), especially for the smaller grain sizes, breakthrough gradually increased. A gradual increase was also visible for the 2 and 1 cm columns (Fig. 1b, c), but to a lesser extent. Such gradual increase could indicate many processes. One of these was the filling of

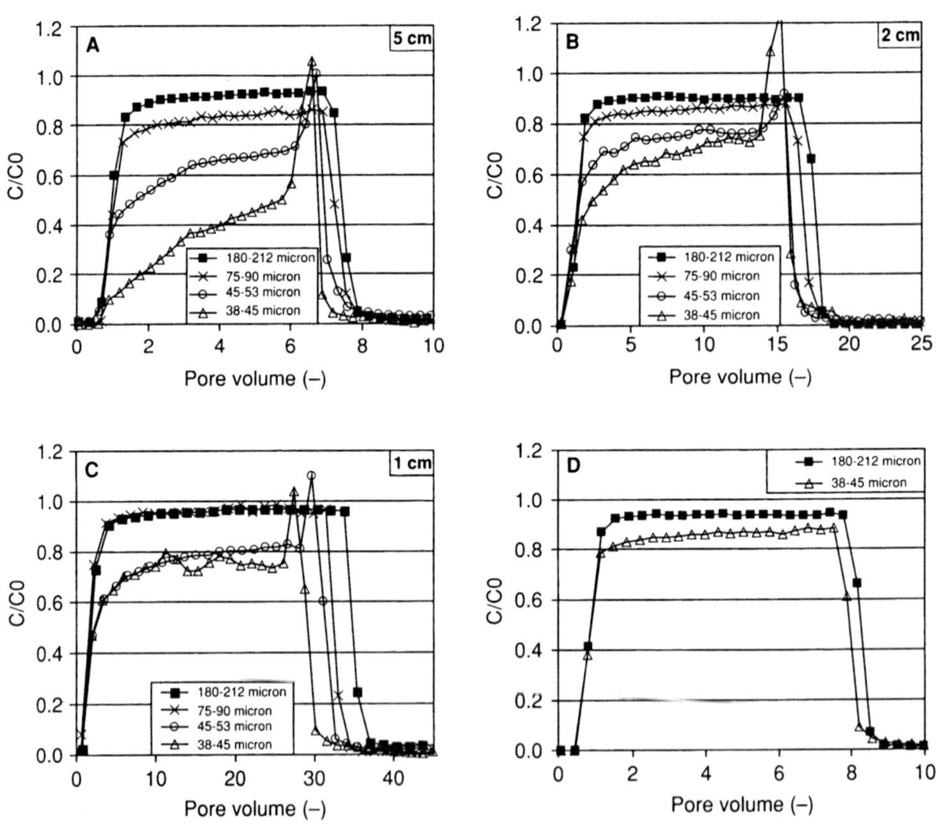

Figure 1 Breakthrough curves for the 4 quartz sand fractions and for column lengths 1, 2 and 5 cm. Lower right (d): breakthrough curves in 5 cm glass bead columns

limited pore space available for straining, in accordance with Foppen et al. (2005). Another possibility was the saturation of sites available for kinetic attachment and a third possibility was a combination of these two processes. The second observation was that the retained bacteria mass increased with decreasing grain size. Attachment and/or straining of bacteria were apparently dependent on the grain size, or, the pore size distribution of the column sediment. If straining was an important process, then, based on a comparison with the glass microspheres breakthrough curves, we concluded that straining did not really depend on grain size, because breakthrough in the glass beads was similar for both the 38–45 μm and the 180–212 μm size fractions (compare Fig. 1a and 1d). Retention of bacteria mass appeared to be highly dependent on not only the size but also the shape and the surface roughness of the grain (irregular quartz and perfect glass spheres) and therefore on the pore size distribution of the sediment. The third observation was the sudden C/C0-increase for the smallest two quartz grain sizes immediately after deloading with demineralised water had started (e.g. Fig. 1a, between pore volume 6 and 7). We did not attribute this to a kind of site saturation (the peaks were absent during pore volume 6–7 in the 2 and 1 cm columns), but to very loosely attached bacteria to the quartz grain surface. Apparently, in these types of systems with very low ionic strength, negatively charged bacteria, negatively charged sediment surface and therefore a high double layer repulsion, already a very small change in conditions (deloading with demineralised water) was enough to cause some detachment of previously attached bacteria.

Flow reversal experiments

Reversed flow effluent concentrations are given in Fig. 2. Bacteria concentration, loading time, pore water flow velocity and sampling frequency in all experiments were similar. Release of bacteria from the columns was instantaneous, and the release process was completed within 1–1.5 pore volumes. Also, the amount of *E. coli* in the reversed flow effluent

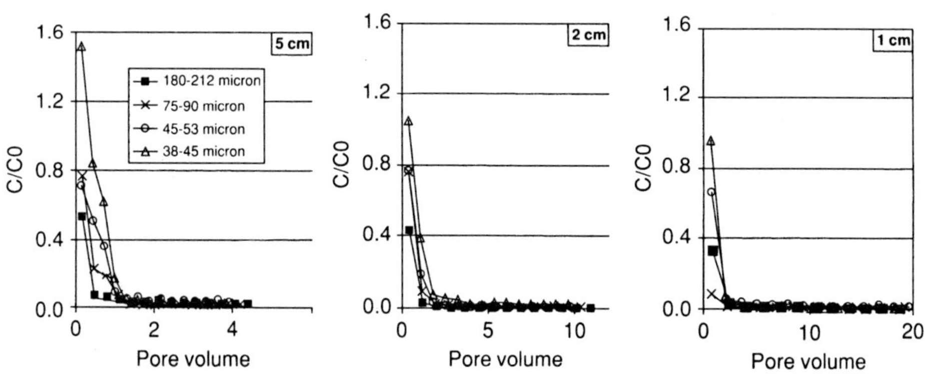

Figure 2 *E. coli* released (expressed as fraction of the influent concentration) as a result of reversing the flow direction for the 5, 2, and 1 cm columns (left, middle and right graph) for the various grain size fractions

increased with increasing column length and decreasing grain size (or rather: pore size). The highest *E. coli* concentrations were found in the reversed flow effluent of the 5 cm 38–45 μm quartz grains columns while the lowest concentrations were in the reversed flow effluent of the 1 cm columns with the largest quartz grain sizes. Based on these observations, we concluded that the released amount of *E. coli* had to be strained in the column and was not attached. However, immediately two questions arose: 1) Was the amount of strained bacteria dependent on the duration of the experiment and 2) Since it was likely that straining and attachment also took place during reversing the flow, what was the magnitude of this secondary retention process? The first question will be answered in the next paragraph. The second question was outside the scope of this paper, because the question could not be answered experimentally. For this, transient mass balance modelling was required in which retained bacteria masses both from modelling and from experiments are compared. For now, we assumed that the amount of *E. coli* that was released due to reversing the flow direction was the minimum amount of strained bacteria.

Duration of the experiment

In Fig. 3a, loading time was 3 minutes, while deloading time varied between 3 and 30 minutes. Minimum straining was 1.5 times higher for the shortest deloading time. However, the minimum strained amount did not significantly change for deloading times between 6 and 30 minutes. In Fig. 3b, the 1 cm columns from Fig. 1c (10 minutes loading and 8 minutes deloading) were compared with experiments in which loading and deloading were 3 minutes only. From these experiments, the minimum amount of straining was 1.8 times higher for the shorter duration experiments. We concluded that the minimum amount of strained bacteria was indeed influenced by the duration of the experiment. For the experiments given in Fig. 1a–c, the minimum amount of strained bacteria that was measured had to be multiplied by a factor of 1.5–1.85 to arrive at a more accurate estimate of the minimum amount of strained bacteria.

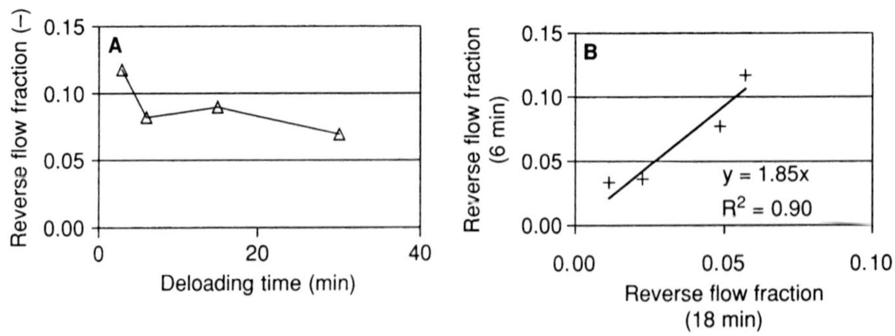

Figure 3 Reverse flow fraction as a function of deloading time in the 38–45 μm quartz sand (a) and relation between 6 minutes experiments and 18 minutes experiments (b)

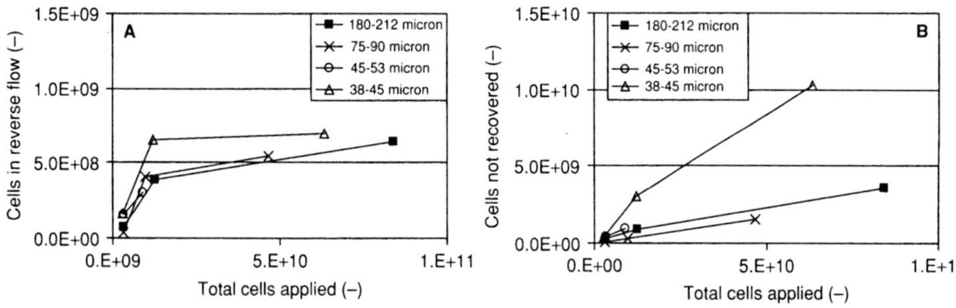

Figure 4 Amount of bacterial cells (a: in reverse flow; b: not recovered) in a 1 cm column as a function of total cells applied to the column

Concentration effect

Upon increasing the bacteria concentration , the amount of minimum strained bacteria was linearly dependent on the *E. coli* concentration of the suspension. However, when concentrations exceeded 2×10^8 cells/mL, the amount of cells in the reversed flow effluent remained constant. This indicated the presence of a maximum pore volume available for straining, which was in accordance with the findings of Foppen et al. (2005) and also fitted in the pore space filling concept to explain the observed gradual increase in breakthrough for the smallest quartz grain size fractions (Fig. 1a–c). The amount of cells not recovered was linearly dependent upon increasing bacteria concentration (Fig. 4b). If these cells were thought to be representative for the amount of cells that were kinetically attached, then, to explain the gradual increase in breakthrough curves, site saturation due to kinetic attachment did not seem to be the case.

If the cells not recovered in Fig. 4b were all considered to be attached, then the slopes of the straight lines were equal to the actual deposition rate, I, and this could be used to determine the dimensionless single collector removal efficiency, η, according to (Ryan and Elimelech, 1996)

$$\eta = \frac{I}{UC_0S} \tag{1}$$

where U is the fluid approach velocity, C_0 is the bulk concentration, and S is the surface area of the column. With η, the sticking efficiency, α, could be calculated according to

$$\alpha = \frac{\eta}{\eta_0} \tag{2}$$

where η_0 is the single collector contact efficiency, determined with the Tufenkji-Elimelech correction equation (Tufenkji and Elimelech, 2004; see Table 1 for parameter values used). Sticking efficiencies were calculated for all quartz grain size fractions (see Table 1). Values ranged between 0.0013 and 0.0099 and were considered to be low. These low sticking efficiency values demonstrated the unfavorable attachment conditions of the system, with negatively charged quartz grains and negatively charged *E. coli*.

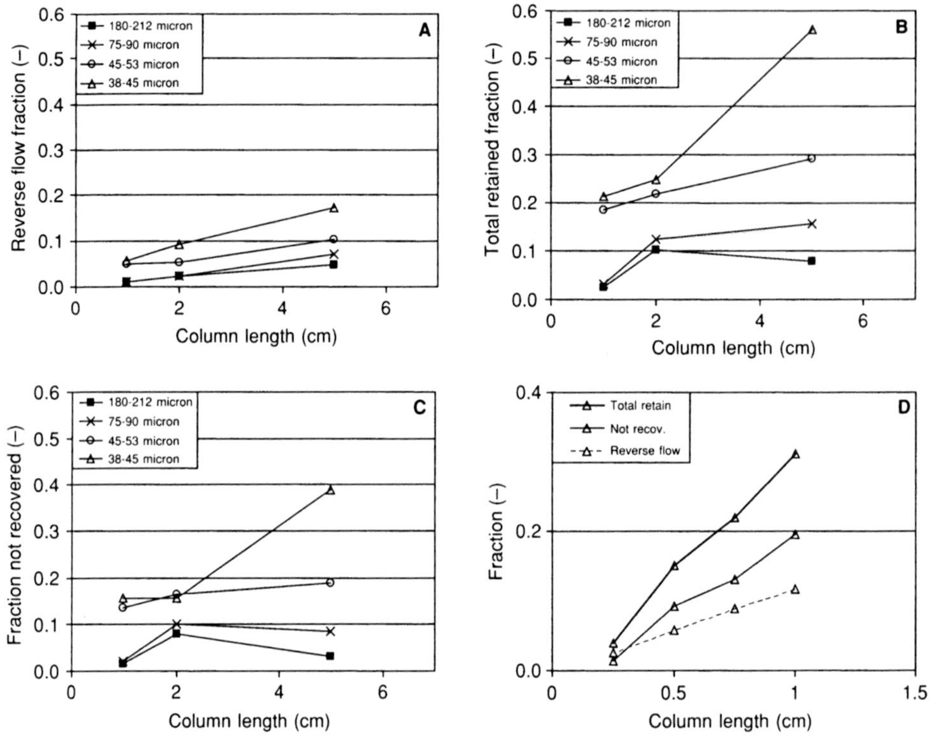

Figure 5 Reverse fraction (a), total fraction (b) and not recovered fraction (c) as a function of column length. Lower right graph (d): Results of 4 mini-columns of 38–45 μm quartz sand

Transport distance

When the bacteria concentration in the reversed flow effluent was expressed as fraction of the total bacteria mass applied to the column (Fig. 5a–d), then we observed that the minimum amount of straining (Fig. 5a) was roughly half of the fraction that could not be recovered (Fig. 5c). The latter was calculated as total amount applied to the column minus the total amounts of bacteria in both the effluent and the reversed flow effluent. Although the fraction not recovered may have contained a secondary straining fraction, we interpreted the not recovered fraction as being indicative for the kinetic attachment of *E. coli*. The ratio of 0.5 underlined the importance of straining in the retention of bacteria during their transport in this type of saturated porous medium. Furthermore, we found that the minimum straining fraction was almost linearly dependent on transport distance. This finding was not in correspondence with the straining model developed by Bradford et al. (2003). These workers found non log-linear retention profiles in their quartz grain columns, which they attributed to straining. This non log-linear behaviour was not present in our experiments. Finally, from Fig. 5b and 5c it seemed as if fractions intersected the vertical axis at values higher than 0.1. This was no true. When the results of total retained, not recovered, and reverse flow fractions for a series of mini-columns (Fig. 5d) were extrapolated with a straight line, obtained with linear regression, to a column length of zero, then the vertical axis was intersected close to zero (values ranged between −0.004 and −0.04, while $R^2_{straight\ line} > 0.98$ in all cases). It should be noted, that also for these mini-columns, the minimum straining fraction (given as reverse flow fraction in Fig. 4d) was linearly dependent on the column length ($R^2 = 0.999$).

Conclusions

The flow reversal method was useful in estimating the amount of straining that occurred during the transport of *E. coli* in saturated porous media. Release of bacteria from the columns was instantaneous, and the release process was completed within 1–1.5 pore volumes. Also, the amount of *E. coli* in the reversed flow effluent increased with increasing column length and decreasing grain size (or rather: pore size). The highest *E. coli* concentrations were in the reversed flow effluent of the 5 cm 38–45 μm quartz grains columns while the lowest concentrations were in the reversed flow effluent of the 1 cm columns with the largest quartz grain sizes. The amount of strained bacteria that we measured, however, was a minimum amount, because 1) for short duration experiments strained bacteria concentrations were 1.5–1.85 times higher than for long duration experiments, and 2) while reversing the flow direction, bacteria could be strained and/or attached again (secondary retention). We could not determine experimentally the amount of secondary retained material. This would require transient mass balance modelling.

In our experiments, we identified the following straining characteristics:

(1) Straining was not dependent on the grain diameter, but -most likely- on the presence of pores that are too small to allow passage of (bio)colloids. It would therefore be useful to describe straining in terms of the pore size distribution of sediments;

(2) The amount of strained bacteria was linearly dependent on the *E. coli* concentration of the suspension. In addition, when *E. coli* concentrations exceeded 2×10^8 cells/mL, the amount of strained bacteria remained constant. This indicated the presence of a certain maximum pore volume available for straining;

(3) In the 0–5 cm column length range we used, straining was almost linearly dependent on column length. This finding was <u>not</u> in correspondence with the straining model developed by Bradford et al. (2002, 2003). These workers found non log-linear retention profiles in their quartz grain columns, which they attributed to straining. This non log-linear behaviour was not present in our experiments;

These observations will certainly assist in developing a better transient mass balance model that includes both straining and kinetic attachment.

References

Bradford, S.A., J.Simunek, M. Bettahar, M.Th. van Genuchten, and S.R. Yates, 2003. Modeling colloid attachment, straining, and exclusion in saturated porous media. *Environ. Sci. Technol.* **37**, 2242–2250.

Foppen, J.W.A., A. Mporokoso and J.F. Schijven, 2005. Determining straining of *Escherichia coli* from breakthrough curves. *J. Contam. Hydrol.* (**76**), 191–210.

Foppen, J.W.A. and J.F. Schijven, 2005. Transport of *E. coli* in columns of geochemically heterogeneous sediment. *Wat. Res.* (**39**), 3082–3088.

Ryan, J. and M. Elimelech, 1996. Review: Colloid mobilization and transport in groundwater. *Colloids and Surfaces A: Physicochemical and Engineering Aspects* **107**, 1–56.

Tufenkji, N. and Elimelech M., 2004. Correlation Equation for Predicting Single-Collector Efficiency in Physicochemical Filtration in Saturated Porous Media. *Environ. Sci. Technol.* **38**, 529–536.

Part III

SLOW SAND FILTRATION – TREATMENT PERFORMANCE

15 The removal of turbidity in a multistage slow sand pilot-plant under challenging conditions

William B. Anderson*, Jeffrey L. DeLoyde*, Robert A. LeCraw, Michael Galan**, Shawn A. Cleary***, and Peter M. Huck***

*Department of Civil Engineering, University of Waterloo, Waterloo, ON, Canada, N2L 3G1
**MS Filter Inc., 17665 Leslie St., Suite 47, Newmarket, ON, Canada L3Y 3E3
***Golder Associates Ltd., 2390 Argentia Road, Mississauga, ON, Canada L5N 5Z7

Abstract A multistage slow sand filter pilot plant was tested under challenging conditions. The analysis of turbidity data from over a full year of operation showed that multistage filtration (including roughing and slow sand filtration) was very robust over a wide range of temperature (from 0 to ~30°C) and turbidity (up to 80 NTU). The pilot plant was able to operate within the Province of Ontario regulatory standards and Canadian federal guidelines for turbidity in drinking water. During the operation of the pilot plant over lengthy, largely unattended, periods of time that there had to be a recognition that some turbidity data would have to be excluded for reasons related to conditions not resulting from filter operation (e.g. flow interruptions to turbidimeter, turbidimeter cleaning, biofilm sloughing from lines leading to turbidimeter, etc.).

Keywords Slow sand filtration, biological filtration, turbidity

Introduction

It is well established that slow sand filtration is a proven and sustainable technology for drinking water treatment in small communities. Slow sand filters, as stand-alone units, however, are sensitive to lower water temperatures that can lead to decreased biological treatment, and high raw water turbidity levels that can result in premature clogging of the filter and frequent cleaning requirements. Multistage filtration, consisting of roughing filtration followed by slow sand filtration, can overcome these treatment limitations. However, concerns persist with regard to the ability of multistage filtration to treat waters of variable quality which may occasionally experiences episodes of high turbidity, and while it is known that biological activity in the schmutzdecke is reduced as temperatures fall there is relatively little published data available for multi-stage systems operating at or close to freezing.

This study evaluated the turbidity removal performance in a multistage slow sand filtration pilot plant exposed to a range of influent turbidities (1 to >80 NTU) and water temperatures (~0 to 30°C). Except for one brief test, the hydraulic loading rate was held steady at 0.4 m/h which is on the high side of typical rates. Raw water was obtained from the Grand River in Ontario, Canada, whose watershed drains a relatively large catchment area

 Recent Progress in Slow Sand and Alternative Biofiltration Processes edited by Rolf Gimbel, Nigel J.D. Graham and M. Robin Collins. ISBN: 9781843391203. Published by IWA Publishing, London, UK.

that receives runoff from both rural agricultural and urban areas. The pilot system included a pre-ozonation step followed by two parallel trains, each consisting of a roughing and slow sand filter in series. The turbidity data presented here are for the raw water and the slow sand filter (SSF) effluents only. The reported results therefore reflect the effectiveness of these three steps in combination. The ozone step was not operated during periods of extremely cold water temperature for reasons discussed later.

The current Ontario and Canadian drinking water filter effluent turbidity standard/ guideline for slow sand filtration is 'less than or equal to 1.0 NTU in at least 95% of the measurements made or at least 95% of the time each calendar month'; and not to exceed 3.0 NTU at any time (MOE, 2003; Health Canada, 2003). This compares with a 0.3 NTU regulated value for rapid sand filtration and a target of 0.1 NTU (MOE, 2003).

Methods

Pilot System

The two-train multi-stage slow sand filter pilot plant was provided by MS Filter, Inc. (Newmarket, ON, Canada). The plant treated raw water from the Grand River, which occasionally experiences turbidity peaks well over 100 NTU.

The pilot plant consisted of two identical trains (Pilot 1, Trains 1 and 2), each having pre-ozonation and upflow roughing filtration (with 20 cm of fine gravel and a 30 cm GAC cap) followed by slow sand filtration. In the 31.5 cm diameter PVC slow sand filter (SSF) column, water flowed down through 40 to 45 cm of silica sand (0.35 mm effective size, uniformity coefficient 1.7) and 20 cm of underdrain gravel before flowing back to the river. The hydraulic loading rates in the slow sand filters in Trains 1 and 2 were both maintained at 0.4 m/h, except for a brief period in late 2004 (Sept. 27-Oct. 1) when the hydraulic loading was increased in Train 1 to 0.8 m/h for a protozoan spiking test. Ozone was applied at a set (unadjustable) dosage of ~7 mg/L prior to the roughing filter for most the period discussed here (except in January, February, November, and December 2005 as water temperature approached 0°C). Ozone was not applied during these months due to problems associated with having a fixed ozone dose (higher than desirable ozone residuals in water) and occasional freezing of gas discharge tubing outlets preventing proper venting in the pilot plant facility (ozone in ambient air). These problems, although not unique to pilot plants, would be accounted for in various ways at full-scale.

Flow to the pilot plant was initiated in October 2003 and data presented in this paper begins in mid-August 2004 when the filters were biologically mature (approaching 10 months of continuous operation).

Turbidity

Continuous turbidity measurements (every 15 minutes) were obtained using online HACH 1720D turbidity meters (Hach Co., Loveland, CO, USA).

Results and discussion

Turbidity has been continuously monitored since start-up. Figure 1 compares raw water turbidity to that of the two slow sand filter effluent streams for the period from mid August

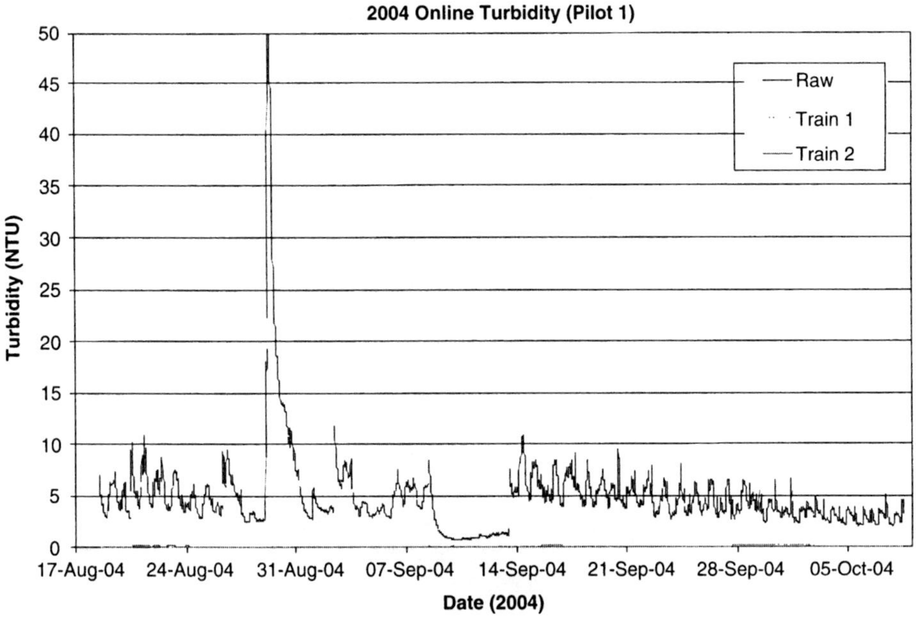

Figure 1 Raw water and slow sand filter effluent turbidities in August/September 2004.

to early October, 2004. Raw water turbidity ranged from about 3 to 10 NTU during this period with the exception of a rain event in late August when turbidity spiked to 50 NTU. Figure 2, which has been scaled to illustrate in more detail the SSF effluent turbidities, shows that both SSF filter effluent turbidities were only very marginally affected by the

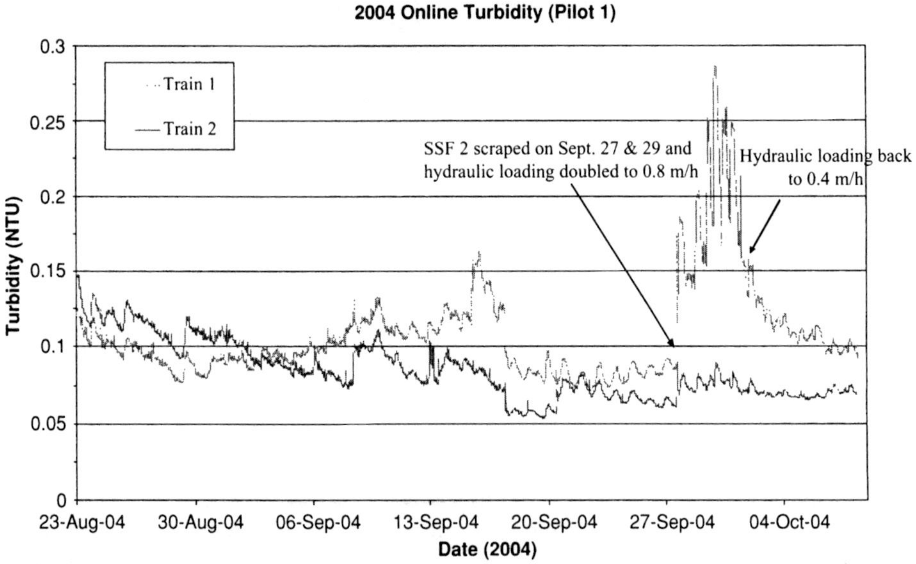

Figure 2 Slow sand filter effluent turbidities in August/September 2004.

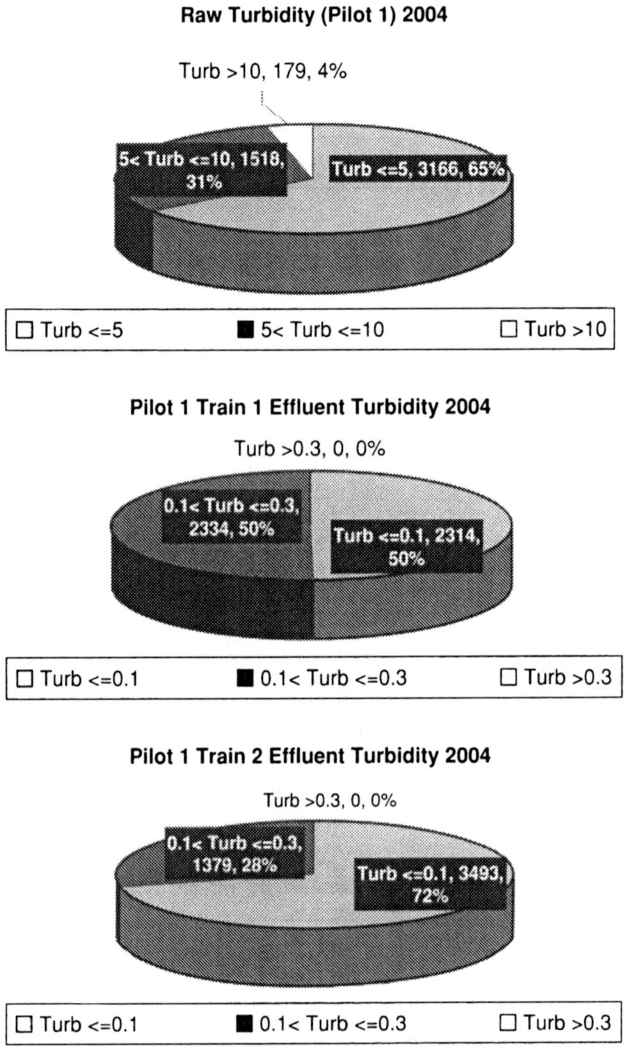

Figure 3 Distribution of turbidity levels for selected ranges in raw water and SSF effluents (data for August and September, 2004).

August rain event (~0.02 NTU increase). In fact the SSF effluent turbidities were typically around 0.1 NTU. The exception to this was during the period from Sept. 27 to Oct. 7, 2004 when a spiking experiment was performed on Train 1. On Sept. 27 SSF 1 was scraped and the hydraulic loading was doubled (from 0.4 to 0.8 m/h). On Sept. 29 SSF 1 was scraped again. This resulted in effluent turbidities spiking upwards toward 0.3 NTU (still less than the regulated value of 1.0 NTU).

To put data for this period into perspective, Figure 3 shows pie charts for the raw and SSF effluent streams. A total of 4,863 data points were analyzed for the raw water and 4,648 and 4,818, respectively, for Train 1 and 2 SSF effluents. During this period the raw

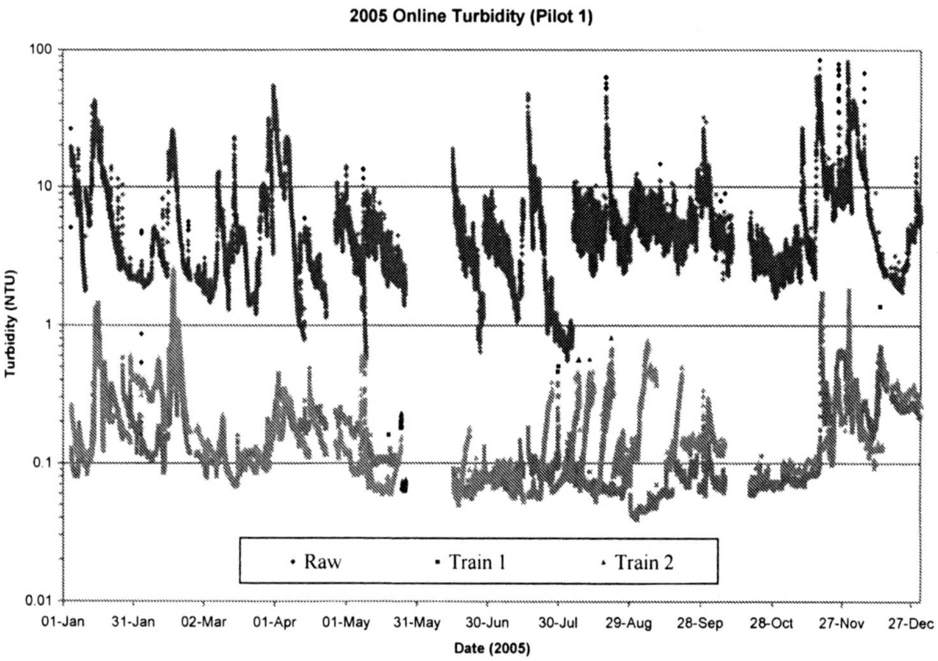

Figure 4 Raw water turbidity *vs.* SSF effluent turbidity for 2005 (note log scale).

water turbidity was less than 10 for 96% of all measurements and above that for 4% of measurements. Pie charts for the SSF effluent turbidities show that for Train 1, 50% of measurements were less than 0.1 NTU and 100% were below 0.3 NTU. For Train 2, 72% of SSF measurements were less than 0.1 NTU and all were less than 0.3 NTU. Train 2 was expected to perform better because it was the control train, with experiments being conducted on Train 1 only.

Data for the entire calendar year 2005 were then examined for trends. The pilot plant had been running for over a year at the beginning of this period and there were sufficient numbers of 'events' to truly observe plant performance. Figure 4 shows the raw water and SSF 1 and 2 effluent turbidity data on a log plot to facilitate comparison of effluent spikes to those in raw water. It can be seen that there were about 17 separate occasions when raw water turbidities spiked to over 20 NTU with a maximum of about 80 NTU being recorded in November. In general, SSF 1 and 2 effluents tracked similarly and responded with spikes at the same time in which raw water spikes were observed. There is some deviation between SSF effluents in August. The reason for this is uncertain but it may be related to the fact that experiments were being conducted using Train 1 and not Train 2.

The data presented in this paper has been vetted for known events that artificially cause turbidity spikes. Table 1 summarizes data before and after this process. Some reasons for removing data included: documented artificial increases in turbidity induced by sampling, disturbing tubing leading to turbidimeters, air bubbles, sample lines plugging, cleaning/flushing of turbidimeters, power interruptions, stopping flow for short periods to move the pilot trailer, routine plant and turbidimeter maintenance, turbidimeter malfunctions, algal/bacterial

Table 1 Comparison of turbidity before and after removal of data known to have been influenced by factors affecting turbidimeter operation/performance

			Uncensored Data				
	Turbidity (NTU) =	**≤5**	**>5 to ≤10**	**≥10**			**Total**
Raw	# of observations	19,006	7,640	4,417	–	–	31,063
	% of observations	61.2	24.6	14.2	–	–	100
	Turbidity (NTU) =	**≤0.1**	**>0.1-≤0.3**	**>0.3-≤1.0**	**>1.0-≤3.0**	**>3.0**	
SSF 1	# of observations	14,211	12,949	2,604	657	641	31,062
	% of observations	45.8	41.7	8.4	2.1	2.1	100
SSF 2	# of observations	9,631	15,863	4,989	201	375	31,059
	% of observations	31.0	51.1	16.1	0.6	1.2	100
	After Removing Data Known to be affected by Events not Related to Filter Operation						
	Turbidity (NTU) =	**≤5**	**>5 to ≤10**	**≥10**			**Total**
Raw	# of observations	18,967	7,615	4,366	–	–	30,948
	% of observations	61.3	24.6	14.1	–	–	100
	Turbidity (NTU) =	**≤0.1**	**>0.1-≤0.3**	**>0.3-≤1.0**	**>1.0-≤3.0**	**>3.0**	
SSF 1	# of observations	14,136	12,582	2,457	374	0	29,549
	% of observations	47.8	42.6	8.3	1.3	0	100
SSF 2	# of observations	9,607	14,934	4,958	177	0	29,676
	% of observations	32.4	50.3	16.7	0.6	0	100

growth in turbidimeters or feed lines, etc. Occasionally data were removed following upflow cleaning of the slow sand filters (up to an hour afterwards) but this was not the rule. For the most part, data were not removed if they were associated with filter scraping, topping up a filter with sand, or backwashing/scraping of the preceding roughing filters.

Most of these reasons for removing data are attributable to the fact that this work was conducted in a pilot scale plant which is extremely susceptible to disturbances which would not normally affect a full-scale plant and its associated monitoring equipment. Nonetheless the turbidity effluent levels were quite low as can be seen in Figure 5. In 2005, 61% of raw water turbidity measurements were less than 5 NTU, 25% were between 5 and 10 NTU, and 14% were greater than 10 NTU (n = 30,948). SSF effluent turbidities were less than 0.1 NTU, 48 and 32% of the time in Trains 1 and 2, respectively. They were between 0.1 and 0.3 NTU, 43 and 50% of the time for Trains 1 and 2, respectively. For both SSF effluents, 99% of all measurements were below the regulated level of 1.0 NTU. The remaining 1% of values were all lower than 3 NTU. According to the wording of the current Ontario regulations these plants would have been in compliance with Ontario Drinking Water Regulations (MOE, 2003) when documented turbidity exceedances from pilot plant issues discussed above were removed. These findings demonstrate the resiliency of the process which was operated at times under extremely challenging conditions, down to 0°C, at the upper end of typical hydraulic loading (0.4 m/h), and at turbidities often exceeding 20 NTU and occasionally approaching 80 NTU.

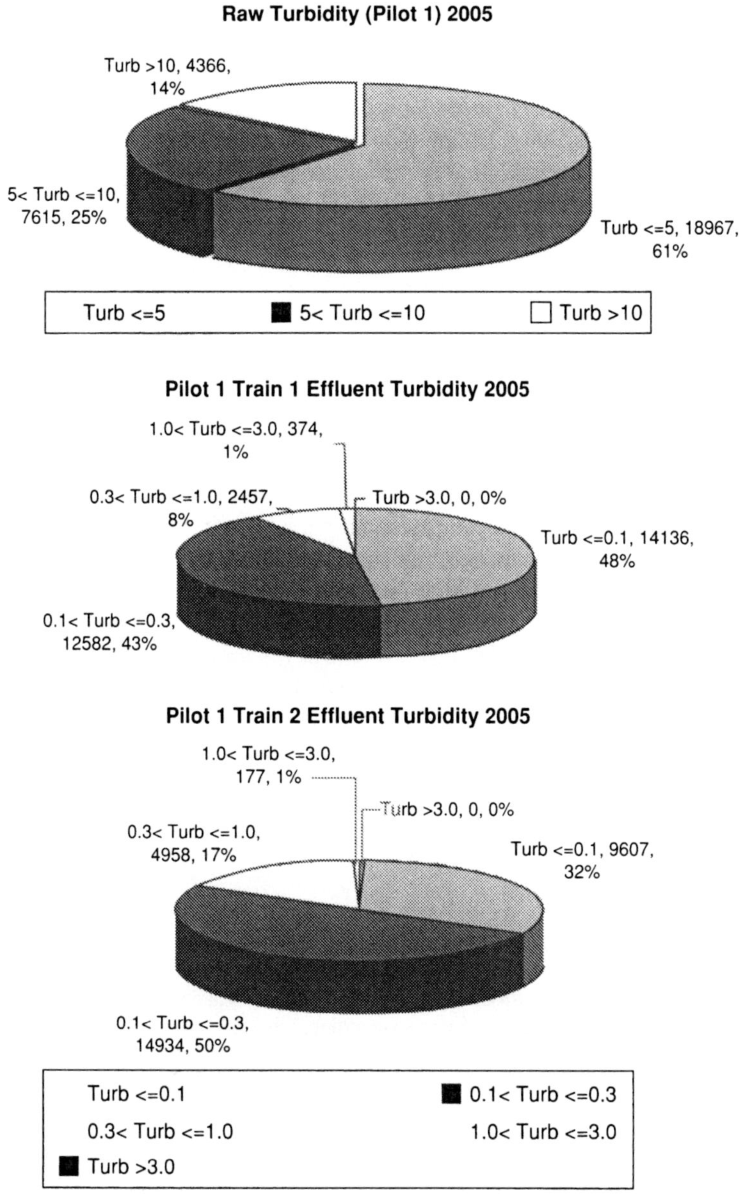

Figure 5 Distribution of turbidity levels for selected ranges in raw water and SSF effluents (after removal of data)

While a thorough discussion of all turbidity spikes is not possible due to restrictions on the length of this article a few observations are worth noting (see Figure 4). First, the filters performed well and generally did not respond to spikes in turbidity under 10 NTU, confirming what has been widely reported. Raw water turbidity spikes up to 30 NTU during spring runoff that were caused solely by melting snow and not rain events did not substantially affect slow sand filter performance (SSF 1 and 2; Mar. 14, 27, 29, 30). On the other hand, rain events particularly in the fall (SSF 1 and 2; Nov. 9, 11, 25, 29) and winter (SSF 1 and 2; Jan. 15-17), were more likely to result in corresponding increases in slow sand filter effluent turbidity. Scraping of the slow sand filters during or very shortly following rain events, particularly in very cold water (<5°C), had longer term deleterious effects in filter performance and should be avoided if possible (e.g. on Nov. 17 for SSF 1 and Jan. 29 for SSF 2).

Conclusions

- This evaluation showed that a pilot scale multistage slow sand filter unit was capable of producing water which was within regulated standards and guidelines under conditions of temperature and turbidity challenge.
- Although the pilot system included a pre-ozonation step, it was not used in the coldest winter months for technical reasons. This did not, however, affect the performance of the system with respect to meeting standard/guideline turbidity values.
- Spring runoff caused by snowmelt, and not accompanied by rain, had a smaller impact on slow sand filter effluent turbidity than rain-induced turbidity resulting in similar raw water turbidities.
- Scraping of slow sand filters should be avoided at very low water temperatures (<5°C).
- When conducting long-term pilot scale research studies it is important to note that turbidity will be affected for reasons other than the performance of the filter(s). Researchers should attempt to anticipate what these may be and maintain detailed log books to provide scientific justification for the potential rejection of data at a later point.

References

Health Canada (2003) Guidelines for Canadian Drinking Water Quality: Supporting Documentation- Turbidity. Water Quality and Health Bureau, Healthy Environments and Consumer Safety Branch, Health Canada, Ottawa, Ontario, Canada. http://www.hc-sc.gc.ca/ewh-semt/pubs/water-eau/doc_sup-appui/turbidity/chap_1_e.html (Jan. 27, 2006)

MOE (2003) Ministry of the Environment. Procedure for Disinfection of Drinking Water in Ontario. (As adopted by reference by Ontario Regulation 170/03 under the Safe Drinking Water Act). http://www.ene.gov.on.ca/envision/gp/4448e.pdf (Jan. 27, 2006)

16 Removal of *Cryptosporidium* oocysts and *Giardia* cysts by pilot-scale multi-stage slow sand filtration

J. L. DeLoyde*, W. B. Anderson*, S. A. Cleary, S. Ndiongue***, R. A. LeCraw****, M. Galan**** and P. M. Huck***

*Department of Civil Engineering, University of Waterloo, Waterloo, ON, Canada, N2L 3G1
**Golder Associates Ltd., 2390 Argentia Road, Mississauga, ON, L5N 5Z7
***Walkerton Clean Water Centre, P.O. Box 160, 220 Trillium Court, Building 3, Walkerton, ON, NOG 2V0 (University of Waterloo at time of study)
****MS Filter Inc., 17665 Leslie St., Suite 47, Newmarket, ON, L3Y 3E3

Abstract Two slow sand filters were seeded with *Cryptosporidium* oocysts and *Giardia* cysts during six challenge tests carried out on two multi-stage slow sand filtration pilot plants between April 2004 and February 2005. The slow sand filters were operated at hydraulic loading rates ranging from 0.4 to 0.8 m/h. *Cryptosporidium* oocyst removal r anged from 2.0 to >5.2 logs at 0.4 m/h and from 1.5 to 3.7 logs at 0.8 m/h. Removal of *Giardia* cysts ranged from >2.4 to >4.9 logs at 0.4 m/h and from >2.5 to >4.8 logs at 0.8 m/h. In general, removal of microorganisms improved as the filters matured, increased hydraulic loading rates had a slight negative impact on removals, and cold temperatures (~ 3°C) did not effect removals in a 15-month old filter.

Keywords Filtration, slow sand filtration, biological filtration, *Cryptosporidium*, *Giardia*

Introduction

Slow sand filtration is highly effective at removing or inactivating *Cryptosporidium* oocysts and *Giardia* cysts through biological, chemical and physical process. The US EPA (2006) LT2ESWTR assigns a 3.0-log *Cryptosporidium* treatment credit to slow sand filtration when used for primary treatment. In Canada, the "Procedure for Disinfection of Drinking Water in Ontario" (2003) document under Ontario Regulation 170/03 assigns a 2-log removal credit for *Giardia* cysts by slow sand filtration.

Numerous studies in the literature have demonstrated greater than 4-log removal of *Cryptosporidium* by slow sand filtration. Timms *et al.* (1995) reported >99.997% removal of inactivated *Cryptosporidium parvum* oocysts by a slow sand filter with a 50 cm bed depth. Increasing the hydraulic loading rate from 0.3 to 0.4 m/h had no effect. Analysis of the top 30 cm of sand showed that all oocysts were removed in the top 2.5 cm. Schuler *et al.*

 Recent Progress in Slow Sand and Alternative Biofiltration Processes edited by Rolf Gimbel, Nigel J.D. Graham and M. Robin Collins. ISBN: 9781843391203. Published by IWA Publishing, London, UK.

(1991) seeded formalin inactivated *Cryptosporidium parvum* oocysts and *Giardia muris* cysts into a 90 cm deep slow sand filter operated for nine months. Oocyst removals were always >99.99% and cyst removals were consistently >99.99% after two months of filter maturation. Increasing hydraulic loading from 0.15 to 0.40 m/h had no effect on oocyst or cyst removal. Bellamy *et al.* (1985) demonstrated that a slow sand filter with 90 cm bed depth achieved average (from 222 analyses) *Giardia* cyst removal of 99.991% at 0.04 m/h, 99.994% at 0.12 m/h and 99.981% at 0.40 m/h. As shown, cyst removal was approximately 4 logs regardless of hydraulic loading over the range tested. The researchers also showed that dropping the temperature from 15°C to 5°C during the filter run did not affect cyst removal, but noted that the effects of maintaining the filter at 5°C for a longer time should be studied. Bellamy *et al.* (1985) concluded that complete removal of *Giardia* cysts can be expected in a biologically mature slow sand filter regardless of the presence of a schmutzdecke layer, hydraulic loading rate and concentration of influent *Giardia* cysts.

Methods

Pilot systems

Six *Cryptosporidium* and *Giardia* challenge tests were conducted between April 2004 and February 2005 on two multi-stage slow sand filtration pilot plants. Pilot 1 was provided by MS Filter, Inc., and commissioned in October 2003. Pilot 2 was built by researchers at the University of Waterloo and was commissioned in December 2003. Both pilots treated raw water from the Grand River, a large river in Ontario, Canada, that is heavily impacted by rural and urban inputs and experiences turbidity peaks over 100 NTU and water temperature variations in the range of approximately 1 to 30 °C.

Pilot 1 consisted of two identical trains (train 1 and 2), each having pre-ozonation, roughing filtration (with 20 cm of fine gravel and a 30 cm GAC cap) and slow sand filtration in series. In the 31.5 cm diameter PVC slow sand filter column, water flowed down through 40 to 45 cm of silica sand (0.35 mm effective size, UC 1.7) and 20 cm of underdrain gravel. The hydraulic loading rates for train 1 and 2 slow sand filters ranged from 0.4 to 0.8 m/h. No ozone residual was measured in the water by the time it entered the roughing filter.

Pilot 2 employed roughing filtration followed by two slow sand filters in series. The 29.5 cm diameter downflow slow sand filter 1 (SSF1) contained 100 cm of silica sand supported by 15 cm of underdrain gravel. Some of the effluent from SSF1 flowed to slow sand filter 2 (SSF2), which was a 15 cm column (half the diameter of SSF1) with a 50 cm bed of silica sand supported by 15 cm of underdrain gravel. Sand was the same as Pilot 1 (0.35 mm effective size, UC 1.7). Both SSF1 and SSF2 were operated in downflow mode at hydraulic loading rates of 0.4 m/h.

Cryptosporidium and *Giardia* feedstock

Two 50 mL vials of formalin-inactivated *Cryptosporidium parvum* oocysts of bovine origin were obtained from a commercial laboratory (Waterborne, Inc., New Orleans, LA) in March and July 2004. Each vial had a concentration of 20×10^9 oocysts/L. In July 2004, four 10 mL vials containing *Giardia muris* at a concentration of 1×10^{10} cysts/L were purchased from the same laboratory. All vials were stored in the dark at 4°C.

Cryptosporidium and *Giardia* stock vials were vortexed and 23 mL of the *Cryptosporidium* stock and 9 mL of the *Giardia* stock were added to 1 L of raw water on the day of the challenge test. For challenge tests 1 to 3 no *Giardia* was added to the feedstock. Actual feedstock concentrations were obtained from five replicate counts using a hemacytometer and ranged from 4.2×10^8 to 6.3×10^8 oocysts/L and from 8.0×10^7 to 1.0×10^8 cysts/L. Target slow sand filter influent *Cryptosporidium* and *Giardia* concentrations were 10^6 oocysts/L and 10^5 cysts/L, respectively. Using a peristaltic pump, continuously stirred feedstock was injected for six hours into the slow sand filter influent. During challenge tests run at 0.4 and 0.8 m/h, feedstock was injected at 1 and 2 mL/min, respectively.

Sampling procedure

Theoretical detention times for Pilot 2 were 85 minutes (SSF1) and 52 minutes (SSF2), calculated using a sand porosity of 0.45. A chloride tracer test showed hydraulic detention times were 80 and 45 minutes for SSF1 and SSF2, respectively. For simplicity, detention times were assumed to be 120 minutes in SSF1 and 60 minutes in SSF2, which allowed influent and effluent sampling to be better coordinated. The theoretical detention time for the slow sand filter of Pilot 1 train 1 operated at 0.4 m/h was 50 minutes and confirmed by a chloride tracer test. For simplicity, a detention time of 60 minutes was used when collecting effluent samples. At 0.8 m/h, the theoretical detention time (25 minutes) was used as the basis for collecting effluent samples 30 minutes after influent samples. Samples were collected in 250 mL (influent) and 1000 mL (effluent) Wheaton glass bottles rinsed in buffered detergent. Samples were refrigerated in the dark at 4°C and processed and enumerated within three weeks.

Immunofluorescence assay and enumeration

One to 2 mL of each influent sample was processed and 200 to 500 mL of each effluent sample was processed for enumeration. A vacuum filter manifold (Hoefer Scientific, San Fransisco, CA) was used to filter samples through 25 mm 0.4μm polycarbonate membranes (Nucleopore, Corning, Acton, MA) and 25 mm 8μm nitrocellulose support membranes (MF Millipore Membrane Filters, Millipore Canada, Ltd., Nepean, ON).

Two millilitres of 1% bovine serum albumem (BSA) was filtered through the membranes, followed by the sample, then another 2 mL of 1% BSA. When needed, 1X PBS was added to keep the membranes moist. The immunofluorescence assay (IFA) procedure was carried out according to IFA staining techniques (USEPA, 1996) using the Hydrofluor combination *Cryptosporidium* and *Giardia* Kit (Strategic Diagnostics, Newark, DE). After staining, the membranes were placed onto glass microscope slides and enumerated using light microscopy at the University of Waterloo. Cleary (2005) indicated that mean oocyst recovery was approximately 75%. Negative and positive controls were processed at the same time to ensure no contamination of samples and adequate staining, respectively. Oocyst and cyst concentrations, used to calculate removals, were calculated by dividing the number

Table 1 Challenge test experimental conditions

	Challenge Tests					
	Test 1	**Test 2**	**Test 3**	**Test 4**	**Test 5**	**Test 6**
Pilot/train	2	1/1	1/1	1/1	1/1	1/1
Maturation time (months)*	4	7	7	10	11	15
Pathogen**	C	C	C	C & G	C & G	C & G
Hydraulic loading rate (m/h)***	0.4	0.4	0.8	0.4	0.8	0.4
SSF detention time (h)	2 (SSF1) 1 (SSF2)	1	0.5	1	0.5	1
SSF sand depth (cm)	100 (SSF1) 50 (SSF2)	40	40	43	37	44
Time since filter cleaning (days)	26	10	3	6	1	4
Depth of water above sand (cm)	51 (SSF1) 15 (SSF2)	104	60	20	50	36
Water Temperature (°C)$^{+}$	10.8	18.9	21.8	23.6	17.4	3.6
Dissolved oxygen (mg/L)$^{+}$	11.6	6.4	7.0	3.4	5.8	11.0
Turbidity (NTU)$^{+}$	0.6	2.8	2.6	1.1	2.7	1.9

* Pilot 1 and Pilot 2 were commissioned in October 2003 & December 2003, respectively
** *Cryptosporidium* oocysts (C) and *Giardia* cysts (G)
*** Hydraulic loading rate across the slow sand filter
$^{+}$Slow sand filter influent water quality

of cysts or oocysts counted on the membrane by the volume of sample processed, then multiplying by 1000 mL.

Results and discussion

A total of six challenge tests were conducted between April 2004 and February 2005. The first *Cryptosporidium* challenge test was carried out on the slow sand filter (SSF) of Pilot 2, whereas *Cryptosporidium* challenge tests 2 and 3 were performed on the SSF of Pilot 1 train 1. During challenge tests 4 to 6, both *Cryptosporidium* and *Giardia* were injected prior to the Pilot 1 train 1 SSF. Experimental conditions are detailed in Table 1.

Cryptosporidium Oocyst Removal

Challenge Test 1

Cryptosporidium challenge test 1 was undertaken on April 28, 2004 after approximately 4 months of cold-water acclimation (Table 1). Even though the system was relatively biologically immature at the time of challenge test 1,*Cryptosporidium* removal through SSF1 ranged from 2.6 to 3.2 logs, with an average removal of 2.9 logs and standard deviation of

Table 2 Challenge test 1 *Cryptosporidium* oocyst removals

Sample Time (h)*			Oocyst Removal (logs)		
Influent	**Effluent SSF1**	**Effluent SSF2**	**SSF1**	**SSF2**	**Total**
1	3	4	2.6	1.5	4.1
2	4	ns	3.0	–	–
3	ns	6	–	–	3.9
4	6	7	3.2	0.3	3.5
6	8	ns	2.8	–	–
		Average (standard deviation)	2.9 (0.3)	0.9 (0.8)	3.8 (0.3)

* 2 hr detention time through SSF1 and 1 hr through SSF2
ns – not sampled (therefore some removals could not be calculated)

0.3 logs (Table 2). SSF2 removed an additional 0.3 to 1.5 logs of *Cryptosporidium*, with an average of 0.9 logs and standard deviation of 0.8 logs. In total, *Cryptosporidium* removal by both slow sand filters in series ranged from 3.5 to 4.1 logs with an average of 3.8 and standard deviation of 0.3 logs.

Challenge Tests 2 to 6

Challenge tests 2 to 6 were carried out on the SSF of Pilot 1 train 1 at hydraulic loading rates of 0.4 m/h or 0.8 m/h. All water was pre-ozonated except during challenge test 6. *Cryptosporidium* oocyst removals through the Pilot 1 train 1 slow sand filter for these tests are presented in Table 3.

Challenge Test 2

Cryptosporidium challenge test 2 was carried out on May 31, 2004 after 7 months of acclimatization (Table 1). Challenge test 2 *Cryptosporidium* oocyst removals ranged from 2.0 to 4.9 logs (Table 3). Average removal was 2.8 logs with a 1.1 log standard deviation. Although Pilot 1 train 1 SSF bed depth was half that of Pilot 2 SSF1, average removal in test 2 (2.8 logs) was similar to the average removal through SSF1 during test 1 (2.9 logs). This suggests that increased bed depth in a biologically immature filter does not increase removal significantly, which is supported by literature findings that *Cryptosporidium* oocysts are primarily removed in the top few centimetres of a slow sand filter (Timms *et al.*, 1995).

Challenge Test 3

Cryptosporidium challenge test 3 was carried out on the slow sand filter of Pilot 1 train 1 on June 7, 2004 after 7 months of acclimatization (Table 1). Challenge test 3 *Cryptosporidium* oocyst removal ranged from 1.5 to 3.2 logs, averaging 2.5 logs with a standard deviation of 0.6 logs (Table 3). Although the hydraulic loading rate was double that of test 2, the average log removal in test 3 (2.5 logs) was only 0.3 logs less than the average removal in test 2 (2.8 logs). The effect of increased hydraulic loading appeared to be minimal, in

Table 3 Removal of *Cryptosporidium* oocysts through Pilot 1 train 1 slow sand filter

Lapsed time (h) at*		Oocyst Log Removal through Slow Sand Filter				
0.4 m/h	**0.8 m/h**	**Test 2 May 31/04**	**Test 3 Jun 7/04**	**Test 4 Aug 26/04**	**Test 5 Sept 30/04**	**Test 6 Feb 2/05**
0	0	–	–	–	–	–
2	1.5	2.9	2.7	>4.5	3.7	3.5
4	3.5	4.9	2.2	>5.2	3.7	4.1
5	4.5	2.3	2.8	4.7	3.7	4.0
6	5.5	2.6	2.5	4.4	3.4	4.5
7	6.5	2.4	3.2	4.0	2.9	4.1
9	8.5	2.0	1.5	2.7	2.3	>4.5
Average log removal (standard deviation)		2.8 (1.1)	2.5 (0.6)	>4.3**	3.3 (0.6)	>4.1**
Hydraulic loading-m/h		0.4	0.8	0.4	0.8	0.4
Water temperature-°C		19	21.8	23.6	17.4	3.6

* Times shown are effluent sampling times. Corresponding influent samples were taken before effluent samples to account for hydraulic detention time through the SSF (1 h detention time at 0.4 m/h and 0.5 h at 0.8 m/h)

** Standard deviation not calculated because averages had a "greater than" sign

agreement with other studies (Timms *et al.*, 1995; Schuler *et al.*, 1991). However, since the hydraulic loading rate had been increased from 0.4 m/h to 0.8 m/h only one day before test 3, it is possible that changes in filter operation and biological composition had not had enough time to impact filter performance.

Challenge Test 4

Challenge tests 4 to 6 involved spiking both *Cryptosporidium* and *Giardia* into the slow sand filter of Pilot 1 train 1. Challenge test 4 was conducted at 0.4 m/h on August 26, 2004 after 10 months of SSF acclimatization (Table 1). *Cryptosporidium* oocyst removal ranged from 2.7 to >5.2 logs, with an average of at least 4.3 logs (Table 3). Since no oocysts were recovered in the effluent samples corresponding to the 2-h and 4-h influent samples, log removals were given a "greater than" sign. Therefore, the average removal would be at least, and possibly greater than, the 4.3 logs reported. For this reason, a standard deviation was not calculated for this test.

Comparatively, the average *Cryptosporidium* removal in test 4 (at least 4.3 logs, 0.4 m/h) was much greater than those for test 2 (2.8 logs, 0.4 m/h) and test 3 (2.5 logs, 0.8 m/h). This was likely the result greater biological maturity of the filter and warmer water temperatures.

Challenge Test 5

Cryptosporidium and *Giardia* challenge test 5 was conducted at 0.8 m/h on September 30, 2004, after nearly one year of filter maturation (Table 1). *Cryptosporidium* removal ranged from 2.3 to 3.7 logs, averaging 3.3 logs with a standard deviation of 0.6 logs (Table 3). The Pilot 1 train 1 SSF achieved these removals even though the hydraulic loading rate was 0.8 m/h and the filter had been cleaned the day before the test.

The greater average removal observed in test 5 (3.3 logs, 0.8 m/h) compared to test 3 (2.5 logs, 0.8 m/h) suggests that increased filter maturity improved oocyst removal, even without the schmutzdecke. However, it is possible that even greater removals might have been attained if the filter had not been cleaned. Bellamy *et al.* (1985) found that a mature filter was able to remove *Giardia* cysts without the presence of a schmutzdecke layer. Test 5 average *Cryptosporidium* removal was greater than test 2 (2.8 logs, 0.4 m/h), also suggesting improved removal due to filter maturity. However, Test 5 average removal was less than test 4 (at least 4.3 logs, 0.4 m/h), showing that increased hydraulic loading in test 5 likely impaired *Cryptosporidium* removal.

Challenge Test 6

Cryptosporidium and *Giardia* challenge test 6 was conducted on February 2, 2005 in cold water (raw water ~2°C) on the Pilot 1 train 1 SSF that had matured for 15 months (Table 1). *Cryptosporidium* removal ranged from 3.5 to >4.5 logs and averaged at least 4.1 logs (Table 3). Despite the cold water temperature, the average *Cryptosporidium* removal in test 6 was greater than the average removals from test 2 (2.8 logs, 0.4 m/h), test 3 (2.5 logs, 0.8 m/h) and test 5 (3.3 logs, 0.8 m/h) because of greater filter maturity. On the basis of this test, cold water did not appear to significantly effect *Cryptosporidium* removal by a mature slow sand filter.

***Giardia* Cyst removal**

Giardia cysts were seeded into the Pilot 1 train 1 slow sand filter influent along with *Cryptosporidium* oocysts in challenge tests 4 to 6 (Table 1). Cyst removals are presented in Table 4.

As shown in Table 4, *Giardia* removal in challenge test 4 ranged from >2.4 to >4.5 logs with an average of at least 3.6 logs. *Giardia* cyst removal ranged from >2.5 to >4.8 logs (averaging at least 4.2 logs) in challenge test 5 and ranged from >4.0 to >4.9 logs (averaging at least 4.6 logs) in challenge test 6.

The "greater than" sign on some removals in Table 4 indicates that the slow sand filter achieved complete cyst removal (i.e. zero cysts were counted in the effluent volume processed). If it had been economically feasible to seed a higher concentration of *Giardia* cysts into the filter, greater removals may have been observed.

Effect of Sampling Time on Calculated Removals

As shown in Tables 3 and 4, the *Cryptosporidium* and *Giardia* removals corresponding to the 9-hr effluent samples (0.4 m/h) and 8.5-h effluent samples (0.8 m/h) were always the lowest, except for test 6. These apparent lower removals were a result of drastically lower influent oocyst concentrations because oocyst seeding was stopped after six hours. For example, the test 5 *Cryptosporidium* and *Giardia* 8-h influent sample concentrations were approximately two orders of magnitude lower than the 6-hr influent sample concentrations (Table 5).

Table 4 Removal of *Giardia* cysts through Pilot 1 train 1 slow sand filter

Lapsed time (h) at**		*Giardia* Cyst Log Removal		
0.4 m/h	0.8 m/h	Test 4 Aug 26/04	Test 5 Sept 30/04	Test 6 Feb 2/05
0	0	–	–	–
2	1.5	>3.7	>4.5	>4.5
4	3.5	>4.2	>4.8	>4.8
5	4.5	>4.5	3.9	>4.5
6	5.5	3.4	>4.8	>4.8
7	6.5	3.7	>4.8	>4.9
9	8.5	>2.4	>2.5	>4.0
Average Log Removal		>3.6	>4.2	>4.6
Hydraulic loading-m/h		0.4	0.8	0.4
Water temperature-°C		23.6	17.4	3.6

** Times shown are effluent sampling times. Corresponding influent samples were taken before effluent samples to account for hydraulic detention time through the SSF (1 h detention time at 0.4 m/h and 0.5 h at 0.8 m/h)

As shown in Table 5, the influent oocyst and cyst concentrations were within the same order of magnitude from one to six hours. While the effluent samples corresponding to the 1-hr and 8-h influent samples both had 40 oocysts/L, the removals were 3.7 and 2.3 logs, respectively. Similarly, while the effluent samples corresponding to the 1-h and 8-h influent samples both showed complete *Giardia* removal (zero cysts/L), the removals were >4.5 and >2.5 logs, respectively. This shows the effect of the lower influent oocyst and cyst concentrations. Nevertheless, the removals corresponding to the 8-h influent samples were included when calculating average removals for each challenge test. Although this makes

Table 5 Challenge test 5 influent and effluent concentrations

Influent Sample Time (h)	Influent		Effluent Sample Time (h)	Effluent		Log Removal	
	oocysts/L	cysts/L		oocysts/L	cysts/L	*Crypto-sporidium*	*Giardia*
0	430	0	0	0	0	–	–
1	182000	32000	1.5	40	0	3.7	>4.5
3	302000	70000	3.5	55	0	3.7	>4.8
4	205000	42000	4.5	45	5	3.7	3.9
5	267000	59000	5.5	105	0	3.4	>4.8
6	227000	61000	6.5	290	0	2.9	>4.8
8	7130	340	8.5	40	0	2.3	>2.5

the reported average removals somewhat conservative, the *Cryptosporidium* value for hour 8 suggests that log removals may be somewhat lower at lower influent concentrations. However, more than this single data point would be required to make such a statement.

Conclusions

This pilot scale investigation of the removal of *Cryptosporidium* oocysts and *Giardia* cysts through the slow sand filter of two parallel multistage slow sand filtration processes showed that:

- It appears that the shallower bed depth in Pilot 1 did not effect oocyst removal. *Cryptosporidium* removals during challenge test 1 (average removal 2.9 logs, 0.4 m/h, Pilot 2 SSF1) and test 2 (average removal 2.8 logs, 0.4 m/h, Pilot 1) were similar, even though Pilot 2 SSF1 had a bed depth of 100 cm while Pilot 1 train 1 SSF had a bed depth of 40 cm. Although both filters were relatively biologically immature, this trend (similar removals between pilots regardless of bed depth) would be expected to continue as the filters matured over time.
- *Cryptosporidium* removal increased with increasing filter maturity. At 0.4 m/h, average oocyst removal increased from test 2 (2.8 logs) to test 4 (at least 4.3 logs). At 0.8 m/h, average oocyst removal increased from test 3 (2.5 logs) to test 5 (3.3 logs).
- Cold water temperature had little effect on oocyst removal in a fully mature filter. Water temperature was 23.6°C during test 4 and 3.6°C during test 6. Oocyst removals were similar, averaging at least 4.3 logs in test 4 and at least 4.1 logs in test 6. Both tests were conducted at 0.4 m/h. Additional cold water tests would be valuable to increase the confidence in this result.
- Increased hydraulic loading had a slightly negative effect on oocyst removal when the filter was relatively immature. Average oocyst removal in test 2 (0.4 m/h) was 2.8 logs, which was slightly better than 2.5 logs in test 3 (0.8 m/h).
- Increased hydraulic loading may have had a stronger negative effect on oocyst removal when filters were more mature. Average oocyst removal in test 4 (0.4 m/h) was at least 4.3 logs, compared to 3.3 logs in test 5 (0.8 m/h). However, the lower oocyst removals in test 5 may have been partially due to filter cleaning the day before the test.
- In challenge tests 4 to 6, complete removal of *Giardia* cysts was often achieved. Average cyst removals were at least 3.6 logs for test 4 (0.4 m/h), at least 4.2 logs for test 5 (0.8 m/h) and at least 4.6 logs for test 6 (0.4 m/h).
- Thus, in general, these slow sand filters were demonstrated to provide effective removal of *Cryptosporidium* oocysts and *Giardia* cysts.

References

Bellamy, W.D., Silverman, G.P., Hendricks, D.W., and Logsdon, G.S. (1985). Removing *Giardia* cysts with slow sand filtration. *Jour. AWWA* 77(2), 52–60.

Cleary, S. (2005). Sustainable drinking water treatment for small communities using multistage slow sand filtration. M.A.Sc. thesis, Department of Civil Engineering, University of Waterloo.

Ontario Regulation 170/03 (2003). Procedure for Disinfection of Drinking Water in Ontario.

Schuler, P.F., Ghosh, M.M., and Gopalan, P. (1991). Slow sand and diatomaceous earth filtration of cysts and other particulates. *Wat. Res.* 25(8), 995–1005.

Timms, S., Slade, J.S., Fricker, C.R. (1995). Removal of Cryptosporidium by slow sand filtration. *Wat. Sci. Tech.* 31(5-6), 81–84.

U.S. EPA (2006). National Primary Drinking Water Regulations: Long Term 2 Enhanced Surface Water Treatment Rule; Final Rule. Fed. Regist. 71(3), 654–786.

U.S. EPA (1996). ICR Microbiology Laboratory Manual. Office of Research and Development, Washington, D.C.

17 Faecal contamination indicator organisms in slow sand filters

H. Petry-Hansen*, H. Steele, M. Grooters**, J. Wingender***, H.-C. Flemming*,*****

*IWW Centre for Water, Moritzstrasse 26, D-45476 Mülheim an der Ruhr, Germany
*Biofilm Centre, Molekulare Enzymtechnologie, Lotharstraße 1, D-47057 Duisburg, Germany
***Biofilm Centre, Aquatische Mikrobiologie, Geibelstraße 41, D-47057 Duisburg, Germany

Abstract A major purpose of slow sand filtration in drinking water treatment is the improvement of the hygienic water quality. The efficacy of filtration is normally monitored by determination of the removal of faecal contamination indicator organisms. In this study, using experimental filters, the fate of coliform bacteria, *Escherichia coli* and enterococci was investigated by quantification of these bacteria in filter influents and effluents as well as in biofilms on the filter material. Determination was performed by conventional culturing methods which require 20–50 hours for analysis, and a fluorescent microcolony hybridization (FMH) method which is based on fluorescent *in situ* hybridisation (FISH) of microcolonies of the target organisms and yielded results only after 6–16 hours. Water temperature significantly influenced removal efficiency. Enhanced elimination of *E. coli* and enterococci was observed with increasing temperatures from 5°C to 30°C, while coliforms could multiply in filter biofilms in filters operated at temperatures between 20°C and 30°C, leading to higher coliform levels in filter effluents than in influents. Inactivation of indicator organisms in the filter beds could be partially attributed to the activity of bacterivorous protozoa which were found in abundance. Glass granulate as an alternative filter material revealed a removal efficiency similar to sand, while coconut fibre material was less efficient. Using amplified rDNA restriction analysis (ARDRA) of 16 S rDNA clone libraries, a bacterial population analysis of the slow sand filter biocoenosis was performed. It revealed that the populations were diverse and composition was clearly influenced by the raw water populations. In all cases, the microbial communities were dominated by non-pathogenic species which are known utilize a wide spectrum of substrates, indicating the biodegradative potential of the filters.

Keywords Slow sand filtration, indicator organisms, coliforms, *Escherichia coli*, enterococci, biofilms, biocoenosis, fluorescent microcolony hybridisation, ARDRA

Introduction

The primary purpose of water treatment is to provide a microbially safe drinking water. Various filtration processes are used in drinking water treatment which are efficient in the

 Recent Progress in Slow Sand and Alternative Biofiltration Processes edited by Rolf Gimbel, Nigel J.D. Graham and M. Robin Collins. ISBN: 9781843391203. Published by IWA Publishing, London, UK.

elimination of waterborne pathogens (Stanfield *et al.*, 2003). The treatment process of slow sand filtration is well established for countries with a relatively high source water quality and temperate climate conditions (Visscher, 1990; Collins *et al.*, 1991). Optimum treatment performance is achieved in mature slow sand filters, whose efficiency is dependent on well-established biofilms in the form of the schmutzdecke at the surface of the filters, and biofilms on the sand grains of the filter bed below the schmutzdecke (Hijnen *et al.*, 2004). Hier den ersten Absatz einfügen (ist nicht eingerückt).

Microbial indicators are used to assess drinking water quality and can also be employed to evaluate the efficiency of a treatment process. Total coliforms, *Escherichia coli* and enterococci are common bacterial indicators of water contamination; at present, *E. coli* appears to provide the best bacterial indication of faecal contamination in drinking water (Tallon *et al.*, 2005). The routine quantification of these indicators is based on cultivation methods. The main disadvantages of culture-based methods are the length of time taken to get the results and the fact that not all existing cells of the target organisms can be cultured. By the use of detection methods based on molecular techniques these disadvantages are overcome.

The aims of this study were

- to investigate the elimination of bacterial indicator organisms by slow filters operating under a range of environmental conditions,
- to optimize a detection method for important indicator parameters based on a molecular technique and
- to investigate the complex bacterial populations of slow filters which were used for drinking water treatment.

Methods

Investigated filters

Experimental slow filters were operated by the Institute for Water (IfW), Schwerte, Germany and by the IWW Centre for Water, Mülheim, Germany. In laboratory columns (diameter 0.2 m–0.5 m) and lysimeters outside (surface area 4 m^2) different filter materials (sand, glass granulate and coconut fibre) and the continual and intermittent operation were tested. The raw water passing through these experimental filters had a temperature of 5–10°C, 20°C and 30°C. Raw water was surface water from the River Ruhr or treated effluent from a wastewater treatment plant (Dinslaken, Germany). Additionally, the biofilms of a full-scale slow sand filter which was operated as part of the standard surface water treatment process at the Ruhr waterworks Mülheim-Styrum-East (RWW, Mülheim, Germany) was investigated.

Culture methods for detection of coliform bacteria, *E. coli* and enterococci

Water samples from filter influents and effluents, and samples from filter material were investigated. The biomass was detached from the filter material by shaking 10 g filter material in sterile water (ad 100 ml, 30 min at room temperature). The bacteria in the supernatant were then quantified by the same methods as for water samples. Coliform bacteria and *E. coli* were quantified with the Colilert-18 (QuantiTray/2000) system (IDEXX Laboratories). Enumeration of enterococci was done by the membrane filtration method

Table 2.1 Nutrient media and hybridization conditions for FMH

Indicator organism	Medium	Incubation time	Incubation temperature	Probe (reference)
Enterobac-teriaceae	Standard no. I nutrient agar	3–4 h	36°C	ProbeD (Ootsubo *et al.*, 2002)
E. coli	Standard no. I nutrient agar	3–4 h	44°C	Colinsitu (Regnault *et al.*, 2000)
Enterococci	Kanamycin aesculin azide agar	16 h	36°C	ENC131 and EFS130 (Meier *et al.*, 1997)

according to ISO 7899-2 (2000), using Slanetz-Bartley agar for primary culture and bile aesculin azide agar for confirmation.

Fluorescent microcolony hybridization (FMH)

The samples were concentrated by membrane filtration (polycarbonate filter, pore size 0.2 μm). The filter was placed on a suitable medium and incubated for 3–16 h at 36°C or 44°C (Table 2.1). After incubation the bacteria cells were fixed and permeabilised by treatment with ethanol (ethanol absolute for 30 sec) and heat (80°C for 20 min). The hybridization reaction was run in a humid chamber at 46°C for 90 min. The hybridization buffer contained 5 ng/μl oligonucleotide probe, 0.9 M NaCl, 20 mM Tris/HCl (pH 8), 0.01% SDS und 35% formamide. After hybridization, the filter was washed for 5 min at 48°C in a buffer containing 80 mM NaCl und 20 mM Tris/HCl (pH 8), 5 mM EDTA und 0.01% SDS. All cells were additionally stained with DAPI (1 μg/ml). Dried filters were embedded in Citifluor AF2. Enumeration of fluorescent microcolonies was done at 400 times magnification with an epifluorescent microscope.

Fluorescent *in situ* hybridization (FISH)

The samples of filter materials were fixed with 4% formaldehyde and frozen at −20°C until investigation. Bacteria were detached by vortexing (30 sec highest speed) from filter material. Bacteria in the supernatant were then analysed by FISH. The hybridization was run for the probes EUB338, NONEUB338, BET42a and GAM42a as described by Manz *et al.* (1992). For quantifying the α-Proteobacteria, the probe ALF968 was used (Neef, 1997). For ALF968 a formamide concentration of 20% in hybridization buffer and 88 mM NaCl in washing buffer was used. All bacterial cells were additionally labelled with DAPI (1 μg/ml). Dried filters were embedded in Citifluor AF2. The microscopy analysis was done at 1000 times magnification with an epifluorescence microscope.

Analysis of 16S rDNA clone libraries

The total DNA of the schmutzdecke and of biofilms from sand grains recovered from the filter layer in 5 cm depth was isolated by the method of Kilb *et al.* (1998). The 16S rDNA genes were amplified by PCR with the oligonucleotides 27f and 1492r described by Medlin *et al.* (1998) and isolated by cloning with the TOPO TA Cloning Kit (Invitrogen, Karlsruhe, Germany). The isolated 16S rDNA genes were characterized by their restriction patterns with the enzymes *Dde*I and *Msp*I (amplified rDNA restriction analysis = ARDRA). Analysis of the restriction patterns was done by the software BioNumerics (Applied Maths, Austin, Texas, USA) using UPGMA (Unweighted Pair Group Method with Arithmetic Mean).

Detection of protozoa

The "non flooded petri dish method" was chosen for reactivating the protozoa in samples of filter material (Lee and Soldo, 1992). The quantification and determination of the protozoa was done *in vivo* by light microscopy (from 125 up to 1250 times magnification). The systematic follows Levine *et al.* (1980) and Cavalier-Smith (1998).

Results and discussion

The efficiency of the experimental slow filters was determined by measuring the elimination of bacterial indicator organisms, including coliform bacteria, *E. coli* and enterococci. These indicators were quantified by conventional cultural methods and by optimized fluorescent microcolony hybridization (FMH). The quantitative detection of indicator organisms in water samples by the FMH method was significantly faster than the detection by conventional cultural methods. By using the FMH method the quantitative result for coliform bacteria and *E. coli* was obtained after 6 h versus 20 h using the conventional method (Colilert-18 system) and for enterococci the result was obtained after 20 h compared to 48 h using the standard method (ISO 7899-2, 2000). Using FMH, fluorescent microcolonies could be clearly observed and easily enumerated on the membrane filters by epifluorescence microscopy. The statistical evaluation of the quantitative data shows correlation coefficients which vary between the middle and the high area of valuation (Table 3.1). These results show that the FMH method allows the quantitative detection of indicator parameters in water samples in a shorter time than the conventional methods and in an acceptable correlation to the reference methods.

For removal of the indicator parameters used in the present study, continuous operation of the filters proved to be more suitable than intermittent operation. This was indicated by

Table 3.1: Correlation coefficients of the variables "culture methods" and "FMH"

Indicator parameter	Correlation coefficient	Valuation of the correlation coefficient[1]
Coliform bacteria/ *Enterobacteriaceae*	0.679	Middle correlation
E. coli	0.709	High correlation
Enterococci	0.797	High correlation

[1] after Spearman-Rho

the higher portion of all indicator organisms in the filtrate after every interruption of the filter runs (intermittent operation).

The reduction of coliform bacteria was influenced by the water temperature in the experimental filters. Elimination of *E. coli* and enterococci increased with temperature from 5°C to 30°C, with a maximal removal of more than 3 log units at 30°C. Comparing the number of coliforms in the water before and after filtration, a reduction of 1 log step or more was observed at 5–10°C and 20°C. When the water had a temperature of 30°C, there was no detectable removal in coliform numbers in either the laboratory experiments or the lysimeter tests. In the case of the laboratory columns, the concentration of coliforms in the filtrates was even higher than in the water before filtration.

In sand filters 90–99% of microorganisms are retained in the first 5–10 cm of the filter bed (Stevik *et al.*, 1999; van Cuyk *et al.*, 2001). In order to investigate the fate of the indicator organisms, a comparison of the number of indicator organisms which go onto the experimental filters during the whole operation time and the number of indicator organisms in the active layers (schmutzdecke and layer at 5 cm depth of the filter bed) at the end of the experiment was done for filters running at 20°C and 30°C. At both temperatures the total number of *E. coli* which enter the experimental filters was higher than the number in the active layers, while the number of coliform bacteria in the active layers was higher than the total number which entered the filters. Thus, coliform bacteria seem to be able to survive and to multiply in the filter bed at temperatures of 20–30°C under the operating conditions of the filters used in the present study. It is known that coliform bacteria and *E. coli* are part of the natural biocenoecis of soils and waters in tropical and subtropical regions where the conditions may be similar to the operating conditions of the experimental filters in this study (water temperature 20–30°C and availability of sufficient microbiologically degradable nutrients). That is why under tropical and subtropical conditions coliform bacteria and *E. coli* are not suitable as sole indicators for faecal contamination (Winfield and Groisman, 2003; Byappanahalli and Fujioka, 2004).

The comparison of different filter materials showed that glass granulate was approximately as efficient in reducing indicator organisms as the reference material sand. For coconut fibre it was observed that in the first weeks after starting a freshly filled filter the numbers of coliform bacteria and *E. coli* in the filtrate was higher than in the filter influent. The investigation of fresh coconut fibre material showed that this material was already contaminated with a high number of coliforms (5×10^4 coliforms/g dry weight) and some *E. coli* (5×10^0 *E. coli*/g dry weight). Coconut fibre is produced in tropical regions where *E. coli* and coliform bacteria belong to the natural biocenoecis of environmental habitats. It is possible that the contamination of coconut fibre with these environmental organisms happened during the production process. All coliforms and *E. coli* in the coconut fibre material could be eliminated by a heat treatment (minimum 1 h at 70°C).

In this study the bacterial communities of slow sand filters for treatment of drinking water were investigated for the first time by molecular techniques. The taxonomically classified bacteria in the active layers of the investigated slow sand filters (Table 3.2) belong to bacterial groups which have the ability of carrying out diverse physiological processes, like the *Proteobacteria*, or have the ability of carrying out very specific physiological processes which are important for the sequential degradation of nutrients. Organisms of the *Cytophaga-Flavobacteria-Bacteriodes* cluster, for example, possess the ability to degrade

Table 3.2: Distribution of phylogenetic bacterial groups in slow sand filter communities as determined by analysis of 16S rDNA clone libraries.

Bacterial groups	Filters running at 5–10°C	Filters running at 20–30°C
Proteobacteria	27% (50 clones)	13% (46 clones)
Nitrospira	0% (0 clone)	3.5% (12 clones)
Gram positive bacteria with low GC content	1% (2 clones)	2.5% (9 clones)
Gram positive bacteria with high GC content	1% (2 clones)	5% (18 clones)
Cytophaga-Flavobacterium-Bacteroides group	1% (2 clones)	3.5% (12 clones)
Planctomyces	1.5% (3 clones)	1% (4 clones)
Bacteria Division TM6	0% (0 clones)	1% (4 clones)
Chloroflexus	1% (2 clones)	0% (0 clones)
Unclassified bacteria	67.5% (126 clones)	70.5% (251 clones)

polymers and make the products available for other bacteria (van Hannen *et al.*, 1999; Casamayor *et al.*, 2000; Laetitia *et al.*, 2000). Organisms of these important groups are known from similar environmental habitats, like soils, sediments, lakes and rivers. These organisms present in the mature filters can originate from two sources: by the raw water and by the filter material itself. Probably, the organisms which are introduced with fresh filter material are gradually washed out when the operation time is longer and the organisms coming on the filter with the raw water then dominate the population (Pell, 1991; Calvo-Bado *et al.*, 2003). Thus, the organisms identified in the mature slow sand filters primarily represent autochthonous bacteria from the aquatic environment. Another factor which appears to influence the composition of the bacterial population is the temperature of the water which passes through the filter (Table 3.2).

For the most important *Bacteria* group, the *Proteobacteria*, 16S rDNA sequence analysis revealed the presence of representatives of the α-, β- and γ-subclasses. These subclasses were quantified *in situ* by FISH. In all active layers of the investigated slow sand filters bacteria of the β-subclass (25% of all *Bacteria*) dominated the populations; 2.5% of all *Bacteria* belong to the α-subclass and 3.5% to the γ-subclass. This dominance of β-*Proteobacteria* was also reported for drinking water biofilms (Manz *et al.*, 1993; Kalmbach *et al.*, 1997), and lakes and rivers (Alfreider *et al.*, 1996, Weiss *et al.*, 1996, Böckelmann *et al.*, 2000). Among 36 sequenced clones, only one hygienically relevant microorganism, a *Legionella* species, was found in all examined filters .

In addition to bacteria, a high abundance of bacterivorous protozoa (heterotrophic nanoflagellates, *Anisonema* sp., *Bodo* sp.) was observed in the schmutzdecke of the experimental filters. Thus, these organisms seem to be important members of microbial communities in slow sand filters. The activity of bacterivorous protozoa can be assumed to be one of the factors involved in the elimination of the indicator organisms.

Conclusions

The elimination of bacterial indicator organisms (coliforms, *E. coli* and enterococci) by slow filters is mainly influenced by the operational temperature, but also by the filter material and the operation mode (continuous, intermittent). Compared to conventional culture methods, FMH allows the quantification of indicator parameters in shorter times, and thus the more rapid evaluation of treatment efficiency of slow sand filters. It is important that the filter material is not contaminated by hygienically relevant bacteria before filtration is started. In regions with water temperatures of 20–30°C and high amounts of microbially available nutrients in raw water, for example in tropical and subtropical regions, coliform bacteria and possibly *E. coli* may persist and multiply in slow sand filters. For this reason they are not suitable as sole indicators for the evaluation of the removal efficiency of faecal pathogens by slow sand filtration.

The microbial communities of slow sand filter biofilms are dominated by environmental bacteria without any hygienic relevance. These environmental bacteria are able to degrade sequentially the nutrients coming with raw water on the filter. Bacterivory by protozoa seems to be an important elimination process of bacteria in slow filters. It is responsible for elimination of hygienically relevant bacteria and maintains the bacterial population in an young and active state of growth.

Acknowledgement

This study was conducted with financial support of the German Ministry of Education and Research (BMBF) and is part of the joint research project "Adjusted water treatment technologies and water distribution under regional conditions".

References

Alfreider, J.P., Amann, R., Sattler, B., Glöckner, F.O., Wille, A. and Psenner, R. (1996) Community analysis of the bacterial assemblages in the winter cover and pelagic layers of a high mountain lake by *in situ* hybridization. *Appl. Environ. Microbiol.* **62**, 2138–2144.

Böckelmann, U., Manz, W., Neu, T.R. and Szewzyk, U. (2000) Characterization of the microbial community of lotic organic aggregates ('river snow') in the Elbe river of Germany by cultivation and molecular methods. *FEMS Microbiol. Ecol.* **33**, 157–170.

Byappanahalli, M., Fujioka, R. (2004). Indigenous soil bacteria and low moisture many limit but allow faecal bacteria to multiply and become a minor population in tropical soils. *Wat. Sci. Tech.* **50**(1), 27–32.

Casamayor, E.O., Schäfer, H., Baneras, L., Pedrós-Alió, C. and Muyzer, G. (2000) Identification of and spatio-temporal differences between microbial assemblages from two neighboring sulfurous lakes: comparison by microscopy and denaturing gradient gel electrophoresis. *Appl. Environ. Microbiol.* **66**, 499–508.

Cavalier-Smith, T. (1998) A revised six-kingdom system of life. *Biol. Rev.* **73,** 203–266.

Collins, M.R., Eighmy, T.T. and Malley, J.P. (1991) Evaluating modifications to slow sand filters. *J. Am. Water Wks. Assoc.* **83**, 62–70.

Calvo-Bado, L.A., Pettitt, T.R., Parsons, N., Petch, G.M., Morgan, J.A.W. and Whipps, J.M. (2003) Spatial and temporal analysis of the microbial community in slow sand filters used for treating horticultural irrigation water. *Appl. Environ. Microbiol.* **69**, 2116–2125.

Van Cuyk, S., Siegrist, R., Logan, A., Masson, S., Fischer, E., Figueroa, L. (2001) Hydraulic and purification behaviors and their interactions during wastewater treatment in soil infiltration systems. *Wat. Res.* **35**, 953–964.

Van Hannen, E., Zwart, G., van Agterveld, M.P., Gons, H.J., Ebert, J.and Laanbroek, H. (1999) Changes in bacterial and eukaryotic structure after mass lysis of filamentous cyanobacteria associated with viruses. *Appl. Environ. Microbiol.* **65**, 795–801.

Hijnen, W.A.M., Schijven, J.F., Bonné, P., Visser, A., Medema, G.J. (2004). Elimination of viruses, bacteria and protozoan oocysts by slow sand filtration. *Wat. Sci. Tech.* **50**(1), 147–154.

ISO 7899-2 (2000) Water quality – detection and enumeration of intestinal enterococci – Part 2: membrane filtration method. Geneva, Switzerland, International Standard Organization.

Kalmbach, S., Manz, W. and Szewzyk, U. (1997) Isolation of new bacterial species from drinking water biofilms and proof of their *in situ* dominance with highly specific 16S rRNA probes. *Appl. Environ. Microbiol.* **63**, 4164–4170.

Kilb, B., Kuhlmann, B., Eschweiler, B., Preuß, G., Ziemann, E. and Schöttler, U. (1998) Darstellung der mikrobiellen Besiedlungsstruktur verschiedener Grundwasserhabitate durch Anwendung molekularbiologischer Methoden. *Acta Hydrochim. Hydrobiol.* **26,** 349–354.

Laetitia, B., Schäfer, H., Joux, F., Courties, C., Muyzer, G. and Lebaron, P. (2000) Genetic diversity of total, active and culturable marine bacteria in coastal seawater. *Aquat. Microb. Ecol.* **23**, 1–11.

Lee, J.J. and Soldo, A.T. (1992) Protocols in Protozoology. Lawrence, Kansas, USA.

Levine, N.D., Corliss, J.O., Cox, F.E.G., Deroux, G., Grain, J., Honigberg, B.M., Leedale, G.F., Loeblich, A.R., Lom, J., Lynn, D., Merinfeld, E.G., Page, F.C., Poljanski, G., Sprague, V., Vavra, J. and Wallace, F.G. (1980): A newly revised classification of the Protozoa. *J. Protozool.* **27**, 37–58.

Manz, W., Amann, R., Ludwig, W., Wagner, M. and Schleifer, K.H. (1992) Phylogenetic oligonucleotide probes for the major subclasses of *Proteobacteria:* problems and solutions. *Syst. Appl. Microbiol.* **15**, 593–600.

Manz, W., Szewzyk, U., Ericsson, P., Amann, R., Schleifer, K.-H. and Stenström, T. (1993) *In situ* identification of bacteria in drinking water and adjoining biofilms by hybridization with 16S and 23S rRNA-directed fluorescent oligonucleotide probes. *Appl. Environ. Microbiol.* **59**, 2293–2298.

Medlin, L., Elwood, H.J., Stickel, S. and Sogin, M.L. (1998) The characterization of enzymatically amplified eukaryotic 16S-like rRNA coding regions. *Gene* **71**, 491–499.

Meier, H., Koob, C., Ludwig, W., Amann, R., Frahm, E., Hoffmann, Obst, U. and Schleifer, K.H. (1997) Detection of enterococci with rRNA targeted DNA probes and their use for hygienic drinking water control. *Wat. Sci. Tech.* **35**(11-12), 437–444.

Neef, A. (1997) Anwendung der *in situ*-Einzelzell-Identifizierung von Bakterien zur Populationsanalyse in komplexen mikrobiellen Biozönosen. Ph.D. Thesis. Technische Universität München, Munich, Germany.

Ootsubo, M., Shimizu, T., Tanaka, R., Sawabe, T., Tajima, K., Yoshimizu, M., Ezura, T. and Oyaizu, H. (2002) Oligonucleotide probe for detecting *Enterobacteriaceae* by *in situ* hybridization. *J. Appl. Microbiol.* **93**, 60–68.

Pell, M (1991) Microbiology and nitrogen transformations in sand-filter systems for treatment of household septic-tank effluents. Ph.D. thesis, Swedish Univerity of Agricultural Science, Uppsala.

Regnault, B., Martin-Delautre, S., Lejay-Collin, M., Lefèvre, M. and Grimont, P.A.D. (2000) Oligonucleotide probe for the visualization of *Escherichia coli/Escherichia fergusonii* cells by in situ hybridization: specificity and potential applications. *Res. Microbiol.* **151**, 521–533.

Stanfield, G., LeChevallier, M. and Snozzi, M. (2003) Treatment efficiency. In: Assessing microbial safety of drinking water. Improving approaches and Methods. OECD/WHO, OECD Publications, France, 159–178.

Stevik, T.K., Ausland, G., Jennsen, P.D., Siegrist, R.L. (1999) Removal of E. coli during intermittent filtration of wastewater effluent as affected by dosing rate and media type. *Wat. Res.* **33**, 2088–2098.
Tallon, P., Magajna, B., Lofranco, C., Leung, K.T. (2005) Microbial indicators of faecal contamination in water: a current perspective. *Water Air Soil Pollut.* **166,** 139–166.
Visscher, J.T. (1990) Slow sand filtration: design, operation, and maintenance. *J. Am. Water Wks. Assoc.* **82**, 67–71.
Weiss, P., Schweitzer, B., Amann, R. and Simon, M. (1996) Identification *in situ* and dynamics on limnetic organic aggregates (lake snow). *Appl. Environ. Microbiol.* **62**, 1998–2005.
Winfield, M.D. and Groisman, E.A. (2003). Role of nonhost environments in the lifestyles of *Salmonella* and *Escherichia coli*. *Appl. Environ. Microbiol.* **69**, 3687–3694.

18 Microbiological conditions before and after cleaning in slow sand filters under tropical conditions

L.D. Sánchez*, J. Latorre*, G. Galvis and J.T. Visscher*****

**Universidad del Valle, Instituto Cinara, A.A 25157, Cali, Colombia
**Los Pinos 259, Urbanización Camacho, Lima, 12, Perú
***jtvisscher@yahoo.com

Abstract This study examines the microbiological conditions before and after cleaning in a slow sand filter (SSF) for algae, protozoa and heterotrophic bacteria. The effect of scraping and harrowing techniques used to clean the SSF were analysed in full scale water treatment plants with a mature filter media and at pilot scale plant with new filtering media. The experimental data obtained for algae and protozoa behaviour show reductions in algae populations between 0.58 and 5.68 Log units/cm3 of sand in SSF cleaned by scraping, and between 0.47 and 1.53 Log units/cm3 of sand in pilot SSF units cleaned by harrowing. For protozoa populations these reductions were respectively between 0.91 and 4.82 Log units/cm3 of sand, and between 0.19 and 4.82 Log units/cm3 of sand. According to experimental data, the bacteria population does not have the same recovery pattern as algae and protozoa. The following recovery periods were recorded after SSF pilot unit cleaning by scraping: algae between 44 and 275 hours; protozoa; between 68 and 320 hours; bacteria between 48 and 137 hours and after cleaning by harrowing: algae between 128 and 260 hours; protozoa between 200 and 210 hours; bacteria between 177 and 229 hours. Full scale plants cleaned by scraping, operating at a filtration rate between 0.11 and 0.12 m/h, recorded maturation periods for faecal coliforms between 0 and 39 hours and 17 hours for turbidity. The effect of cleaning on the effluent caused reductions in efficiency of faecal coliforms removal between 0 and 3 logarithmic units. Turbidity showed a maximum reduction of 18%.

Keywords Slow sand filtration, maturation period, filter scraping, filter harrowing, biomembrane, algae, protozoa and bacteria populations.

Introduction

SSF is a very suitable water treatment technology, but one limiting factor is that a maturation period is needed after filter cleaning to obtain a good quality effluent. The best-known and most used cleaning technique in SSF systems is scraping off the top two or three centimetres of the filter surface when operation is stopped after the maximum head-loss has been reached. According to Lloyd (1974), filter bed cleaning affects the functional micro

 Recent Progress in Slow Sand and Alternative Biofiltration Processes edited by Rolf Gimbel, Nigel J.D. Graham and M. Robin Collins. ISBN: 9781843391203. Published by IWA Publishing, London, UK.

fauna and flora distribution in the filter bed. When the filter bed dries, a large number of bacteria, particularly spores which are resistant to chlorination, appear in the effluent as the filter is put into operation after cleaning.

Collins *et al.*, (1992) reported another method of cleaning called filter harrowing. Water above the sand is drained slowly to a depth of 15 inches or so. The upper layer of sand is raked or harrowed while water is being drained off above the filter bed. The harrowing releases fine suspended particles that have accumulated. These particles are washed off by the water as it drains through valves above the filter bed, Several authors suggest that with this technique, filter interruptions are minimised and maturation periods, the "time needed after cleaning to reach a stable condition of a relevant variable in the system's effluent water". are shortened as the bacterial population is not completely removed during cleaning. (Cullen *et al.*, 1985 and Collins *et al.*, 1991). Also the time consuming removal and washing of the sand is avoided. However, in countries like Colombia, this cleaning technique is not used.

The research reported upon in this paper focused on 1) Studying the effect of cleaning by respectively harrowing and scraping on the algae and protozoa populations in an SSF, 2) Measuring the time these populations take to recover after filter cleaning, and 3) Identifying possible water quality problems due to filter cleaning.

Slow sand filter maturation

Bellamy *et al.*, (1985), established that the microbiological conditions that govern the effectiveness of a slow sand filter are: 1) the degree of formation of the bio-membrane and 2) the microbiological maturation of the sand bed. Four types of filter conditions can be identified: 1) filters with fresh sand that are just initiated, 2) filters in which the sand was replaced, 3) filters that have been cleaned (removal of the biomembrane), and 4) mature filters that are in stable operation. These conditions effect the biological populations in the filter and therewith the treatment efficiency.

Collins et al., (1992), have demonstrated that filter maturation refers to the age of the microbiological development of the biomembrane and is generally quantified by reductions in turbidity and bacterial content in the effluent water. Cullen et al., (1985), reported maturation periods that varied between 6 hours and 12 weeks after scraping the filter. It is still not clear why or exactly how maturation develops. One of the most widely accepted theories is that a completely developed biomembrane is needed before maximum removal efficiency can be achieved. During the period that the biomembrane matures, part of the particles and microorganisms may not be retained and may travel into the filter media and pass into the effluent, thus entailing a sanitary risk for the consumers.

Different indicators are used by different authors in order to quantify the biological population at the filter surface: Lloyd (1974) reported 1,500 protozoa/cm2 in a column of sand from a mature filter and Dibernardo (1995) reported 2,400 Bacillariophyta/gram of sand in the first 5 centimetres of sand. According to Krebs (1985), a population of organisms that is present in a favourable environment will increase in number. However, populations do not continuously increase geometrically. The density increases little when a population lives and grows in a limited space. The growth curve which represents a population of this type is sigmoid or 'S' shaped. Ellis (1993) and Lloyd (1974) have suggested that bacteria removal is due above all to the action of predators, especially protozoa and rotifers. According to Rittman (1993), the accumulation of bacteria is controlled by four processes: growth, deposition, decline and separation.

Methods

The study was carried out at Cinara Research and Technology Transfer Station in Puerto Mallarino, Cali, Colombia which draws water from the highly polluted Cauca River. Two pilot SSF units preceded by upflow roughing filters in layers (URFL) and dynamic roughing filters (DyRF) were evaluated for filter bed cleaning by respectively scraping and harrowing. In addition, five full-scale plants in the city of Cali were reviewed which are being cleaned by scraping. These are located at: La Parcelación Cañasgordas, El Colegio Colombo Británico, La Ribera, Shalom and La Sirena and all receive water from the moderately contaminated Pance River. In all systems, the SSF units were preceded by DyRF and URFL.

Sampling

The pilot units operated at a filtration rate of 0.15 m/h and the real scale plants at between 0.11 and 0.12 m/h. The pilot filters operated for 110 days and a total of two cleanings per unit were evaluated. In the full-scale plants, only one cleaning of the filter was evaluated. Before and after each cleaning, water samples were taken simultaneously at the different sampling points in the pilot units and the full-scale plants – the affluent, the filter bed and the effluent. The sampling was carried out through the following procedure: For the algae and protozoa count, the filter was divided into four equal parts and a sample of the surface sand was extracted from each part using a plastic flask with a capacity of 1 cm^3. Later, in the laboratory, the four samples were mixed in a 100 ml sterile flask to integrate the whole sample and subsequently separated again in four portions. Each of the portions of 1 cm^3 of sand was put in a sterile flask of 100 ml, to which 25 ml of protozoa cultivator was added. The sample of sand with cultivator was stored at a temperature of 4°C. It was shaken for 3 minutes, to homogenize and unstick the organisms that have adhered to the grains of sand. The parameters and variables in the study are presented in Table 1.

Pilot filters

Two pilot SSF units where built in Puerto Mallarino both preceded by DyRF and URFL treatment. The units worked in parallel making it possible to compare cleaning by filter scraping and by filter harrowing. The pilot filters had the following characteristics: building material - Ferro cement, shape – circular, diameter –2 m, total height –2.1 m, filter bed height –1 m, support bed height –0.25 m, available charge loss –0.7 m, and free border –0.15 m. The filtration rate used was 0.15 m/h and the characteristics of the filter sand were: effective size (d10) 0.15 – 0.30 mm; uniformity coefficient 2.4 and porosity 0.48.

Full-scale plants

The recovery process of algae and protozoa and the performance of the SSF in terms of removal of turbidity and faecal coliforms after scraping were first explored in the Cañasgordas

Table 1 Study parameters and variables

Cleaning Technique	Parameters	Unit	Independent Variables	Independent Variable Values
	Turbidity Faecal coliforms Flow	NTU CFU/100 ml l/s	Filtration rate Filtering bed height Scraping thickness Effective bed size Filtering media age	0.15 m/h 1.0 m 2–3 cm 0.15 mm Pilot filter: 1 month
Scraping - Harrowing	Algae Protozoa Bacteria populations	No. organisms/ cm^3 wet sand	**Harrowing** Surface wash speed Deep wash speed Supernatant water height for cleaning. Harrow penetration depth in the bed	Real scale plants: > 1 year $0.15\ m^3/m^2/h$ $0.10\ m^3/m^2/h$ 0.30 m 5 cm

Note: The sampling began at the moment when the filter stopped operation for cleaning. The sampling was stopped at the moment when the removal efficiency for faecal coliforms was restored to the level before cleaning.

and Colombo Británico plants. With these results as a base, the measuring method was applied to three more plants: La Ribera, Shalom and La Sirena also comparing the performance of these parameters before cleaning and during the filter runs. The Colombo Británico treatment plant in Cali supplies water to a school with 1,200 students. The system was designed for a flow of 1.0 l/s. The slow filters have a total area of 24 m^2, the filtering media has a depth of 1.2 m and a filtration rate of 0.15 m/h. The plant in Cañasgordas is designed for a flow of 10.5 l/s. The SSF units have a total area of 249 m^2, the filter sand is 1.0 m deep and has a filtration rate of 0.15 m/h. The plant in La Sirena operates at a flow of 12 l/s, La Ribera, 10 l/s and Shalom, 1.5 l/s. The filtration rate in these three systems ranges between 0.12 m/h and 0.20 m/h. The average temperature of the water was 20°C.

Results and discussion

Algae and protozoa in the filtering bed

In three cases in the pilot filters no algae and protozoa were found after cleaning by scraping method, which shows that it can have a dramatic effect (Table 2). The fact that in the fourth case algae and protozoa were found may be the result of a removal of a thinner layer of the top of the sand in the scraping. In the full-scale plants, the impact of cleaning on the algae and protozoa level was less severe, although the scraping method was used, perhaps because a thinner layer of sand was removed, as may have been the case in one of the pilot units as well. The algae population in all tests tended to be greater than the protozoa population. The full-scale plants had higher populations of algae and protozoa. The latter may be the effect of the better water quality and lower turbidity of the Pance River in comparison with the Cauca River. Of the two treatment plants, the Colombo Británico plant had higher algae and protozoa populations. This could be due to differences in the pre-treatment systems installed, as the Cañasgordas plant uses pre-filters in series. More retention of particles or

Table 2 Algae and protozoa conditions in the filter surface before and after cleaning

			Algae Count - Cleaning Unit. Log/cm³ sand			Protozoa Count Unit. Log/cm³ sand		
Cleaning Tech.	**Filter Evaluated**	**Test No.**	**Before**	**After**	**Reduction Logs.**	**Before**	**After**	**Reduction Logs.**
Scraping	SSF 1 Pilot	1	5,68	0	5,68	4,82	0	4,82
		2	5,94	4,56	1,38	4,5	0	4,5
Harrowing	SSF 2 Pilot	1	6,17	4,64	1,53	4,49	3,57	0,92
		2	6,33	5,86	0,47	4,97	4,78	0,19
Scraping	SSF 2 Colombo Británico	1	7,51	6,64	0,87	7,29	6,48	0,81
	SSF 3 Cañasgordas	1	6,61	6,07	0,54	6,12	5,20	0,92

micro-organisms could be occurring, thus affecting the number of organisms that could colonise in the SSF.

Filter cleaning affects the algae (primary producers) and protozoa populations (principal primary consumers) in the filter (Table 2). This confirms the statement of Lloyd (1974) that cleaning causes a reduction in the functional micro fauna in the filter. After cleaning by scraping, there are populations that are virtually removed, especially protozoa and algae. This was more significant in the pilot SSF, test 1. In the second test in the same filter, the protozoa were extracted and there was a remaining algae population of 3.8 logs/cm3 sand. In the pilot filters that where cleaned by harrowing the protozoa and algae populations suffered less alteration than with the cleaning by scraping. This is in line with the finding by Collins et al., (1991) who reports that cleaning by harrowing leaves better biological conditions.

The results of the different tests indicate that recovery of algae and protozoa after cleaning correlates with a sigmoid or 'S' shaped "growth or colonisation" pattern (Figures 1 and 2).

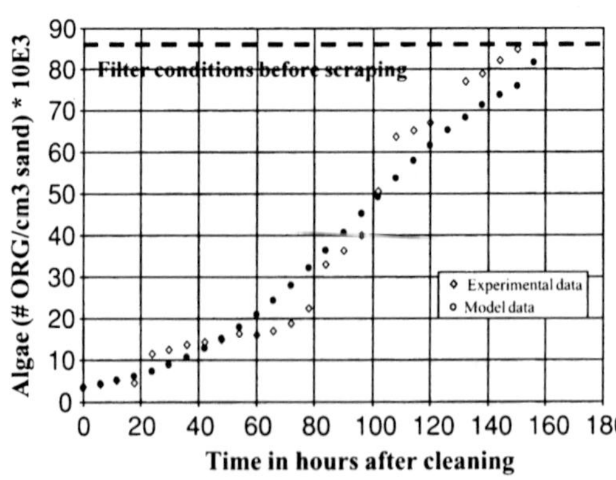

Figure 1 Algae behaviour after cleaning by scraping (SSF 1 Pilot Test 2)

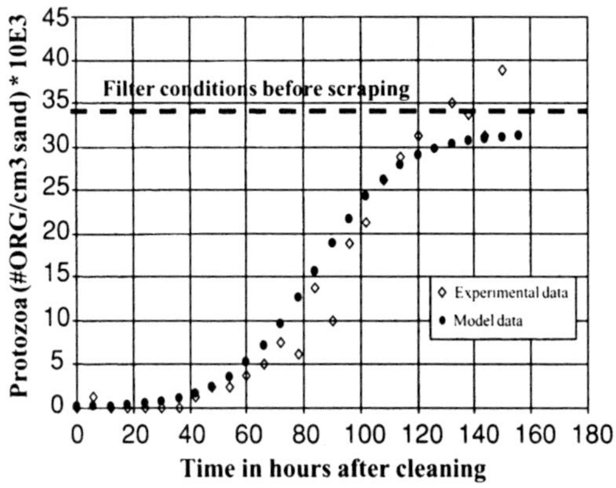

Figure 2 Protozoa behaviour after cleaning by scraping (SSF 1 Pilot Test 2)

The correlation coefficients (r) obtained on calculating the parameters in the different tests varied between 0.70 and 0.97, indicating a good match between the proposed growth model and the experimental data.

The mathematical growth model is similar to that described by Krebs (1985) for natural populations. The upper asymptote tends to be equal to the level of the population in the filter before cleaning. At this point, the filter can be considered to be in a 'stable' or 'mature' state, where the population density in the space studied becomes saturated with organisms and the biomass may be in balance with the available nutrients, as suggested by Bellamy et al., (1985). The population of organisms after cleaning is the intercept of the figure with the y-axis. The recovery pattern of the algae and protozoa populations, more than representing a growth pattern, seems instead to be a process of "colonization" that, according to Ritman (1993), is controlled by four processes: growth, deposition, decline and separation. Unsticking may be the mechanism causing the greatest loss of both algae and protozoa. This can be influenced by three factors: cleaning method, number of organisms adhering to the filtering media before cleaning and hydrodynamic forces caused by the filtration rates. The remaining population in the filter after cleaning is important, given the organism population's dependence on the maturation period. Therefore the cleaning technique needs to affect the populations as little as possible, especially the protozoa. This is also important for maintaining the greatest possible number of micro-organisms afterwards (Huisman et al., 1974; Lloyd, 1974 and Collins, et al.,1991).

Heterotrophic bacteria behaviour

Bacteria populations also showed significant reductions in SSF cleaned by scraping. The populations reduced by between 0.59 and 1.66 Log units/cm^3 of sand, and in pilot SSF with cleaning by harrowing, by 0.63 to 0.93 Log units/cm^3 of sand. The lower reduction in biological populations was found in the full-scale plants with more mature filter beds as shown by the data recovery patterns of bacteria, algae and protozoa population differ.

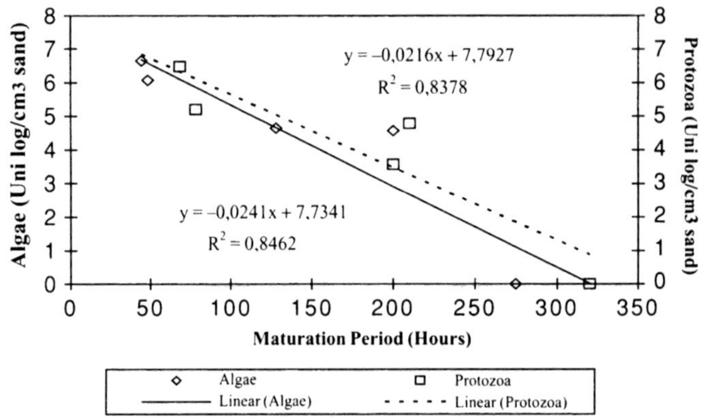

Figure 3 Maturation period in function of the remaining algae and protozoa after SSF cleaning

Maturation period

In SSF cleaned by scraping the recovery periods were as follows: For algae between 44 and 275 hours; for protozoa; between 68 and 320 hours; and for bacteria between 48 and 137 hours. For SSF cleaned by harrowing: For algae between 128 and 260 hours; for protozoa between 200 and 210 hours; and for bacteria between 177 and 229 hours. The correlation of the maturation period and the remaining population of algae and protozoa after is shown in Figure 3 and shows that there is dependency ($r^2 = 0.85$ for algae and protozoa). Algae seem to have a shorter recovery period than protozoa but also in this case a correlation exists between both populations after cleaning (Figure 3). Lloyd (1974), established that in the SSF food chain, the algae play an important role as they are the primary producers, and the food source of the protozoa.

Another factor that can affect the maturation period is filtering media age, as the pilot SSF were in a condition corresponding to a new bed which has just begun to be used (4 logs/cm3 sand for protozoa; 5 logs/cm3 sand for algae) and does not have a stable microbiological condition (Bellamy et al., 1985). The full-scale plants operated over a year, which suggests a more balanced biological population (between 5–6 logs/cm3 sand for protozoa and 6 logs/cm3 sand for algae).

The maturation period of the different systems after cleaning was established for both faecal coliforms and turbidity. This period ranged for faecal coliforms between 0 and 142 hours, for raw water qualities between 142 and 1500 UFC/100 ml. For turbidity the range was between 17 and 160 hours, with lower values when the inflowing water was below 2.5 NTU for 90% of the time. The periods for turbidity are in general shorter than those reported by Cullen et al., (1985) who recorded periods of between 6 hours to 2 weeks in full-scale plants that operated with turbidity less than or equal to 5 NTU. The system in Cañasgordas needs further review as here maturation was immediate. SSF cleaning by scraping as well as harrowing caused reductions in the removal efficiency of faecal coliforms at the beginning of the filtering run of between 0 and 3 logarithmic units. Turbidity did not record significant reductions after cleaning (maximum reduction of 18%) and its removal efficiency is restored much before that of faecal coliforms, hence it cannot be used as

Table 3 Algae and protozoa populations in full-scale plants

MSF Plant	**Sirena**		**Ribera**		**Shalom**	
SSF Populations (No. org/cm^3 wet sand)	Algae	Protozoa	Algae	Protozoa	Algae	Protozoa
Average	1,87E+06	5,00E+03	6,94E+04	2,50E+03	9,14E+05	3,75E+03
Maximum	2,38E+06	7,50E+03	1,89E+05	3,75E+03	1,09E+06	5,00E+03
Minimum	1,30E+06	2,50E+03	2,88E+04	1,25E+03	7,86E+05	2,50E+03
STD	5,07E+05	3,54E+03	7,96E+04	1,77E+03	1,56E+05	1,25E+03

Note: La Ribera Plant has covered SSF.

an indicator to put the system back in operation again, unless the raw water is moderately contaminated and chlorination is applied. The maturation period is not only a function of the affluent water quality (Letterman et al., 1985), but rather that it depends on the relationship between algae, remaining protozoa after cleaning and the technique or means of cleaning the filtering media. Similar findings have been reported by Lloyd (1974), Bellamy et al., (1985) and Collins et al., (1991), in filters evaluated in the USA and Europe.

Follow-up in full-scale plants on algae and protozoa populations and water quality

Measurements of algae and protozoa on the sand surface have been carried out in other plants that have been operating for between 12 and 18 years (Table 3). According to the data in Table 3, the plant with the largest biological population of algae and protozoa in the first centimetres of SSF sand was La Sirena. The La Ribera plant had the lowest biological population in the first few centimetres of sand, which could be an effect of having covered filters. The water quality behaviour in these plants is shown in Table 4.

The La Sirena plant had the highest risk in its SSF affluent, but it recorded the highest removal efficiency for turbidity and faecal coliforms, which is associated with a larger biological population of algae and protozoa in the filter media. This fact is reflected in the La Ribera plant SSF, which is treating the water with the lowest risk, but has the lowest level of faecal coliform removal, which coincides in turn with the smaller populations of algae and protozoa in the filtering media. Filter run length differs between the filters and is affected by the quality of the water that is treated. The SSF at the La Sirena plant had an average run of 66 days, whilst for Shalom the run lasted 81 days and for La Ribera, 70 days.

Conclusions and recommendations

After cleaning, when SSF units are put back into operation, there is a trend for algae and protozoa to have an ‘S’ shaped growth pattern, which can be represented by a logistical model. This behaviour was similar in the SSF cleaned by scraping and by harrowing. The biological populations recorded the following recovery periods after SSF cleaning: scraping; algae, between 44 and 275 hours; protozoa, between 68 and 320 hours and harrowing; algae, between 128 and 260 hours; protozoa, between 200 and 210 hours.

Table 4 SSF inlet and outlet quality in full-scale plants (January to November 2005)

	Affluent		Effluent	
Plant: Sirena	**Turbidity (NTU)**	**Faecal Coliforms (CFU/100 ml)**	**Turbidity (NTU)**	**Faecal Coliforms (CFU/100 ml)**
Maximum	365	2.000	3	14
Minimum	120,0	224,0	0,10	0,00
STD	173,24	1255,82	1,5	6,0
Average	243	1112	0,8	5,8
Plant: Ribera				
Maximum	4,00	300	3,00	56
Minimum	0,45	0	0,10	0
STD	1,22	94	0,93	23
Average	2,43	34	0,68	12
Plant: Shalom				
Maximum	2,1	600	5	28
Minimum	0,2	32	0,1	0
STD	0,83	204	1,61	9,0
Average	0,85	179	0,64	5,67

After cleaning by scraping, the pilot SSF operating at a filtration speed of 0.15 m/h had maturation periods for faecal coliforms of 120 and 142 hours and of 120 and 160 hours for turbidity. The pilot SSF cleaned by harrowing, which operated at the same filtration speed, registered shorter maturation periods: 116 and 86 hours for faecal coliforms, and 124 and 72 for turbidity. The full-scale plants with cleaning by scraping, operating at a filtration rate of between 0.11 and 0.12 m/h, recorded considerably lower maturation periods for coliforms of between 0 and 39 hours. It was also found that there was a direct relationship between the water quality of the treated water and the maturation periods, but maturation also depends on the populations of organisms remaining in the filter after cleaning, the cleaning technique, how cleaning is carried out, filtering media age, (lower biological populations in the pilot SSF with beds one month old and higher populations in the full-scale SSF plants that had been in operation for more than a year) and the filtration rate (lower maturation periods in the full-scale plants operating between 0.11 and 0.12 m/h).

The selected pre-treatment system should guarantee that the SSF can operate properly. This implies that the turbidity level of the water reaching the SSF unit should not exceed 10 NTU for more than 50% of the time. Furthermore it is essential that the cleaning of the pre-treatment units should never coincide with cleaning of the SSF units. If considerable fluctuation in water quality occurs, than the filtration rate in the pre-treatment systems should be reduced.

It is important to continue with this research and include also testing of harrowing in full scale systems as it may improve the operational efficiency of the systems and reduce the workload of the operators. The explorations should also look at the potential risk of accumulation of impurities in filter that are being harrowed as this might in the longer term lead to earlier clogging and thus more frequent cleaning.

References

Bellamy, W. D., Silverman, G. P. and Hendricks, D. W. (1985) Filtration of Giardia Cysts and Other Substances. Slow Sand Filtration. EPA-600/2-85/026. Water Engineering Research Laboratory. Cincinnati, Ohio, USA, **2**, 304.

Cullen, T.R., and Letterman, R.D. (1985) The effect of Slow Sand Filter Maintenance on Water Quality. *A.W.W.A.* (12).

Collins, R.M, Eighmy, T.T., and Malley, J.P. (1991) Evaluating Modifications to Slow Sand Filters. *A.W.W.A.* 83(09), 62–70.

Collins, M. R., Eighmy, T.T., Fenstermacher, J. M. and Spanos, S. K. (1992) Removing Natural Organic Matter by Conventional Slow Sand Filtration. *A.W.W.A.* (05).

Dibernardo, L. (1995) Algas e Suas. Influências na Qualidade das Aguas e nas Tecnologias de Tratamento. Associacâo Brasileira de Engenharia Sanitária e Ambiental, ABES, Río de Janeiro, Brazil, September.

Ellis, K. V. and Aydin, M. E. (1993) A Study of Three Slow Sand Filters at Various Flow Rates with Constant Temperature. *Wat. Sci Tech.* **2**, 88–96.

Huisman, L. and Wood, W. E. (1974) Slow Sand Filtration. World Health Organization (WHO). Geneva.

Krebs, CH. J. (1985) Ecología. Estudio de la Distribución y la Abundancia. Editorial Harla, Mexico. Second edition.

Lloyd, B. (1974) The Functional Microbial Ecology of Slow Sand Filtration. PhD. Thesis. University of Surrey, England. September.

Rittman, B. (1993) The Significance of Biofilms in Porous Media. Water Resources Research, **7**(07), 2195–2202.

19 Biofiltration of microcystin toxins: an Australian perspective

L. Ho*, D. Hoefel*, T. Meyn, C.P. Saint* and G. Newcombe***

*Cooperative Research Centre for Water Quality and Treatment, Australian Water Quality Centre, SA Water Corporation, PMB 3 Salisbury SA 5108
(E-mail: *lionel.ho@sawater.com.au*)
**Institute of Environmental Engineering, Technical University of Berlin, Strasse des 17. Juni 135, Berlin D-10623, Germany

Abstract Granular activated carbon (GAC) and slow sand filters were evaluated for their ability to remove microcystin toxins through biological degradation processes. The lag period prior to degradation commencing, common for most biological filters, was shown to be eliminated when the filter biofilms had been previously exposed to microcystin. Using polymerase chain reaction (PCR) *mlrA*, a gene involved in microcystin degradation, was detected within the biofilm of a sand filter, providing supporting evidence that biodegradation of microcystin was the primary removal mechanism through the filters.

Keywords Biodegradation; biological filtration; granular activated carbon (GAC) filter; microcystin; polymerase chain reaction (PCR); slow sand filter

Introduction

Microcystins are a group of algal toxins which are prevalent in water bodies worldwide. They are potent hepatotoxins which have been responsible for the death of livestock and in more recent times, humans; the most documented case occurred in Brazil in 1996 where fifty patients in a dialysis centre died as a result of microcystin poisoning (Dunn, 1996; Jochimsen *et al.*, 1998). Consequently, the complete removal of microcystins from drinking water is paramount.

In dissolved form, microcystins are not effectively removed by conventional water treatment processes (Hoffman, 1976; Himberg *et al.*, 1989; Lahti and Hiisvirta, 1989). However, studies have shown that microcystins are readily degraded by aquatic bacteria in reservoirs and lakes (Jones and Orr, 1994; Rapala *et al.*, 1994; Cousins *et al.*, 1996; Christoffersen *et al.*, 2002); therefore, the accumulation of these degrading bacteria in biological filters offers an attractive mechanism for the removal of these problematic compounds from drinking water. One of the advantages of biological filtration is its ability to remove contaminants without the addition of other chemicals that may have potential to produce undesirable by-products in drinking water. However, for the confident application of biological filtration for the removal of algal toxins, it is essential that the

 Recent Progress in Slow Sand and Alternative Biofiltration Processes edited by Rolf Gimbel, Nigel J.D. Graham and M. Robin Collins. ISBN: 9781843391203. Published by IWA Publishing, London, UK.

optimum conditions are known, and the complete removal of these toxic compounds is guaranteed.

The most difficult aspect of any biological process is the time required for degradation to commence, also known as the ripening period or lag period – the biofiltration of microcystin is no exception. This lag period is a major hindrance for the application of biological filtration processes, particularly for the removal of transient contaminants such as algal toxins. If the lag period can be reduced or even removed, then it should be possible to optimise conditions for the successful biological filtration of these contaminants. By performing batch studies, Rapala *et al.* (1994) showed more rapid degradation of microcystins using bacterial inocula collected during or after a microcystin bloom, compared with inocula collected from places where no blooms had occurred. Similarly, using bacteria which had previously been exposed to microcystin, Christoffersen *et al.* (2002) and Holst *et al.* (2003) documented biodegradation of microcystin in natural waters with no lag period. Each of these three studies was conducted in natural waters; however, to date, no studies of this kind have been conducted using biological filters.

Furthermore, few studies have either isolated or identified bacteria capable of degrading microcystin toxins in water (Bourne *et al.*, 1996, 2001; Park *et al.*, 2001; Saito *et al.*, 2003; Harada *et al.*, 2004; Ishii *et al.*, 2004). Recently, Saito *et al.* (2002) designed a polymerase chain reaction (PCR) assay which amplifies a fragment of *mlrA*, a gene involved in microcystin degradation, within a microcystin degrading bacterium. It is envisaged that such an assay will be useful for screening biological filters for their potential microcystin degrading ability prior to a contamination event.

The major objective of this study was to reduce and/or eliminate the lag period prior to microcystin degradation occurring in granular activated carbon (GAC) and slow sand filters which were functioning predominantly in biological mode. A further objective was to utilise the aforementioned PCR assay for detection of the *mlrA* gene within bacteria attached to a slow sand filter matrix, therefore providing additional evidence that removal occurred via biodegradation. In addition, the PCR assay will aid in subsequent isolation and identification of microcystin degrading bacteria from within the sand filter biofilm.

Methods

Materials and reagents

Microcystin-LR (MCLR) and -LA (MCLA) were isolated from a natural bloom of *Microcystis aeruginosa* that occurred in the Torrens Lake in South Australia during the summer of 1998–99. The isolation procedure involved freeze-thawing the bloom material in water and methanol, followed by preparative reverse phase flash chromatography and preparative high performance liquid chromatography (HPLC).

Sample water was collected from the Myponga Reservoir, South Australia, and immediately filtered through a 1 μm filter cartridge to remove particulate matter. The GAC employed in this study was Picazine, a wood based-chemically activated carbon (PICA Carbons, Australia). This carbon is highly mesoporous and has a high surface area, characteristics which have previously been shown to be favourable for microcystin adsorption

(Donati *et al.*, 1994; Ho, 2004). Sand was obtained from the filter beds at the Morgan Water Treatment Plant (WTP) in South Australia where pre-chlorination is not practiced.

Laboratory filter experiments

Granular activated carbon (GAC) filter

Prior to use in this study, the GAC had been in commission in a pilot scale filter at the Myponga WTP (South Australia) for a period of 6 months where it was constantly fed with treated water, prior to chlorination. During this time the GAC pilot filter was periodically spiked with a mixture of MCLR and MCLA. Full details of this pilot plant have been described previously (Ho, 2004).

At the completion of the pilot plant trials samples of the used GAC were loaded into a laboratory glass column (I.D. of 2.5 cm) at a bed height of 15 cm. The laboratory filter was constantly fed with Myponga treated water (MTW) for nine months, during which time a series of five intermittent trials were conducted where a microcystin solution containing MCLR and MCLA was spiked into the column (at target concentrations of 10 $\mu g\ L^{-1}$ each) during each of the trials. Duplicate samples were taken from the column influent and effluent at regular intervals for microcystin analyses. The empty bed contact time (EBCT) of the filter was targeted at 15 minutes, which was identical to the pilot scale filter at the Myponga WTP. The experiment was conducted at room temperature (20±2°C).

Slow sand filters

Filter sand was collected from the Morgan WTP and packed into two separate laboratory glass columns (I.D. of 2.5 cm) at a bed height of 15 cm. The columns were fed with Myponga Reservoir water (MRW) spiked with a microcystin solution containing MCLR and MCLA (at target concentrations of 20 $\mu g\ L^{-1}$ each). Duplicate samples were taken at regular intervals for microcystin analysis. The columns were run under slow sand filtration conditions (EBCT of 30 minutes, filtration rate of 0.3 m hr^{-1}) at room temperature (20±2°C). Prior to the commencement of this study, one of the columns had been in operation for approximately 12 months where it was continuously fed with MRW with periodic spiking of MCLR and MCLA (only for the first 6 months). During this time the column was able to effectively remove both microcystin analogues (unpublished data).

At the completion of the column experiments, sand was removed from the pre-exposed column and the bacteria detached from the sand filter biofilm by periodic vortexing for 15 minutes. The supernatant was collected and any carry over of sand was pelleted by slow speed centrifugation at 1,000 × g for 30 seconds. The sand filter biofilm associated bacteria, now within the supernatant, were assessed by PCR for the presence of the *mlrA* gene (described below).

Analysis of microcystin

Prior to HPLC analysis, microcystins were concentrated from sample waters by C18 solid phase extraction according to the methods described by Nicholson *et al.* (1994). A HPLC system consisting of a 600 pump controller, 717plus autosampler and 996 photodiode array detector (Waters Pty Ltd, Australia) was employed. Sample volumes of 50 μL were injected into a 150 × 4.6 mm Luna C18 column (Phenomenex, Australia) at a flow rate of 1 mL min^{-1}. Two mobile phases were used for the gradient run (30 % acetonitrile/0.05 % trifluoroacetic acid and 55 % acetonitrile/0.05 % trifluoroacetic acid). Concentrations of MCLR were determined by calibration of the peak areas (at 238 nm) with that of external standards (Sapphire Bioscience Pty Ltd, Australia). All MCLA concentrations were expressed in terms of MCLR equivalents. The HPLC method has a detection limit of 0.025 μg L^{-1}. Microcystin recoveries were greater than 95 % with a relative precision of 10 %.

Molecular techniques

DNA from the sand filter biofilm was extracted and prepared using the InstaGene™ Matrix DNA Preparation Kit according to the manufacturer's instructions (Bio-Rad, Australia). The DNA was then used as a template for PCR targeting the *mlrA* gene using the specific oligonucleotide primer set MF, 5′-GACCCGATGTTCAA GATACT-3′ and MR, 5′-CTCCTCCCACAAATCAGGAC-3′ (Saito *et al.*, 2003; Ho *et al.*, in press).

PCR reactions were performed in 50 μL volumes, containing 1 pmol of each primer, 2.5 mM $MgCl_2$ (Perkin Elmer, Australia), 1 × PCR Buffer II (Perkin Elmer, Australia), 200 μM deoxynucleoside triphosphates (dNTPs) (Roche Diagnostics Pty Ltd, Australia), 1.25 units of Ampli *Taq* Gold DNA Polymerase (Perkin Elmer, Australia) and 10 μL of genomic DNA template. A GeneAmp® 2400 PCR System (Perkin Elmer, Australia) was used for the amplifications with the following thermal cycling conditions: 94°C for 10 minutes; 45 cycles of 94°C for 20 seconds, 60°C for 10 seconds and 72°C for 30 seconds; and 72°C for 10 minutes. PCR products were separated by 1 % agarose gel electrophoresis and visualised on a Dark Reader illuminator (Clare Chemical Research, USA). Procedures for the cloning and sequencing of PCR products were performed as described previously (Ho *et al.*, in press). DNA sequences were analysed using the SeqMan™ II and EditSeq™ software programs (DNASTAR, USA). Identification of DNA sequences was performed using the Nucleotide-nucleotide BLAST® network service on the NCBI GenBank® database (http://www.ncbi.nlm.nih.gov/BLAST).

Results and discussion

GAC filtration of microcystin

An attempt was made to reduce the lag period for microcystin degradation in a laboratory GAC filter by conducting a series of intermittent microcystin spike experiments. During the abatement of the microcystin spikings, MTW was continuously passed through the GAC filter. Figure 1 shows five spiking trials over a period of nine months. In each of the spiking trials the lag periods are shown in parentheses (Note: error bars were omitted for clarity, the results were within ± 1.0 μg L^{-1} of the mean).

In the first trial, a lag period of 21 days was evident before microcystin degradation occurred, comparable to findings by Newcombe *et al.* (2003) who observed a lag period of 16 days in a similar column experiment. During this lag period the GAC was functioning in predominantly adsorption mode as indicated by the significant breakthrough of

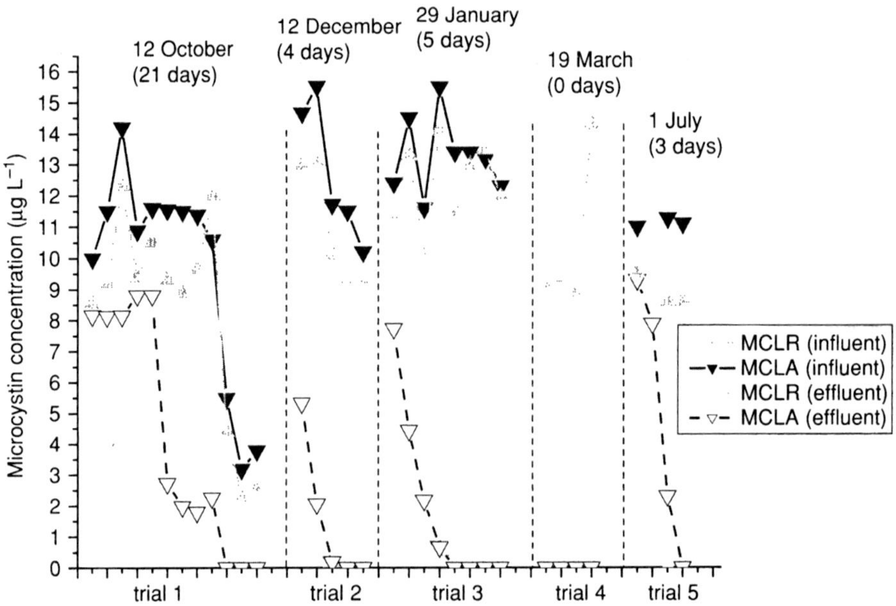

Figure 1 Spiking trials for microcystin-LR (MCLR) and -LA (MCLA) removal through a laboratory granular activated carbon (GAC) filter.

both analogues, with MCLR much more readily adsorbed than MCLA. The differences in the adsorption of both analogues have been observed previously and were attributed to electrostatic interactions reducing the adsorption of MCLA (Newcombe *et al.*, 2003; Ho, 2004).

The second trial, undertaken two months later, resulted in a lag period of only 4 days, suggesting that the bacteria responsible for the degradation may have acclimatised to both microcystin analogues. The third trial conducted approximately one and a half months later resulted in a lag period of 5 days, similar to trial 2. No lag period was evident in the fourth trial, conducted almost two months after trial 3. Although it appears no MCLA was detected in the influent during this trial, MCLR was detected. Since the microcystin spike solution contained both analogues in approximately equal concentrations, then it is likely that MCLA was present and that there was an error in the detection of MCLA.

The final trial was conducted approximately three and a half months after trial 4 with an observed lag period of 3 days. The reappearance of the lag period can be attributed to the time between the trials which was considerably longer than the time frame between the previous trials. However, the final lag period was considerably shorter than the 21 day lag period of the first trial conducted approximately eight months prior to the final trial beginning. These results suggested that once acclimatised, bacteria within a biologically active GAC filter are able to effectively degrade microcystin toxins without a lag period. Moreover, these results indicate that GAC filters do not have to rely solely on adsorption and that the biodegradation mechanism can increase the operational bed life of the filter making it practically feasible for the removal of microcystin.

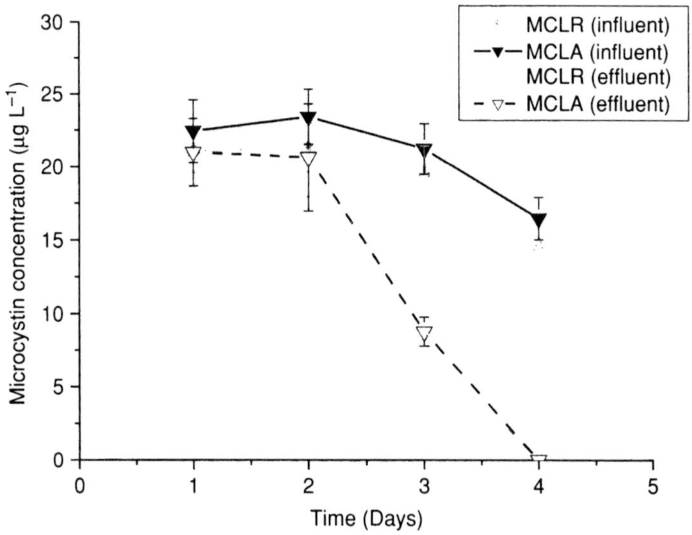

Figure 2 Microcystin-LR (MCLR) and -LA (MCLA) removals through the "new" laboratory sand filter. Error bars represent standard deviations from duplicate measurements.

Slow sand filtration of microcystin

Two laboratory slow sand filters were commissioned to assess their ability to remove MCLR and MCLA. One of the filters (designated as "old") had been previously exposed to periodic microcystin spiking over a year prior to this study commencing; in contrast to the "new" filter which was commissioned at the beginning of this study. Figures 2 and 3 show removals of both microcystin analogues through the new and old filters, respectively. Both filters utilised the same influent water during the experiment.

Breakthrough of both microcystin analogues was initially observed in the first few days of the new filter, providing strong evidence that minimal physical microcystin losses could have occurred during the experiment. Degradation of MCLR and MCLA commenced on day 3 with no detection of either microcystin analogues in the effluent of the new filter by day 4. Subsequent column experiments using a combination of different filter sands and influent waters have yielded similar results (data not shown). These results suggested that the time required for a biofilm to establish and acclimatise to microcystin was quite short, in contrast to the GAC filter which required a lag period of 21 days prior to degradation of microcystin commencing. The differences between the results obtained for the sand filter and GAC filter experiments may be attributed to the influent water as the sand filter was fed with reservoir water (MRW) which would be expected to have higher bacterial numbers than the treated water (MTW) used to feed the GAC filter. In addition, the longer EBCT (30 minutes) employed with the sand filter compared with the GAC filter (15 minutes) could have also enhanced degradation. Solvent (acetonitrile/methanol) extraction of the apparatus of the new filter, including the sand, at the completion of the experiments yielded no microcystin, confirming that no physical removal occurred.

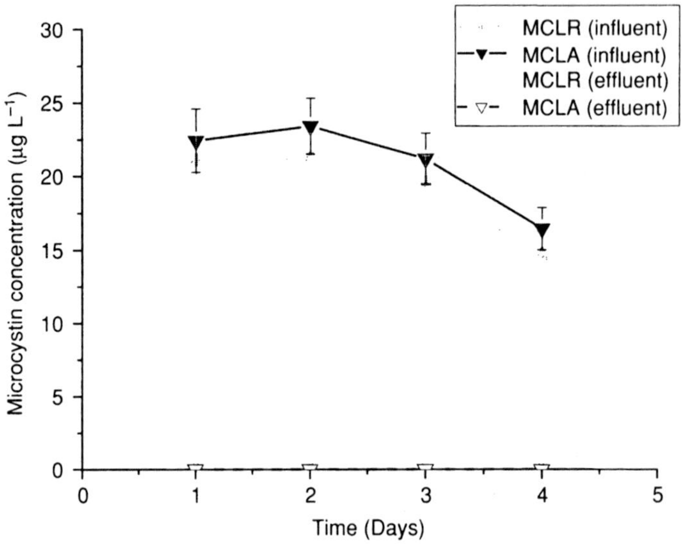

Figure 3 Microcystin-LR (MCLR) and -LA (MCLA) removals through the "old" laboratory sand filter which had previously been exposed to both microcystin analogues. Error bars represent standard deviations from duplicate measurements.

Throughout the experiment no microcystin was detected in the effluent of the old sand filter, suggesting that pre-exposure of the biofilm to microcystin was able to eliminate the lag period prior to degradation commencing. However, this could also have been due to the more advanced bacterial community within the biofilm of the old filter, compared with that of the new filter which had not been acclimatised to the experimental conditions in this study. Nevertheless, the old filter was still able to effectively remove microcystin even after no toxins had been spiked for six months prior to this study. The absence of microcystin breakthrough in the old filter suggests that biologically active slow sand filters are a viable treatment option for effective microcystin removal.

Detection of the *mlrA* gene within bacteria detached from sand filter biofilm

A PCR assay was used to identify the presence of the *mlrA* gene within bacteria previously attached to the biofilm of the old sand filter. DNA from *Sphingomonas* sp. ACM-3962, a microcystin degrading bacterium originally isolated from an Australian water body by Jones *et al.* (1994), served as a positive control for the PCR assay since the MF-MR primer set was designed from its *mlrA* gene sequence (Saito *et al.*, 2003). This generated a PCR product of approximately 800 base pair (bp) in size (data not shown). A product of equivalent size to the positive control was evident for the DNA template extracted from the biofilm of the sand filter. This product had a DNA sequence of 98 % similarity to an 807 bp nucleotide sequence of the *mlrA* gene from *Sphingopyxis* sp. C-1. Experiments are currently underway in an attempt to isolate and identify the bacterium which contains this homologous *mlrA* gene with preliminary results indicating that the bacterium also belongs to the sphingomonad group.

The detection of the *mlrA* gene within the old sand filters suggests that the removal of microcystin through this filter most likely followed a degradation pathway similar to one previously reported by Bourne *et al.* (1996, 2001) where the enzyme encoded by the *mlrA* gene cleaved the cyclic structure of microcystin. Furthermore, this study demonstrates the potential of the PCR assay to be used as a screening tool to assess the microcystin degrading capability of biologically active sand filters prior to an impending episode of microcystin.

Conclusions

Biologically active GAC and slow sand filters were shown to be viable treatment options for the effective removal of microcystin toxins. Furthermore, the lag period prior to degradation commencing was eliminated in both filter types, highlighting their applicability for microcystin removal. A homologous *mlrA* gene was detected within the bacterial fraction of the sand filter biofilm providing additional evidence that biodegradation of microcystin was the primary removal mechanism through the filters.

Acknowledgements

The authors would like to thank Najwa Slyman for her assistance in maintaining the GAC filter. Alexandra Keegan is also duly acknowledged for her assistance with the development of the PCR assay.

References

Bourne, D.G., Jones, G.J., Blakeley, R.L., Jones, A., Negri, A.P. and Riddles, P. (1996) Enzymatic pathway for the bacterial degradation of the cyanobacterial cyclic peptide toxin microcystin LR. *Appl. Environ. Microbiol.* **62**(11), 4086–4094.

Bourne, D.G., Riddles, P., Jones, G.J., Smith, W. and Blakely, R.L. (2001) Characterisation of a gene cluster involved in bacterial degradation of the cyanobacterial toxin microcystin LR. *Environ. Toxicol.* **16**(6), 523–534.

Christoffersen, K., Lyck, S. and Winding, A. (2002) Microbial activity and bacterial community structure during degradation of microcystins. *Aquat. Microb. Ecol.* **27**(2), 125–136.

Cousins, I.T. Bealing, D.J., James, H.A. and Sutton, A. (1996) Biodegradation of microcystin-LR by indigenous mixed bacterial populations. *Wat. Res.* **30**(2), 481–485.

Donati, C., Drikas, M., Hayes, R. and Newcombe G. (1994) Microcystin-LR adsorption by powdered activated carbon. *Wat. Res.* **28**(2), 1735–1742.

Dunn, J. (1996) Algae kills dialysis patients in Brazil. *BMJ* **312**, 1183–1184.

Harada, K.-I., Imanishi, S., Kato, H., Masayoshi, M., Ito, E. and Tsuji, K. (2004) Isolation of Adda from microcystin-LR by microbial degradation. *Toxicon* **44**(1), 107–109.

Himberg, K., Keijola, A.-M., Hiisvirta, L., Pyysalo, H. and Sivonen, K. (1989) The effect of water treatment processes on the removal of hepatotoxins from *Microcystis* and *Oscillatoria* cyanobacteria: A laboratory study. *Wat. Res.* **23**(8), 979–984.

Ho L. (2004) The removal of cyanobacterial metabolites from drinking water using ozone and granular activated carbon. PhD Dissertation, University of South Australia, Adelaide, Australia.

Ho, L., Meyn, T., Keegan, A., Hoefel, D., Brookes, J., Saint, C.P. and Newcombe, G. Bacterial degradation of microcystin toxins within a biologically active sand filter. *Wat. Res.* (in press).

Hoffman, J.R.H. (1976) Removal of *Microcystis* toxins in water purification processes. *Wat. SA* **2**(2), 58–60.

Holst, T., Jørgensen, N.O.G., Jørgensen, C. and Johansen, A. (2003) Degradation of microcystin in sediments at oxic and anoxic, denitrifying conditions. *Wat. Res.* **37**(19), 4748–4760.

Ishii, H., Nishijima, M. and Abe, T. (2004) Characterization of degradation process of cyanobacterial hepatotoxins by a Gram-negative aerobic bacterium. *Wat. Res.* **38**(11), 2667–2676.

Jochimsen, E.M., Carmichael, W.W., An, J., Cardo, D.M., Cookson, S.T., Holmes, C.E.M., Antunes, M.B.C., Filho, D.A.M., Lyra, T.M., Barreto, V.S.T., Azevedo, S.M.F O. and Jarvis, W.R. (1998) Liver failure and death after exposure to microcystins at a hemodialysis center in Brazil. *N. Engl. J. Med.* **338**(13), 873–878.

Jones, G.J., Bourne, D.G., Blakeley, R.L. and Doelle, H. (1994) Degradation of the cyanobacterial hepatotoxin microcystin by aquatic bacteria. *Nat. Tox.* **2**(4), 228–235.

Jones, G.J. and Orr, P.T. (1994) Release and degradation of microcystin following algicide treatment of a *Microcystis aeruginosa* bloom in a recreational lake, as determined by HPLC and protein phosphatase inhibition assay. *Wat. Res.* **28**(4), 871–876.

Lahti, K. and Hiisvirta, L. (1989) Removal of cyanobacterial toxins in water treatment processes: Review of studies conducted in Finland. *Wat. Supp.* **7**, 149–154.

Newcombe, G., Cook, D., Brooke, S., Ho, L. and Slyman, N. (2003) Treatment options for microcystin toxins: Similarities and differences between variants. *Environ. Technol.* **24**(3), 299–308.

Nicholson, B.C., Rositano, J. and Burch, M.D. (1994) Destruction of cyanobacterial peptide hepatotoxins by chlorine and chloramine. *Wat. Res.* **28**(6), 1297–1303.

Park, H.-D., Sasaki, Y., Maruyama, T., Yanagisawa, E., Hiraishi, A. and Kato, K. (2001) Degradation of cyanobacterial hepatotoxin microcystin by a new bacterium isolated from a hypertrophic lake. *Environ. Toxicol.* **16**(4), 337–343.

Rapala, J., Lahti, K., Sivonen, K. and Niemalä, S.I. (1994) Biodegradability and adsorption on lake sediments of cyanobacterial hepatotoxins and anatoxin-a. *Lett. Appl. Microbiol.* **19**(6), 423–428.

Saito, T., Okana, K., Park, H.-D., Itayama, T., Inamori, Y., Neilan, B.A., Burns, B.P. and Sugiura, N. (2003) Detection and sequencing of the microcystin LR-degrading gene, *mlrA*, from new bacteria isolated from Japanese lakes. *FEMS Microbiol. Lett.* **229**(2), 271–276.

20 Removal of cyanobacterial toxins (microcystins) during slow sand and bank filtration

G. Grützmacher*, G. Wessel*, I. Chorus* and H. Bartel*

*Umweltbundesamt (UBA), German Federal Environmental Agency, Versuchsfeld Marienfelde, Schichauweg 58, D-12307 Berlin.

Abstract Mass occurrences of cyanobacteria – often called blue-green-algae – can pose a threat to human health as many are capable of producing toxic substances (so-called cyanotoxins). In case of drinking water production from surface waters frequently infested by cyanobacteria secure elimination of toxins has to be ensured. In an interdisciplinary research project managed by the Berlin Competence Center for Water (KWB) field scale experiments addressing this topic in biofiltration systems (slow sand filtration, artificial groundwater recharge, bankfiltration) were conducted on the test site for aquatic simulation run by the German Federal Environmental Protection Agency (UBA). Cell-bound and extracellular, hepatotoxic microcystins (MCYST – the most frequently found group of cyanotoxins) – were applied to the water reservoir of slow sand filters and smaller scale enclosures. Toxin concentrations were measured in different depths and in the effluent after 80 cm to 130 cm of sand passage. The results confirmed the importance of biodegradation for MCYST elimination in sandy material and yielded generally high degradation rates (0.04 h^{-1} to 0.17 h^{-1}) under optimal conditions. Optimal conditions can be defined as the following i) previous conditioning of the filter for more than 4 days, preferably for several weeks, ii) aerobic conditions, iii) moderate temperatures ($> 10\,^{\circ}$C).

Keywords Slow sand filtration, bank filtration, cyanobacterial toxins, microcystins, biodegradation, retardation.

Introduction

Cyanobacterial toxins are toxic substances produced by cyanobacteria which occur in surface waters world wide. At elevated nutrient levels mass developments of cyanobacteria can occur that can lead to hazardous toxin concentrations exceeding the provisional guideline value published by the WHO for Microcystin-LR of 1 μg/L for drinking water (WHO 1998). If surface water infested by cyanobacteria is used as source for drinking water (eg. by slow sand or bank filtration) secure removal of these toxins, of which the group of hepatotoxic microcystins (MCYST) is the most frequent, has to be ensured. Previous investigations have shown that MCYST is likely to be degraded during sand or sediment contact. This was shown in laboratory experiments (Lahti *et al.* 1998, Miller, 2000) and field observations (Grützmacher *et al.* 2002). However, cases of toxin breakthrough in bank filtered water

edited by Rolf Gimbel, Nigel J.D. Graham and M. Robin Collins. ISBN: 9781843391203. Published by IWA Publishing, London, UK.

have also been observed and published (eg. Lahti *et al.* 1998, Eynard *et al.* 2000). It was therefore the aim of investigations conducted in the course of an interdisciplinary research project NASRI (Natural and Artificial Systems for Recharge and Infiltration) to quantify elimination under most realistic conditions and to find out in which cases toxin breakthrough is possible. In order to simulate worst-case conditions, experiments with extra-cellular microcystins were carried out at technical scale on the UBA's center for aquatic simulations in Berlin, Marienfelde.

Methods and materials

Microcystins

Extra-cellular microcystins were extracted from a mass culture of *Planktothrix agardhii* (HUB76) by repeated freeze thawing of the concentrated culture, thus releasing the cell-bound toxin. The crude extract used for the experiments was obtained after separation from the particulate cell-debris by centrifuging.

Experimental site marienfelde

The site consists of a storage reservoir (about 3,500 m^3 volume) with an adjacent artificial aquifer (4 m depth, area: 2,000 m^2) built up of fine and coarse gravel. Additionally the surface water can be conducted into 4 infiltration basis each with an average surface area of 60 m^2. Two of these basins are sealed by concrete at the bottom with a sand depth of 0.8 m so they act as slow sand filters, and two have direct contact to the underlying sediments and can be operated as aquifer infiltration ponds. For smaller scale experiments three enclosures with a surface area of 1 m^2 and medium grained sand filling (80 cm) with a gravel drainage layer of 50 cm depth are installed in one of the infiltration basins. The enclosures used for the experiments described below had either two or four additional sampling ports in 40 cm and 80 cm (for experiments E4, E5 and E9) or 20 cm, 40 cm, 60 cm and 80 cm depth (for experiments E2 and E3).

The whole site is separated from the surrounding aquifer by a layer of clay thus forming an independently operable hydraulic system so that experiments with toxic substances can be carried out without adverse effects on the environment (for details see Bartel & Grützmacher 2002).

Enclosure and slow sand filter experiments

Extracellular microcystins were applied to the water reservoir in a pulsed application. The filtration velocities varied between 0.6 m/d and 2.4 m/d and in some cases the uppermost centimetre of the filter bed was removed before the experiment in order to compare the effect of presence and absence superficial "schmutzdecke" on toxin removal. NaCl was applied as a tracer simultaneously to the toxin and detected by electrical conductivity measurement. Care was taken not to raise the electrical conductivity by more than 10 %, in order to avoid massive changes in hydrochemistry. Toxin and tracer concentrations were measured in the water reservoir and in the effluent (slow sand filter) as well as in different depths (20 cm, 40 cm,

60 cm and 80 cm in the enclosure experiments). For one enclosure experiment anaerobic conditions were induced by adding biodegradable DOC (acetic acid) continuously with a resulting concentration of 0.3 mmol/L additional DOC (resulting DOC concentration in water reservoir: 9.6 mg/L). After 4 days of continuous dosing oxygen could not be detected in 40 cm depth; after 9 days, no more oxygen was found in the effluent (Grützmacher *et al.* in press).

For all slow sand filter and enclosure experiments with pulsed application the elimination rate was calculated as the MCYST recovered ($MCYST_{rec}$) in the effluent according to equation (1) in relation to the total MCYST applied (initial concentration multiplied by volume of the water reservoir).

$$MCYST_{rec} = \sum c_i * Q^* \Delta t$$

c_i: concentration of MCYST at time i
Q: flow rate
Δt: corresponding time interval.

Microcystin analysis

Microcystin analyses were carried out by ELISA (Enzyme-Linked ImmunoSorbent Assay) according to Zeck *et al.* (2002) and HPLC (High Performance Liquid Chromatography) with a photodiode array detector (Lawton *et al.* 1994). Usually samples were first tested for their microcystin content using the ELISA and selected ones were subsequently analyzed by HPLC to verify the results and to distinguish the different microcystin variants.

Modelling

The experiments with pulsed MCYST application were modelled with the computer program VisualCXTFIT (Nützmann *et al.* 2005) based on the parameter estimation code CXTFIT (Toride *et al.*, 1995). For determination of pore velocity and dispersivity tracer experiments were carried out simultaneously with each MYST experiment. These parameters were then used to model the effluent curves of MCYST for determination of retardation coefficients (R) and degradation rates (λ).

Results and discussion

Slow sand filter experiments

Table 1 gives the initial conditions for three slow sand filter experiments as well as the results of the elimination calculations (Grützmacher *et al.* in press). The results show that the experiment conducted with conditions assumed to be quite unfavourable for degradation, i.e. virgin sand, no visible schmutzdecke and high flow velocity (2.5 m/d), did not show reduced elimination of MCYST compared with the experiments SSF 5 and SSF 6 under conditions more conducive to bacterial degradation, i.e. preconditioned sand and lower flow velocities. In contrast, the highest elimination was observed under these presumably less favourable conditions.

Table 1 Characteristic initial parameters and results of three slow sand filter experiments with MCYST (v_f: filtration velocity, c_0: initial concentration).

Experiment no.	v_f [m/d]	c_0 [μg/L]	Visible schmutz-decke?	Previous MCYST-contact?	T min - max] [°C]	Aerobic?	Elimination [%]
SSF 2	2.5	10.5	No	No	8–19.5	Yes	77.5
SSF 5	1.4	8.6	Yes	Yes	15–25	Yes	73.5
SSF 6	0.6	6.0	Yes	Yes	4.8–13	Yes	62.8

Table 2 Retardation coefficients (R) and degradation rates (λ) obtained by modelling for three slow sand filter experiments (Grützmacher *et al.* in press).

Experiment no.	R	λ [h^{-1}]
SSF 2	2.6	0.17
SSF 5	1.3	0.04
SSF 6	1.5	0.05

Table 3 Characteristic initial parameters and results of five enclosure experiments with MCYST (values in brackets: uncertain, due to smaller database).

Experiment no.	v_f [m/d]	c_0 [μg/L]	Visible schmutz-decke?	Previous MCYST-contact?	T min – max [°C]	Aerobic?	Elimination [%]
E2	1.25	6.1	No	No	23.6–31.6	Yes	97.3
E3	1.12	6.2	Yes	Yes	16.4–23.1	Yes	90.2
E4	1.09	9.0	Yes	No	0.6–4.7	Yes	(99.5)
E5	1.22	7.6	No	Yes	6.3–11.2	Yes	90
E9	0.62	6.0	Yes	Yes	23–27	No	85

Retardation coefficients and degradation rates obtained by modelling the MCYST experiment taking into account flow velocities and dispersion rates from simultaneous tracer experiments are shown in Table 2 (Grützmacher *et al.* in press). These results show that the higher elimination observed for the experiment with highest flow velocity and virgin sand (SSF 2) is due to a clearly higher degradation rate (0.17 h^{-1}) compared to the following experiments (0.04 h^{-1} and 0.05 h^{-1} for SSF 5 and SSF 6 respectively). This might be due to reactions (e.g. irreversible sorption) that take place only during first contact with MCYST.

Enclosure experiments

An overview of the characteristic initial parameters and percentages of elimination during five enclosure experiments under different conditions is given in Table 3. The elimination

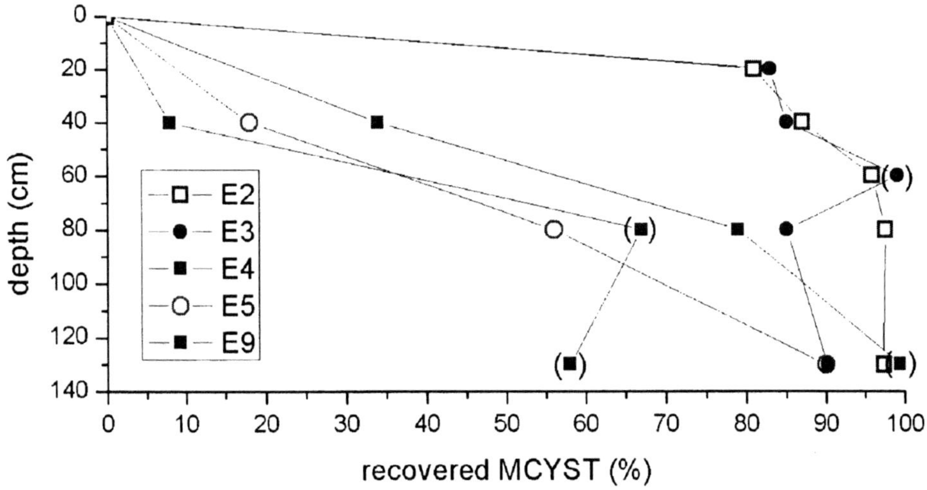

Figure 1 Elimination rates in different depths calculated for the enclosure experiments (symbols in brackets refer to uncertain values due to insufficient data).

was calculated for the concentrations of MCYST measured in the effluent after 130 cm of filter passage. These results seem to support a clear difference between aerobic and anaerobic conditions, with anaerobic conditions leading to lower total elimination rates than aerobic conditions (85 % as compared to 90 %–99.5 %). Taking the SSF results into account it seems unlikely that this is due only to different flow velocities, which amount to 0.6 m/d in the anaerobic and 1.1 m/d to 1.25 m/d in the aerobic experiments.

No previous MCYST contact (i.e. "virgin" sand) seems to lead to slightly higher elimination rates regarding aerobic experiments only. This is in agreement with the results of the SSF experiments. Due to a longer filtration length (130 cm in the enclosures compared to 80 cm in the SSF) the total elimination rates obtained in the aerobic enclosure experiments exceed those in the SSF experiments by about 20 % to 30 %.

Regarding the development of MCYST elimination over the filter depth as shown in Figure 1, the elimination rates in 80 cm depth during the enclosure experiments (56 % to 98 %) are in the same range as those measured in the SSF experiments (63 % to 78 %) but show higher variations.

The elimination rates calculated for the first 40 cm show even higher variations, ranging from 8 % (E9) to 85 % (E2). In this part of the sand filter the difference between aerobic and anaerobic conditions does not seem as pronounced as when regarding the total depth. This is probably due to the fact that the infiltrating water is saturated with oxygen so that even in the anaerobic experiment a short aerobic passage in the uppermost centimetres has to be assumed.

For the aerobic experiments there is a distinct difference between the results of E2 and E3 on the one hand and E4 and E5 on the other hand. Whereas E2 and E3 show elimination rates over 80 % in the first 40 cm, E4 and E5 only reach values between 18 % and 34 %. The following differences can be determined between these two sets of experiments: i) the experiments were conducted on different enclosures with different pretreatment: E2 and E3 were conducted in enclosure III (4 sampling ports, continuous operation since July 2003, four weeks prior to first experiment) whereas E4 and E5 were carried out in enclosure II (2 sampling ports, continuous operation since November 2003, four days prior to first experiment) and ii) the experiments cover different temperature ranges: E2 and E3 were carried out in August and September with temperatures between 16.4 °C and 31.6 °C, whereas E4 and E5 were carried out in November with temperatures between 0.6 °C and 11.2 °C.

These results support the hypothesis that the existence of a visible, superficial clogging layer or "schmutzdecke", is not of decisive importance concerning biodegradation of MCYST. Temperature, redox conditions as well as the pretreatment of the filter as a whole (conditioning or maturation) play a much more important role. Ellis (1975) states that the "schmutzdecke" or organic filter adds substantially to the effectiveness of the straining process. Biological activity, however, is observed up to a depth of 40 cm or more, and thus the application of Ellis' observation is not limited to the superficial "schmutzdecke".

The most important conclusion for water supply ist that for the settings simulated (aerobic and anaerobic infiltration into sandy material) travel times of only a few days (< 5) proved sufficient to reduce maximum extracellular MCYST concentrations of 100 μg/L to values lower than the provisional WHO guideline value of 1 μg/L MCYST-LR (WHO 1998).

Acknowledgements

We thank the Berliner Wasserbetriebe (BWB) and Veolia Water for sponsoring the NASRI research project during which the experiments were conducted. We would also like to thank I. Flieger for the HPLC analyses as well as HW. Althoff, T. Starzetz and T. Köhler for their help during the technical scale experiments.

References

Bartel, H. and Grützmacher, G. (2002) Elimination of microcystins by slow sand filtration at the UBA's experimental field. In *Riverbank filtration: Understanding Contaminant Biogeochemistry and Pathogen Removal*, (ed. C. Ray), pp. 123–133, Kluwer Academic Publishers.

Grützmacher, G., Böttcher, G., Chorus, I., Knappe, A. and Pekdeger, A (2002) Cyanobacterial toxins in bank filtered water from Lake Wannsee, Berlin. In *Management of Aquifer Recharge for Sustainability* (ed. P.J. Dillon) Proceedings of the 4th international symposium on artificial recharge of groundwater, ISAR-4, Adelaide, South Australia 22–26 Sept. 2002.

Grützmacher, G., Wessel, G., Chorus, I. and Bartel, H. (in press) Are there limits to cyanobacterial toxin (microcystin) elimination by sand passage? In *Proceedings 5th International Symposium on Management of Aquifer Recharge (ISMAR 5)*, June 10–16, Berlin, Germany.

Lahti, K., Vaitomaa, J., Kivimäki, A.L., Sivonen, K. (1998) Fate of cyanobacterial hepatotoxins in artificial recharge of groundwater and in bank filtration. In *Artificial Recharge of Groundwater*, (eds. Peters *et al.*), Balkema, Rotterdam.

Lawton, L.A., Edwards, C. and Codd, G.A. (1994) Extraction and high-performance liquid chromatographic method for the determination of microcystins in raw and treated waters. *Analyst*, **119**, 1525–1530.

Miller, M.J. (2000) Investigation of the removal of cyanobacterial hepatotoxins from water by river bank filtration. PhD Thesis, Flinders University.

Nützmann, G., Holzbecher, E., Strahl, G., Wiese, B., Licht, E. and Knappe, A. (in press) Visual CXTFIT – a user-friendly simulation tool for modelling one-dimensional transport, sorption and degradation processes during bank filtration. In *Proceedings 5th International Symposium on Management of Aquifer Recharge (ISMAR 5)*, June 10–16, Berlin, Germany.

Toride, N., Leij, F.J., and Van Genuchten, M. Th. (1995) *The CXTFIT code for estimating transport parameters from laboratory or field tracer experiments*, Research Report No. 137, U.S. Salinity Lab., Agric. Res. Service, US Dep. of Agric., Riverside (CA).

World Health Organisation (1998) *Guidelines for drinking water quality*, 2nd edition, Addendum to Volume 2, Health criteria and other supporting information, WHO, Geneva.

Zeck, A., Eickenberg, A., Weller, M.G. and Niessner, R. (2001) Highly sensitive immunoassay based on an monoclonal antibody specific for [4-arginine]microcystins. *Anal. Chim. Acta* **441**, 1–13.

21 Removal of *microcystis aeruginosa* and microcystins by slow sand filtration: A pilot scale study

J. C. Sá*, C. C. S. Brandão*

*University of Brasília, Civil and Environmental Engineering Department, Brasília – Brazil (cbrandao@unb.br)

Abstract To evaluated the influence of the filtration bed characteristics (sand effective size and bed depth) and filtration rate, as well as the influence of the concentrations of *M. aeruginosa* and dissolved microcystins in the raw water, on the performance of slow sand filtration a pilot scale study was carried out with living cells of *Microcystis aeruginosa* and microcystins. Four experimental stages were conducted. Altogether, the results indicate that sand effective size and bed depth influence the removal of *Microcystis* and microcystins, but the filtration rate does not seem to affect significantly the efficiency of the filter. However, the degree of maturation of the filter and its acclimatisation to the toxin play the most important roles in the effectiveness of the treatment of water with microcystins by slow sand filtration.

Keywords *Microcystis aeruginosa*, microsystins, bed characteristics, sand size, bed depth

Introduction

Cyanobacterial blooms have been occurring in many reservoirs, rivers and lakes worldwide, and may contaminate water sources used for drinking-water production. Cyanotoxins are intracellular compounds produced by various cyanobacterial genus and species. The genera *Microcystis* may produce, among its secondary metabolites, microcystins, toxins that can cause liver damage or even death. In 1998 WHO, recognizing the risks involved in human consumption of water contaminated by microcystins, recommended a drinking water guideline of 1 μg/L. Brazil was the first country to enforce this guideline at national level.

Water treatment processes more indicated for the removal of cyanotoxins are the chemical oxidation, the biological oxidation and activated carbon adsorption (Donati *et al.*, 1994; Hart *et al.*, 1998; Mouchet and Bonnélye, 1998; Newcombe, and Nicholson, 2002 and 2004). Generally, if optimized, the conventional treatment can be efficient in the removal of viable cells but has a limited removal efficiency of cyanotoxins (Hrudey *et al*, 1999).

Slow sand filtration has been suggested as a water treatment technology able to remove both cyanobacteria and cyanotoxins. The first study in this area is described by Keijola *et al.* (1988). In their experiments, two pilot slow sand filters were supplied by raw water containing cyanotoxins produced by *A. flos-aquae*, by *M. aeruginosa* and by *O. agardhii*. The reduction of cyanotoxins was higher than 70% for toxins from *A. flos-aquae* and *M. aeruginosa* blooms and was lower for toxins from *O. agardhii* culture.

In the work of Grutzmacher *et al.* (2002), two full-scale experiments with slow sand filters were conducted. The first experiment was carried out with dissolved microcystins (dosed as short-term, single pulse application, filtration rate of 0.8 m/day) and yielded high removal rates (>95%). The second, that consisted of a long-term exposure of the SSFs to living cells of *Planktothrix aggardhii* (intracellular microcystins), carried out at a filtration rate of 0.2 m/day, showed a removal higher than 85% during the first 11 days. The lowest daily removal of total microcystins was 67%. The authors reported that in moderate temperatures, with an intact *schmutzdecke* with previous contact with microcystins, slow sand filtration is an effective treatment for eliminating microcystins from drinking water.

Similar results regarding dissolved microcystins were reported by Sá (2002) in experiments carried out in a pilot slow sand filters operated at a filtration rate of 3 m/day. Total removal (100%) of dissolved microcystins was observed when toxin concentration in the raw water was in the range of 58 to 127 μg/L for 2 days. The removal was lower, about 85%, when the concentration of dissolved microcystins in the raw water was kept very high (~420 μg/L) for a period of 3 days. It was observed that the removal of coliform bacteria was negatively influenced by the presence of dissolved microcystins in the raw water.

However, Sá (2002) reports that when the filters were firstly exposed, during three days, to raw water containing living cells of *M. aeruginosa* (~130 μg/L of intracellular microcystins), the overall removal of microcystins (intra plus extracellular) was almost nil, although the removal of cells was high, around 87%. The low overall removal was due to the release of dissolved microcystins from lysing cells retained in the filter beds. In the second exposure of the filters to cells, the overall removal efficiency rose to 43%, and reached 75% in the third exposure, when the filtration rate was also lowered to 2 m/day. The results suggest that the filter bed biological maturity have a positive influence on the overall removal efficiency and that previous exposition of the filter to microcystins is also important, as quoted by Grutzmacher *et al.* (2002).

In this scenario, the aim of this work is to evaluate the efficiency of slow sand filtration in removing viable cells of *M. aeruginosa* and dissolved microcystins and the occurrence of cells lysis inside filter bed. To accomplish this main objective, this study evaluated the influence of the filtration bed characteristics (sand effective size and bed depth) and filtration rate, as well as the influence of the concentrations of *M. aeruginosa* and dissolved microcystins in the raw water, on the performance of the filtration process.

Methods

Pilot-scale experiments were conducted using 3 slow sand filter columns operated in parallel (Figure 1).

The raw water was taken from Paranoá Lake (Brasília-DF-Brazil), a mesotrophic reservoir, and it was periodically spiked during few days with a toxic strain *M. aeruginosa* cultivated in laboratory and dissolved microcystin obtained by the lysis of cultivated cells. The experimental work was carried out in the four stages presented below.

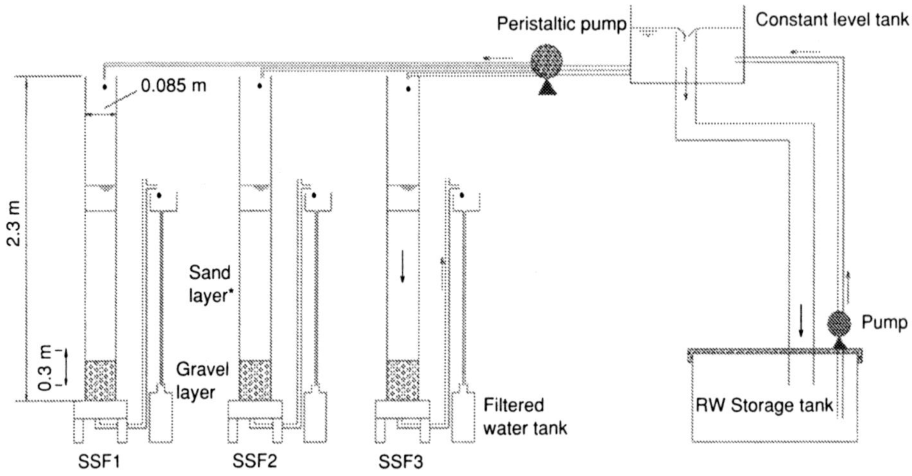

Figure 1 Schematic arrangement of experimental apparatus

Experimental Stage 1: Each of the 3 filter columns was filled with 0.9m of sand with different effective sizes, SSF1-0.35 mm, SSF2-0.28 mm and SSF3-0.22 mm. The SSFs were operated with a filtration rate of 0.125 m/h. After a ripening period, the raw water was spiked during 3 days with a toxic strain of *M. aeruginosa* to obtain a final concentration of 10^5cell/mL. Following this phase the filters were monitored for evaluating the occurrence of breakthrough of viable cells of *M. aeruginosa* and cell lysis in the filter bed.

Experimental Stage 2: The filter bed depth was varied, SSF1-0.6 m, SSF2-0.9 m and SSF3-1.1m. The effective size and the filtration rate were kept constant (0.125 m/h). The effective size adopted was the one that led to the best filtration performance in the stage 1. Similarly to stage 1, after ripening, the SSFs were submitted to two periods of exposure to *M. aeruginosa* (3days, 10^5cell/mL) followed by periods of monitoring (raw water without *M. aeruginosa*).

Experimental Stage 3: Each SSF in the pilot plant was operated at a different filtration rate: SSF1-0.083 m/h; SSF2-0.167 m/h; and, SSF3-0.125 m/h. Sand with the effective size and bed depth optimized in the previous stages was used in all SSFs. After ripening period, the SSFs were fed with raw water either spiked with *M. aeruginosa* (10^5cell/mL) or with dissolved microcystins. Each spiking period was followed by a monitoring period.

Experimental Stage 4: In this stage, the 3 SSFs were operated at the filtration rate select from results obtained in the stage 3 and the raw water was spiked with higher concentrations of *M. aeruginosa* as well as with 5 different concentrations of dissolved microcystins. The same sand characteristic of stage 3 was adopted.

It is important to point out that between experimental stages the filter beds (sand) were removed from each filtration column, washed, dried, re-sieved (when necessary) and re-installed. So, at the beginning of each experimental stage the filter sand beds were like new.

To evaluate the performance of the slow sand filters, water head in points along the filter beds were measured and water quality parameters such as pH, total coliforms, *E. coli*,

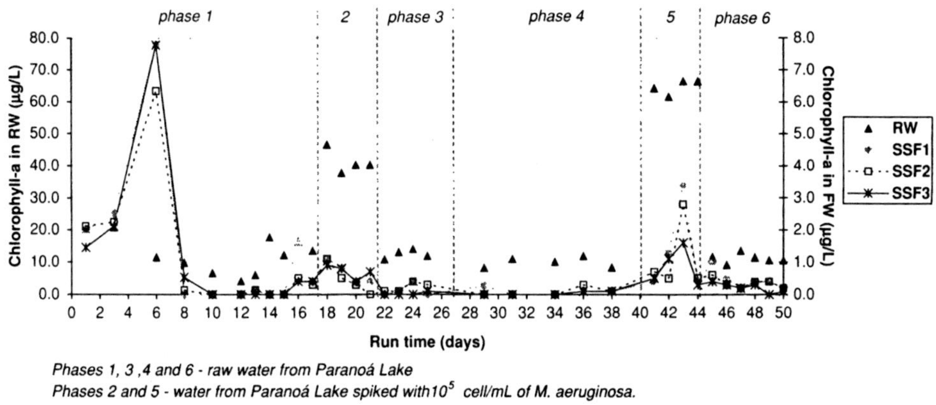

Figure 2 Chlorophyll-a in the raw and filtered water during experimental stage 1

chlorophyll-a, intracellular and extracellular microcystins were analysed. During spiking and monitoring periods, 24 hours composite samples of filtered water of were daily analized. The routine adpted for the chlorophyll-a analysis is described by Lloyd and Tucker (1988). For the analysis of microcystins it was used an enzyme-linked immunosorbent assay - ELISA kit (Envirologix®, USA), which has a range of detection of 1.6 to 2.5 µg/L.

Results and discussion

Experimental stage 1

Figure 2 presents the behaviour of the raw and filtered water during the experimental stage 1 regarding the parameter chlorophyll-a. As can be seen, the chlorophyll-a in the filtered water was not highly influenced by the effective size of the filter media. Statistical analysis confirms this observation. Similar behaviour was observed for turbidity (data not shown).

The average removal of chlorophyll-a by filters was 98.1%, 98.3% and 98.5%, respectively, for filters SSF1 ($\varnothing_{eff}$-0.35 mm), SSF2 ($\varnothing_{eff}$-0.28 mm) and SSF3 ($\varnothing_{eff}$-0.22 mm). In all filters it was observed a degradation of the quality of filtered water in the days during and following the periods of cell spiking, suggesting some breakthrough of viable cells.

The pilot filters were operated during 50 days and they were cleaned and restarted on the 26th day. Figure 3 shows the behaviour of the total head loss during experimental stage 1.

The rate of head loss development was higher for the filter SSF3 ($\varnothing_{eff}$-0.22 mm) and was similar in the two other filters (SSF1, $\varnothing_{eff}$-0.35 mm and SSF2, $\varnothing_{eff}$-0.28 mm), particularly during the second half of the experiment. The head loss development in the filters was higher in the first half of the experiment than in the second half, probably due to the existence of a very thin layer of fine sand on the top of the filters bed as a result of an inadequate procedure of filling the filter column with sand. After scraping 2 cm in day 26, the rate of head loss development decreased.

Considering the similarity of filtered water quality produced by the three SSFs, and the differences in head loss development, it is suggested that effective size of 0.28 mm is more adequate for the removal of *M. aeruginosa* cells, and was used in the following stages.

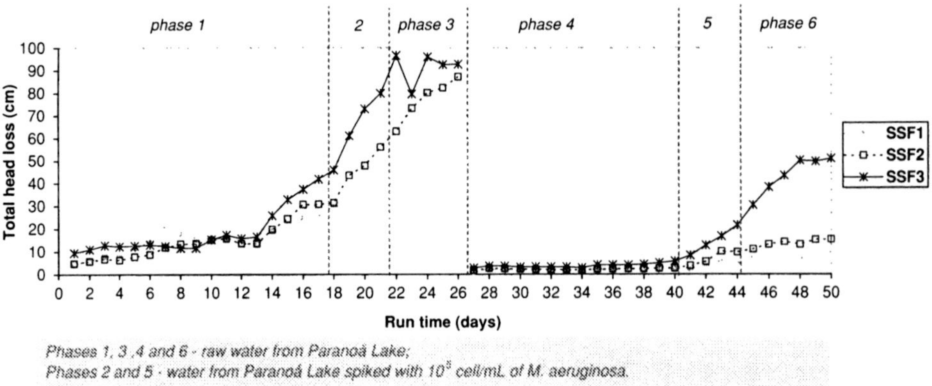

Figure 3 Head loss development during experimental stage 1

Experimental stage 2

In the experimental stage 2 the influence of the bed depth was evaluated. Figures 4 and 5 present, respectively, the behaviour of chlorophyll-a and intracellular microcystins in the raw and filtered water. The results suggest that the removal of *M. aeruginosa* was influenced by the filter depth. Filter SSF1 (h_{bed}-0.6 m) presented the lower chlorophyll-a removal (94.4%) as the filter SSF3 (h_{bed}-1.1 m) was able to remove, in average, 99.5% of the chlorophyll-a during spiking periods. The chlorophyll-a of the spiked water ranged from 76 to 118 μg/L.

Although Figures 4 and 5 show that highest cell concentration occurred in the effluent of SSF1, the highest content of dissolved toxin was detected in the effluent of SSF3, as can be observed in Figure 6.

Figures 4 and 5 show that during and following spiking periods the filtered water contains significant amounts of cells - intracellular microcystins was undetectable in the filtered

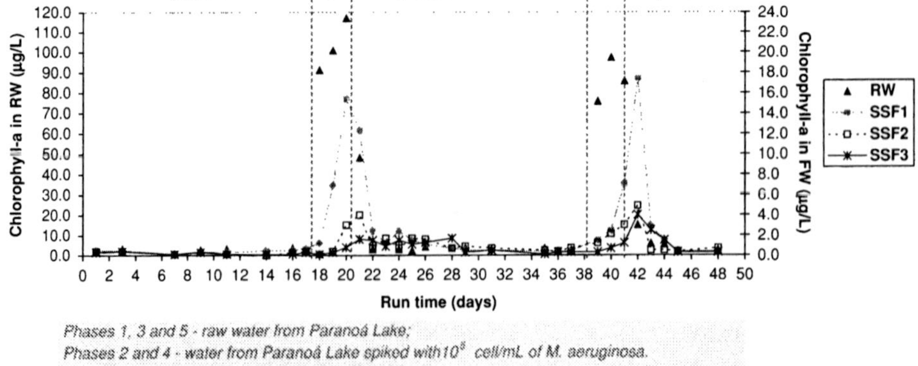

Figure 4 Chlorophyll-a in the raw and filtered water during experimental stage 2

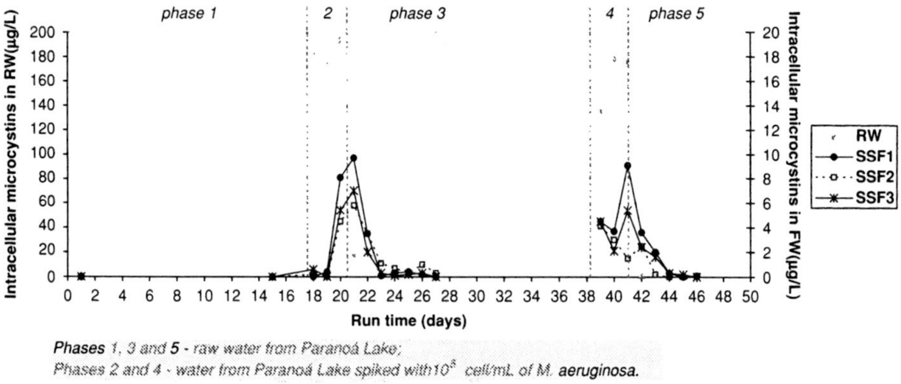

Figure 5 Intracellular microcystins in the raw and filtered water during experimental stage 2

water only about 6 days after ceased the spiking periods. The highest chlorophyll-a and intracellular microscystins concentrations occurred in the last day of spiking or in the first day of monitoring, suggesting that some of the previously retained cells were carried through the filter bed. This phenomenon was more intense in the shallower filter bed (SSF1).

Figures 5 and 6 clearly show that the retained cells were subjected to both carriage through the filter bed and lysis. Nevertheless, the overall removal of microcystins (intra plus extracellular) was high, around 93%, and similar in the 3 SSFs. As extracellular microcystins can be easier, and safer, oxidised than intracellular microcystins, with lower oxidant demand, the authors suggest that deeper beds should to be used to treat water containing *Microcystis aeruginosa*, even because shallower beds poses higher risks of pathogen breakthrough.

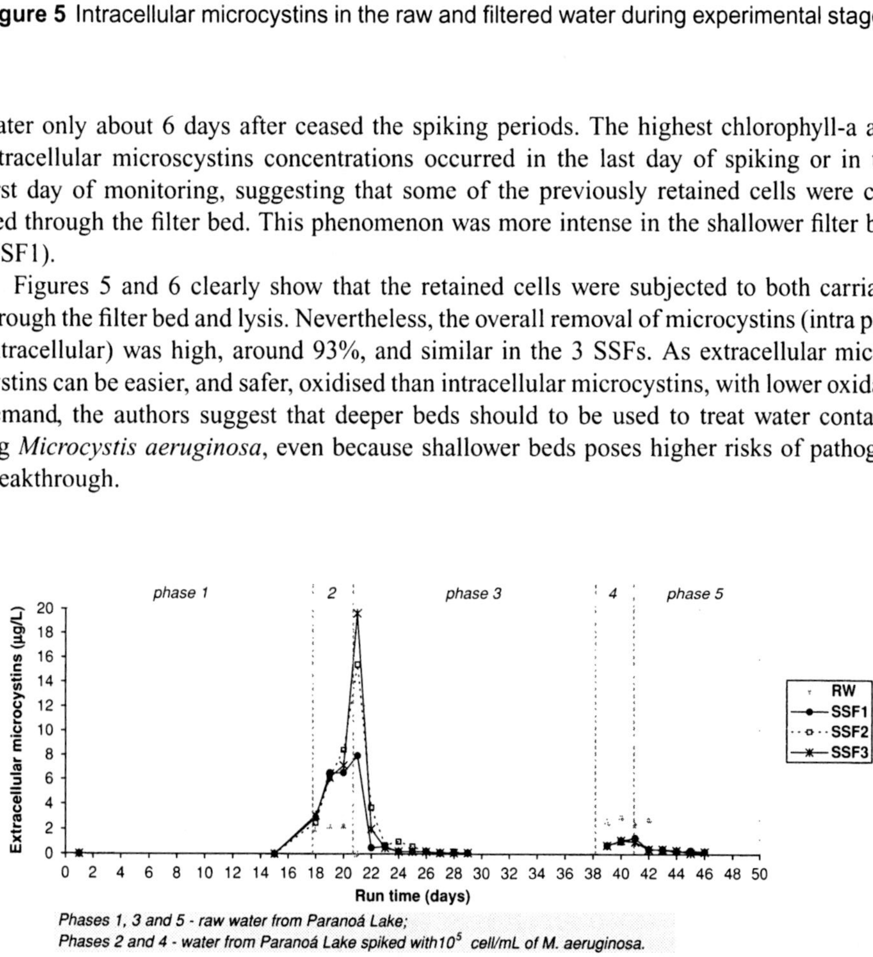

Figure 6 Extracellular microcystins in the raw and filtered water during experimental stage 2

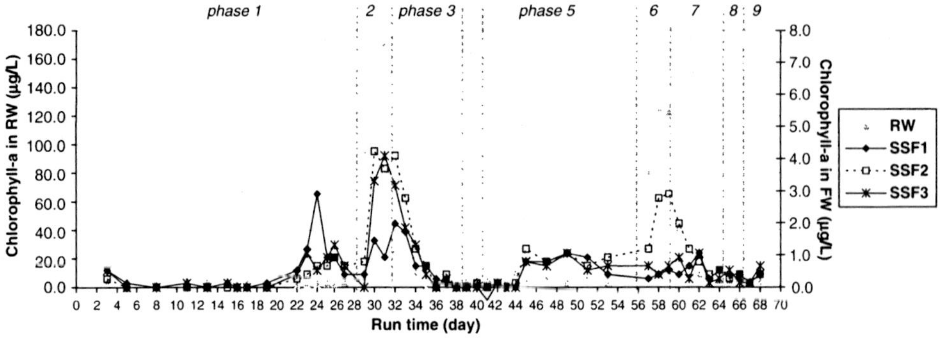

Phases 1, 3, 5, 7 and 9 -raw water from Paranoá Lake; Phases 2 and 6 - water from Paranoá Lake spiked with 10^5cell/mL of M.aeruginosa; Phases 4 and 8 - water from Paranoá Lake spiked with around 20 μg/L of dissolved microcystins.

Figure 7 Chlorophyll-a in the raw and filtered water during experimental stage 3

Experimental stage 3

In stage 3, attention was focused to the influence of filtration rate (SSF1-0.083 m/h; SSF2-0.167 m/h and SSF3-0.125 m/h). All 3 pilot filters had the same bed configuration- sand, 0.9 m depth and effective size of 0.28 mm. Figures 7, and 8 present the behaviour of the raw and filtered water for, respectively, chlorophyll-a and extracellular microcystins.

In Figure 7 it can be seen that the SSF1, operated with the smallest filtration rate, presented the best removal of chlorophyll-a during the spiking periods and following days.

Differently from the behaviour observed in experimental stage 2, after the spiking periods, phases 4 and 8, the concentrations of intracellular microcystins detected in the filtered water (data not shown) was very low, <1 μg/L, indicating that few cells were carried through the filter beds. This probably can be explained by the low concentration of intracellular microcystins in the spiked water (about 35 μg/L), as compared with stage 2.

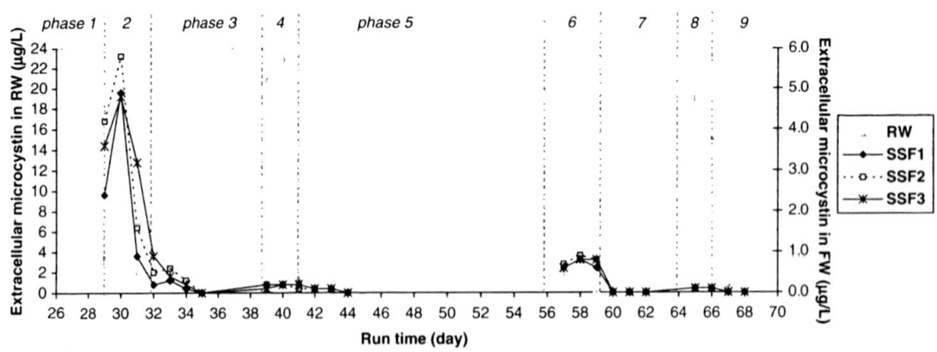

Phases 1, 3, 5, 7 and 9 - water from Paranoá Lake; Phases 2 and 6 - water from Paranoá Lake spiked with 10^5 cell/mL of M.aeruginosa; Phases 4 and 8 - water from Paranoá Lake spiked with around 20 μg/L of dissolved microcystins.

Figure 8 Extracellular microcystins in the raw and filtered water during experimental stage 3

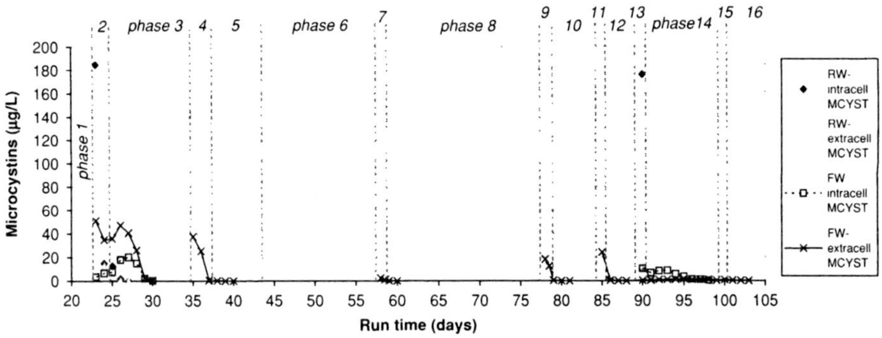

Phases 1, 3, 5, 6, 8, 10,12, 14 and 16 - water from Paranoá Lake; Phases 2 and 13 - water from Paranoá Lake spiked with 10^6 cell/mL of M.aeruginosa (intracellular microcystins) and some amount of dissolved microcystins; Phases 4, 7, 9, 11 and 15 - water from Paranoá Lake spiked with different concentrations of dissolved (extracellular) microcystins.

Figure 9 Microcystins in the raw and filtered water during experimental stage 4

The release of dissolved microcystins due to retained cells lysis in the SSFs was clear in phases 2 and 3 (concentration of dissolved microcystins higher in the filtered water than in affluent water) but was not detected in phases 6 and 7 (Figure 8). This fact may be explained by the degree of maturation of the SSFs after almost 2 months of continuous operation.

When the raw water was spiked with dissolved microcystins during 48 hours (phases 4 and 8) the removal of microcystins was very high with concentrations in the filtered water lower than 0.5 µg/L. The removal of both intra and extracellular microcystins did not seem to be influenced significantly by the filtration rate.

Experimental stage 4

All pilot filters were similar and were operated at the same filtration rate (0.125 m/h) during the fourth experimental stage. Figure 9 presents the mean values of microcystins calculated from the values detected in the effluent water of the three pilot filters operated in parallel.

In Figure 9 it can be observed that in period after the raw water was spiked during 24 hours with approximately 180 µg/L of intracellular microcystins, phases 3 and 14, the filtered water contained significant amounts of both intra and extracellular microcystins. The overall removal, considering both intra and extracellular microcystins, was almost nil during the 23rd and 30th day period, whereas it was about 70% during the 90th and 99th day period. These results highlight the importance of the degree of maturation of the filter bed, as well as its acclimatisation to microcystins, for effective treatment of waters containing *Microcystis* and microcystins, as previously observed by Grutzmacher *et al.* (2002) and Sá (2002).

The removal of the dissolved microcystins when they were directly applied to the SSFs was in the range of 65 to 90%, depending on the degree of maturation of the filter and concentration in the raw water. These values were obtained considering the amount loaded to the filter over spiking period and the amount of dissolved toxins detected in the filtered water during spiking and monitoring periods (until microcystins were not detected in the effluent).

The results obtained suggest that the dissolved microcystins are more easily degraded when it is already dissolved in the raw water than when they are released for the water as a result of retained cells lysing. Dissolved microcystins present in the affluent water tend to have more contact with the *schmutzdecke* and with the first top centimetres of the sand layer, where the biological activity is concentrated. On the other hand, cells of *M. aeruginosa*, due to their shape and size, can be carried to deeper layers of the bed and then lysed as result of hostile environmental conditions. In deeper layers the biological activity is scarce and no appreciable degradation is likely to takes place.

During all experimental stages it was systematically observed that whenever dissolved microcystins were present in the raw water in significant amounts, the removal of coliform bacteria and *E. coli* dropped. Depending on the dissolved microcystins concentration (high concentrations), the removal of coliform bacteria was almost nil. The filters usually recovered their ability to perform high removal of coliform between 1 and 3 days. This phenomenon was already observed by Sá (2002) and it needs to be further investigated.

Final considerations

The results obtained in the four experimental stages indicate that sand effective size and bed depth influence the removal of *Microcystis* and microcystins, but the filtration rate (in the range studied) does not seem to affect significantly the quality of the filtered water regarding the parameters analysed. In this study, the head loss was a decisive factor in the performance of the filters only when the effective size was evaluated. However it is important to bear in mind that the periods of exposure to high concentration of cells in the experiments were relatively short, and not necessarily represents the natural conditions of raw water.

It was observed that high amounts of microcystins (intra e extracellular) can be removed by the slow sand filter, however it largely depends on the degree of maturation of the filter and its acclimatisation to the toxin. These two aspects play the most important roles in the effectiveness of the treatment of water with microcystins by slow sand filtration.

References

Donati, C., Drikas, M., Hayes, R. and Newcombe, G. (1994) Microcystin-LR adsorption by powdered activated carbon. *Water Research* **28**(8), 1735–1742.

Grützmacher, G., Böttcher, G., Chorus, I. and Bartel, H. (2002) Removal of Microcystin by Slow Sand Filtration". *Environmental Toxicology*, **17**, 386–394.

Hart, J., Fawell, J.K. and Croll, B. (1998) The fate of both intra- and extracellular toxins during drinking water treatment. *Water Supply*, **16**(1/2), 611–616.

Keijola, A-M., Himberg, K., Esala, A.L., Sivonen, K. and Hiisvirta, L. (1988). Removal of cyanobacterial toxins in water treatment processes: Laboratory and pilot scale experiments. *Toxicity Assessment: An international Journal*, **3**, 643–656.

Loyd, S.W. and Tucker, C.S. (1988). Comparison of three solvent systems for extraction of chlorophyll a from fish pond phytoplankton communities. *World Aquaculture Society*, USA.

Mouchet, P. and Bonnélye, V. (1998). Solving algae problems: French expertise and world-wide applications. *Journal of Water SRT – Aqua"*, **47**(3); 125–141.

Newcombe, G. and Nicholson, B. (2002) Treatment options for the saxitoxin class of cyanotoxins. *Water Science and Technology: Water Supply*, **2**(5–6), 271–275.
Newcombe, G. and Nicholson, B. (2004) Water treatment options for dissolved cyanotoxins. *Journal of Water Supply: Reseach and Technology – Aqua*, **53**(4), 227–239.
Sá, J.C. (2002). *Remoção de Microcystis aeruginosa e microcistina pelo processo de filtração lenta.* MSc Thesis. University of Brasília, 115 p. Brasil.

22 Behavior of selected drugs during slow sand filtration

B. Kuhlmann*, N. Zullei-Seibert*,, J. Nolte***, M. Grote******

*IfW Institut für Wasserforschung GmbH, Zum Kellerbach 46, D-58239 Schwerte, Germany
**WWU Westfälische Wasser- und Umweltanalytik, Willy-Brandt-Allee 26, D-45891 Gelsenkirchen, Germany
***ISAS Institute for Analytical Sciences, Bunsen-Kirchhoff-Str. 11, D-44139 Dortmund, Germany
****Universität Paderborn, Fakultät für Naturwissenschaften, Department Chemie, Warburger Str. 100, D-33098 Paderborn, Germany

Abstract Fate and behavior of selected drugs were investigated in laboratory filter columns simulating slow sand filtration. During several weeks ibuprofen, carbamazepine, sulfamethoxazol and diclofenac were added continuously to the influent of filter columns filled with sand. The filters were operated with natural ground or surface water. The concentration of the drugs was regularly analysed in the filter effluent. No degradation could be observed in the filter operated with groundwater. In the filters run with surface water from the river Ruhr, only Ibuprofen was metabolized, but degradation did only take place when no other carbon source was available. Even when the filters were operated for up to 9 weeks to give more time for microbial adaptation processes, no removal of the selected drugs was detected. Degradation of ibuprofen, sulfamethoxal and diclofenac was detected in a filter column run as a closed cycle resulting in longer contact times and a higher concentration of the test substances.

Keywords Slow sand filtration, drugs, microbial degradation

Introduction

Pharmaceutical products are used world wide in a great amount and enter the environment through different pathways. Surplus drugs are often directly disposed to sewage, drugs used by humans are partly discharged to sewage treatment plants with urine and feces. It is well known that only a small part of the medicinal substances that reach sewage treatment plants in unchanged form or as metabolites is eliminated by physical, chemical or microbiological processes. For that reason the main pathway of drugs into surface waters is via the effluents of sewage treatment plants. (Halling-Sörensen et al. 1998, Kümmerer 2004, Klopp 2005)

When surface water - contaminated with drug residues - is used as raw water for drinking water production, the drugs or their metabolites may reach the drinking water. Looking at slow sand filtration as a typical surface water treatment technique, its effectiveness for

 Recent Progress in Slow Sand and Alternative Biofiltration Processes edited by Rolf Gimbel, Nigel J.D. Graham and M. Robin Collins. ISBN: 9781843391203. Published by IWA Publishing, London, UK.

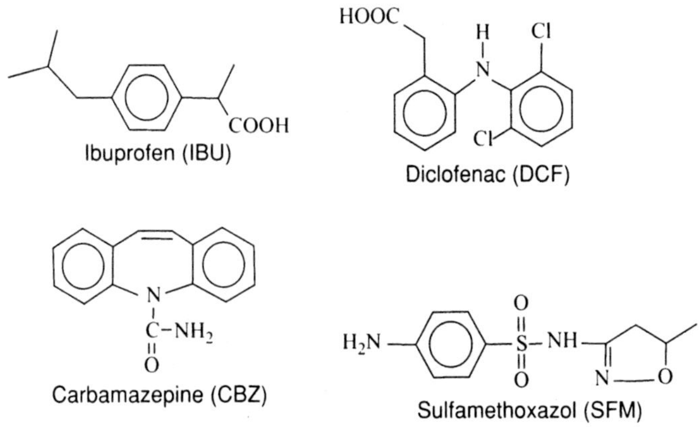

Figure 1 Chemical structures of the selected drugs.

removal of drugs is therefore of important concern. Fate and behavior of selected pharmaceuticals in slow sand filters is determined mainly by sorption and degradation processes, that depend on several factors like temperature, pH, organic carbon content or redox situation. Scope of the work presented here is to investigate the behavior of selected pharmaceutical compounds under defined conditions.

As test substances carbamazepine, sulfamethoxazol, diclofenac and ibuprofen were chosen. They belong to different chemical classes of active drugs and represent important groups of pharmaceutical applications. Their chemical structures are shown in figure 1.

Ibuprofen and diclofenac are analgetic agents, carbamazepine is used against epilepsy and sulfamethoxazol is an antibiotic. All four compounds are produced and used in great amounts and are frequently found in surface waters (Stumpf et al. 1996, Heberer und Stan.1998, Ternes 2001,). Monitoring activities of the water works show that the four test substances are regularly detected in the river Rhein (ARW 2004).

Methods

To investigate the behavior of the selected drugs during slow sand filtration under defined conditions, a model system was used. It consisted of glass columns filled with sand and was operated with a natural groundwater or surface water from the river Ruhr (figure 2). Table 1 gives some technical details.

As in practice of slow sand filtration in the central Ruhr area, the surface water was first filtered through a gravel filter.

Table 1 Technical details of the model system

Material of the filter columns:	Glass
Hight:	100 cm
Internal diameter:	22 cm
Filter hight:	80 cm
Flow velocity:	5 cm/h
Direction of flow	Bottom to top

Figure 2 Model system for slow sand filtration

During an adaptation period the sand filters were only run with groundwater or gravel filtered surface water. Then the test substances were continuously dosed to the filter influents for 4 weeks in the first and 9 weeks in the second experiment. Spiked concentration was 100 μg/L for each drug.

Another filter column was operated as a closed cycle. During the first 9 days about 5 mg of each test substance were added, during the following 13 days additionally 0.9 mg/day. The experiment was run for 9 weeks althogether.

Samples were taken regularly from the filter effluents and analyzed for the test substances. Drugs were determined by HPLC-UV, LC-MS or GC-MS (Baitaneh et al. 2006). For chemical characterization, pH, conductivity, temperature, oxygen, DOC, nitrate, nitrite, ammonia, chloride and sulfate were regularly controlled in the filter effluent. Colony forming units (CFU) and esterase activity were determined to investigate number and activity of microrganisms and PCR/DGGE was used to look at microbial diversity.

Results and dicussion

During the first tests, no degradation of carbamazepine, sulfamethoxazol and diclofenac could be observed. In the filter operated with surface water, ibuprofen was degraded after a lag period of about 2 weeks (figure 3).

In the filter run with groundwater, no removal of the test substances could be observed. This corresponds to the lower microbial number and activity found in the groundwater.

To enhance microbial adaptation processes, in another experiment a filter run with surface water was continuously dosed with the test substances for nine weeks. The other experimental conditions were the same as in the first test. No degradation could be observed within nine weeks for all pharmaceuticals applied.

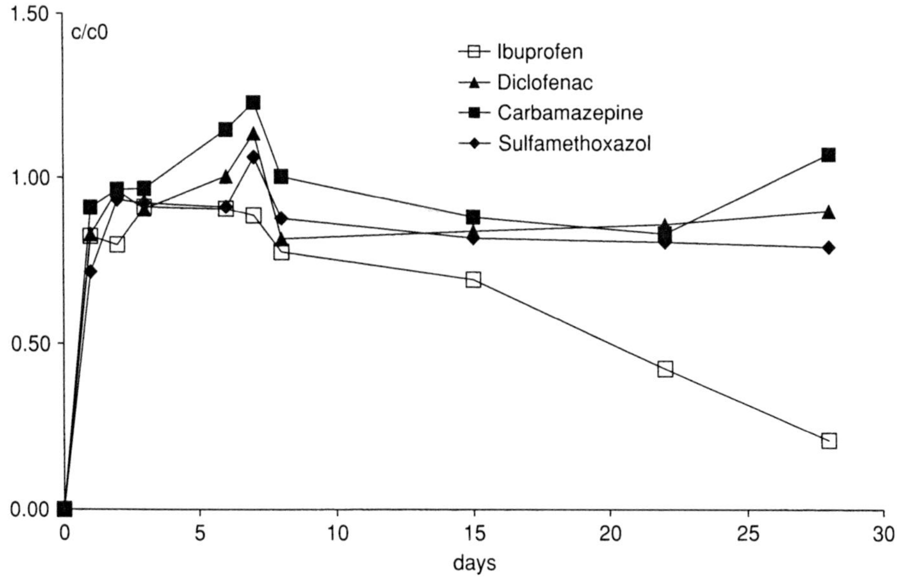

Figure 3 Relative concentrations of test compounds in the filter operated with surface water as a function of time

These results seem to be contradictory to the data of the first filter run. However, they are conclusive when considering certain other parameters measured in the filter effluent, indicating the reduction of nitrate accompanied by a formation of nitrite and ammonia. Evidence is given that in the filter an unknown organic substance was favorable degraded and the tested drugs were not co-metabolized.

In the filter operated as a closed cycle, removal of ibuprofen, sulfamethoxazol and diclofenac started after 7 weeks, sulfamethoxazol was degraded even below the detection limit (figure 4). Carbamazepine remained persistent.

The test substances are known to be not or only partly degraded in sewage treatment systems and sorption does not play a role in their removal (Ternes et al 2004). This is stressed by the various positive findings in surface waters. Our investigations show that under normal conditions of slow sand filtration, the concentration of the test substances is not decreased. With the exception of carbamazepine which seems to be very persistent, a removal of ibuprofen, sulfamethoxazol and diclofenac is possible, but depends on certain local environmental conditions. Only high contact times and high substrate concentrations in the experiment with the column operated as a closed cycle did lead to a reduction of these three compounds. Long adaptation periods and a lack of other carbon sources in the flow-through filter columns resulted only in degradation of ibuprofen.

Conclusions

Contamination with pharmaceuticals poses a severe problem for drinking water production form surface water. Findings of pharmaceuticals in groundwater that is influenced

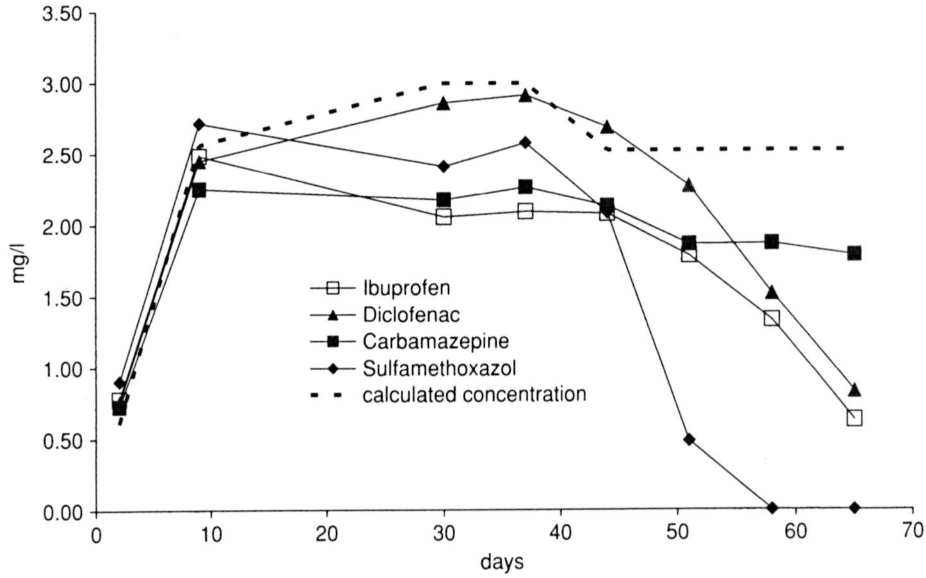

Figure 4 Concentrations of test compounds in the filter operated as a closed cycle

by polluted surface water stress that natural processes are often not able to reduce these compounds (Heberer, et al. 1998).

Under the typical conditions found in the practice of slow sand filtration, it can be assumed that carbamazepine, ibuprofen, sulfamethoxazol and diclofenac are almost persistent and not reduced by sorption or microbial degradation, the main processes to remove organic pollutants during slow sand filtration.

To decrease the risk of the occurrence of drugs in drinking water, more sophisticated and therefore as well more expensive water treatment techniques like ozonization and active carbon filtration may be used (Ternes et al 2004). But it seems to make more sense to reduce the entry of drugs into surface water by optimizing their reduction in the sewage plants or - even more efficient - to reduce the sewage burden by alternative measures.

Acknowledgements

The authors gratefully acknowledge financial support by the MUNLV NRW.

References

ARW (Arbeitsgemeinschaft der Rheinwasserwerke) (2004) 60. Bericht der ARW. Jahresbericht 2003. Eigenverlag, Köln

Bataineh, M., Nolte, J. Kuhlmann, B., Zullei-Seibert, N. and Grote, M. (2006) Degradation behavior of selected pharmaceuticals and their main metabolites in model systems. Part I: Development of the Analytical Procedures. In press

Halling-Sörensen, B., Nielsen, S.N., Lanzky, P.F., Ingerlev, F., Lützhoft, H.C.H. Jorgensen, S.E. (1998) Occurence, Fate and Effects of Pharmaceutical Substances in the Environment - A review. *Chemosphere* **36**(2), 357–393.

Heberer, T., and Stan, H.J. (1998) Arzneimittelrückstände im aquatischen System. *Wasser+Boden*, **50**, 20–25.

Heberer, T., Schmidt-Bäumler, K. and Stan H.-J. (1998) Occurence and Distribution of Organic Contaminants in the Aquatic System in Berlin. Part I: Drug Residues and other Polar Contaminants in Berlin Surface and Groundwater: *Acta hydrochim.hydrobiol.*, **26**, 272–278.

Klopp, R. (2005) Gefährliche Stoffe in kommunalen Kläranlagen. *GWA*, **200**, 6/1–6/15

Kümmerer, K. (2004) Herkunft, Vorkommen, Verhalten und Verbleib von Arzneimitteln in der aquatischen Umwelt. *GWA*, **23**, 1–11

Stumpf, M, Ternes, T.A., Haberer, K., Seel, P. and Baumann, W. (1996) Nachweis von Arzneimittelrückständen in Kläranlagen und Fliessgewässern. *Vom Wasser* **86**, 291–303.

Ternes, T (2001) Vorkommen von Pharmaka in Gewässern. *Wasser+Boden*,. **53**, 9–14.

Ternes, T.; Janex-Habibi, M.-L.; Knacker, T.; Kreuzinger, N. and Siegrist, H. (2004) Assessment of Technologies for the Removal of Pharmaceuticals and Personal Care Products in Sewage and Drinking Water Facilities to Improve the Indirect Potable Water Reuse. POSEIDON Project Report EU Contract No. EVKI-CT-2000-00047

23 Down-scaled study of slow sand filtration of secondary effluents

Gilad Ben Ari* and Avner Adin**

*Environmental Studies Program, The Hebrew University of Jerusalem, Israel. gil-b-a@012.net.il.
**Soil and Water Sciences Dept., The Hebrew University of Jerusalem, Israel. adin@vms.huji.ac.il. To whom correspondence should be addressed.

Abstract Slow sand filtration (SSF) can serve as a tertiary wastewater treatment process particularly but not only in developing countries. Research objectives were to characterize the biological activity level during the process, to study the effect of concentrated organic matter (OM) on the filtration process and to construct, operate and examine experimental units of small, unorthodox scale. Results show that he biological removal is responsible for one third to half of the total contaminant removal performed by SSF of secondary effluents. Chlorination as a pre-treatment to SSF increases the filtration efficacy by promoting flocculation high OM concentrations cause a considerable reduction in filter run providing poor quality filtrate. The resulting increased biological activity did not reduce the total quantity of organic mater that entered the sand segment of the filter, while larger quantities of microorganisms passed through, thus replacing OM with larger biomass. Frequently larger quantities of OM passed through the filter than originated in the wastewater supply. The filtration units used in this work were unusual in their small sizes. The reliability of the system was proven by the correlation between the results obtained during this research and to the results from previous experiments using considerably larger units.

Keywords slow sand filtration, tertiary treatment, biological filtration, wastewater reuse.

Introduction

The principles of slow sand filtration are not new and it has been in use for a number of years for treating potable water. The treatment is based on a slow flow of fluids through a porous sand media. The filtration unit consists of a column of water followed by a column of sand. The filtration removes contaminating particles; organic and inorganic materials and microorganisms. Part of the removal is performed by physiochemical mechanisms (O'Melia & Stumm, 1967) causing the particles to be adsorbed to the grains of the filtration media. Additional contaminant particles are removed by microorganisms developing in the filter's water column and in the filtration media (mainly on the top five centimetres of the media column (Hendel *et al*, 2001). The microorganisms use the decomposable organic material in the filter water as a food source. There are two stages in the filtration cycle: The first stage,

 Recent Progress in Slow Sand and Alternative Biofiltration Processes edited by Rolf Gimbel, Nigel J.D. Graham and M. Robin Collins. ISBN: 9781843391203. Published by IWA Publishing, London, UK.

known as the "filter ripening", is characterized by a lower quality filtrate than desired. The quality improves as the filtration media absorbs particles and the microorganism population in the filter grows. The ripening stage is relatively long, during which the quality of treated water is below the required levels (Letterman, 1987). The second stage of the filtration cycle is known as the "work stage" during which the filter provides the desired quality filtrate. The stage ends when the filter clogs up. In this research, slow sand filtration of secondary effluents from an activated sludge system was studied. The goal of the tertiary filtration using slow sand filtration is to receive effluents in an adequate quality, sufficient for agriculture irrigation and other utilities, with out damaging public health and the environment. The prime advantages of this treatment are low cost and the simplicity of the operation. Those advantages, with the demand for large operational area, make the SSF particularly attractive for developing countries and small communities (Adin, 2003).

Research objectives were threefold: (a) to characterize the level of biological activity during the filtration process, (b) to study the effect of wastewater with a high level of OM on the filtration process in order to improve the process, the ripening stage in particular, and (c) to construct, operate and examine experimental units of small, unorthodox scales.

Methods

Secondary effluent samples were brought to the laboratory from the Shafdan (Tel-Aviv) activated sludge plant every 4 days and stored in 4°C refrigerator. Before use, the effluents were brought to room temperature. The filtration system is presented in Fig. 1. Column filters, 0.9 m or 0.6 m long high with 30 mm internal diameter and bed depth of 0.6 m or 0.2 m were used. Effective grain size was 0.64 and uniformity coefficient was 1.5 and sand porosity was 0.42. Filtration rate was 0.15 m/h. Head loss was determined by piezometers.

The work consisted of four main experiments. The first experiment was conducted using two filters, examining the filtration of dissolved matter either from primary or secondary wastewater effluents. The second experiment was conducted with secondary wastewater using two filters. A chlorine solution was added to one of the filters to compare the filtration of chlorinated wastewater in which there was no biological activity. The second filter was fed with non-chlorinated secondary wastewater. The third experiment consisted of two separate filters. Conditions of high levels of OM were created in the supply tank of one of the filters. The second filter was supplied with usual secondary wastewater. In the fourth experiment, the two filters were supplied from the same tank. The supply pipes of both filters had a sugar solution injected thus creating conditions of high OM levels. Feeding and sampling systems: Two 10 litre tanks were placed 1.5 m above the filters. The effluents flow was gravity driven to the filters. Sampling valves were drilled into the filters 7cm and 55 cm below the sand surface. The sampling valves penetrate approximately 1 cm into the filter bed. At the height of 40 cm, a hose was installed in order to drain excessive effluents. A sampling valve was placed in the base of each filter. A second valve was connected to a flowmeter. A third valve was connected to a piezometer to measure head loss. Particle size distribution was determined using a laser diffraction analyser.

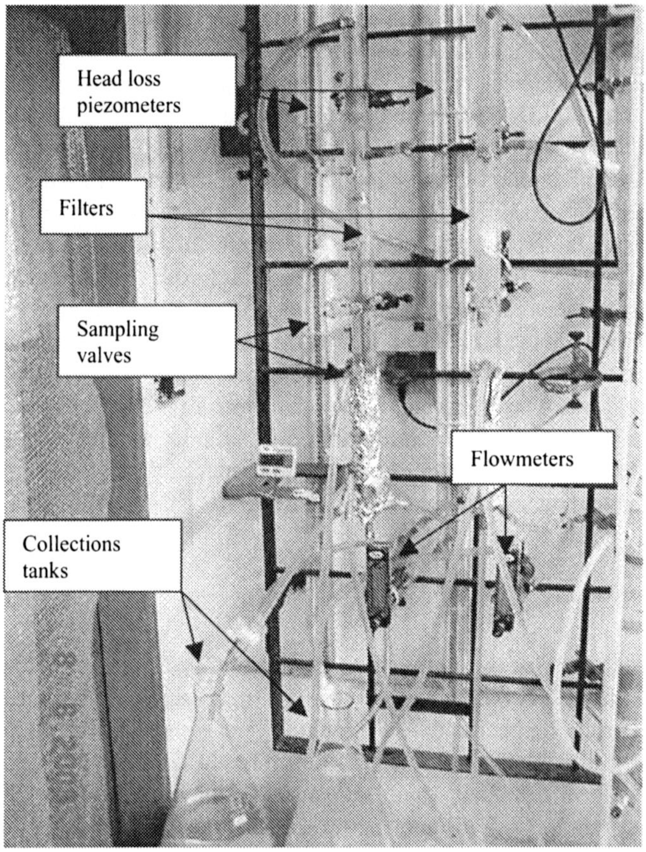

Figure 1 Experimental system setup

Results and discussion

In Experiment 1 the filters were fed with wastewater that had been pre-treated with a 1.2 micron GFC filter. One filter was supplied with primary wastewater and the other with secondary wastewater. A significant difference was found between the filters; however neither followed the common filtration cycle. The maximum turbidity removal of the filter supplied with primary wastewater was 30%. After the 20th day the removal rate gradually decreased until it reached 5% by the end of the experiment. The average turbidity reduction was 15%. The instability of the system as seen in Fig. 2 indicates that the biological layer (Schmutzdecke) development in the filter media was sporadic. We presume that the pre-treatment considerably reduced the available nutrients and biomass in the inlet effluent therefore disturbing the microbial development in the media. The experiment demonstrated the combined effect of both reduced biological activity and the diminished physiochemical mechanisms on the filtration efficiency. The filtration coefficient λ (Fig. 3) was a magnitude lower than expected in efficient treatments (Gelman, 1997). These results can be compared to Gelman's (1997) in which the average turbidity removal was 50%. Our result of 27% removal

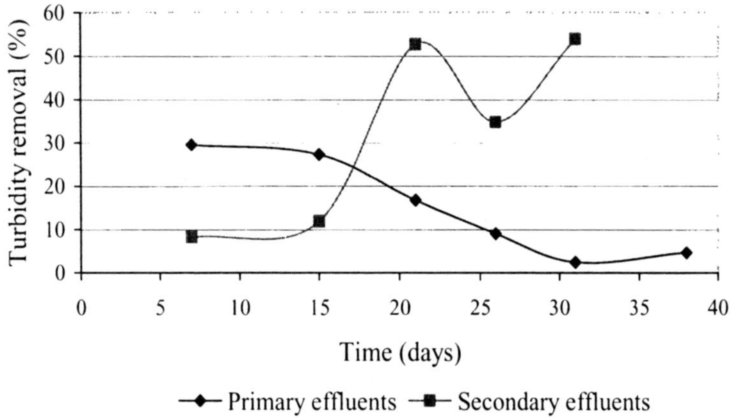

Figure 2 Turbidity removal during experiment 1.

using secondary wastewater indicates that approximately half of the removal achieved during SSF is due to biological activity. Comparing the results to primary wastewater indicates less than a third of the removal was due to biological activity.

Experiment 2 had to be terminated after 32 days, before completing the full filtration cycle due to the destruction of the laboratory during a snow storm.

The average turbidity removal in the non-chlorinated filter was 63% similar to the results obtained by Ellis (1987). The filter fed with chlorinated water was considerably more effective with an average turbidity removal of 74% (Fig. 4). This result is consistent with the findings of Farook *et al* (1993) after filtering chlorinated inlet effluent achieved removal in the region of 90%. The improved efficacy of the chlorinated wastewater filtration can be explained if flocculent properties are attributed to the chlorine. The oxidation of the OM in the inlet effluent by the chlorine destabilises the particles allowing contact between particles and the creation of larger flocks that settle and are then subsequently trapped in the filter media (Rav Acha *et al*, 1995). Smaller particles have larger relative surface areas

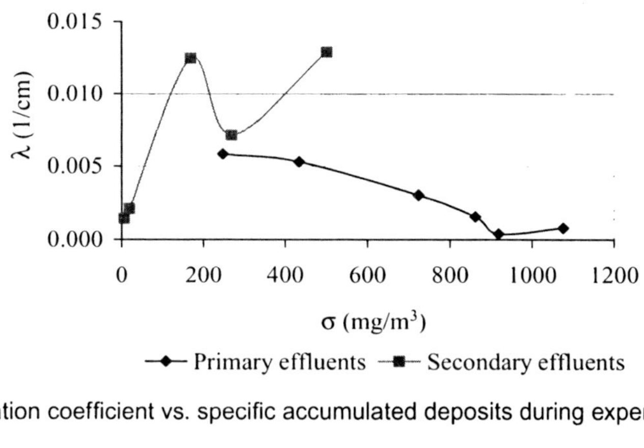

Figure 3 Filtration coefficient vs. specific accumulated deposits during experiment 1

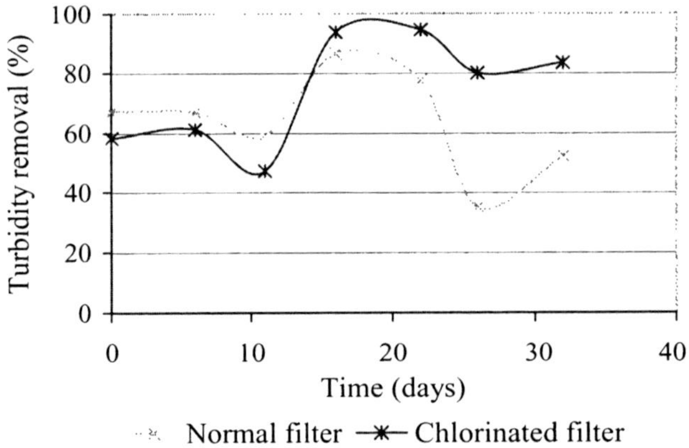

Figure 4 Change in turbidity removal over time in experiment 2

and therefore the destabilisation is more severe, possibly explaining the enhanced removal of the smaller particles. Due to the increased coagulation potential, the dominant removal mechanism is particle-particle interactions and to a lesser degree surface filtration. A jar test was conducted to compliment this experiment and confirmed that chlorine concentrations of 20 mg/l have a flocculative effect on secondary wastewater. The hypothesis that experiment 2 initially intended to examine was that chlorination would be significantly detrimental to turbidity removal. In practice, the opposite was established; the chlorination increased the turbidity removal. We propose that a trade off occurred between the positive flocculative effect of the chlorine and the detrimental decrease of biological activity in the filter system. An alternative explanation is that the biological activity in the water column and on the surface of the filter simply replaces inlet effluent biomass with live biomass. Therefore the disinfection has no direct effect on the filtration as the inactive biomass has the same filtration effect. The flocculative effect of the chlorine therefore increases the filter efficacy.

Figure 5 describes the change in the bulk specific deposit (σ) in both filters, and is seen to be almost identical. The rate of change of the bulk specific deposit can be compared to the model suggested by Ojha & Graham (1994). Allowing for a number of basic assumptions, the model proposes that the rate is time constant. The high correlation between the experimental results and the linear prediction confirms that the model is appropriate for this experiment and indeed that rate of change of the bulk specific deposit is constant.

Figure 6 describes the removal of all particles between 2–80 μm in experiment 2. The non-chlorinated filter removed 77% of the particles, the chlorinated filter 84%. The non-chlorinated filter did not exhibit any change in trend over time. The chlorinated filter showed ripening from the 16th day seen in the increased average turbidity removal of 95%. This change can be seen in the two diagrams describing the particle size distribution, from the first day of the experiment (Fig. 7) and the 22nd day (Fig. 8). Figure 9 demonstrates the high correlation (R^2=0.89) between turbidity removal and particle removal in chlorinated filter.

TOC removal values (Fig. 10) of the chlorinated filter (average 56%) were significantly higher than the values of the normal filter. The normal filter behaved as expected of a SSF. The chlorinated filter revealed the dual action of the chlorine; direct breakdown of the OM

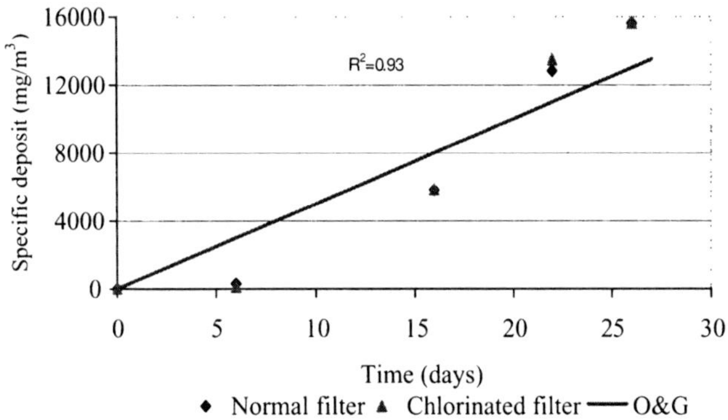

Figure 5 Change in bulk specific deposit in experiment 2 and comparison to Ojha & Graham model.

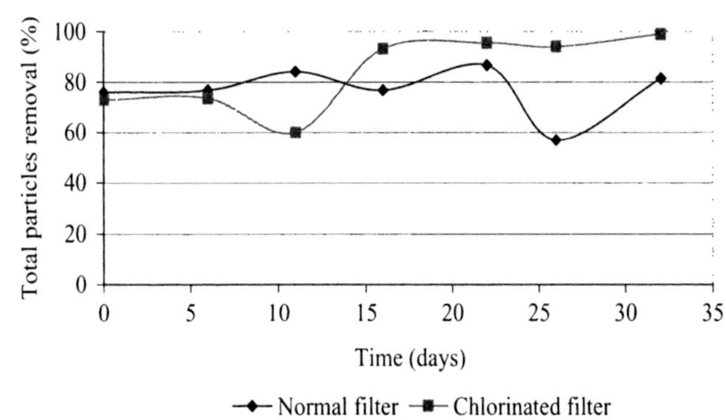

Figure 6 Change in particle removal of all particles between 2–80 μm in experiment 2.

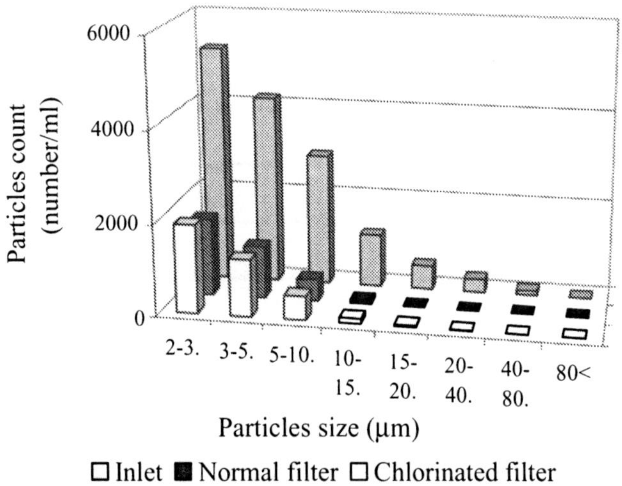

Figure 7 Particle size distribution of all particles between 2–80 μm in experiment 2, day 0

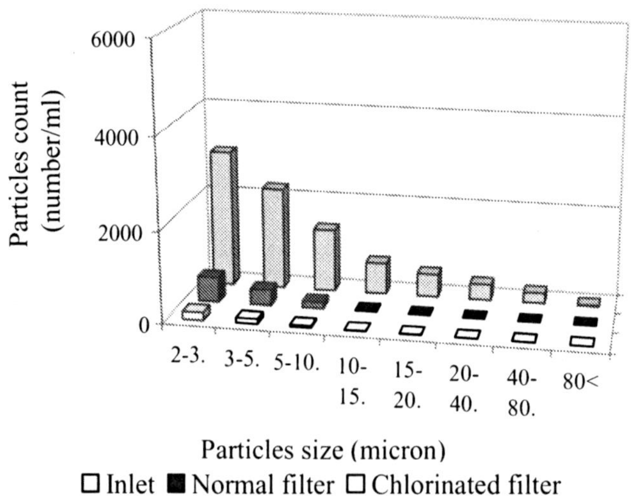

Figure 8 Particle size distribution of all particles between 2–80 μm in experiment 2, day 22

and the prevention of the biomass accumulation in the filter water column. On the 26th day a TOC breakthrough occurred in the normal filter. This breakthrough is not equivalent to the turbidity breakthrough (due to shear forces exceeding adhesive forces) found in rapid sand filter cycles, but an indication of the instability of the filter due to the lack of ripening.

In Experiment 3, the feed tank of one filter system was loaded with a sugar solution in addition to the secondary wastewater, creating high organic pressure. Due to the intense microbial activity, anaerobic conditions were rapidly achieved. The feed tank of the control system was supplied with secondary wastewater. No head loss was noted in the control system during the ten days of the experiment. This was expected as head loss usually occurs

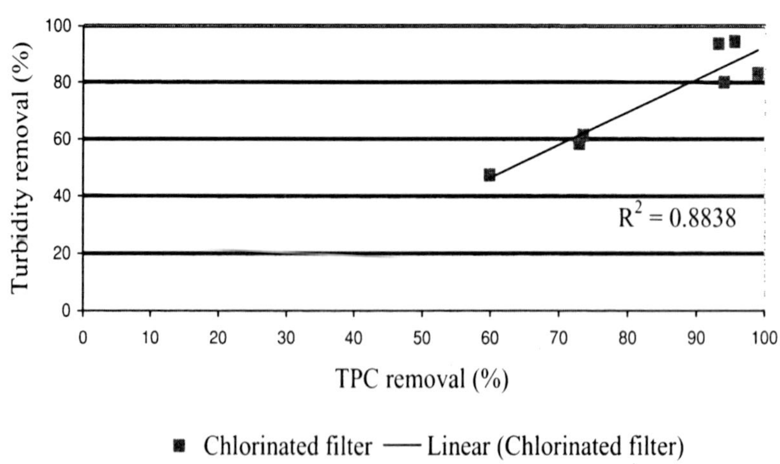

Figure 9 Turbidity removal vs. total particle removal in experiment 2

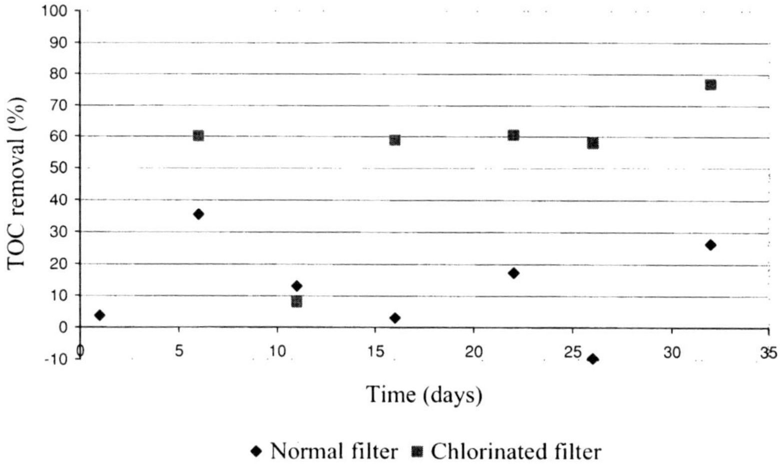

Figure 10 Percentage of TOC removal in filters during experiment 2 (outlier points removed)

after 30 days. The filter with high organic load developed exponential head loss indicating surface filtration. High correlation (Fig. 11) can be found with the surface filtration head loss model developed by Boucher (1947). This correlation supports the hypothesis that the dominant mechanism was surface filtration and not deep filtration.

For turbidity removal follow up in Experiment 4, both filters were fed with the same effluents. The organic load was about 200 BOD. The inlet turbidity was 3.5 NTU. The average turbidity in the filters water column (10 cm above the bed surface) was 12.4 NTU in filter 1, and 7.2 NTU in filter 2. The average turbidity in filter 1 outlet was 5 NTU, and 7.2 NTU in filter 2. It is a notable fact that the turbidity in the water fraction of the filter, was higher (except day 6 in filter 1) than the turbidity in the inlet effluent, occasionally 3–13 times higher (Fig. 12).

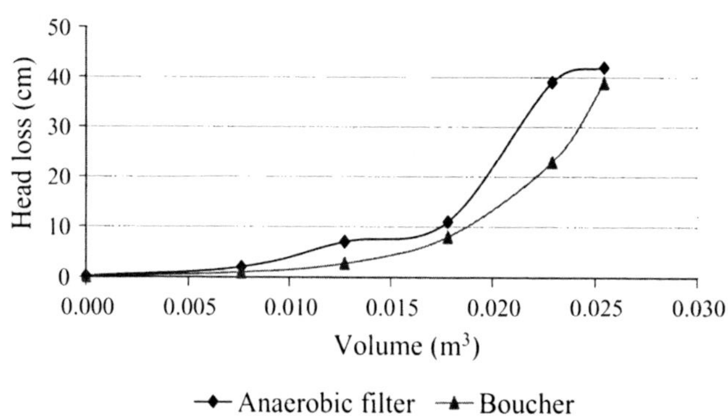

Figure 11 Development of head loss vs. both volume of inlet effluent in the filter with high organic pressure and according to the model of Boucher (1947): $y = 0.2e^{207x}$

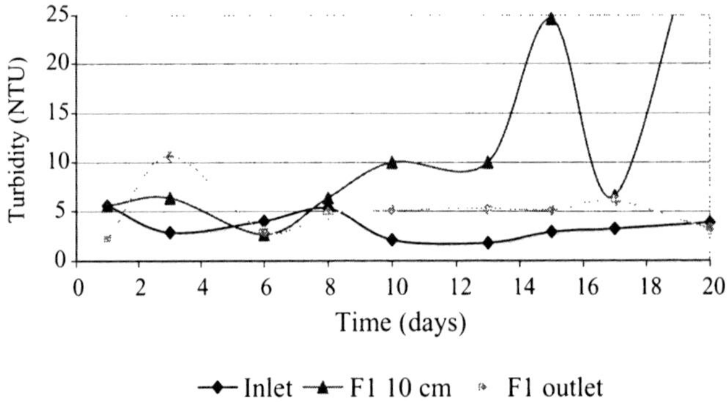

Figure 12 Change in turbidity over time in column inlet effluent, water column and filtrate. Filter 1, experiment 4

Those high values probably reflect an intensive microbiological activity that flourish because of the large amount of organic substance (200 BOD), generally soluble (The organic loud was made of a sugar solution), and from addition of dissolved oxygen from the effluents dripping to the filter. In fact, the filter bed filtered a much higher levels of turbidity than the levels that were originally measured in the inlet effluents.

TOC removal with time (by percent), from filter 1 water column and outlet, can be seen in Fig. 13. The contrast between filter 1 water column removal curve and the filter outlet removal curve, demonstrates the difference in the removal processes in the two parts of the filter. In the filter water column, the concentration of organic matter changed dramatically: in filter 1 TOC removal in days 3, 8, 20 was 80%– 90%, in day 6 an increase of 10% TOC, and a 40% increase in day 10. In filter 2 the pattern was similar. The dramatic changes are most likely due to microbiological activities as mentioned above. A restrained outcome of those phenomena was sometimes seen in the outlet of the filters (day 10 filter 1) as well as additional reduction of the organic matter. In the first three days of the experiment, in

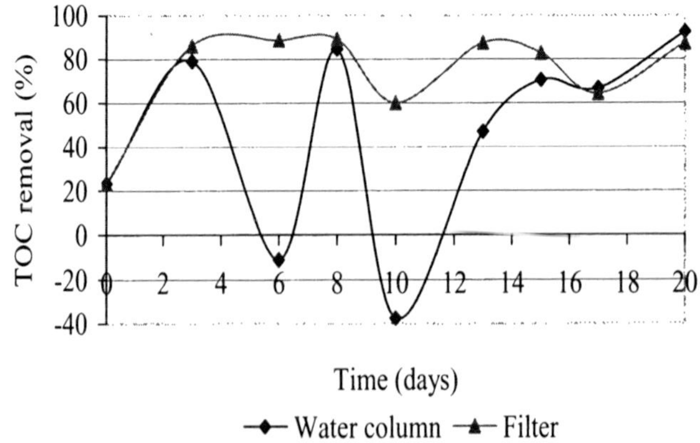

Figure 13 TOC removal over time in water and filtrate. Filter 1, experiment 4

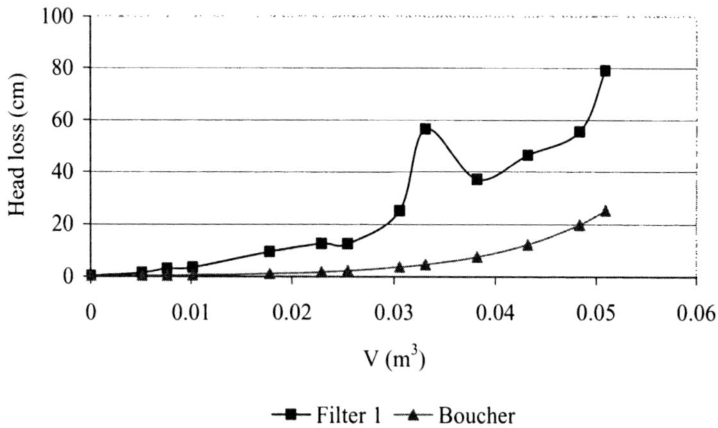

Figure 14 Change in head loss vs. inlet volume. Filter 1, experiment 4. experiments. Comparison to model of Boucher: $y = 0.2e^{95x}$

day 8 and in the last five days, similar TOC removal levels were detected in the filters' water columns and in the filters' outlets. During those days, TOC removal in the filter's water columns was satisfactory. During the other days, in which TOC removal from the filter's water columns was ineffective, the filter's bed removed 80%–90% of the TOC. Our conclusion is that the physiochemical removal mechanisms in the filters' beds were effective even if there wasn't TOC removal in the filters' water columns. Both TOC removal and turbidity, demonstrated that in the case of effluents with high organic levels, the biological activities in the water column of the filter didn't contribute to the quality of the filtrate, and sometimes filter bed faced higher levels of TOC than the levels that were first measured in the inlet effluents. The biological activities changed the characteristics of the organic matter from dissolved to colloidal, increasing the removal efficiency of the filter bed. A different observation was seen in experiment 1, in which dissolved matter wasn't efficiently removed, probably because the biological growth was not developed properly.

The head loss measured in the first filter developed exponentially. The surface filtration head loss model proposed by Boucher (1947) also describes exponential head loss (Fig. 14), however the results in our experiment did not completely overlap the predicted values.

Literature dealing with SSF and specifically with SSF of wastewater effluents cites head loss development according to the surface filtration model (Gelman, 1997 and more). These investigations indicated that the almost all the head loss build up was at the top 5 cm of the sand media. In the first filter, positive correlation was found between bulk specific deposit and the development of head loss. A similar relationship was found between the TOC removal and the head-loss development.

Comparison of bulk specific deposit (BSD) in the various experiments

Comparing the BSD in Experiments 1, 2 and 4 indicates three trends (Fig. 15). In experiment 1, there was very little BSD. The pre-treatment of the wastewater in this experiment removed

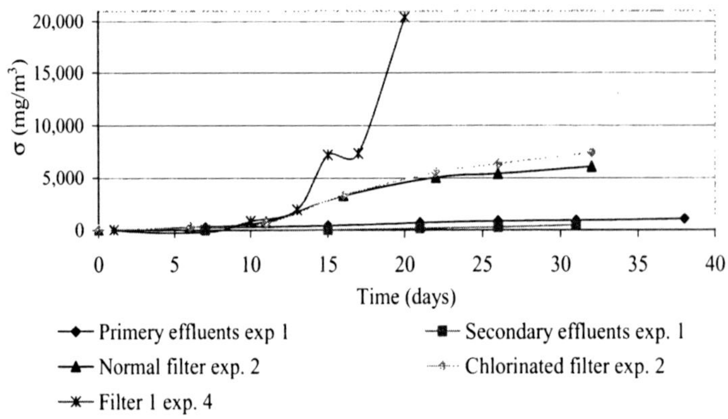

Figure 15 Comparison between all four effluent Time vs. bulk specific deposit

all particles over 1.2 μm. This pre-treatment reduced the efficiency of the physiochemical filtration mechanisms. Furthermore, the removal of nutrients and microorganisms inhibited the development of the schmutzdecke thus reducing the biological filtration. In experiment 2, the BSD accumulation was similar in both filters. This finding can possibly be explained by the trade off between the beneficial flocculative behaviour of the chlorine and the detrimental reduction of microbial development in the filter water column and media. In experiment 4 to which dissolved organic matter had been loaded to a level of BOD 200, a mirror image of Experiment 1 was found. Drastically increased microbial growth predominantly in the filter water column produced a large number of particles, partly filtered by the filter media.

Conclusions

(1) Quantification of the biological factor in SSF is important. Isolation of the biological factor was achieved by removing the suspended particles from the inlet effluent. Comparison to typical literature values for SSF of secondary effluents revealed that the biological removal accounts for half to a third of the total particle removal.

(2) SSF ripening time can be shortened by adding organic pressure to the inlet effluent during the first days of the filtration cycle.

(3) Chlorination as a pre-treatment to SSF increases the filtration efficiency. Chlorine's flocculative characteristics were proven. Wastewater treatment plants using SSF should consider pre-oxidation in addition to disinfection at the end of the filtration process.

(4) The increased microbial activity in the filter water column, due to the loading of high values of dissolved organic matter directly into the filter inlet, did not cause a reduction in the quantity of OM in the filter water column. However, the microbial activity changed the nature of the OM from dissolved to particulate thus allowing improved removal by the filter media.

(5) The head loss measured in experiments 3 & 4 developed exponentially and agreed with the surface filtration head loss model proposed by Boucher (1947). In Experiment 4 a correlation was found between the bulk specific deposit and the head loss

build up. A similar relation was established between the TOC removal and the head loss development.

(6) In experiments 2 & 4, the bulk specific deposit was compared to the model of Ojha & Graham (1994). In experiment 2, high correlation was found between the model and the bulk specific deposit in both filters. The bulk specific deposit change rate was found to be time constant. In experiment 4, a 70% correlation was found between the bulk specific deposit and the predicted results according to the model.

(7) The experimental system was found to be stable and reliable. The system's advantages are its simple construction, ability to locate in a laboratory, low maintenance, low costs, and the small quantities of materials required (sand, effluents and additives).

Acknowledgment

This paper is part of M.Sc. requirements of Gilad Ben-Ari. USAID for their research funding, the family of Ian Karten for their generous scholarship, Ms. Rivka Kalbo and Ms. Luba Rubenstein of the Water Treatment Technology Laboratory, the Israeli Ministry of Environment for its assistance and Dr. Cheikh Diop from Dakar University, Senegal, for his cooperation should be greatly acknowledged.

References

Adin, A. (2003). Slow Granular Filtration for Water Reuse, *Water Supply*, **3** (4), 123–130.

Boucher, P.L., (1947). A new measure of the filterability of fluids with applications to water engineering. *Jour. Inst. Civil Eng.*, **27**, 415–421.

Ellis, K.V. (1987). Slow sand filtration as technique for the tertiary treatment of municipal sewages. *Wat. Res.*, **21**, 403–410.

Farook, S. & Yousef, A.K., (1993). Slow Sand Filtration of secondary effluent. *Jour. of Env. Eng.*, **19**(4), 615–630-I.

Gelman, A. (1997) *Slow Granular Filtration as a pre-treatment to secondary wastewater*. Master's Thesis. The Hebrew University of Jerusalem, Jerusalem, Israel. In Hebrew.

Hendel, B., Marxsen, J., Fiebig, D., & Preu, G. (2001) Extracellular Enzyme Activities During Slow Sand Filtration In A Water Recharge Plant. *Jour. Wat. Res.*, **35**(10), 2484–2488.

Letterman, D. R. (1987). An Overview of Filtration. *Jour. AWWA*, **79**(12), 26–32.

Ojha, C.S.P., & Graham, N.J.D. (1994). Computer-Aided simulation of Slow Sand Filter Performance. *Jour. Wat. Res.*, **28**(5), 1025–1030.

O'Melia, C. R.. & Stumm, W. (1967). Theory of water filtration. *Jour. AWWA*, **59**(11), 1393–1412.

Rav-Acha, Ch., Kummel, M., Salomon, I. and Adin, A. The effect of chemical oxidants on effluent constituents for drip irrigation. Water Research, 29:1:119–129 (1995).

24 Performance Evaluation of Three Slow Sand Filters

A. Alicea*, J. C. Robles, H. Guillont** and R. A. Rios****

*University of Turabo, Environmental Sciences Department, School of Science and Technology, Gurabo, Puerto Rico 00778
**University of Puerto Rico, Environmental Health Department, Medical Sciences Campus, San Juan, Puerto Rico 00936

Abstract This manuscript describes the evaluation of three slow sand filters operating in different configurations: unit #1 was a classic slow sand filter; unit #2 used a fine gravel pre-filter and unit #3 uses a geotextile cloth. The units were evaluated for fecal coliforms and turbidity removal. Filter #2 and filter #3 were effective in the removal of turbidity and fecal coliforms. However, this did not occur in filter #1 because the sand was not well graded and the inadequacy of the raw water had a negative effect on the schmutzdecke.

Keywords Slow Sand Filters, geotextile, gravel pre-filter, turbidity, fecal coliforms, schmutzdecke.

Introduction

Slow Sand Filtration is an inexpensive and efficient way to obtain potable water for countries in the process of development and in isolated communities. A classic slow sand filter is composed of gravel, fine gravel, sand, a container and a faucet to extract the purified water.

This work describes the evaluation of three slow sand filters operating in different configurations. The overall purpose of the study was the development of appropriate technology to be used in the small (approximately 100–300 persons) rural water systems of Puerto Rico. Two of the units were built for this study in 2004 and the third unit has been operating since 1994. Unit #1 was a classic slow sand filter, unit #2 used a fine gravel pre-filter for turbidity removal and unit #3 used a geotextile cloth for the development of a second schmutzedecke. The units were evaluated for fecal coliforms and turbidity removal.

Methods

The three slow sand filters evaluated in this work were designed using different filtering approaches and methods. The materials used in the construction of the units were common and inexpensive, using both metal or plastic containers and pipes.

Unit #1 consisted of a polyvinyl chloride container with the classic $^3/_4$ inch gravel, $^1/_4$ fine gravel and sand arrangement. The container was 3 feet in diameter and 5 feet in height.

 Recent Progress in Slow Sand and Alternative Biofiltration Processes edited by Rolf Gimbel, Nigel J.D. Graham and M. Robin Collins. ISBN: 9781843391203. Published by IWA Publishing, London, UK.

This unit was operated for 10 years uninterruptedly. This filter was connected by galvanized pipes to three water fountains from which a colony of 1,200 free ranging primates drank the purified water. Unit #2 was a slow sand filter with an external fine gravel pre-filter consisting in a metal container with gravel inside it. The fine gravel was $^1/_4$ inch in size and evenly distributed inside the container. The purpose of the pre-filter was to reduce the turbidity before the raw water entered the filter. Unit #3 used a geotextile cloth as a pre-filter, integrated inside the slow sand filter design. This cloth was 6 inches above the sand to allow for oxygen diffusion to create a proper schmutzdecke in the sand. It was submerged in water (1 foot deep) inside the filter to create a second schmutzdecke to determine if this was a factor in improving water quality. Also, this design was intended to determine if the two schmutzdeckes differed in their types of microorganisms. Both unit # 2 and unit #3 main slow sand filters were constructed using a 2 feet diameter metal container with a 3 feet height.

Results and discussion

Results

The sand used in the three filters was from different sources. This was done to take into account the different array of river sands available. Also, the different types of sands could produce possible differences in water quality due to the effective size and uniformity coefficient of the sand.

The sand in unit #1 had an effective size of 0.17 mm and a uniformity coefficient of 4.53, the one in unit #2 had 0.26 mm and 7.69, and the one in unit #3 had 0.21 mm and 6.19, respectively. The concavity coefficient for units #2 and #3 was close to 1, indicating well-graded sand. However, unit #1 was not well graded and contained a significant amount of silt. The analysis results are shown in Figure 1.1.

In 1994, unit #1 was evaluated for fecal coliforms and turbidity removal. The results were encouraging because it removed a maximum of 73% of the turbidity and an average of 93.3% of the fecal coliforms. The sand had an effective size of 0.33 mm and a uniformity coefficient of 3. The only difference during these ten years was that the sand was changed.

In 2004, unit #1 was tested to determine if its efficiency had changed and to compare its actual performance with the results of 1994. The outcome showed a significant negative difference. There was no turbidity removal and the fecal coliforms in the "treated" water actually increased. The granulometry showed that the sand had a significant amount of fine particulates passing sieve 200. To explain this, a hydrometry test was done, which showed that this fine particulate (0.100 mm to 0.002 mm) was silt. This explained the actual negative values for turbidity and fecal coliform removals because silt can be filtered and fecal coliforms can be attached to them. Silt could also sediment and obstruct the slow sand filter.

Unit #2 showed a higher efficiency of turbidity removal when the levels exceeded 6 NTU in comparison to unit #3. Overall, unit #2 showed an 88% turbidity removal. This indicates that the gravel pre-filter is well suited for high turbidity raw water. In addition, unit #2 reached 100 % removal of fecal coliforms. The organisms found in the schmutzdecke were ciliates, annelids, nematodes and organisms of the Phylum Chrysophyta. These organisms reached a population of 105 organisms per mm^3.

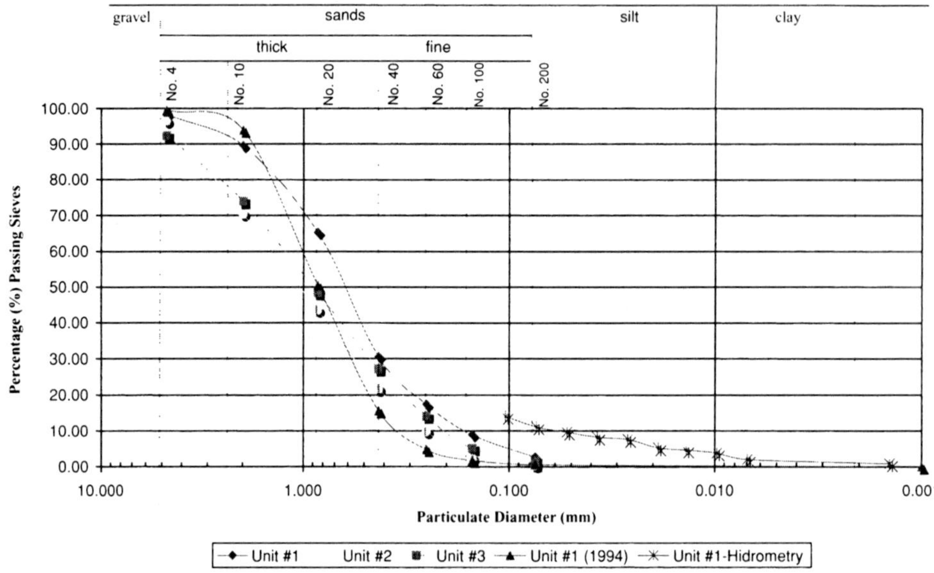

Figure 1.1 presents the granulometry for units #1, #2, #3 and the results of unit #1 in 1994 and the hydrometry for unit #1.

Unit #3 removed a maximum of 80% of the turbidity and a maximum of 100% of the fecal coliforms. On the other hand, the schmutzdecke of the geotextile exhibited populations of green, red and golden algae and organisms of the Phylum Oomycota. This last species requires additional analysis to see if it has something to do with the quality of the effluent because it may be feeding on the decay of organic matter such as the algae. The schmutzdecke of the sand presented populations of red and golden algae. It didn't show green algae populations possibly because the geotextile above it gave the surrounding environment a lasting darkness and only little, if any, photosynthesis could take place. The schmutzdecke of the sand also presented rotifers and nematodes. The rotifer population was about 1250 organisms per mm^3 and the nematode population was about 750 organisms per mm^3. Water quality in slow sand filters depends on a greater number of superior life form organisms (nematodes) above the number of lower life form organisms (rotifers). The number of rotifers was superior to the number of nematodes, which is indicative that the filter was in an early stage of maturity. Even so, the filter was capable of removing a maximum of 100% of fecal coliforms, which indicates that the schmutzdecke of the geotextile was effective (with the aid of the schmutzdecke of the sand) in removing the fecal coliforms completely and a maximum of 80% in the removal of turbidity. The results for turbidity and fecal coliform removals for units #1, #2 and #3 are shown in Figures 1.2 and 1.3.

Discussion

Unit #1 did not achieve acceptable removals for turbidity and fecal coliforms. This could be due to the inadequacy of the effective size and uniformity coefficient of the sand. Also, there

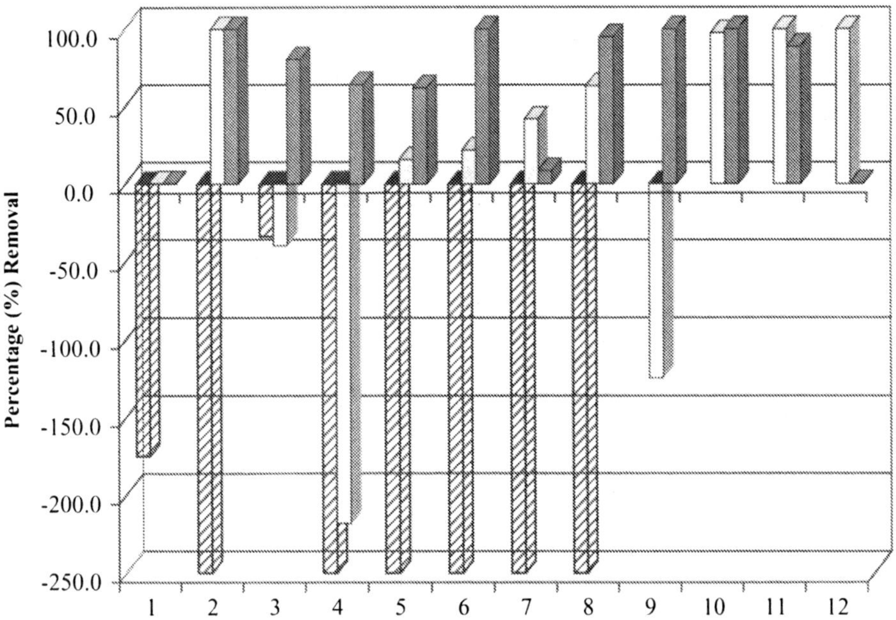

Note: The maximun value for Unit #1 is -oo (>627.8%).
Unit #1 was evaluated for eight samples.

Sample Number

■ Unit #1 □ Unit #2 ■ Unit #3

Figure 1.2 Turbidity removals.

was a considerable amount of silt, which could be filtered with fecal coliforms attached to it. Hence, the increased number of fecal coliform colonies in the filtered water. In addition, the long galvanized piping from the filter to the water fountains could have adversely affected the turbidity. It must be taken into account that unit #1 had worked for ten years. In several occasions during its service, the filter had been dry, which could have affected the stability of the schmutzdecke, influencing negatively the quality of the treated water. A good quantity of microorganisms (especially nematodes) in the schmutzdecke is generally associated with good quality of the treated water.

Unit #2 and unit #3 received raw water from the same location. The pre-filter of unit #2 showed an increased amount of turbidity removal (88%) when the NTU values exceeded 6. On the contrary, unit #3 showed only a maximum of 80% turbidity removal, which could indicate that the geotextile pre-filter works best at low turbidity raw waters. In terms of fecal coliform removals, both units removed a maximum of 100% of the colonies. This showed that both systems of pre-filtration can be used to obtain potable water in places were there is no potable water treatment plants. The success of unit #2 and unit #3 over unit #1 could be attributed to the pre-filters which help lower the turbidity and to the correct effective size and uniformity coefficient of the sand although it was obtained from different sources. Also, the schmutzdecke in unit #2 and unit #3 was always submerged in water, which contributed to the stability of the microorganism populations and the overall quality of the filtered water.

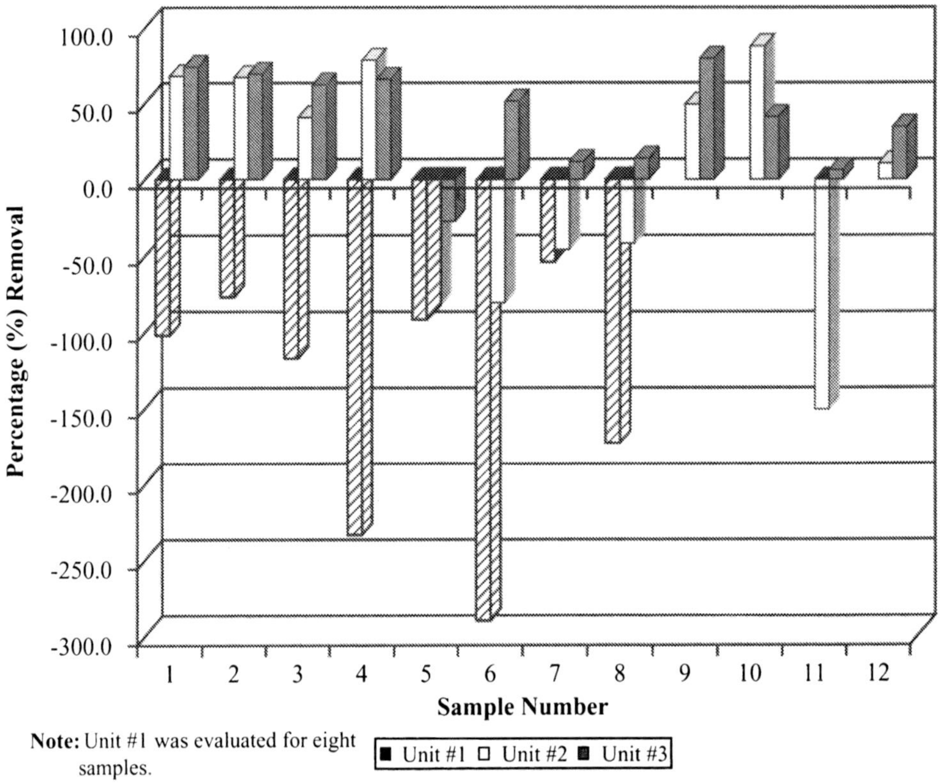

Figure 1.3 Fecal coliform removals.

Conclusions

Filter #2 and filter #3 were effective in the removal of turbidity and fecal coliforms. This was due to external and internal pre-filter systems and to the adequacy of the effective size and uniformity coefficient of the sand. Nevertheless, this did not occur in filter #1 because the sand utilized did not meet the generally accepted specifications. Also, the instability of the water supplied to filter #1 affected the microorganism population of the schmutzdecke which consequently affected adversely the quality of the treated water.

To obtain the best efficiency in a slow sand filter, it is important to use good quality sand with the appropriate effective size and uniformity coefficient, free of silt or any other contaminant and to maintain a constant flow of water. This will help achieve a relative stability in the overall system and the microorganism population in the schmutzdecke, which will contribute to a high quality water.

References

Amblard, C., Bourdier, G., Carrias, J.F., Maurin, N. and Quillier, C. (1996) Seasonal Evolution of Microbial Community Structure in a Drinking Water Reservoir. *Water Research.* **30**, pp. 3, 613–614.

Anderson, O.R. (1988) *Comparative Protozoology; Ecology, Physiology, Life, History*. Springer - Verlog, Inc. New York. pp. 131–145.
Basu, O. and Cleary, S. (2003) EWB Workshop. Biosand/Slow Sand Filtration. University of Waterloo, Waterloo, Ontario, Canada.
Bowles, J.E. (1992) *Engineering Properties of Soils and their Measurement*. 4th Edition. Boston: Irwin McGraw-Hill. pp. 43–55, 57–69.
Hickman, C.P., Roberts, L. and Larson, A. (1997) *Integrated Principles of Zoology*. 10th Edition. Chapters 16, 18, 19.
Huisman, L. and Wood, W.E. (1974) Slow Sand Filtration. *World Health Organization*, Geneva, Switzerland.
Marcantoni C., Maldonado E. and Cotto, M. (1996) Determinación de las densidades poblacionales microbiológicas existentes en el schmutzdecke de un filtro de arena lento y su relación con la calidad del efluente. Master Thesis, Environmental Health Department, Public Health School, University of Puerto Rico, Río Piedras, Puerto Rico.
Prescott, L. M. (1990) *Fundamentals of Microbiology*. 3rd Edition. Mc-Graw Hill. pp. 851–870.
Raven, P.H., Evert, R.F. and Eichhorn S.E. (1999) *Biology of Plants*. W. H. Freeman and Company Worth Publishers. pp. 371–374.
Riesenberg, F., Walters, B., Steel, A. and Ryder, R.A. (1995) Slow Filters for a Small Water System. *Journal of the American Water Works Association*.
Yung, K. (2003) Biosand Filtration: Application in the Developing World. CE 401 Project prepared for the University of Waterloo. pp. xxi–xxiii.

Part IV

SLOW SAND FILTRATION – PROCESS DEVELOPMENTS

25 Impacts of chemical pre-treatment on slow sand filtration

C. C. Dorea*, B. A. Clarke**

* Centre for Environmental Health Engineering (CEHE), School of Engineering (C5), University of Surrey, Guildford, Surrey, GU2 7XH, UK (Email: caetanodorea@hotmail.com)
** Centre for Environmental Health Engineering (CEHE), School of Engineering (C5), University of Surrey, Guildford, Surrey, GU2 7XH, UK (Email: b.clarke@surrey.ac.uk)

Abstract The chemical enhancement of gravel filtration with aluminium sulfate, *i.e.* direct (gravel) filtration, has been proposed as a pre-treatment alternative for slow sand filters. Most studies have frequently focused on the *efficiencies* of the pre-filters in terms of reduction percentages and the *effectiveness*, *i.e.* extension of slow sand filter runs, is not usually cited or even evaluated. Also, the presence of Al residuals could have a negative (*i.e.* toxic) effect on the biological treatment offered by the downstream slow sand filters and need to be considered. A pilot-scale evaluation showed that albeit *efficient*, the chemically-enhanced gravel pre-filters were not more *effective*. At certain levels, residual Al hydroxide precipitates caused rapid headloss development and limited the effectiveness of the pre-treatment. There was no evidence of an effect of Al on the biological slow sand filtration process.

Keywords Aluminium sulfate; coagulation; gravel (roughing) filtration; slow sand filtration.

Introduction

The simple operational and maintenance requirements of slow sand filtration (SSF), coupled with its high treatment efficiency makes it an attractive technology. The main limitation of the process is its vulnerability to high suspended solids loadings. Pre-treatment in such instances can be achieved by gravel (or roughing filtration). Yet, reported cases of underperformance of gravel pre-filtration have been attributed to significant fractions of influent suspended particles in the colloidal range (Lloyd *et al.*, 1991; Visscher *et al.*, 1996; Ingallinella *et al.*, 1998); which are theoretically less efficiently removed than other particles in gravel filters (Boller, 1993). Moreover, maximum limits of turbidity for the application of gravel pre-filters have also been suggested and the general consensus in the literature is the "rule-of-thumb" value of 10 Nephelometric Turbidity Units (NTU).

A simple intervention to overcome such water quality constraints can be through the dosing of a coagulant (aluminium sulfate – AS) upstream of the pre-filtration stage; enhancing the pre-treatment efficiency, in what could be defined as "direct" or "contact" (gravel) filtration. Previous studies have investigated its use as a pre-treatment for SSF (Ahsan; 1995;

 Recent Progress in Slow Sand and Alternative Biofiltration Processes edited by Rolf Gimbel, Nigel J.D. Graham and M. Robin Collins. ISBN: 9781843391203. Published by IWA Publishing, London, UK.

Ingallinella *et al.*, 1998; Carvalho *et al.*, 2001; Mwinga *et al.*, 2004); however, the results emphasise pre-filtration treatment *efficiency* in terms of particulate (or other contaminant) removal efficiency and do not consider the *effectiveness* of the pre-treatment in protecting the slow sand filters from premature clogging.

Coagulation with AS can result in increased levels of Al (*i.e.* Al residuals) relative to the Al levels naturally present in source waters. Dissolved Al in its free (hydrated) Al ionic form is considered its most toxic species to aquatic life (Florence and Batley, 1980). As such, there would be an inherent potential for the Al residuals from the pre-treatment to affect the biological treatment offered by aquatic (micro)organisms of the slow sand filter, if the Al residuals are in a toxic form. In general, it is difficult to make analytical determinations of individual Al species (Van Benschoten and Edzwald, 1990; Reiber *et al.*, 1995). Yet, by fractionation methods, groups of Al species can be isolated by operationally defined procedures (*e.g.* membrane filtration, ion exchange) and analytically quantified.

The three major groups into which Al can be divided, typically by membrane filtration, are the particulate (Al_P), total dissolved (Al_D; with acid digestion of sample), and dissolved monomeric (Al_M; without acid digestion) forms. The sum of the Al_D and Al_P makes up the total Al (Al_T). By means of a strongly acidic cationic ion exchange column (Van Benshoten and Edzwald, 1990) the organically-bound Al (total – Al_O and monomeric – $Al_{O\text{-}M}$) can be isolated from Al_D. By difference the total inorganically-bound Al ($Al_I = Al_D\text{-}Al_O$) and the monomeric inorganically-bound Al ($Al_{I-M} = Al_M - AL_{O\text{-}M}$) can be calculated.

The objectives of this study were: (1) assess the *efficiency* as well as the *effectiveness* of chemically-enhanced up-flow gravel filters in series (UGFS) as pre-treatment for SSF and (2) to study the impact of aluminium residuals on the treatment performance and potential effects on biological activity.

Materials and methods

Pilot-scale multi-stage filtration (MSF) systems comprising two parallel lines of triple stage UGFS connecting to slow sand filters were setup to evaluate the performance of a chemically-enhanced system (dosed with AS – BDH, UK) in comparison with an undosed (control) system (Figure 1). Different raw water turbidities and filtration rates were tested on the

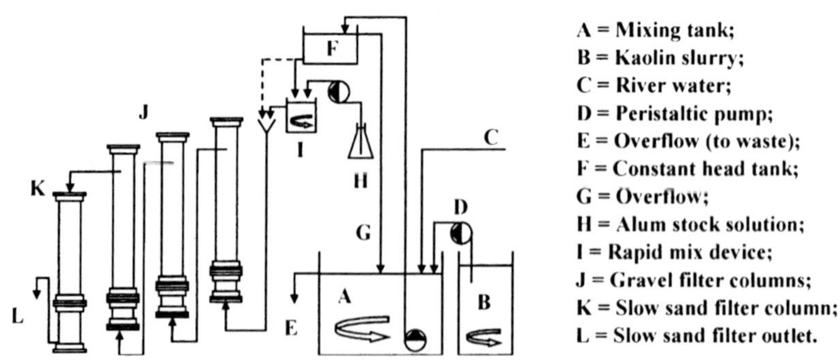

Figure 1 Schematic diagram of experimental system (second filter series omitted for clarity).

Table 1 Summary of experimental conditions of slow sand filtration column trials.

Run no.	Filtration rate (m^3/m^2h)	Bed depth (mm)	Observation
1	0.150	300	Preliminary testing of experimental system and low Al dosing (50 μg/L) on day 24.
2	0.150	300	Al dosing at 50, 100, and 200 μg/L on days 6, 10, and 14. New sand used.
3	0.075	100	Al dosing at 50, 100, and 200 μg/L on days 4, 8, and 12. New sand used.
4	0.150	100	Al dosing at 50 μg/L on day 5. New sand used.
5	0.075	100	Al dosing at 100 and 200 μg/L on days 4 and 8. New sand used.

experimental MSF system during the two slow sand filter runs that were carried out; further details of this experimental setup and its operation are available elsewhere Dorea and Clarke, 2005; Dorea, 2005). Filters were challenged with kaolin (WhitChem, UK) to simulate high turbidities. Performances were measured mainly in relation to headloss development and turbidity reductions (2100P HACH, USA), dissolved organic substances (*i.e.* ultraviolet absorption at 254 nm wavelength – UVA-254; Eaton, 1995), and faecal coliforms (FC – ISO, 1990). Al residuals (particulate and dissolved) were also measured and further fractionated (Van Benschoten and Edzwald, 1990) using a strongly acidic cationic ion exchange column (Amberlite IR-120 PLUS – Sigma-Aldrich, UK) to assess chemical speciation and potential bioavailability of dissolved Al on the aquatic (micro)organisms of a slow sand filter.

On a separate experimental rig, two lab-scale slow sand filter columns (Al-dosed and undosed control), run identically with pond water, were used to evaluate the impact of simulated Al residuals on the treatment performance of SSF and possible effects on biological activity. In order to simulate treatment residuals, Al was added as aluminium nitrate (Spectrosol-BDH, UK) on the assumption that upon addition the Al would be in a more reactive (more bioavailable) form as studied by Gardner and Comber (2003). Performance indicators were measured mainly as particulate (turbidity) and FC removals. Dissolved oxygen (DO – DO 175, HACH, USA.) uptake by the columns was used to assess biological activity and was measured according to the procedures established by the manufacturer of the instrument. A total of five runs were carried out under different conditions (Table 1). Filter bed depths were reduced from 300 mm to 100 mm, as it was thought that larger bed depths could "mask" possible effects on the biological treatment.

Results and discussion

Gravel pre-filtration reduction efficiencies for turbidity, UVA-254 absorption, and FC levels were all significantly higher in the Al-dosed (2 to 8 mg/L as Al) series than in the undosed control (Table 2). The pre-treatment efficiencies achieved in the pilot-scale MSF system were in agreement with other similar studies (*e.g.* Ahsan, 1995; Mwinga *et al.*, 2004). The effluents of the slow sand filter from the Al-dosed series were of better quality than the undosed control (Table 2) due to the higher efficiency of the Al-dosed pre-treatment.

Table 2 Runs 1 and 2 performance data summary (series: A = Al-dosed; B = control).

		Series	Raw water	Pre-filter effluent	SSF effluent
Turbidity (NTU)	Run 1	A	124 [32.6; 354]	3.04 [1.09; 10.5]	0.61 [0.42; 0.95]
		B	124 [32.6; 354]	28.5 [2.48; 79.8]	0.85 [0.32; 1.25]
	Run 2	A	194 [132; 356]	4.89 [2.74; 13.7]	0.57 [0.26; 0.99]
		B	194 [132; 356]	50.7 [3.44; 88.6]	6.84 [1.28; 45.0]
UVA-254 (cm^{-1})	Run 1	A	0.087 [0.071; 0.130]	0.046 [0.031; 0.063]	0.044 [0.030; 0.061]
		B	0.087 [0.071; 0.130]	0.091 [0.066; 0.131]	0.080 [0.054; 0.126]
	Run 2	A	0.106 [0.073; 0.149]	0.048 [0.030; 0.065]	0.047 [0.030; 0.063]
		B	0.106 [0.073; 0.149]	0.102 [0.070; 0.131]	0.100 [0.068; 0.127]
FC (cfu/100 mL)	Run 1	A	1368 [350; 8750]	71 [3; 270]	9 [<1; 61]
		B	1368 [350; 8750]	411 [124; 2775]	41 [4; 212]
	Run 2	A	1060 [520; 1820]	17 [4; 33]	5 [2; 11]
		B	1060 [520; 1820]	326 [140; 497]	73 [21; 262]

Note: values are run averages; range: [min.; max.].

Treatment efficiencies have been discussed in greater detail elsewhere (Dorea and Clarke, 2005).

Although SSF-A had an influent that was apparently more "acceptable" (*i.e.* < 10 NTU) than SSF-B in both runs, the duration of the SSF-A run was shorter than that of the control filter. During run 1, SSF-A had reached its terminal headloss (*i.e.* supernatant overflowed) at day 46 whilst SSF-B continued to run until day 53. The second run for SSF-A lasted only 23 days due to the development of excessive headloss; SSF-B operated until day 28 when its run was discontinued without having reached its maximum headloss.

The influent turbidities for SSF-A were an order of magnitude less than that of SSF-B (Table 2), suggesting that the nature of the turbidity-causing particulates in the Al-dosed pre-treatment was causing the observed headlosses in SSF-A. The effluent turbidities of the Al-dosed gravel pre-filters (1A, 2A, and 3A) had a strong positive correlation with particulate Al levels (Figure 2). Particulate Al mainly comprised of insoluble Al-hydroxide precipitates, as only a small fraction of the measured total residual was from the Al contained in the kaolin particles (Dorea and Clarke, 2005).

The relationship between the slow sand filter influent turbidities and the daily headloss increase showed two distinct patterns (Figure 3). The turbidity-causing particulates of SSF-A (*i.e.* mainly Al-hydroxide precipitates), over a narrower range had a greater potential to develop headloss than did those turbidity-causing particulates of control filter SSF-B (*i.e.* kaolin particles). This was thought to be due to factors relating to the properties of the particulates in suspension (*i.e.* degree of destabilisation, size, and mechanical properties) and the design and operating conditions of the slow sand filters (*i.e.* smaller sand grain size and lower filtration rates compared to rapid sand filtration).

It is apparent, at least for the conditions tested in this study, that even a chemically-enhanced pre-treatment target of 3 NTU as proposed by Ahsan (1995), could be unsuitable for SSF. Based on Figure 3, a *maximum* of 2 NTU would be a safer target to ensure the *effectiveness* of chemically-enhanced pre-treatment. Such a limit would most likely require a target *average* turbidity below this value, possibly even less than 1 NTU. If it weren't

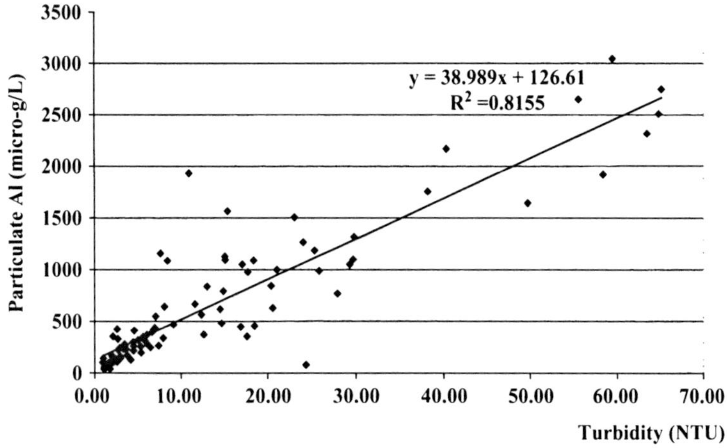

Figure 2 Correlation between Al-dosed gravel filter turbidities and particulate Al.

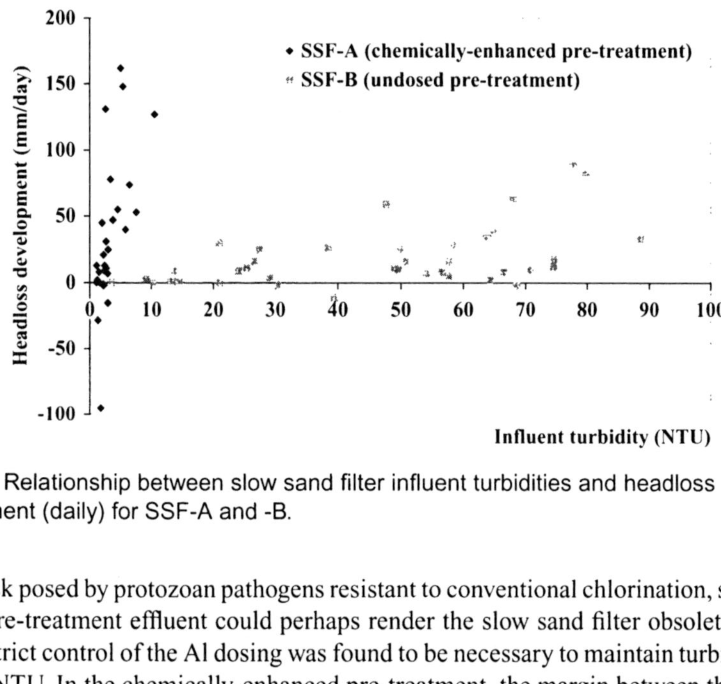

Figure 3 Relationship between slow sand filter influent turbidities and headloss development (daily) for SSF-A and -B.

for the risk posed by protozoan pathogens resistant to conventional chlorination, such a high quality pre-treatment effluent could perhaps render the slow sand filter obsolete. Furthermore, a strict control of the Al dosing was found to be necessary to maintain turbidity levels below 2 NTU. In the chemically-enhanced pre-treatment, the margin between the best that can be achieved and worst that can be tolerated in terms of effluent turbidities, in order to have an *efficient* and *effective* pre-treatment, is very fine.

Average Al concentrations for the different fractions measured at each sampling point are summarised in Table 3. In general, the concentrations of each fraction decreased through each filtration stage. The dissolved fraction increased in comparison to the raw water due to the use of the Al-based coagulant; the Al_D consisted mainly of inorganically-bound Al. Al_I,

Table 3 Average Al fraction concentrations obtained from raw water and filter effluents.

Dissolved Al	Sampling point (Al concentration in μg/L)				
fractions	RW	1A	2A	3A	SSFA
Al_T	240	1040	643	167	41
Al_P	224	973	589	127	8
Al_D	18	68	54	40	33
$^*Al_M$	10	61	44	31	30
Al_I	9	59	48	30	22
$^*Al_{I\text{-}M}$	8	58	42	28	27
Al_O	9	9	7	10	11
$^*Al_{O\text{-}M}$	2	3	2	2	3

Note: n = 8, except those marked * where n = 7.

Al_M, and $Al_{I\text{-}M}$ were present in similar concentrations. It is likely that most of the monomeric dissolved Al is of inorganic origin.

Most of the dissolved Al residual from the pre-filtration effluent (*i.e.* slow sand filter influent) was of inorganic nature (Al_I). This fraction was on average 73 % of the total dissolved Al (Al_D). Lower Al_D concentrations resulted in lower Al_I, however the dataset was too limited to establish a correlation. The relationship between these two fractions is dependent on the presence of various ligands, including the hydroxyl ion, F, sulfate and the functional groups of dissolved organic carbon that compete for the Al cation (Van Benschoten and Edzwald, 1990). For most fractionation test samples, the Al_I fraction consisted mainly of inorganically-bound monomeric Al species ($Al_{I\text{-}M}$). The recorded concentrations of $Al_{I\text{-}M}$ ranged between 0 to 76 μg/L. The $Al_{I\text{-}M}$ is considered to be the potentially more bioavailable Al species to aquatic life (Florence and Batley, 1980) and hence could possibly have a toxic effect. By reducing the total dissolved Al residuals, the inorganically-bound monomeric species should also be reduced; this is possibly the only practicable mitigation to prevent a possible effect on the biological treatment offered by the slow sand filter.

It must be borne in mind that most of the literature on the impact of Al on aquatic life considers higher life forms and the potential impact of dissolved Al species on them. The ecology of the slow sand filter also includes micro- and macro-organisms that feed on particles. These organisms contribute to the biological purification offered by the slow sand filter and could be affected by the ingestion of particulate Al. Hence, as a precautionary measure, the Al residuals should be kept to a minimum, both particulate and dissolved. Our results suggest that the negative impact of headloss development due to particulate Al outweighed the significance of this Al fraction in relation to the biological purification mechanisms.

As the two slow sand filters in the pilot MSF system received influents with different water quality characteristics (*i.e.* turbidity, nutrients, organics, pH), it was not possible to compare both SSF units with regards to any effect on the biological treatment. Such comparison was done with the SSF column trials, which based on a two-tailed paired t-test ($\alpha = 0.01$) of daily averages. From the turbidity (Figure 4) and DO (Figure 5) profiles the similar behaviour of the two SSF units with regards to trends in performance was apparent.

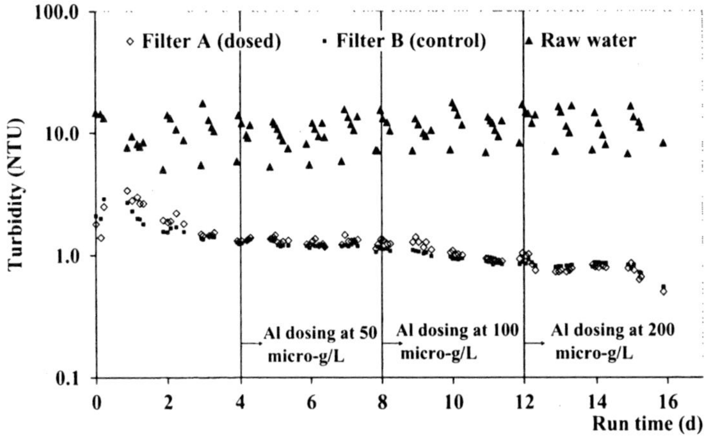

Figure 4 Turbidity profile (Run no. 3; $f = 0.075\ m^3/m^2h$; d = 100 mm).

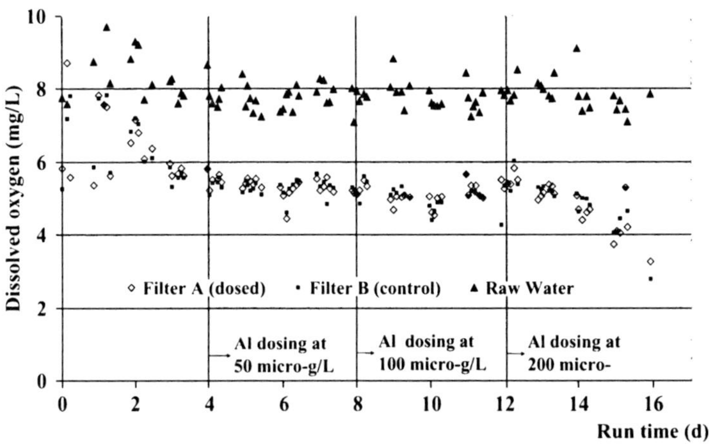

Figure 5 Dissolved oxygen profile (Run no. 3; $f = 0.075\ m^3/m^2h$; d = 100 mm).

Drops in DO levels with filtration run times indicated that biological respiration processes were occurring. As a whole, the statistical comparison of all SSF column trials revealed no significant difference between treatment performance parameters (Table 4) during the runs. The differences in terms of performance were also not statistically significant when considering only the days in which Al was dosed (Dorea, 2005), as during the first days of each run Al was not dosed.

Conclusions

The chemical enhancement of the gravel pre-treatment with AS improved the overall treatment *efficiency* of the multi-stage filtration system in comparison to the control filter series.

Table 4 Comparison* summary for treatment performance parameters for SSF column trials.

	Run number	Turbidity (NTU)	Dissolved oxygen (mg/L)	Faecal coliforms (cfu/100mL)	UVA-254 (cm^{-1})
	Filter A mean (n)	0.42 (29)	5.38 (29)	29 (28)	0.057 (22)
1	Filter B mean (n)	0.39 (29)	5.28 (29)	31 (28)	0.056 (22)
	Significant difference at $\alpha = 0.01$? (p-value)	No (0.023)	No (0.029)	No (0.343)	No (0.100)
	Filter A mean (n)	0.92 (19)	5.18 (19)	1 (18)	0.051 (12)
2	Filter B mean (n)	0.93 (19)	5.17 (19)	1 (18)	0.052 (12)
	Significant difference at $\alpha = 0.01$? (p-value)	No (0.732)	No (0.818)	No (0.478)	No (0.438)
	Filter A mean (n)	1.27 (17)	5.32 (17)	<1 (16)	0.067 (5)
3	Filter B mean (n)	1.19 (17)	5.31 (17)	<1 (16)	0.068 (5)
	Significant difference at $\alpha = 0.01$? (p-value)	No (0.175)	No (0.795)	No (0.352)	No (0.178)
	Filter A mean (n)	1.72 (9)	5.75 (9)	11 (8)	Not monitored
4	Filter B mean (n)	1.75 (9)	5.70 (9)	9 (8)	
	Significant difference at $\alpha = 0.01$? (p-value)	No (0.239)	No (0.276)	No (0.347)	–
	Filter A mean (n)	0.82 (15)	5.04 (15)	4 (14)	Not monitored
5	Filter B mean (n)	0.79 (15)	5.08 (15)	2 (14)	
	Significant difference at $\alpha = 0.01$? (p-value)	No (0.019)	No (0.484)	No (0.066)	–

* Two-tailed paired t-test (1 % significance level).

Pre-treatment efficiencies were dramatically increased in terms of turbidity and UVA-254 reduction, as well as thermotolerant faecal coliform removal. However, the headloss in the slow sand filters of the chemically pre-treated series developed quicker than the control due to residual Al hydroxide precipitates. That is, the chemical enhancement of the pre-filtration stage did not prove to be *effective*. Hence, dosing of coagulants in upstream processes must be carried out with caution. The proposed SSF pre-treatment objectives for similar alum-enhanced pre-filtration studies reported in the literature (*i.e.* 3 NTU) should be reconsidered in light of the results of this investigation. The dosing of simulated Al residuals did not have an effect on the biological activity of the slow sand filter; but *no evidence of an effect* should not be interpreted as *evidence of the lack of an effect*.

Acknowledgments

The authors are grateful for the support provided by RWE Thames Water.

References

Ahsan T. (1995). *Process Analysis and Optimization of Direct Horizontal-Flow Roughing Filtration.* PhD dissertation, International Institute for Infrastructural, Hydraulic and Environmental Engineering (IHE), Delft University of Technology, The Netherlands, A.A. Balkema, Rotterdam.

Boller M. (1993). Filter mechanisms in roughing filters. *J. Water SRT – Aqua*, **42**(3), 174–185.

Carvalho R. P. M., Cezar M. C. M. and Brandão C. C. (2001). Aplicabilidade da pré-filtração ascendente em pedregulho com coagulação química, como pré-tratamento para a filtração lenta de águas com presença de algas (in Portuguese). Paper presented at the *21º Congresso Brasileiro de Engenharia Sanitária e Ambiental*, João Pessoa, Brasil.

Dorea, C. C. (2005). *Chemically-Enhanced Gravel Pre-Filtration for Slow Sand Filtration.* PhD Thesis, University of Surrey, Guildford, UK.

Dorea, C. C. and Clarke, B. A. (2005). "Chemically-enhanced gravel pre-filtration for slow sand filters: Advantages and pitfalls." In *Proc. Particle Separation 2005*, Seoul, 1–3 June.

Eaton, A. (1995). Measuring UV-absorbing organics: a standard method. *J. AWWA*, **87**(2), 86–90.

Florence, T. M. and Batley, G. E. (1980). "Chemical Speciation in Natural Waters." *CRC Crit. Rev. Anal. Chem.*, **9**(3), 219–296.

Gardner, M. J. and Comber, S. D. W. (2003). "Aluminium speciation in effluents and receiving waters." *J. Environ. Monitor.*, **5**, 902–05.

Ingallinella A. M., Stecca L. M. and Wegelin M. (1998). Up-flow roughing filtration: Rehabilitation of a water treatment plant in Tarata, Bolivia. *Wat. Sci. Tech.*, **37**(9), pp. 105–112.

International Standards Organisation (1990). ISO 9380–1:1990[E], *Water quality – Detection and enumeration of coliform organisms, thermotolerant coliform organisms and presumptive* Escherichia coli *– Part 1: Membrane filtration method*, Geneva, Switzerland.

Lloyd B. J., Galvis G. and Eudovique R. (1991). Evaluation of multiple barrier drinking water treatment systems for surface water sources. Paper presented at the *20th Caribbean Water Engineer's Conference*, Cayman.

Mwinga G., Setlhare B. and Loewenthal R. E. (2004). Impacts of coagulation on upflow roughing filtration in layers. Paper presented at the *30th WEDC International Conference*, Lao PDR.

Reiber, S., Kukull, W., and Standish-Lee, P. (1995). "Drinking water aluminium and bioavailability." *J. AWWA*, **87**(5), 86–100.

Van Benschoten J. E. and Edzwald J. K. (1990). Measuring Aluminum During Water Treatment: Methodology and Application. *J. AWWA*, **82**(5), 71–78.

Visscher, J. T., Galvis, G., and Latorre, J. (1996). "Filtración en multiplas etapas-FIME: bondades e limitaciones." (In Spanish) In *Anais da Conferencia Internacional de Mejoramiento de la Calidad del Agua*, Santiago de Cali, Colombia.

26 Removal of humic substances in slow sand and in slow sand/activated carbon filtration using ozone and hydrogen peroxide as pre-oxidants

Luiz, Di Bernardo, Edson, Pereira Tangerino****

*Av. Trabalhador São-Carlense, 400; CEP: 13566-590; Escola de Engenharia de São Carlos – USP; São Carlos, Brazil; Email:bernardo@sc.usp.br.
** Alameda Bahia, 550, Ilha Solteira; CEP 15385-000; Faculdade de Engenharia de Ilha Solteira – UNESP; Ilha Solteira, Brazil; Email: edtang@dec.feis.unesp.br

Abstract This research work was carried out in order to mainly investigate the removal of true colour, UV absorbance 254 nm and total organic carbon by oxidation with ozone and hydrogen peroxide followed by slow filters composed of sand and sand/granular activated carbon as media. It was observed that the best train for the treatment of water with true colour between 50 and 100 HU was by using ozone alone or ozone followed by hydrogen peroxide and slow filters with media consisting of sand and granular activated carbon.

Keywords Multistage filtration, slow sand/activated carbon filtration, ozone; hydrogen peroxide, humic substances

Introduction

The multistage filtration (FiME) is considered to be an appropriate alternative for water treatment to small communities. However, the efficiency of removal of true color due to the presence of humic substances has been reported to be very low, generally less than 25 %. Langlais et al. (1991) reported that for river Sein, the use of peroxon (ozone + hydrogen peroxide) was more efficient than hydrogen peroxide + ozone and much more efficient than the use of ozone alone in removing the absorbance 254 nm. Malley et al. (1994) carried out a pilot plant study to examine the effects of using ozonation prior to slow sand filtration on the overall performance of the treatment. The authors observed that ozone had a marked positive influence on the performance of slow sand filter treatment, microbiology and natural organic matter (NOM) transformations. Although the resulting ozonation was beneficial with regards to the removal of NODOC, BDOC and THMFP (trihalomethane formation potential), the filter run times was observed to decrease presumably due to increase clogging of the pores via biofilm development. This problem could be alleviated by using a larger effective size filter media. By-products of ozonation, primarily as formaldyede, were produced in the pilot slow sand filters (acetaldehyde and glyoxal) were readily removed in the filters. In a compiled study of pilot plants carried out in Pennsylvania State University and

 Recent Progress in Slow Sand and Alternative Biofiltration Processes edited by Rolf Gimbel, Nigel J.D. Graham and M. Robin Collins. ISBN: 9781843391203. Published by IWA Publishing, London, UK.

in the University of New Hampshire, Dempsey and Fu (1994) reported that, in general, the practice of applying medium-level ozone to raw water before it reaches the slow sand filters contributes to promote biological activity, increasing the biofilm thickness, thus resulting in greater removal of THMFP, turbidity, and organic contaminants. Cable and Jones (1996) studied the removal of THM precursor in the two trains (ozonated and non-ozonated) of a full scale plant of slow sand filtration treating of raw water from an upland raw water in the UK. They observed an average of 28 % and 50 % of the removal of THMFP respectively in the non-ozonated and in the ozonated-stream. Furthermore, good linear correlations have been observed between THMFP and true colour and THMFP and TOC (total organic carbon). Yordanov et al. (1996) conducted a study to assess changes in the schmutzdecke resulting from the application of ozonated, nutrients poor, coloured water to slow sand filters. The authors verified that ozonation/slow sand filtration was shown to decrease the colour levels below the maximum admissible value and is suitable to upgrade existing slow sand filtration plants. However, rapid clogging of filter media may occur due to the increased amount of more rapidly biodegradable materials, causing an increase in headloss and shorter filter run times. Bauer et al. (1996) developed a GAC sandwich slow filter which has a sand and granular activated carbon as media. Ozonation was applied prior to the GAC sandwich slow filter. The concept was implemented to a full scale plant which has been achieving removal of organics (colour, taste and odor), ozonation of by-products (THM's) and pesticides.

The present research work was carried out in order to evaluate the removal of humic substances on the slow filtration. Five experiments were performed using four slow filters in parallel: two filters with sand and granular activated carbon as media and the other two filters with sand as medium. Pre-oxidation with ozone and hydrogen peroxide (and combinations of these oxidants) were studied in order to improve the removal of humic substances in slow filtration.

Methods

The pilot plant was comprised of a dynamic roughing gravel filter-DRGF (length = 1 m; width = 0.75 m; height = 1.0 m) containing three gravel sub layers of size between 4.8 mm (top) and 25.4 mm (bottom), total gravel layer thickness equal to 0.9 m, an upflow roughing gravel filter-UFRGF (internal diameter = 0.8 m; total height = 3.0 m) composed of five sub layers with gravel size between 1.41 mm (top) and 50.0 mm (bottom), total gravel layer equal to 2.1 m, both used as pretreatment step and four slow filters. The slow filters had the same support layer with a thickness of 60 cm and consisted of five sub layers with gravel size varying between 1.21 mm (top) and 50.0 mm (bottom). Table 1

Table 1 Characteristics of sand and activated carbon used in the slow filters

Slow Filter	Grain size (mm)	Sub layer thickness (cm)
F1 and F3	Sand: 0.08–1.0	10
	Granular activated carbon: 0.30–0.84	30
	Sand: 0.08–1.0	30
F2 and F4	Sand: 0.08–1.0	70

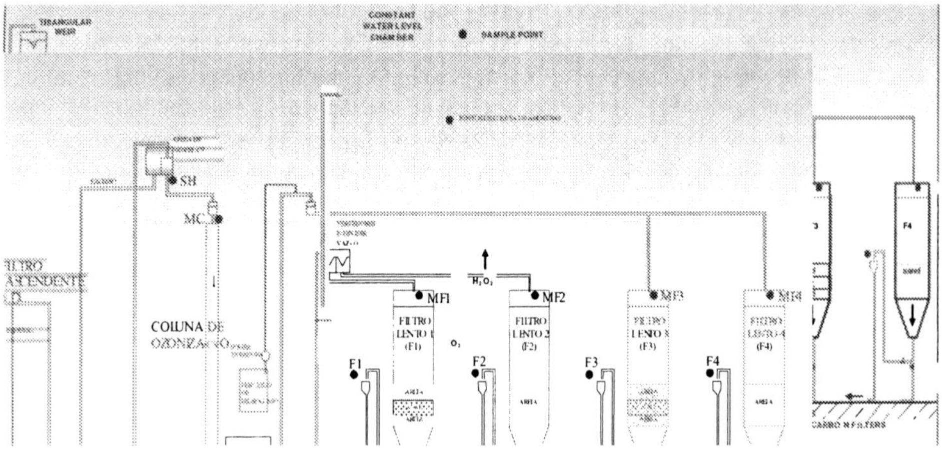

Figure 1 Schematic illustration of the pilot plant

presents the composition of the media of each slow filter. Raw water was abstracted from the local city water supply pipe and pumped to the inlet of the DRGF. To guarantee a superficial rate of 0.75 m/h and an overflow rate of 20 % of the total incoming flow rate in this unit, the flow rate was measured by using a rotameter. Effluent water from this unit was collected through perforated pipes and conducted to a box from which the flow rate that feeds the UFRGF was adjusted by using a rotameter in order to obtain a filtration rate of 0.5 m/h and the excess was wasted to the drainage system. The effluent from the UFRGF was conducted to a flow splitting box from which the influent to the filters was conducted through individual pipes provided with rotameters in order to obtain a filtration rate of 0.125 m/h in each of the four slow filters. A solution of humic substances was pumped to the outlet box of the UFRGF and the water on the study was conducted to a tank provided with a floating valve from which this water was pumped to a box maintained at constant level. From this box, the water on the study was passed through a column in which ozone could be introduced with a applied dosage of 1.0 mg/L with a retention time of 10 min. The effluent pipe was divided in two lines, one feeding the filters F1 and F2 and the other feeding the filters F3 and F4. The line feeding the filters F3 and F4 could receive a solution of hydrogen peroxide with a dose of 0.5 mg/L. Figure 1 shows a general scheme of the pilot plant used in the experimental work. Thus, in those runs with pre-oxidation, this process was performed on the influent water on the study to the slow filters. Arrangements were done to consider the different conditions of the runs performed, as presented in Table 2.

Humic substances used to increase the true colour of the raw water were extracted from samples of black-coloured soil with high concentration of charcoal residues. The following procedure was adopted to extract the humic substances from the soil: i) extraction in 0.5 mol/L solution of potassium hydroxide in a mixing tank during 4 h; ii) soil/extracted solution relationship equal to 1:20 (mass per volume); iii) sedimentation for 48 h and collection of the supernatant; iv) dialysis with hydrochloric acid carried out until a negative test to chloride using silver nitrate was observed. The resulting solution of humic substances was used to

Table 2 Runs Performed in the Pilot Plant

	F1		F2		F3		F4	
Run	True Color (Hu)	Pre-oxidant	True Color (Hu)	Pre-oxidant	True Color (Hu)	Pre-oxidant	EPT (*)	Pre-oxidant
1	EPT (*)	O_3	EPT	O_3	EPT (*)	No Preox	Out of Operation	
2	EPT	$H_2O_2 + O_3$	EPT	$H_2O_2 + O_3$	EPT	H_2O_2	EPT	H_2O_2
3	50–70	$H_2O_2 + O_3$	50–70	$H_2O_2 + O_3$	50– 70	H_2O_2	50–70	H_2O_2
4	50–70	O_3	50–70	O_3	50–70	$O_3 + H_2O_2$	50–70	$O_3 + H_2O_2$
5	EPT	$H_2O_2 + O_3$	EPT	$H_2O_2 + O_3$	50–70	No Preox.	50–70	No Preox

(*) effluent from pretreatment units

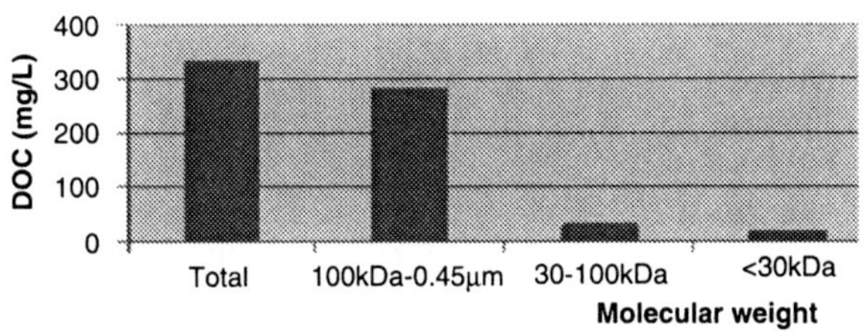

Figure 2 Concentration of DOC versus molecular weight of different fractions of the humic substances

increase the true colour of the raw water. Some characteristics of this solution are shown in Figure 2 (concentration of dissolved organic carbon – DOC versus molecular weight) and Table 3 (elementar analysis).

The following parameters were monitored during the experimental work for different sample points in the pilot plant: pH, absorbance 254 nm, alkalinity, non purgeable dissolved organic carbon, total coliformes, *E. coli*, conductivity, apparent colour, true colour, iron, manganese, particle countings (medium size between 2 and 48 μm), ozone (gaseous phase - continous), residual ozone (continous), residual hydrogen peroxide, suspended solids and turbidity.

Table 3 Humic substances composition and atomic relationship of the elements

Mass (%)				Atomic Relationship		
C	H	N	O	H/C	O/C	C/N
44.34	3.95	1.50	50.21	1.07	0.85	34.5

Table 4 Maximum, minimum and average values of true colour in the five runs

Sample Point		True Colour (Hu)				
		Run 1	Run 2	Run 3	Run 4	Run 5
Efluent UFRGF	Maximun	57	38	169	109	64
	Average	32	23	52	43	23
	Minimum	20	14	25	23	15
After Humic Substances Addition	Maximun	–	–	165	100	85
	Average	–	–	67	63	57
	Minimum	–	–	40	46	46
Upstream of the Ozonation Column	Maximun	59	38	161	100	64
	Average	32	23	67	63	23
	Minimum	10	16	39	46	15
Downstream of the Ozonation Column	Maximun	53	36	120	70	23
	Average	34	22	50	43	15
	Minimum	14	11	20	31	10
Effluent F1	Maximun	38	17	75	53	15
	Average	18	11	30	27	9
	Minimum	4	7	9	17	5
Effluent F2	Maximun	41	25	77	57	26
	Average	18	13	38	30	13
	Minimum	8	4	21	20	9
Effluent F3	Maximun	35	14	58	49	42
	Average	15	7	32	26	32
	Minimum	3	1	11	18	21
Effluent F4	Maximun	–	31	87	53	50
	Average	–	18	45	29	39
	Minimum	–	11	19	20	31

Results and discussion

Since the main objective of this research work was to study the removal of true colour, only the data relating to this parameter are presented herein. Table 4 shows the maximum, average and minimum values of true color in all runs. Figures 3 and 4 show the total removal of true colour for different conditions of pre-oxidation in slow sand filters and slow sand/granular activated carbon filters from which it can be seen that the slow sand/granular activated carbon filters are more efficient than the slow sand filters.

Table 4 confirms that, for influents with and without the addition of humic substances, filters F1 and F3 (sandwich filters) generally are more efficient in removing colour than filters F2 and F4. This result is attributed to the use of activated carbon in filters F1 and F3. It can also be perceived from this table that the ozone has a marked positive influence, aiding in the reduce of true colour of the influent to the filters.

Although not presented herein, a considerable reduction of UV absorbance 254 nm, dissolved organic carbon, turbidity, suspended solids, number of particles with medium size between 2 and 48 μm, total coliformes and *E. coli* was detected mainly in the effluent of the filters with activated carbon.

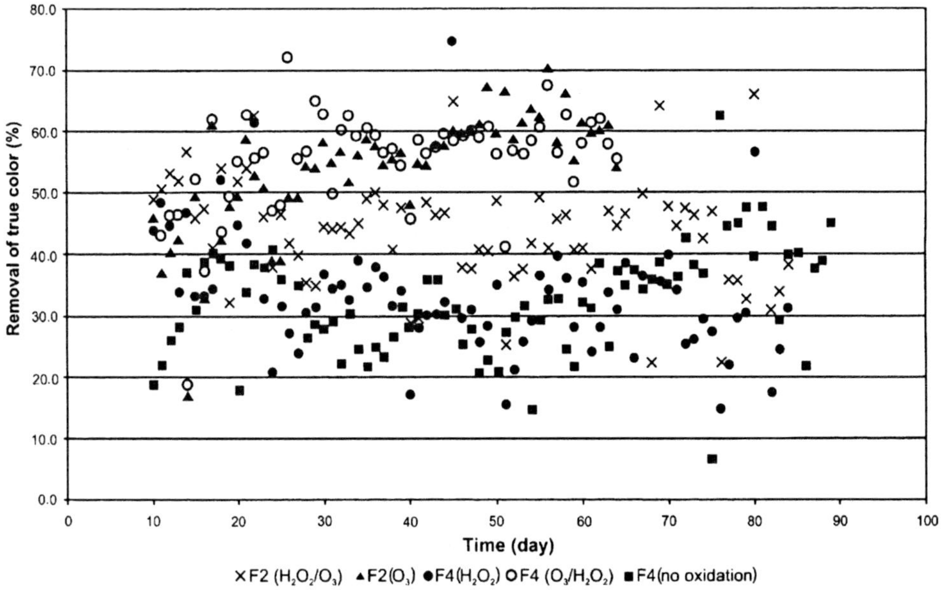

Figure 3 Removal of true colour in slow sand filters for different conditions of pre-oxidation

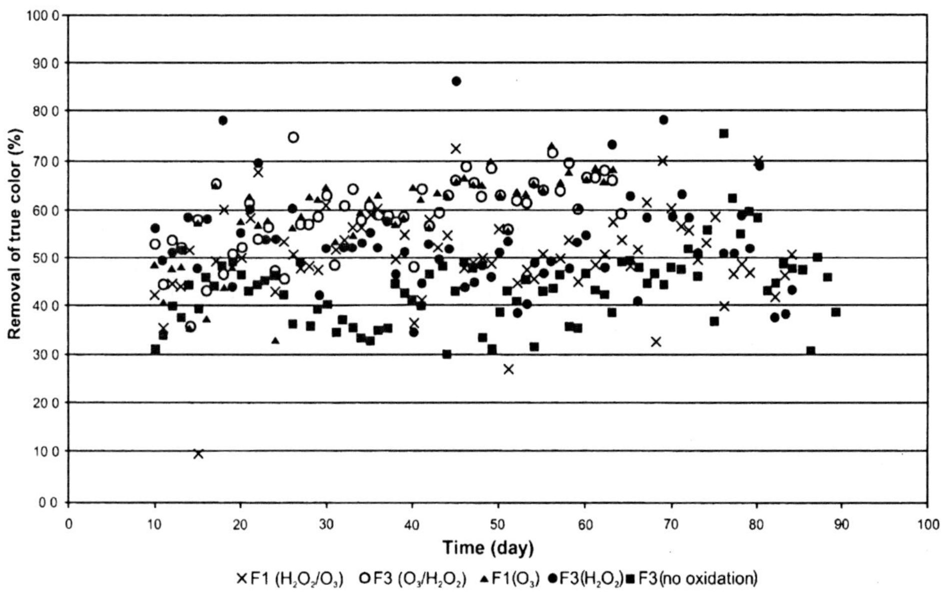

Figure 4 Removal of true colour in slow sand/granular activate carbon filters for different conditions of pre-oxidation

Conclusions

Based on the experimental investigation carried out, the following conclusions can be drawn:

a) for humic substances, the removal of surrogate parameters such as true colour, total dissolved organic carbon and UV absorbance 254 nm was shown to be higher in those slow filters containing granular activated carbon in the media, receiving pre-oxidized influent by ozone or ozone followed by hydrogen peroxide; for the last pre-oxidation condition, the removal of true colour attained results of approximately 64 %;

b) removal of true colour in the filters receiving influent pre-oxidized with hydrogen peroxide alone or with hydrogen peroxide introduced upstream of the ozonation column is unsatisfactory; the best results were obtained when hydrogen peroxide was applied downstream of the ozonation column;

c) residual hydrogen peroxide concentration in the influent to the filters, higher than 0.1 mg/L, caused damage to the biological layer and reduced the removal efficiency of colour, even in those slow filters with granular activated carbon in the media;

d) the slow sand filters receiving natural pre-filtered water presented an average true colour removal of approximately 33 % while in the slow filters with granular activated carbon, the average reduction of true colour observed was equal to 63 %; when receiving natural pre-filtered water oxidized with ozone or ozone/hydrogen peroxide, the removal of true colour by the slow sand filters was higher, varying between 21.5 and 53 %, while for the slow filters with granular activated carbon, this parameter varied between 44 and 68 %;

e) higher removal of UV absorbance 254 nm was observed when the influent was pre-oxidized with ozone/hydrogen peroxide; the overall removal attained was 40–60 % in the filters with granular activated carbon and 20–40 % in the slow sand filters

f) as observed in literature, the ozonation of the influent to the slow filters causes an increase in the rate of head loss during the runs.

References

Bauer, M. J., Colbourne, J. S., Foster, D. M., Goodman, N. V., Rachwal, A. J. (1996). GAC enhanced slow sand filtration (GAC sandwich). In: Advances in Slow Sand and Alternative Biological Filtration. Graham, N. J. D., & Collins, R. ed. John Wiley & Sons Ltd, England. p. 223–232.

Cable, C. J., Jones, R. G. (1996). Assessing the effectiveness of ozonation followed by show sand filtration in removing THM precursor material from an upland raw water. In: Advances in Slow Sand and Alternative Biological Filtration. Graham N.J.D & Collins, R. ed. John Wiley & Sons Ltd, England. p. 29–37.

Dempsey, B. A., Fu, L. T. (1994). Pilot studies of preozonation and slow sand filtration. Slow Sand Filtration – An International Compilation of Recent Scientific and Operational Developments. Ed. Collins, M. R. & Graham, N. J. D., AWWA, USA.

Langlais, B., Reckhow, D.A., Brink, D.R. (1991). Ozone in water treatment. Application and engineering, Lewis Publishers, Inc., Chelsea, Micg. 569 p.

Tangerino, E. P. (2003). Removal of humic substances with ozone and hidrogen peroxide and slow filtration Doctor's Thesis presented at the Department of Hydraulics and Sanitation of the School of Engineering of Sao Carlos, Brazil (in Portuguese).

Yordanov, R.V., Lamb, A.J., Melvin, M.A.I. (1996). Biomass characteristics of slow sand filters receiving ozonated water. In: Advances in Slow Sand and Alternative Biological Filtration. Graham, N.J.D & Collins, R. ed. John Wiley & Sons Ltd, England. p.107–118.

27 Covering slow sand filters; qualitative and operational aspects

J. Abrahamsson* and P. Dromberg*

*Stockholm Water Co, S-106 36 Stockholm, Sweden joab@stockholmvatten.se

Abstract One slow sand filter has been covered and evaluated and compared to uncovered filters for three and a half years. Online monitoring as well as surface algae and floc build-up, water quality and sand profiles has been sampled and analysed throughout the trial period. By covering the filter, the filter runtime has increased markedly and no cleaning at all has been necessary during the trial period. During the same period the uncovered filter has been cleaned 7 times. In both filters freezing has been applied during cold temperatures. Data shows that covering the filter has not changed the water quality in any negative way. Algae are present in the covered filter, but only sparse due to limited light access. From sand profiles it can be concluded that in uncovered filters neither the concentrations of organic matter nor the concentration of aluminium increases during the trial period. In the covered filter on the other hand, where no cleaning took place during the trial period, a gradual increase in these parameters can be observed on the surface. In deeper sand layers, there is no change over time, which indicates that it is not necessary to remove more sand when cleaning the covered filter despite its prolonged runtime.

Keywords Drinking water, slow sand filter, covered filter, sand profiles

Background

Stockholm Water Company operates 2 waterworks that supplies Stockholm with drinking water. Both waterworks use conventional coagulation/sedimentation followed by rapid and slow sand filtration, see figure 1. The slow sand filters at the two waterworks covers an area of 70 000 and 40 000 m^2 respectively. The slow sand filters are cleaned twice a year on a regular basis, the first cleaning of the year taking place in spring when ice cover has melted. During summer the head loss increases due to algae growth, and in the fall the filters are cleaned again before temperature drops to freezing degrees. In case head loss increases rapidly during winter, freezing is applied. During freezing, the filter is drained and the surface is left to freeze for 80–100 (freezing degrees Celsius × hours). Freezing usually has a drastic effect in reducing head loss, and the filter can be put back into operation.

Covering the slow sand filters is not common in Sweden, but several European cities have covered slow sand filters. Covering the slow sand filters in Stockholm has been discussed for a number of reasons. Among these can be mentioned increased filter run times and improved water quality. Problems with animals and birds in and around filters will also be

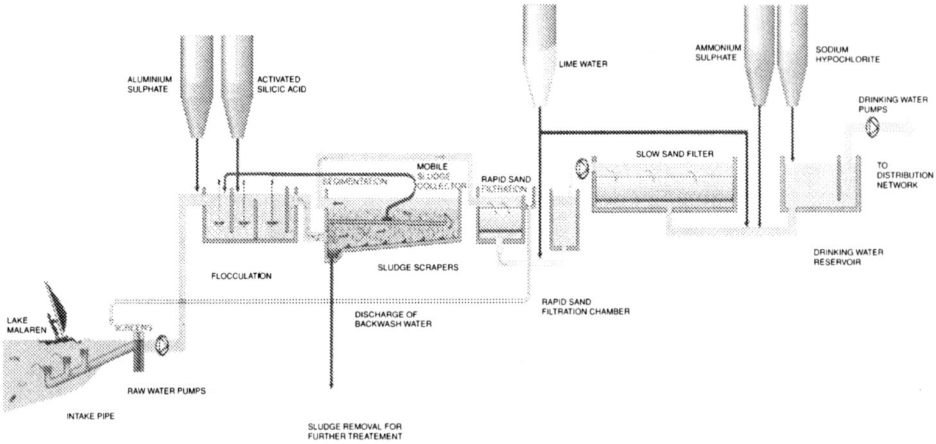

Figure 1 Treatment process overview, drinking water production in Stockholm.

eliminated as well as deposition from rain and/or air. Due to low winter temperatures, ice covers the filters during parts of the year, which makes it impossible to clean the filters during several months – covering the filter could possibly make cleaning procedure more flexible. During the last few years security has also been an emerging issue.

To investigate the aspects of covering slow sand filters, a study was initiated. The comparative study was initiated in April 2002 and is still running January 2006.

Material and methods

Previous studies in Stockholm have shown that covering slow sand filters is best performed in full scale. For that reason, one of the slow sand filters was covered with a steel building for evaluation purposes. One identical filter was chosen as a reference. Both filters contain the same type of sand, similar sand depth and have the same area (1500 m^2). The same surface load has been applied to both filters throughout the experimental period.

One full filter cycle for the covered filter was to be enclosed in the study in order to determine the total filter run time. Since no cleaning has been necessary so far, the trial period reported here is three and a half year. The filters have been monitored continuously online (flow, temperature, head loss), by ocular inspections, standard water quality analysis and algae analysis as well as by sand sampling and analysis.

Water sampling and analysis of standard parameters were performed regularly throughout the trial period in both reference and covered filter. During the first year, weekly sampling was applied, but later the frequency of sampling was decreased. Water was analysed with respect to alkalinity, UV-absorbance, TOC, turbidity and pH. Random samples were analysed with respect to AOC.

Ocular inspections were performed regularly, particularly during summer. During inspections the filter surface was studied and algae samples were extracted from the surface. The algae samples were preserved and studied with microscope.

To evaluate the clogging process, the sand in the filters was examined twice a year. Sampling was always performed immediately before cleaning of the uncovered filter. When sampling for sand, the filter was freshly drained and 4 random samples were taken in each filter. At each sampling point, 200 gram sand was extracted from 5 depths in the sand bed; 0–2 cm, 2–5 cm, 5–10 cm, 10–15 cm and 15–20 cm. The deepest sampled sand, 15–20 cm, is considered background values for each filter. All samples were analysed with respect to clogging factors (aluminium and TOC as indicators of floc build-up) and microbiological activity (standard heterotrophic plate counts cultivated for 3 and 7 days respectively, coliform bacteria and ATP). The samples intended for microbiological analysis were treated for further analysis by shaking 10 g of sand with 10 ml of sterilised water for one minute. The obtained slurry was then processed further depending on analysis. Samples intended for analysis of clogging factors were centrifuged with water before analysis. Data was analysed with respect to a) differences between uncovered and covered filters b) temporal variations and effects of cleaning intervals and c) variation with sand depth.

Results and discussion

Seasonal variation

Raw water temperature varies markedly with season. Maximum temperature during summer reaches 20°C, while the water is just above freezing during winter months. Due to the variations in temperature, the sand filter efficiency also varies. This can be exemplified by the breakdown of TOC in the slow sand filter, see figure 2.

Headloss development and filter run time

The reference filter had to be cleaned twice a year as expected. Due to rapid head loss increase, freezing had to be implemented during the first winter, see figure 3b.

The covered filter has not been cleaned at all during the trial period, and the filter run time has exceeded all expectations. During winter, rapid head loss increase has been observed. The first winter head loss increased markedly followed by a spontaneous gradual reduction

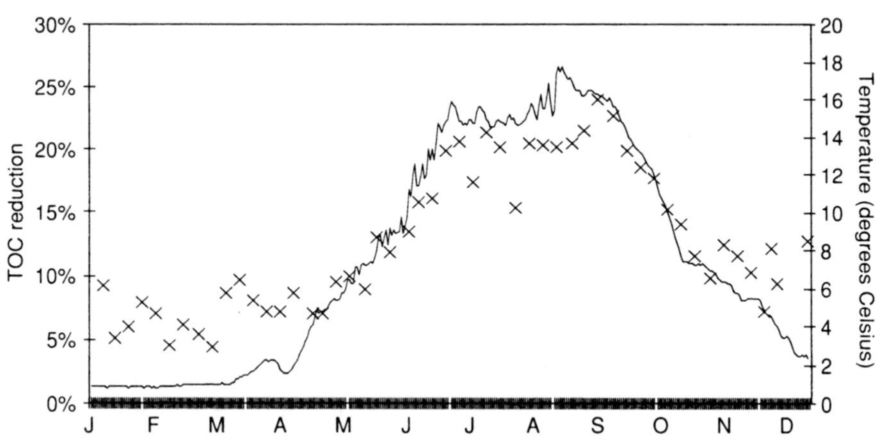

Figure 2 SSF performance variations with season (temperature), exemplified by average TOC reduction in open slow sand filters.

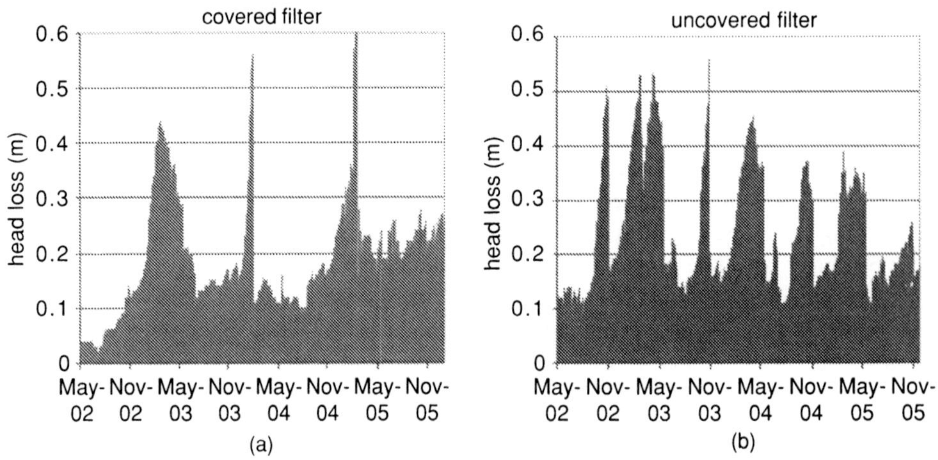

Figure 3 Pressure drop over slow sand filters. (a) The covered filter has not been cleaned at all, but freezing took place in January 2004 and in January 2005. (b) The uncovered filter has been cleaned November 2002, May 2003, October 2003, May 2004, October 2004 and May 2005. Freezing took place in March 2003.

in head loss during spring. The second winter, head loss development was very dramatic in the early winter and the filter had to be stopped for freezing. Despite unfavourable freezing conditions, head loss decreased markedly and did not increase again during the rest of the winter. The procedure was repeated with successful results the third winter, see figure 3a. From figure 3a, it is also obvious that head loss does increase over time, only slower than in uncovered filters. It is likely that the filter run will have to be terminated during 2006.

Head loss in slow sand filters is believed to be explained mainly by two factors; algae growth and residual floc build-up. In open filters, summer head loss increase is expected to be a combination of these two, with algae cover being the most dominant feature. In wintertime, when light access is limited and temperature low, residual floc is believed to be the main contributor to head loss build-up. This does not explain why head loss increase is more rapid during winter in both covered and uncovered filters, when very little algae are present in the filters.

The fact that the covered filter is even more affected than uncovered filters by the rapid head losses during low temperature indicates that this is somehow related to the residual floc that accumulates in the top layer of the filter bed. In open filters, the residual floc is removed regularly, while the covered has accumulated residual floc over the trial period. It is proposed that the structure of the residual floc changes with decreasing temperature, causing decreased permeability at low temperatures. This is supported by the fact that freezing, which breaks down the sludge-like residual floc to particulate matter, totally reduces the head loss. Whether the rapid head loss increase during colder periods are related to residual floc structure is being researched during the winter 2005/2006.

Water quality

Frequent water quality analyses have not revealed any statistical differences in water quality. Turbidity, alkalinity, TOC, AOC, heterotrophic bacteria and coliforms has been monitored regularly in filtrate.

Physio-/chemical parameters

Temperature varies greatly in the source water, with summer temperatures around 20°C and close to freezing during winter. Covering the filter has lowered the temperature in summer and increased the temperature in fall and early winter. Unfortunately, the cover-up does not prevent the filters from freezing.

Other parameters of the filtrate are not influenced at all by the cover with one exception; alkalinity varies with sunlight availability in open filters while alkalinity stays constant in covered filters due to absence of photosynthesis.

Microbiological quality

Neither the covered nor the uncovered filter has had any pathogenic bacteria in filtered water. Heterotrophic plate counts vary with season and there is no statistical difference between the covered and the uncovered filter see figure 4a, but at least it can be concluded that covering has no negative effect on bacterial reduction. The only parameter where covering the filter actually seems to have an effect is when it comes to *Aeromonas* in filtered water, see figure 4b. *Aeromonas* hydrophilia dominates, and the occurrence is very closely linked to water temperature. The reason for the increased growth of *Aeromonas* has not been established. Older data indicates that the now covered filter had a lot of *Aeromonas* even before the cover was added, which may indicate that *Aeromonas* is not an effect of covering, but rather varies greatly from filter to filter.

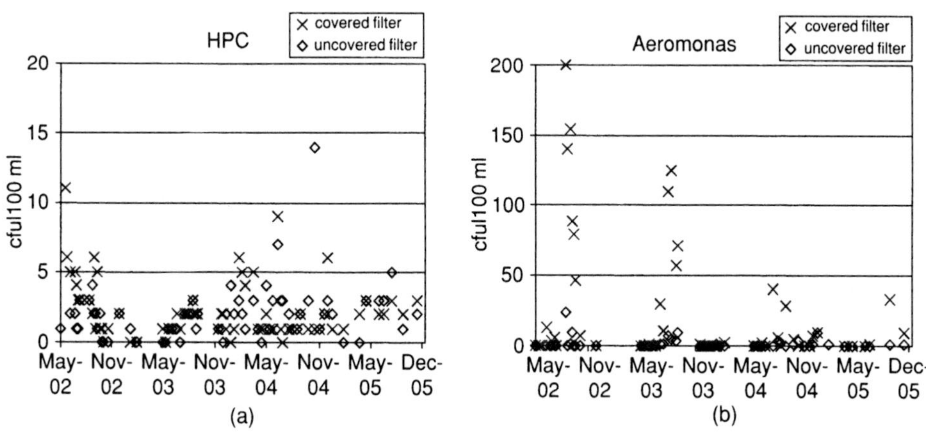

Figure 4 (a) Heterotrophic plate counts in filtered water, comparison between covered and uncovered filter and (b) Comparison of *Aeromonas* occurrence in covered and uncovered filter.

Filter surface and algae growth

Ocular observations reveal that there are a lot of algae growing in uncovered filters. In the covered filter, there is no visible algae growth and no schmutzdecke at all has been established. Algal analysis shows that the same types of algae are present in the covered filter as in uncovered filters, but the occurrence is sparse due to limited light access. It has also been observed that the surface of the covered filter gradually darkens and becomes more compacted during the filter run. With the absence of photosynthesis, very little algae production takes place, and with time the surface of the covered filter has developed a flora and fauna different from those observed in uncovered filters. At sand sampling it has been observed that aquatic organisms such as freshwater isopods and nematodes are common at the surface of the covered filter. Filtrates of the covered and uncovered filters show no difference in abundance of organisms.

Sand profiles

Floc build-up

Concentrations of aluminium are highest in the surface sample (0–2 cm) and decrease gradually with depth in both the uncovered and the covered filter, see figure 5a and b.

At the first sampling time in October 2002, the covered and the uncovered filters had very similar distribution of aluminium in the sand profile. With time, the aluminium concentration in the two top layers of the covered filter increase gradually. It would have been expected to find constant concentrations of aluminium in the uncovered filter, but for unknown reasons the concentration declines over time. Aluminium sulphate dose, pH as well as the thickness of sand removed at cleaning affect the floc that loads the filter.

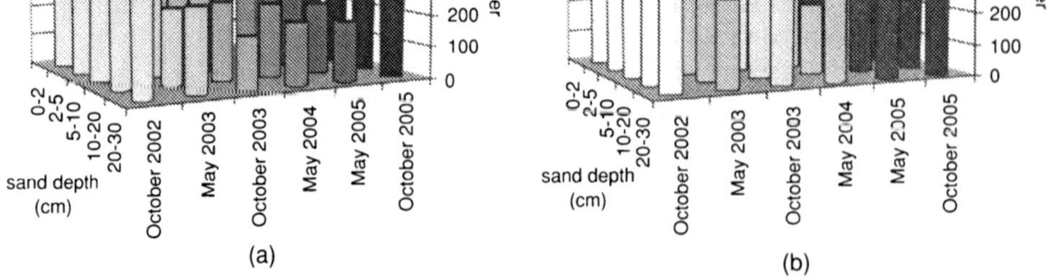

Figure 5 Aluminium content, variation with time in sand profiles. (a) covered filter and (b) uncovered filter

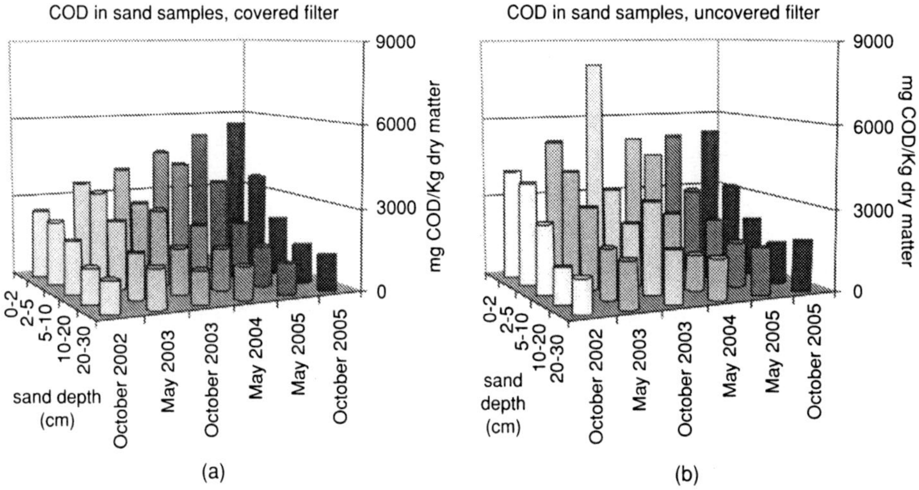

Figure 6 COD, variation with time in sand profiles. (a) covered filter and (b) uncovered filter

The concentrations of organic matter in the uncovered filter are much higher in the top layer while the background concentrations in both filters are very similar and stable over time, see figure 6a and b. With time, the organic matter in the top layers of the covered filter accumulates since no cleaning has taken place.

Microbial activity

Microbial activity in slow sand filters measured as HPC shows that the activity in both filters is high see figure 7a and b. Background activity (HPC in the deepest sampled sand layer) is

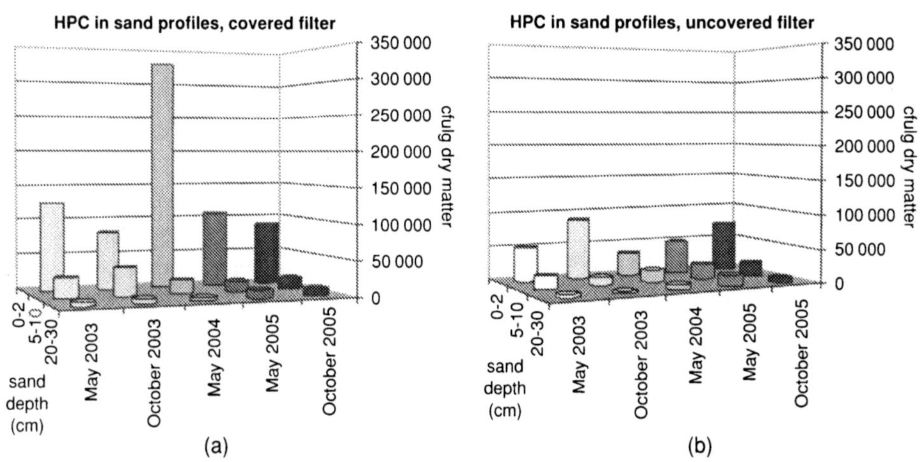

Figure 7 Heterotrophic bacteria in sand profiles, cultivated for 3 days at 22°C. (a) Covered filter and (b) uncovered filter.

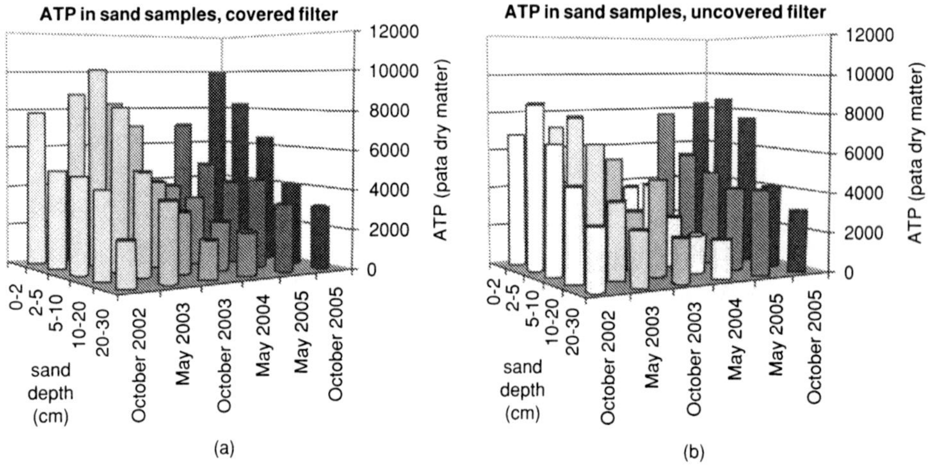

Figure 8 ATP in sand profiles. (a) covered filter and (b) uncovered filter

very similar in both filters and does not vary much over time. Surface samples shows that slowly growing HPC are much more abundant in the covered filter. In the covered filter, HPC is most abundant during spring sampling, while in the uncovered filter, HPC is most abundant at fall. These observations are believed to depend on the presence of UV-light in the uncovered filter, which has a bactericidal effect and is more dominating in spring.

Random samples showed that the only presumptive pathogen present in the surface sample is coliform bacteria, which are 10–100 times more frequent in the uncovered filter. At a depth of 20 centimetres no pathogenic bacteria can be detected in any filter.

Microbial activity measured as ATP confirms that the activity in the top sand layers is higher in the covered filter than in the uncovered, see figure 8a and b. There is no indication that the activity increases with time. Background concentrations remain more or less stable over time and there is no difference between the open and the covered filter.

Conclusions

Since the filter runtime has not yet ended, results are still indicatory. The most surprising observation so far is the filter run time, which is much above expectation. Judging from the slow, gradual increase in head loss, it is believed that cleaning will be necessary during 2006, giving the filter an expected filter runtime of over four years, which should be compared to the standard six months in uncovered filters. Similar intervals between filter cleanings have been reported from other waterworks with covered slow sand filters, for example Amsterdam and Zürich (personal contacts). The dominating cost and labour intensive part of slow sand filter operation are cleanings, and a prolonged filter runtime can greatly affect the cost effectiveness of the slow sand filters.

Other conclusions that can be drawn from this study are that despite the different appearances of the filter surfaces, for example the lack of schmutzdecke on the covered filter as well as the establishment of aquatic organisms, it is not possible to establish any differences in terms of water quality when comparing covered and uncovered filters. The effectiveness in reducing TOC, taste and odour causing substances as well as the barrier function is not affected in any negative way. The only observation that differ the covered filter from the uncovered one is the number of *Aeromonas* present in the covered filter. Whether the increased *Aeromonas* concentrations are a consequence of covering the filters has not been established, there are major differences in between individual slow sand filters when it comes to *Aeromonas* growth.

Temperature effects are only temporary with colder water during spring and warmer water during fall. Unfortunately, the differences are too small to prevent the water surface from freezing which would have made cleaning possible at all times of the year.

From the sand profiles it can be concluded that both the uncovered and the covered filter has decreasing concentrations of both impurities in the form of aluminium (from residual floc) and organic matter with increasing filter depth as well as microbial activity measured as HPC and ATP. In the uncovered filter, exposed to regular cleanings, no temporal difference can be established. In the covered filter, where all impurities are allowed to accumulate over a long period of time, a gradual increase can be observed in the top five centimetres. No changes occur deeper in the filter, which means that it is not necessary to remove more than five centimetres when cleaning the filter in the future. Despite the lack of freshly produced organic matter (algae) in the covered filter, the activity in the top layers is higher than in the uncovered filter. This effect is thought to be positive, since a higher activity in the filter generally yields better filter function. In the deeper layers of the filter there is no difference in activity, indicating no increased risk of the filter leaking bacteria of any kind to the filtered water.

Yet to be explained is the behaviour of the covered filter during cold water temperatures, which is being analysed. Whether or not covering the slow sand filters in Stockholm is a good option is totally dependent on how the rapid head loss during wintertime can be handled practically and/or solved. In case such a solution will be identified in the near future, covering the slow sand filters can drastically increase the filter run times and secure the water production at all times of the year. On top of that, covering of the filter would mean increased security and decreased risk of contamination from animals, by rain and by deposition from air.

28 Robotic cleaning of slow sand filters improves filter quality

J. Back

WEDA Water Service AB, Sweden

Abstract A new method of rinsing slow sand filters using a submersible underwater cleaning robot has proved to have several advantages over traditional rinsing methods. A study comparing traditional rinsing by a compact track loader (BobCat) with an underwater cleaning robot shows that the risk for break through of bacteria is reduced radically with the underwater robot. The underwater cleaning robot as a rinsing method also shows better results concerning the filter's ability to reduce algae content in the water. An underwater cleaning robot is also being tested at Eau de Paris water works in France where traditionally the filters are rinsed by hand. The aim with the test is to improve the working environment, reduce algae growth and secure a stable production capacity.

Keywords Rinsing, slow sand filtration, cleaning, maintenance, algae, bacteria

Introduction

A slow sand filter relies on its flora of micro-organisms to work efficiently. When it works efficiently a slow sand filter achieves 90–99% reduction in bacteria and viruses. This flora of micro-organisms is found in the "schmutzdecke", the mat of biological matter that forms on the sand surface where particles are trapped and organic matter is biologically degraded.

After a certain time and as the throughput of the filter decreases due to clogging, the surface of the filter has to be rinsed. Normally 2–3 cm of the filter surface is removed.

The removal of this top layer means a drastic decrease in the amount of micro-organisms which means that it no longer works as a bacteriological barrier and thus bacteria can pass through the filter. It takes several days for the bacteriological barrier to re-establish in the filter. During this time, it can be several days up to weeks, it is recommended that the filter is not used in the drinking water production and put to waste.

Rinsing methods with prior filter draining

There are mainly two traditional methods for rinsing slow sand filter: Manual cleaning using spades and rakes (see figure 1.1) or machine-rinsing using compact track loaders, e.g. BobCat (see figure 1.2). Both methods implies draining of the filter prior to cleaning. Filter draining means a substantial disturbance in the filter's flora of micro-organisms. This increases the time necessary for the flora of micro-organisms to re-establish in the filter.

 Recent Progress in Slow Sand and Alternative Biofiltration Processes edited by Rolf Gimbel, Nigel J.D. Graham and M. Robin Collins. ISBN: 9781843391203. Published by IWA Publishing, London, UK.

Figure 1.1 Manual cleaning of slow sand filters using high-pressure water hoses

Figure 1.2 Machine park used for cleaning slow sand filters in a Water Works in the UK

Figure 1.3 Underwater cleaning robot ready for immersion in the slow sand filter at Ivry-sur-Seine (Eau de Paris)

To minimize the time that the filter's capacity is deteriorated it is important to disturb the filter's micro-flora as little as possible (Jabur and Mårtensson (2003)). This can be done by minimizing the time the filter is drained and by using light machinery to avoid that bacteria is pushed down in the sand.

Underwater robotic cleaning – without filter draining

In a study by Jabur and Mårtensson (2003) the traditional way of rinsing has been compared to a new way using an electrical submersible underwater cleaning robot. The robot is radio remote controlled and equipped with a 1 m wide suction nozzle and a powerful submersible pump. A discharge hose connected to the pump outlet is used to discharge the sludge to the designated point of discharge. The suction nozzle can be lifted or lowered in order to control the depth of rinsing (see figure 1.3). By using a submersible robot the basin is never drained which means that the disturbance to the filters micro-organisms is minimal (see figure 1.4). By comparison using a compact track loader means draining the filter for 6–12 hours.

Advantages of using an underwater cleaning robot

- □ No need to drain the slow sand filter
- □ Minimal disturbance to the slow filters micro-organisms

Figure 1.4 Immerged robot rinsing a filter with heavy algae growth. The sediment is pumped away through a 3" discharge hose.

- The filter is not exposed to air during the rinsing so there is no need to fill up the filter from the bottom
- Pressure and vibrations on the sand surface are reduced due to the small weight (approximately 150 kg in water) of the robot compared to a compact track loader (2000 kg)
- The powerful pump removes all the dirt without causing turbidity

Method

Currently trials of a submersible robot is performed at the Eau de Paris water works in Ivry-sur-Seine. The water works with a production capacity of 300000 m^3/day has a total of 30 slow sand filters with a total surface of 22000 m^2. Traditional cleaning is performed in teams of 4 persons using high-pressure water hoses and rakes to clean off the sediment and algae. Improvements in the working environment, a possible feminization of the hard work of cleaning, increase in production capacity, decrease in algae growth and securing a stable water quality in the filtrate are among the key factors for the evaluation of the test (Back, Chamoux, Niay and Rocher (2005)). It was recently decided to expand the trials and also incorporate another Eau de Paris Water works in Joinville in the trial. It is yet to early to say what the results of these trails will be.

Jabur and Mårtensson (2003) has done practical tests to compare and evaluate robotic rinsing of slow sand filters to traditional methods. The comparison was performed at Lugnet Water Works in Nässjö and at Häggeberg Water Works, Jönköping, Sweden. Filter surfaces varied from 500–1000m^2.

Two filters at each Water Work were cleaned three times, one filter using a submersible robot, one filter using the traditional method using a compact track loader of type BobCat. Measures were taken concerning water quality, risk of break through of pathogenic bacterias after rinsing, time to re-establish the purifying capacity of the filter, filter capacity, possible penetration of particles in the lower layers of the sand and its effects, the effect of different rinsing depths and cost analysis.

Comparison of rinsing equipment

Submersible robot

- Length: 115 cm
- Width: 100 cm
- Height: 90 cm
- Weight in air: 200 kg
- Weight in water: 150 kg
- Ground pressure: 500 kg/m^2
- Pump capacity: 100 m^3/h
- Speed: 0,5 to 10 m/minute
- Rinsing depth: Variable 0–4 cm
- Power supply: 3-phase 400 V, 5.5 kW
- Equipped with underwater cameras
- Possibility to put back removed sand by equipping the machine with an onboard hydro-cyclone for cleaning of sand

Compact track loader (BobCat)

- Length: 3 m
- Width: 1,6 m
- Height: 2 m
- Weigth: 2223 kg
- Ground pressure: 1900 kg/m^2
- Powered by a 32,5 kW Diesel engine

Results and discussion

Jabur and Mårtensson (2003) show that there is a risk of break-through of bacteria after rinsing of a slow sand filter. The results indicate that this risk is higher when using a compact track loader than when using the underwater cleaning robot. After the summer cleaning substantially more heterotrophic bacteria where found in the filtrate from the filter cleaned using the compact track loader. Some fecal indicator bacteria where also found. The filtrated water the hours after rinsing would not pass the norm for drinking water. This was especially noticeable in the filter that had very fine sand.

The study further shows that the filter's capacity for removing algae is not affected when using the submersible robot. The results from the filters rinsed using the BobCat shows that the filters capacity of removing algae is decreased.

The considerable weight of the compact track loader also means that there is a risk that dirt is mixed into the lower layers of the filters when the BobCat operates on the filter surface. The results suggests that the risk of this is higher the finer sand used in the filter.

The underwater cleaning robot has a much lower pressure on the sand surface that reduces the risk of pushing dirt lower into the sand. Since the underwater cleaner uses a pump to remove the "schmutzdecke" the dirt is extracted directly from the place where it lays and is not moved around.

Even though the actual rinsing takes longer with the submersible robot compared to a compact track loader a reduction in filter downtime can be obtained. Using the compact track loader the filter was put out of production during an average period of 24 hours. The actual rinsing took 3,5–5 hours. After the rinsing the filter was replenished with water from underneath which took 18–20 hours. For the submersible robot the time the filter is put out of production is reduced to the time of rinsing which was 7–10 hours.

When using a compact track loader a layer of approximately 4,5 cm were removed. The rinsing depth of the underwater cleaning robot can be adjusted and was during the tests set at 3 cm. This means that the filter can be used for a longer period of time before sand has to be added using the submersible robot.

The cost of rinsing varies depending on size of sand filters and how often they need to be cleaned. The robot can be operated by only one person whereas manual cleaning is very labour-intensive. Traditional cleaning using compact track loaders requires heavy investment in machinery since you usually need more than just a compact track loader to clean the surface. Equipment for removing algae, making the surface flat after rinsing etc is also needed. For the Water Works in the study by Jabur and Mårtensson (2003) a cost reduction of 40% was possible.

Conclusions

- Rinsing with an underwater cleaning robot reduces the risk of break-through of bacteria.
- A slow sand filter's ability to filter algae is not reduced after rinsing with a submersible robot. Cleaning with a compact track loader reduces the filter's ability to remove algae.

- When taking into account the time to replenish the filter with water after draining the time that the filter is put out of production can be reduced by 50% using an underwater cleaning robot.
- The time between deep-rinsing can be reduced using the underwater cleaning robot
- Savings in time and money is possible by using submersible cleaning equipment

References

Jabur, H. and Mårtensson, J. (2003) Rinsing methods of slow sand filters. VA-Forsk 2003–39

Back, J., Chamoux, J-F., Niay, R. and Rocher, F. (2005) Robotic cleaning of underwater floors by in place suction. L'eau, l'industrie, les nuisances. 286, 125–127.

29 Comparison between traditional and under-water rinsing methods of slow sand filters

H.S. Jabur, J. Mårtensson*

*SWECO VIAK, Jönköping, Sweden

Abstract In a study traditional rinsing of slow sand filters by a tractor have been compared with a semi-submerged machine. The results show that the under-water rinsing can reduce the risk for breakthrough of microorganisims. The costs for filter rinsing can in longer perspective be reduced considerably by submerged rinsing.

Keywords Breakthrough, bacteria, operation time, costs.

Introduction

Rinsing is an essential process to the proper and economical operation of slow sand filters. The principal factor during filter rinsing is that the disturbance of the filter bed and especially the bacteriological layer should be minimised. The effect of rinsing will not be significant if the filter bed is mature and the disturbance of the filter bed limited. It is likely that the effect of the rinsing is influenced more by the length of time the filter is drained than by the removal of the surface layer.

The use of under-water machines with no need of dewatering may limit the disturbance of the bacteriological layer considerably and also the deterioration of the filtrate water quality. Rinsing by under-water machines will improve the working conditions and reduce the labour cost. Further reduction of labour work can be achieved by automatic operation of the process. The time when filters are out of service will be reduced as well as the amount of filtrate water lost.

The purpose of the research work was to compare different water quality parameters and how the two methods affects the operation conditions as time consumption for the rinsing, time to empty and fill up, first filtrate etc. The final report includes both a summary of the conclusions made during the project and recommendation for future use of such machines (Jabur and Mårtensson 2003).

Experiments

The study was carried out at Häggeberg Water Works in Jönköping and Lunget Water Works in Nässjö in central Sweden from 2001 to 2003. At each water works, two full-scale slow sand filters were selected for the test, which were made during three operation periods and with rinsing at about the same time.

 Recent Progress in Slow Sand and Alternative Biofiltration Processes edited by Rolf Gimbel, Nigel J.D. Graham and M. Robin Collins. ISBN: 9781843391203. Published by IWA Publishing, London, UK.

Figure 1 Rinsing at Lugnet WW, Nässjö, with a Bobcat, note the tracks.

A Bobcat tractor rinsed one of the filters, while in the other filter a semi-submerged machine was used as shown in fig1 and fig 2.

A comprehensive sampling and analysing programme have been conducted to the tests considering chemical and bacteriological parameters and algae. Samples was also taken after rinsing from different depth in the filter sand for the four filters and analysed regarding bacteria, glowing rest etc.

Figure 2 Rinsing at Lugnet WW, Nässjö, using a semi submerged machine.

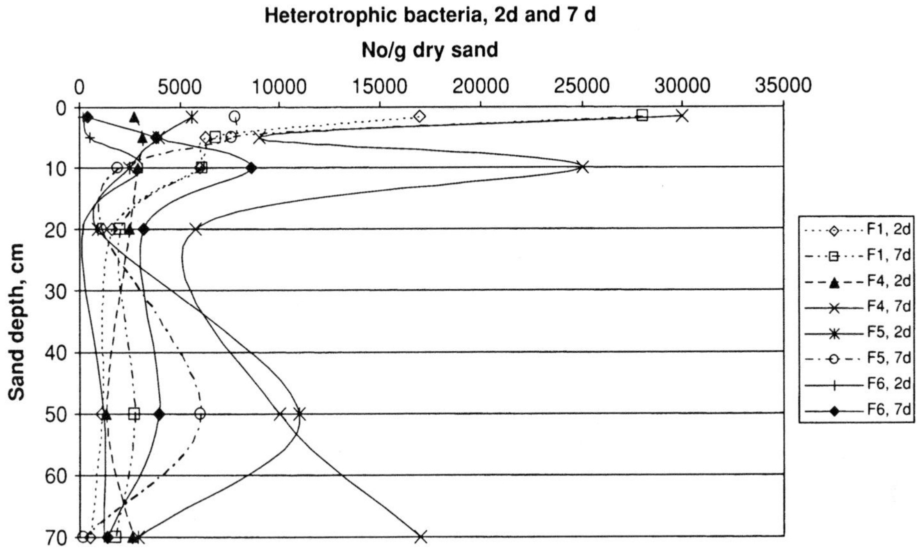

Figure 3 Heterotrophic bacteria in the filter sand at Lugnet and Häggeberg waterworks.

Results and discussion

At traditional rising with Bobcat the supernatant water level is lowered to about 10 cm below the sand surface or more. After rinsing the filter is refilled slowly from bottom using filtrate water to prevent air entrapment. The filter is put out of service for about 24 hours before operation to waste.

Using an under-water machine the filter is put out of service only for the time during rinsing. This took 7–10 hours for each filter with about 100 m^2/h machine rinsing capacity.

The top layer sand removed by a Bobcat was in average 4.5 cm, while the under-water machine was more precise and removed about 3.0 cm as intended.

Considering the filtrate water quality, deterioration could be seen after rinsing especially for bacteriological parameters. The risk for breakthrough of bacteria, also pathogen bacteria, is thus increased considerably during this period, which could be seen by analysis of the heterotrophic plate count bacteria (HRC) and selected indicator organisms (coliform bacteria, E-coli, faecal streptococci, Clostridium perfringens). At Swedish conditions this period takes 6–7 days.

Under-water rinsing may reduce the risk for breakthrough of microorganisims, since the disturbance of the sand is much larger by tractor rinsing and since lowering of the water surface itself affect the micro-flora. This could be seen in the results concerning bacteria, and algae analyses.

Algae samples were taken 24–48 hours after rinsing. The results show that the capacity of the filters to remove algae was not reduced by using under-water rinsing, but it was reduced to half at Näjssö WW and from 99% to 97% at Häggeberg WW when using a Bobcat.

Bacteriological analyses from different depths in the filter sand suggests that the bacteria has been brought down in an irregularly manor. This is most obvious in relatively fine and single graded sand by tractor rinsing (Fig 3). This shows that the magnitude of the disturbance of the biological layer depend on the characteristic of the sand.

An increase of the turbidity could be observed during the first day after rising in all the filters while the capacity of the filters to oxidise organic matter expressed as COD was affected in a very small extent.

The cost for filter rinsing can in a longer perspective be reduced considerably (up to 40%) by submerged rinsing.

Conclusions

The research results confirm that the filter rinsing in general lead to a certain negative effect on filtrate water quality and there is a risk of breakthrough of bacteria including resistant bacteria during the first days after rinsing.

The under-water rinsing may reduce the risk of breakthrough of bacteria and algae since the disturbance of the filter sand is much bigger by tractor rinsing.

The under-water rinsing will reduce the amount of filtrate water lost and improve the working conditions.

Reference

Jabur, Husam S. and Mårtensson, J. (2003) Rinsing methods of slow sand filters. *Svenskt vatten AB, VA-Forsk 2003-39.*

30 Extension of slow sand filter running times by protection layers

H.-J. Mälzer*, R. Gimbel*,**

*IWW Rhenish-Westphalian Institute for Water Research, Moritzstr.26, 45476 Muelheim a. d. Ruhr, Germany (a.maelzer@iww-online.de)
**University of Duisburg-Essen, Institute of Energy and Environmental Engineering/Water Technology, Bismarckstr. 90, 47048 Duisburg, Germany (gimbel@uni-duisburg.de)

Abstract The running time of slow sand filters is limited by the increase of pressure drop which is mainly caused by the surface filtration of suspended solids. Surface filtration mechanisms on the top of filter sand layer can be avoided e.g. by protection layers. Choosing suitable materials for protection layers a considerable extension of slow sand filter running times can be achieved.

Keywords Slow sand filter, protection layer, running time extension

Introduction

By slow sand filtration suspended solids are removed and dissolved organic compounds are degraded by microorganisms. These processes mainly take place on the surface and in the first centimetres of the filter and cause an increase of the pressure drop in this layer due to the removal of particles and due to the growth of microorganisms. If the pressure drop exceeds the maximum available height the surface of the slow sand filter must be scraped to remove the first few centimetres and a new filtration cycle can be started. One possibility to extend the running time is the use of protection layers to avoid a fast increase of the pressure drop and to expand the removal mechanisms into the protection layers. With the financial support of the German Federal Ministry of Education and Research (BMBF) the suitability of different natural and synthetic materials for the running time extension of slow sand filters has been investigated (Mälzer and Gimbel, 1999; Mälzer, 2005).

Experimental plant and tested materials

Four effluent controlled filters (Filter F1 ... F4) of 0.5 m diameter have been operated in parallel with the effluent of a waste water treatment plant. The effluent of a waste water treatment plant was used as raw water to simulate the raw water abstraction from a river which is polluted by domestic wastewater but still has a certain self cleaning capacity.

 Recent Progress in Slow Sand and Alternative Biofiltration Processes edited by Rolf Gimbel, Nigel J.D. Graham and M. Robin Collins. ISBN: 9781843391203. Published by IWA Publishing, London, UK.

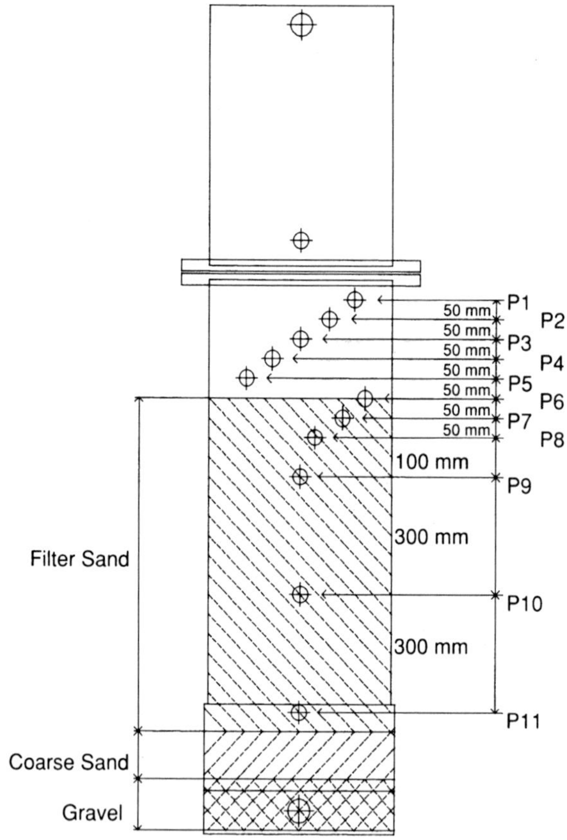

Figure 1 Design of the filter columns

The design of the filter columns is shown in Figure 1. The protection layers have been always applied on the surface of the filter sand. Therefore, the total height of the filter bed (filter sand plus protection layer) was increased according to the application of the protection layers. Filter F3 was always operated as a reference filter without protection layers. The sampling points at the filter columns which are listed in Table 1 had been of special interest for the investigations.

In a first set of experiments carried out in the year 1998, the following synthetic materials have been tested for the extension of the filter running times: geotextiles, porous foam mats (porosity: 60 pores per inch) and so-called permeable synthetic collectors (PSC), which are small cylinders made of reticulated foam (porosity: 60 pores per inch). Two types of PSC have been tested: large PSC (diameter: 10 mm, height: 10 mm) and small PSC (diameter: 5 mm, height: 5 mm). Both types are shown in Figure 2. The foam mats and the PSC have been placed in layers of different heights (5, 10 and 15 cm) on the surface of the filter sand, the heights of the geotextile layers had been 3.2, 6.4 and 9.6 mm. The grain size of the filter sand was 0.4 – 0.8 mm. The filter velocity was kept constant at 0.4 m/h and the height of supernatant was kept constant at about 0.9 m above the sand layer.

Table 1 Location of sampling points of special interest at the filter columns

Sampling point	Location
P2	20 cm above filter sand surface
P3	15 cm above filter sand surface
P4	10 cm above filter sand surface
P5	05 cm above filter sand surface
P6	surface of the filter sand
P7	05 cm below filter sand surface
P11	80 cm below filter sand surface

In a second set of experiments carried out in the years 2003/2004 the following natural materials have been tested as protection layers: gravel (grain size 3.1 – 5.6 mm), pumice stone (grain size 1.0 – 4.0 mm) and coconut fibres (mean fibre length 2 cm, mean fibre diameter 0.09 mm). The protection materials have been placed in layers of 20 cm height on the surface of the filter sand. The grain size of the filter sand was 0.2 – 2.0 mm. The filter velocity was kept constant at 0.2 m/h.

Results and discussion

First set of experiments

As an example for the results of the first set of experiments, the increase of pressure drop in the filters with the small PSC as protection layers are presented. Figure 3 shows the development of the pressure drop in the small PSC layers and in different layers of the filters. The height of the PSC layers on the top of the filter sand was 10 cm in filter F1, 15 cm in filter F2 and 5 cm in filter F4. Filter F3 was operated without a protection layer. The pressure drop in the first 5 cm (between P6 and P7) of the filter sand in filter F3 increased within a few days and the first 1–2 cm of the sand had to be scraped, cleaned and refilled

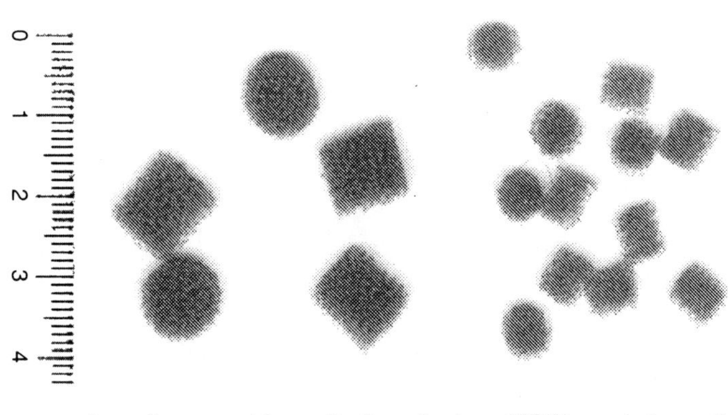

Figure 2 Large and small permeable synthetic collectors (PSC), scale in centimetres

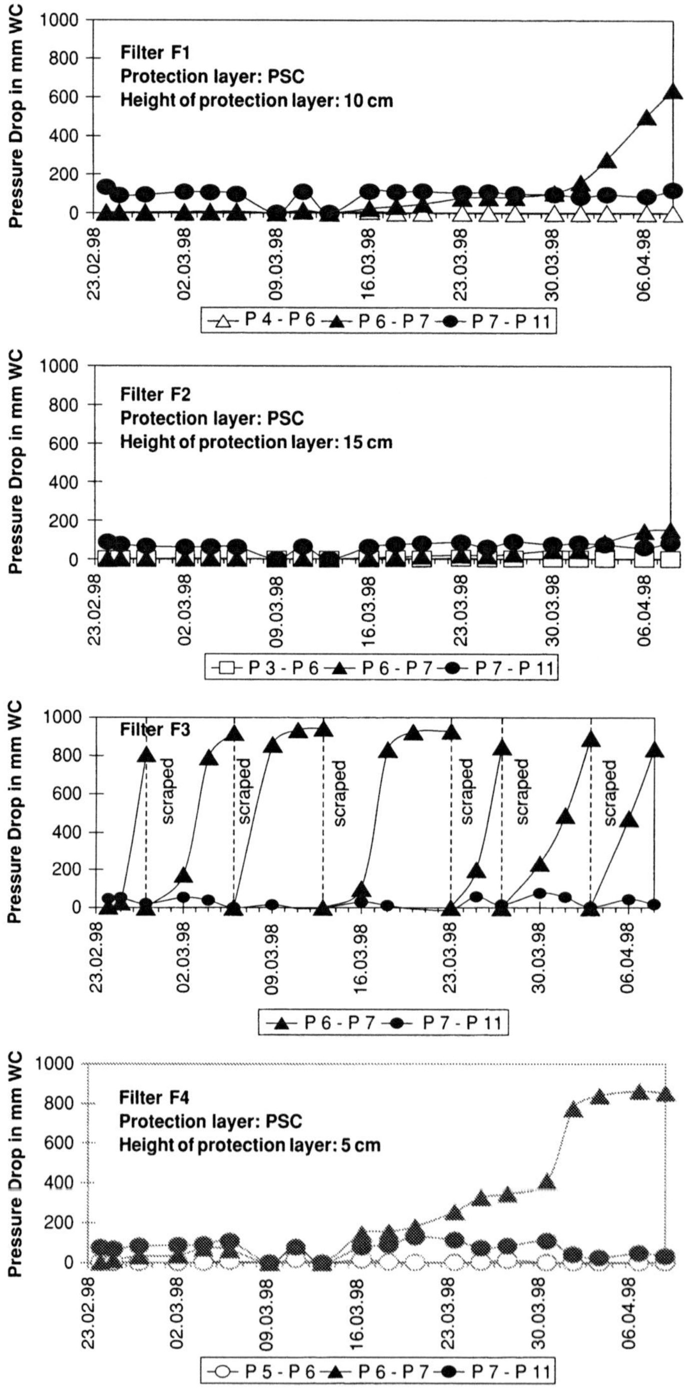

Figure 3 Pressure drop in mm water column (WC) between different sampling points of the filters covered with different heights of small PSC protection layers (first set of experiments)

before the filter F3 was restarted. In deeper filter layers no increase of the pressure drop could be observed because the suspended solids have been removed mainly by surface filtration mechanisms. With increasing height of the PSC layer the increase of the pressure drop between the sampling points P6 and P7 became slower because particles which may block the surface of the filter sand have been removed by the PSC protection layer. The higher the protection layer the longer was the filter running time. The pressure drop in the PSC layers showed nearly no increase. Particles have been deposited between the collectors as well as inside the collectors which led to high particle loads of the PSC layers and only to a very low increase of the pressure drop.

Second set of experiments

The development of the pressure drop in the filters covered with the natural protection materials is shown in Figure 4. In filter F1, a layer of 20 cm of gravel (3.1 – 5.6 mm) was applied on the surface of the filter sand, the surface in filter F2 was protected by a layer of 20 cm of pumice stone (1.0 – 4.0 mm) and filter F4 was protected by a layer of 20 cm of coconut fibres. Filter F3 was operated without a protection layer. As already observed in the previous test, the pressure drop in the first 5 cm (between P6 and P7) of the filter sand in filter F3 increased within a few days and the filter had to be scraped. In Filter F1, a slow increase of the pressure drop could be observed after a running time of about 5 weeks.

The increase occurred in the first 5 cm of the filter sand and indicated that particles broke through the protection layer and have been removed on the surface or within the first 5 cm of the sand layer. In the protection layer no significant increase of the pressure drop occurred. In Filter F2, an increase of the pressure drop could be observed after a running time of about 8 weeks. The increase occurred in the protection layer and indicated, that the protection layer was finally blocked by the retained particles. In Filter F4, an increase of the pressure drop could be observed after a running time of about 12 weeks. Again, the increase occurred in the protection layer.

Corresponding to the increase of the pressure drop in the gravel protection layer of filter F1 an increase of the turbidity at the sampling point F1P6 (between filter sand and protection layer) has been observed. This increase also corresponded with an increasing breakthrough of particles though the gravel protection layer. The other protection layers of natural materials showed only a slight increase or even a decrease of the particle concentration with increasing running time (Mälzer and Gimbel, 2005).

In contrary to the slow sand filter layers, which act mainly as a surface filter, the protection layers act mainly like a deep bed filter. The behaviour of the protected filters is finally determined by the behaviour of the protection layer. A breakthrough of particles through the protection layer causes the increase of the pressure drop in the filter sand layer and the termination of the filter run. Does the removal of particles in the protection layer cause an increase of pressure drop in the protection layer, the filter run will be terminated as well.

Calculation of running time extension

For further evaluations of the results the following procedure was applied, which is described in detail by Mälzer and Gimbel (2002). The increase of the filter resistance in the total slow

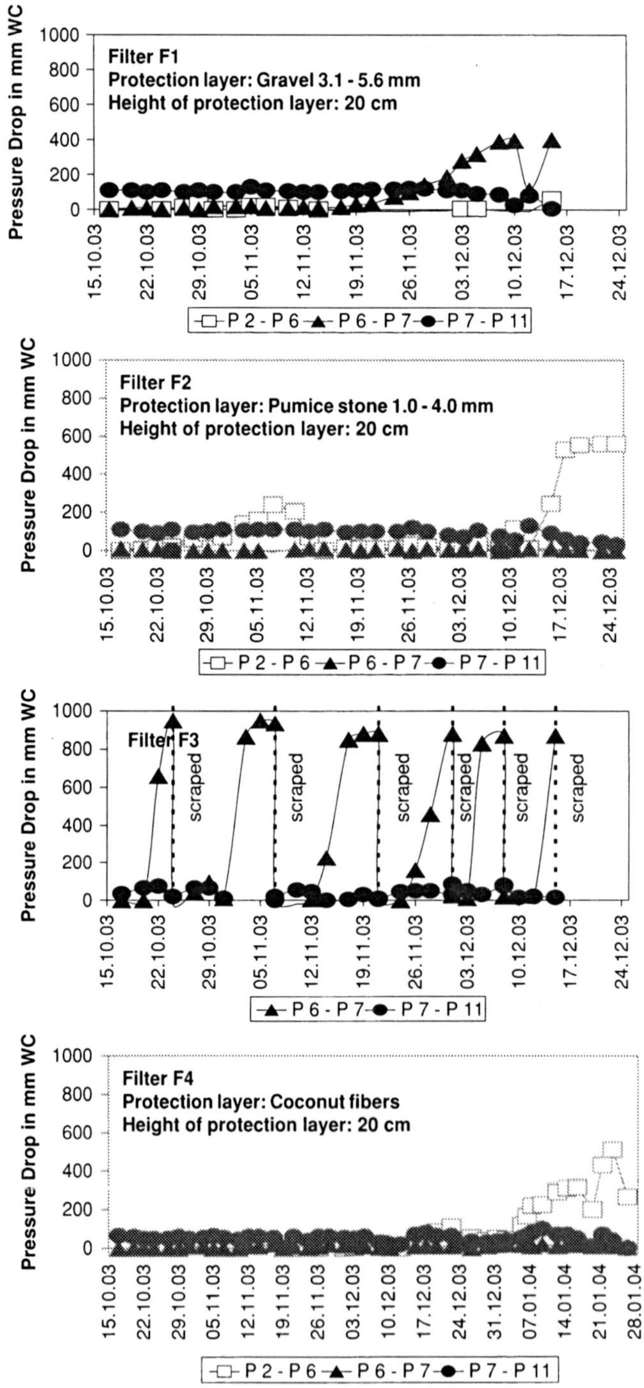

Figure 4 Pressure drop in mm water column (WC) between different sampling points of the filters covered with different natural materials as protection layers (second set of experiments)

sand filters (filter sand layer plus protection layer) due to the removal of suspended solids was calculated by the following equation:

$$\Delta R = \Delta p/v/\eta$$

ΔR = increase of filter resistance in 1/m

Δp = increase of pressure drop in N/m^2

v = filter velocity in m/s

η = viscosity of water in kg/m/s

To compare the results from different filter runs and tested protection materials the following standard conditions have been defined:

filter velocity: 0.4 m/h for the first set of experiments
0.2 m/h for the second set of experiments
viscosity of water: $1*10^{-3}$ kg/m/s
influent concentration of suspended solids: 10 mg/l

Furthermore, the "standard filter running time" was defined as the time after which the pressure drop increases under standard conditions by 200 mm water column (respectively the filter resistance increases by $1.77*10^{10}$ 1/m in the first set of experiments and by $3.54*10^{10}$ 1/m in the second set of experiments). The factor of running time extension is the multiple of the running time which could be observed in the filter without protection layer.

A comparison between the extensions of the filter running times which could be achieved by the application of different protection materials is shown in Table 2. Best results of the synthetic protection materials showed the 15 cm protection layer of small PSC by which a running time extension by the factor of more than 20 was possible. By the use of geotextiles a running time extension only by a factor of about 2 was possible and a formation of a filter cake on the top of the geotextile protection layers could be observed. The coconut fibres have been the best of the natural materials for the running time extension.

Removal of selected water compounds

The removal of selected water compounds by different slow sand filters (partly covered by protection layers) is shown in Table 3 and Table 4. Concerning the removal of suspended and dissolved water compounds no significant difference between the filters with and without protection layers could be observed. The degradation of dissolved organic substances was only poor and a high oxygen consumption was observed which was partially caused by the biodegradation of the accumulated organic suspended solids on the sand surface respectively in the protection layers.

Conclusions

The use of protection layers for the extension of slow sand filter running times can be recommended under the condition that biodegradation processes of dissolved organic compounds and nitrification have been almost completely happened in the waste water treatment plant.

Table 2 Minimum running time extensions of slow sand filters covered by different protection layers

Protection layer material: Small permeable synthetic collectors (small PSC)				
Height of protection layer in cm	5	10	15	-
Height of pr Factor of running time extension	6	14	20	-
Protection layer material: Large permeable synthetic collectors (large PSC)				
Height of protection layer in cm	5	10	15	-
Factor of running time extension	3	7	7	-
Protection layer material: Foam mats				
Height of protection layer in cm	5	10	15	-
Factor of running time extension	2	5	5	-
Protection layer material: Geotextiles				
Height of protection layer in mm	5	10	15	-
Factor of running time extension	about 2	about 2	about 2	-
Protection layer material: Gravel 3.1–5.6 mm				
Height of protection layer in cm	-	-	-	20
Factor of running time extension	-	-	-	4
Protection layer material: Pumice stone 1.0–4.0 mm				
Height of protection layer in cm	-	-	-	20
Factor of running time extension	-	-	-	4
Protection layer material: Coconut fibres				
Height of protection layer in cm	-	-	-	20
Factor of running time extension	-	-	-	10

Table 3 Arithmetic means of influent concentrations and removal of selected water compounds by the slow sand filters during the first experimental set

Parameter	**Influent**	**Removal in % (between influent and P11)**			
		Filter F1 protect. layer: 10 cm Small PSC	**Filter F2 protect. layer: 15 cm Small PSC**	**Filter F3**	**Filter F4 protect. layer: 5 cm Small PSC**
Turbidity	2.8 FNU	88.2	85.2	86.4	89.3
O_2	5.0 mg/l	88.7	88.2	86.4	85.4
DOC	8.4 mg/l	9.5	8.3	3.7	7.0

Table 4 Arithmetic means of influent concentrations and removal of selected water compounds by the slow sand filters during the second experimental set

Parameter	**Influent**	**Removal in % (between influent and P11)**			
		Filter F1 protect. layer: 20 cm Gravel	**Filter F2 protect. layer: 20 cm Pumice stone**	**Filter F3**	**Filter F4 protect. layer: 20 cm Coconut fibres**
Turbidity	10.2 FNU	92.3	93.2	91.1	92.1
O_2	5.8 mg/l	90.9	78.7	87.1	87.3
DOC	8.6 mg/l	16.4	18.6	14.3	18.0

Acknowledgement

The authors thank the German Federal Ministry of Education and Research (BMBF) for funding of the investigations (project codes 02WA9511 and 02WT0282). The authors are responsible for the content of this publication.

References

Mälzer, H.-J.; Gimbel, R. (1999). Untersuchungen zur Laufzeitverlängerung von Langsamsandfiltern durch kontinuierliche Teilregenerierung und/oder Einsatz von Schutzschichten. Schlussbericht zum BMBF-Forschungsvorhaben 02 WA 9511, Mülheim a.d.R., Germany.

Mälzer, H.-J; Gimbel, R. (2002). Protection layers for the extension of slow sand filter running times in wastewater reuse. *Water Science and Technology: Water Supply*, **2**(1), 305–309.

Mälzer, H.-J. (2005). Exportorientierte Forschung und Entwicklung auf dem Gebiet der Wasserver- und –entsorgung. Teil 1: Trinkwasser. Teilprojekt A2: Optimierung und Erweiterung des Einsatzes von Langsamfiltern durch spezielle Auflageschichten. Schlussbericht zum BMBF-Forschungsvorhaben 02 WT 0282, Mülheim a.d.R., Germany.

Mälzer, H.-J; Gimbel, R. (2005). Protection layers for the extension of slow sand filter running times in wastewater reuse. In *Proceedings of the IWA International Conference on Particle Separation 2005, June 1–3, Seoul, Korea.*

31 Study on the application of alternative filter materials using slow sand filtration

U. Hütter, F. Remmler

Institut für Wasserforschung GmbH, Zum Kellerbach 46, 58239 Schwerte, Germany

Abstract The low-cost use of suitable filter sands with reproducible characteristics for slow sand filtration is usual in Europe but not naturally practicable in other developed or developing countries. The aim of this study was to investigate alternative low-cost filter materials (crushed recycled glass, coconut fibre) in laboratory scale experiments as well as in larger scale experimental facilities for their application using slow filtration. The conducted experiments simulated climatic conditions (temperature 5–10°C, 20°C and 30°C) and raw water compositions (with a high content of DOC and ammonium) deviating from Central European circumstances to check whether slow filtration can be adapted to different conditions in other countries. Relevant operating conditions (filtration rate, continuous or intermittent operation) were included in the investigations. The results of the investigations showed that – after taking into account a start-up period comparable to a standard filter sand – the use of recycled crushed glass or coconut fibre for slow sand filtration, in general is possible. The alternative filter materials recycled crushed glass and coconut fibre were characterized by specific advantages and disadvantages compared to standard filer sand used in Central Europe. These aspects have to be considered for their use for drinking water production under local conditions.

Keywords Ammonium, coconut fibre, crushed recycled glass, DOC, sand, slow sand filtration, temperature

Introduction

For more than 100 years slow sand filtration (SSF) has been recognized in Europe as a natural and cheap method of water treatment, and in its many years of use it has become a tested and scientifically well researched procedure. SSF belongs to the established and reliable treatment procedures in drinking water production. Beside the low energy requirement, the low waste production, the low maintenance requirements and the use of locally available materials, the ability to customize the procedure to individual requirements and conditions is advantageous. However, the low-cost use of suitable filter sands with reproducible characteristics for SSF is usual in Europe but not naturally practicable in other developed or developing countries.

This paper reports on investigations made in the subproject "Limitations of slow sand filtration, possibilities of technical modifications and adaptation to local conditions", which

edited by Rolf Gimbel, Nigel J.D. Graham and M. Robin Collins. ISBN: 9781843391203. Published by IWA Publishing, London, UK.

is part of the joint research project "Adjusted water treatment technologies and water distribution under regional conditions". One aim of this study was to investigate alternative low-cost filter materials in laboratory scale experiments as well as in larger scale experimental facilities for their application using slow filtration. The experiments simulated climatic conditions (temperature 5–10°C, 20°C and 30°C) and raw water compositions (with a high content of DOC and ammonium) deviating from Central European circumstances to check whether slow filtration can be adapted to different conditions in other countries. Relevant operational conditions (filtration rate, continuous or intermittent operation) were investigated as well.

Materials and methods

After preliminary examinations the alternative filter materials "recycled crushed glass" and "coconut fibre" were chosen for further experiments. Both materials are readily available in many parts of the world. Water filtration using coconut fibres in combination with burnt rice husks is used e.g. in Southeast Asia and Pakistan (Frankel, 1977 and 1981; Barnes and Mampitiyarachichi, 1983; Heber, 1985; Mughal, 2000). Recycled crushed glass was already investigated in different studies for its performance in a potable water treatment application (CWC, 1995; Evans *et al.*, 2002; Rutledge and Gagnon, 2002; Dryden Aqua Ltd., 2004a and 2004b).

In laboratory scale experiments the effects of temperature and filter material on the purification efficiency were investigated (figure 2.1, left part). The laboratory filter columns were designed to simulate slow filtration conditions (continuous operation with constant supernatant, filtration rate 2.2 m/d). The applied filter materials are characterized in table 2.1.

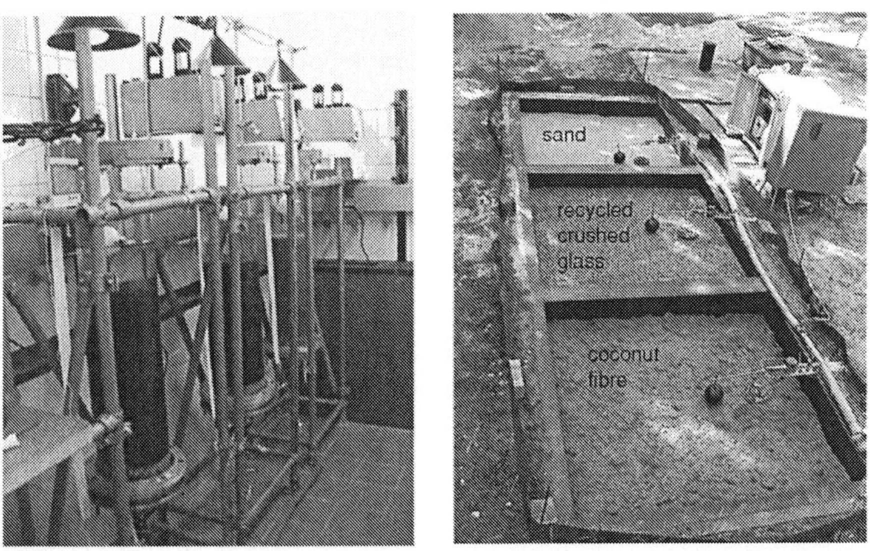

Figure 2.1 Pilot test plants: Filter columns in an air-conditioned room (left part) and pilot plant filter basins in the field (right part)

Table 2.1 Characteristics of the applied filter materials

	sand medium	recycled crushed glass medium*	coconut fibre medium**
Media identification	River sand	brown and green hollow glass	shredded coconut fibre
Particle size distribution [mm]	0.2–2	0.2–5	–
Effective size [mm]	0.28	0.55	–
Unconformity coefficient [-]	2.3	5.2	–

* Preliminary treatment: washing and disinfection (2 h at 180°C)
** Fibre: average length ca 2 cm, average diameter ca 0.09 mm

The filter influent was tempered and aerated surface water (river Ruhr), which in addition was spiked with DOC and ammonium. The experiments were carried out in an air-conditioned room at three different temperatures (5–10°C, 20°C and 30°C) and the filter materials were renewed after each test.

The subsequent experiments conducted in a semi-technical field pilot plant focused on the influence of operating conditions (filtration rate, continuous or intermittent operation) on the filter performance (figure 2.1, right part). For the simulation of climatic conditions deviating from Central Europe the surface water of the river Ruhr (= influent) was warmed up to ca 30°C and additionally spiked with DOC and ammonium. During the operation time four different tests were conducted without a change of filter material between each test:

Stage 1: continuous operation, filtration rate 2.2 m/d,
Stage 2: intermittent operation, filtration rate 2.2 m/d,
Stage 3: intermittent operation, filtration rate 3.8 m/d,
Stage 4: continuous operation, filtration rate 1.0 m/d.

Samples of influent and effluent of the filter columns and the semi-technical field filter basins were analysed for their temperature, conductivity, oxygen, oxygen reduction potential, turbidity, SAK_{254}, pH, heavy metals (copper, iron, manganese, nickel and zinc), DOC, TOC, nitrate, nitrite, ammonium and sulphate. Microbiological parameters were analysed and assessed in another subproject of the joint research project. Sampling procedures and determination of chemical parameters were accomplished according to German standard methods (Drinking Water Ordinance, 2001).

Results and discussion

The mean concentrations of DOC and ammonium in the influent of the test plants are listed in table 3.1. With increasing temperature the DOC and ammonium concentrations had to be reduced to avoid anaerobic filter conditions.

Figure 3.1 illustrates the relative DOC and ammonium concentrations in the column effluents (coconut fibres were not investigated at the temperature 5–10°). Concerning the reduction of DOC a comparable development of the relative concentrations was detected for the different filter materials (figure 3.1, left part). The range of the C/C_0 – ratio grew with increasing temperatures. Recycled crushed glass and sand showed rather similar results whereas the coconut fibres were characterised by an organic emission (predominantly

Table 3.1 Mean concentrations of DOC and ammonium in the influent of the test plants

		influent (= spiked surface water)	
test plant and temperature		**mean DOC [mg/L]**	**mean ammonium [mg/L]**
column tests			
	5–10°C	7 mg/l	5.2 mg/l
	20°C	8 mg/l	2.9 mg/l
	30°C	5 mg/l	1.9 mg/l
semi-technical field tests			
	30°C	5 mg/l	1.4 mg/l

humates). This effect increased with higher temperatures and was even more intense in the start-up stages of the tests.

The degradation of ammonium was similar for all tested filter materials (figure 3.1, right part). The results proved the strong dependence of this process on the temperature. An optimum for nitrification is given at a temperature between 28 and 36°C, below 12°C this

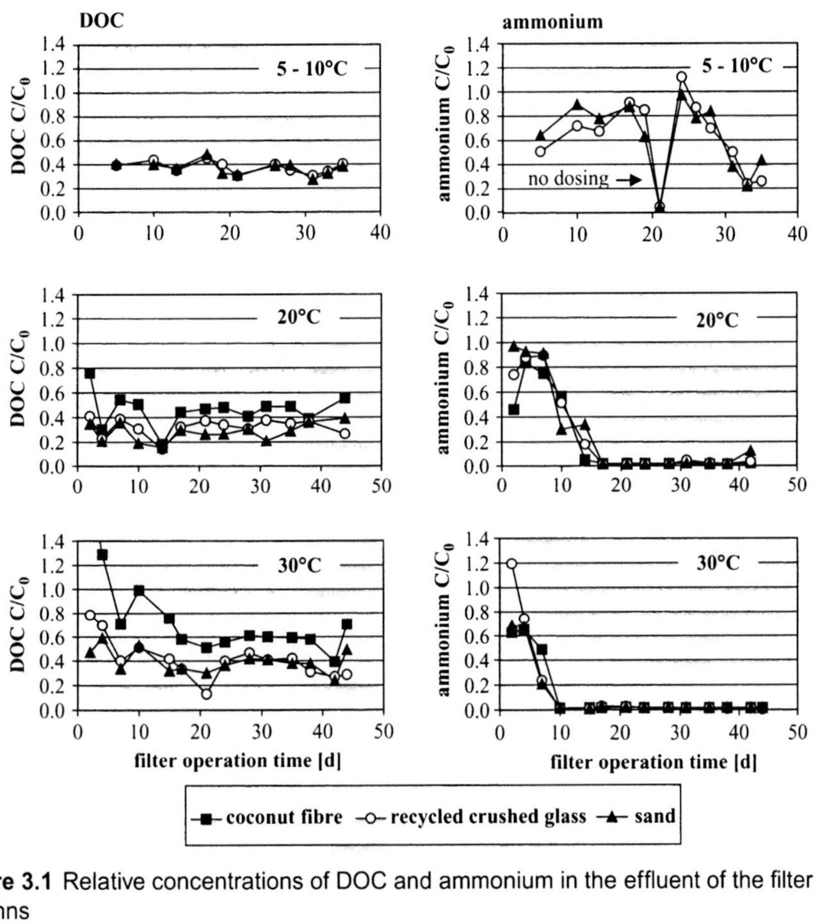

Figure 3.1 Relative concentrations of DOC and ammonium in the effluent of the filter columns

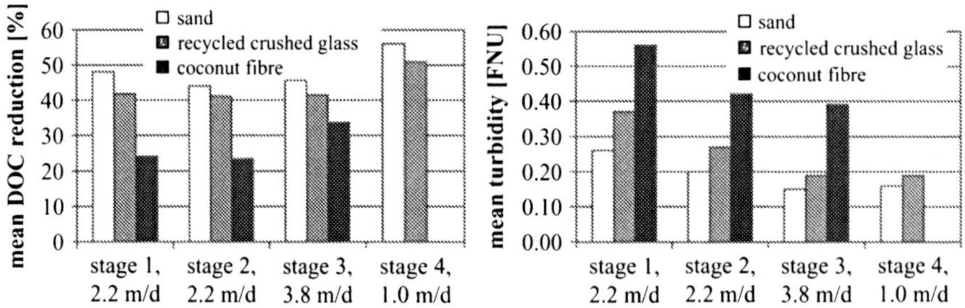

Figure 3.2 Mean DOC reduction in the filters (left part) and mean turbidity of the filtrates (right part) depending on the filtration rate (results achieved during the semi-technical field tests)

process slows down and below 8°C it nearly stops. The pH value of the effluents ranged from 7 to 8 which is also an optimum for nitrification. The development of the concentrations of nitrate and nitrite on the other hand was different for the tested filter materials. Compared to the other filter materials for the coconut fibres a significant accumulation of nitrite was determined but only over a shorter period and with a lower concentration. In the effluents of the recycled crushed glass and sand the concentrations of nitrites were about two times higher as in the effluent of the coconut fibres. This effect was also determined at the pilot scale test plant. In contrast to the recycled crushed glass and sand, the coconut fibre filter showed a much lower accumulation of nitrate.

Figure 3.2 illustrates the effect of the filtration rate on the turbidity and the DOC reduction. The filtration rate showed only a little effect on the DOC reduction in the filters (figure 3.2, left part). With a filtration rate of 2.2 m/d and 3.8 m/d the recycled crushed glass filter and the sand filter had approximately the same mean reduction efficiency of 40 – 50%. With the lower filtration rate 1 m/d a mean DOC reduction above 50% was detected. In all, the coconut fibre filter was characterized by a lower DOC reduction. This effect was influenced by the already mentioned organic emission of this filter material. The mean turbidity showed little changes during the submerged phases (figure 3.2, right part). The mean concentrations of the stage with the highest and the lowest filtration rate were almost equal. As expected the turbidity reduction enhanced with increasing filter operation time.

The observed effects of the filtration rate on the purification capacity of the filter materials are summarized in table 3.2. The highest degree of purification for the sand filter and the recycled crushed glass filter were detected with a filtration rate of 1 m/d. Hereby it has to be considered that this stage (stage 4, continuous operation) represents the period with the longest filter operation time and the observed purification effect could be superimposed by the fact of a longer filter run.

In table 3.3 the achieved mean DOC and turbidity reduction in the different filter materials under the defined conditions at the test plants are given. The filtrates showed no great differences in turbidity. All filtered water turbidity levels were within typical target levels for drinking water treatment processes, except in start-up stages or shortly in restart stages during intermittent operation. The results summarised in table 3.3 prove for the

Table 3.2 Influence of the filtration rate on the purification capacity of the filters (semi-technical field tests)

	sand			recycled crushed glass			coconut fibre		
Filtration rate [m/d]	**1.0**	**2.2**	**3.8**	**1.0**	**2.2**	**3.8**	**1.0**	**2.2**	**3.8**
DOC	+	o	o	+	o	o	n.a.	o	+
turbidity	o	o	o	o	o	o	n.a.	o	o
ammonium	o	o	o	–	o	o	n.a.	o	–
nitrite	o	o	–	–	o	–	n.a.	o	o
nitrate	+	o	o	+	o	o	n.a.	o	o
heavy metals	o	o	o	o	o	o	n.a.	o	o
-		purification			capacity		below		average

n.a. = not available
\+ purification capacity above average
o no change in purification capacity.

recycled crushed glass and the sand that nearly similar reductions for DOC and turbidity were detected. The coconut fibre filter showed a little lower DOC reduction.

Coconut fibres and sand showed a comparable reduction efficiency concerning the removal of iron and zinc. For the recycled crushed glass a distinct lower elimination for zinc in general and for iron at the temperature 5–10°C was detected. In all, the results of the investigations proved a slightly better reduction efficiency with increasing temperatures.

Especially in the start-up stages the filtrate of the coconut fibre filter was characterized by a pH value below 7. This organic filter material also showed at higher temperatures a higher oxygen reduction during the filtration process compared to the other filter materials.

The sand showed the highest head loss during all tests because of the particle size distribution (table 2.1). The sand filter and the coconut fibre filter had to be cleaned during the pilot scale field tests (high algae vegetation) whereas the recycled crushed glass in all cases showed the longest filter operation times. Evans *et al.* (2002) investigated a glass medium that was similar to that of the tested typical sand medium (concerning effective size and unconformity coefficient). In this study the glass medium had the benefit of taking 10–15%

Table 3.3 Mean DOC and turbidity reduction of the filter materials

filter material	mean DOC reduction	mean turbidity reduction	temperature
sand	60	60	5–10°C
	70	80	20°C
	60/45–55*	70/90–95*	30°C
recycled crushed glass	60	65	5–10°C
	70	70	20°C
	60/40–45*	65/90*	30°C
coconut fibre	55	65	20°C
	25/30–40*	25/85*	30°C

* Semi - technical field pilot plant

longer than the sand to reach particle breakthrough, and it also appeared to accumulate head loss in most runs at a slightly lower rate than the sand.

The pilot scale field investigations proved that intermittent filter operation (stage 2 and stage 3) causes a high mobilisation of substances during restart stages. This effect was much more intense at the higher filtration rate in stage 3.

In spite of the suitability of the investigated filter materials, on principle, the chosen test conditions (raw water quality, filter depth, temperature, filter operation, filtration rate) in some cases (ammonium, nitrite, manganese) led to an exceeding of the relevant advisory values given by the Drinking Water Ordinance (2001) and the WHO-Guidelines (2004). The semi-technical investigations in the field proved that only the continuously operated coconut fibre filter was able, after the start-up stage, to produce a drinking water quality which met the WHO-Guidelines (2004).

Conclusions

The results of the investigations showed that – after taking into account a start-up period comparable to a standard filter sand – the use of recycled crushed glass or coconut fibre for slow sand filtration, in general is possible. The alternative filter materials recycled crushed glass and coconut fibre were characterized by specific advantages and disadvantages compared to standard filter sand used in Central Europe (recycled crushed glass: longer filter run / lower elimination of heavy metals, coconut fibre: small formation of nitrite and degradation of nitrate / organic emission in start-up stages, higher oxygen reduction in the filter). These aspects have to be considered for their use for drinking water production under local conditions.

The presented investigations assessed the purification capacity of the sole treatment step slow filtration. However, in Central Europe slow filtration is always a part of a modular system of suitable pre- and post-treatment steps depending on the local conditions and quality demands. In principle, to guarantee a constant drinking water quality, a permanent further development and local optimisation of the multi-barrier-system needs to be ensured.

Acknowledgements

This study was conducted with subsidy from the German Ministry of Education and Research (BMBF) and is part of the joint research project “Adjusted water treatment technologies and water distribution under regional conditions”.

References

Barnes, D. and Mampitiyarachichi, T.R. (1983) Water filtration using rice hull ash. Waterlines **2**, 21–23.

CWC Clean Washington Center (1995) Evaluation of crushed recycled glass as a filtration medium in slow sand filtration. Report No. GL-95-4. (http://www.cwc.org /glass/gl_htm/ gl954rpt.htm).

Drinking water ordinance (2001) *Bundesgesundheitsblatt* Teil I, Nr. 24, 959–980.

Drydren Aqua Ltd. (2004a) *Project to examine the use of recycled glass as a filtration media in the treatment of water and wastewater*, Project GLA2-020, Published by The Waste & Resources Action Program (WRAP), Banbury.
Drydren Aqua Ltd. (2004b) *Treatment of drinking water using recycled glass*, Project GLA2-020. Interim Report 1, Published by The Waste & Resources Action Program (WRAP), Banbury.
Evans, G., Dennis, M., Cousins, M. and Campbell, R. (2002) Use of recycled crushed glass as a filtration medium in municipal potable water treatment plants. *Water Supply* **2**(5–6), 9–16.
Frankel, R.J. (1977) The coconut fibre/burnt rice husks filter for supplying drinking water to rural communities in Southeast Asia. *Asian Environmental* **1**(2), 111, 12–20.
Frankel, R.J. (1981) Design, construction and operation of a new filter approach for treatment of surface waters in Southeast Asia. *Journal of Hydrology*, 51, 319–328.
Heber, G. (1985) *Simple methods for the treatment of drinking water*, Verlag F. Vieweg & Sohn, Braunschweig.
Mughal, F.H. (2000) A low-cost water treatment system. DAWN – Business 17.07.2000 (http://www.dawn.com/2000/07/17/ebr4.htm).
Rutledge, S. O. and Gagnon, G. A. (2002) Comparing crushed recycled glass to silica sand for dual media filtration. *Journal of Environmental Engineering Sciences* **1**, 349–358.
World Health Organization (WHO) (2004) *Guidelines for drinking water quality*. 3rd edn, viol. 1, Geneva.

32 The operation, flow conditions and microbial reductions of an intermittently operated, household-scale slow sand filter

M.A. Elliott*, C.E. Stauber, F. Koksal, K.R. Liang, D.K. Huslage, F.A. DiGiano, M.D. Sobsey.

*University of North Carolina, CB 7431, Chapel Hill, NC, 27514, USA.

Abstract Nearly one-fifth of the world's population lacks access to safe, reliable sources of drinking water. Point of use (POU) household water treatment technology allows people to improve the quality of their water by treating it in the home. A promising emerging POU technology is the biosand filter (BSF). The BSF is a household-scale, intermittently operated slow sand filter that maintains a wet media bed containing a schmutzdecke and allows periodic water dosing by the user. Step input chemical tracer tests indicated that the BSF operates at near-plug flow conditions. Six-to-eight week longitudinal challenge studies were conducted with daily charges of surface water spiked with *E. coli* strain B bacteria, coliphages MS2 and PRD-1 and human enteric virus echovirus type 12. The BSF ripened in a manner similar to conventional SSFs. Flow rate slowed and microbial reductions improved over time with ripening. *E. coli* reductions were ~90% following filter startup but improved to 95–99.5% over time. Microbial reductions were greater with greater residence time within the filter, especially for water retained in the filter bed overnight. *E. coli* and echovirus 12 reductions were greater than those of coliphages MS2 and PRD-1.

Keywords Slow sand filtration, intermittent, point of use, water treatment, developing countries

Introduction

More than one billion people in the developing world lack access to improved sources of drinking water (WHO, 2004). For those with access to safe water, subsequent contamination during transport or household storage can pose significant health risks. Three-to-five billion cases of diarrhea occur annually and lead to 1.8 million deaths (WHO, 2004) in addition to massive morbidity and economic consequences.

The preferred solution to the problem of unsafe water causing disease is to provide universal access to inexpensive, safe, reliable piped water supplies. However, in many nations that reality is many decades away because authorities can not create the infrastructure and deliver reliable service. In the meantime, families obtain their own drinking water wherever they can, often from contaminated, unsafe sources. Most consume this water

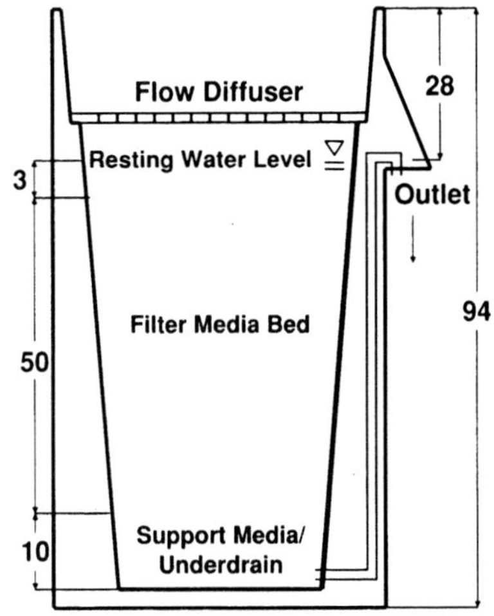

Figure 1.1 Cross-section of the concrete BSF. Lengths are in cm.

untreated because it is their only option. Some boil their water, but boiling requires large amounts of fuel, making it difficult, expensive, and a threat to the sustainability of natural resources.

Point of use (POU) drinking water treatment and safe storage technology allows people without access to safe water sources to improve the quality of their water by treating it in the home (Sobsey, 2002), thereby taking control of the safety of their drinking water. No one POU device is appropriate for all situations and many factors must be considered such as cost, simplicity of use, volume treated, local sustainability, cultural acceptability and treatment efficiency. One of the most promising emerging POU technologies is the biosand filter (BSF), a household-scale, intermittently operated slow sand filter (SSF).

The biosand filter consists of a concrete or plastic chamber filled with sand (Figure 1.1). An elevated discharge tube allows the filter to maintain a layer of water above the sand surface and prevents dewatering. Research has established that full-scale, conventional SSFs remove pathogens from water (Fox et al, 1994) but the effectiveness of this household-scale unit has until recently been uncertain because of different design and operational properties than conventional SSF.

The design and operational features of the BSF and the conventional SSF are compared in Table 1.1. The range of operational parameters listed for the BSF has evolved through rather limited experience in developing countries. The BSF is similar to conventional SSFs in that there is no pretreatment or backwashing and operation is simple, including gravity- rather than pressure filtration. As in conventional SSFs, the sand bed remains wetted throughout operation and a ripening process occurs, during which a schmutzdecke forms and performance improves.

The most important difference between BSF and SSF is the feed flow pattern. The filter is operated intermittently by introducing a single charge of water (perhaps only one per day) rather than continuously. The typical charge is 10-to-20 L after which the

Table 1.1 Recommended range of parameters of conventional SSFs and the range of parameters of the plastic and concrete BSFs.

			Biosand Filter	
		Conventional SSF[1]	Plastic	Concrete
Filtration Rate	m/hr	0.08–0.4	0.0–0.9*	0.0–1.2*
Depth of sand				
Initial	m	0.8–1.2	0.4	0.5
Minimum	m	0.5–0.7	0.4	0.5
Depth of supporting media & underdrains	m	0.3–0.6	0.13	0.1
Depth of Supernatant Water	m	0.9–1.5	0.05–0.17*	0.02–0.28*
Freeboard	m	0.2–0.3	0.0–0.04*	0.0–0.4*

[1]From Pyper and Logsdon, 1991
*Vary within this range during one charge due to intermittent operation.

water level above the filter media declines. The BSF operates, therefore, in the declining rate mode of filtration. The maximum application rate is about 1.1 m/h and at the maximum filtration rate, the filter area of a single BSF unit provides a product water flow rate of about 1.0 L/min. Operation is discontinued when water production (flow rate) is judged insufficient by the user. The filter is manually cleaned and immediately put back in use.

As shown in Figure 1.1, dewatering of the filter in between charges is avoided by a vertical discharge tube that rises from 2- to 5 cm above the height of the filter media. Another unique aspect of the BSF design is to promote uniform drip flow over the sand surface by use of a plastic or sheet metal diffuser above the filter media. This diffuser prevents the charge of water from disturbing the schmutzdecke.

With repeated daily filtration cycles, the schmutzdecke layer eventually develops to cover the media bed and severely limits the filtration rate. At this time, the user disturbs the top few cm of sand to open up pores that have been clogged or blocked by the schmutzdecke and accumulated suspended matter. Sand is not removed intentionally. The dirty supernatant water is then removed and the filter placed in service again.

A number of reports have presented practical aspects of the BSF such as field implementation, user satisfaction, and percentage removal of fecal coliforms or *E. coli* in the field (Murcott, 2002). Over 50,000 BSFs are in use worldwide. Users report satisfaction with the filter and improved health following its introduction into the home (Kaiser et al, 2002). Despite successful field implementation, there has been little systematic, process engineering research to substantiate the effectiveness of the BSF nor to optimize its design and operation. A M.S. thesis addresses bacterial reductions and performance characteristics (Buzunis, 1995) and one peer-reviewed publication reports removal of toxicants and parasites (Palmateer et al, 1999).

The objectives of this research on the BSF are: (1) to characterize its hydraulic operation and (2) to measure the reduction efficiency of enteric bacteria, bacteriophages and viruses under controlled laboratory conditions.

Methods

Filter and media preparation

Plastic, 60-L capacity, filter units were obtained from Davnor Water Treatment Technologies Ltd. (Calgary, Alberta, Canada). Filters contained 5 cm of underdrain gravel, 5 cm of medium-sized support gravel, and 40 cm of sand, the effective size (d10) of which was 0.19–0.22 mm. All filter media were crushed granite gravel available locally. The gravel was sieved according to standard international procedures for the BSF. The media were loaded into the filter in a manner to prevent formation of air pockets. The initial flow rate following the first 20-L charge ranged from 0.8–1.1 L/min. Flow rate was measured daily just after the introduction of a charge.

Some experiments were also conducted with a concrete rather than plastic BSF. These were constructed using a steel mold donated by Samaritan's Purse (Boone, NC). Media was added in the same fashion as for the plastic filters.

Design of microbial challenge studies

Four filtration experiments were conducted in which the daily charge was either 20- or 40 L. Results of four of these experiments are reported herein for which the filter feed conditions (type of filter, total length of filter operation, daily charge volume and feed water) are listed in Table 1.2.

Feed water comprised surface water from one of three local water supply reservoirs (Cane Creek Reservoir in Chapel Hill, N.C., University Lake in Chapel Hill, N.C., or Lake Michie in Durham, N.C.) to which was added up to 4% by volume of pasteurized primary effluent from a local wastewater treatment plant (OWASA, Chapel Hill, N.C.). The purposes of this amendment were to increase the supply of organic matter for development of a biofilm layer and to reproduce the presence of fecal contamination that may be more typical of the feed water in a developing country situation.

On each day of operation, the local surface water supply, which had been allowed to reach room temperature (20°C) was amended with pasteurized sewage and an aliquot of stock solutions of the challenge microorganisms as listed in Tables 1.2 and 1.3.

Table 1.2 Characteristics of longitudinal filter challenge experiments.

Experiment number	BSF type	Length (days)	Daily charge (liters)	Source water*	% sewage pasteurized
2	Plastic	44	40	CC, UL	0%
3	Plastic	54	40	CC	4%
4	Plastic	50	20	LM	1%
5	Concrete	43	20	CC, UL	1%

*CC = Cane Creek Reservoir (Orange County, NC, USA); UL = University Lake (Orange County, NC, USA); LM = Lake Michey (Durham, NC, USA)

Table 1.3 Target concentrations of microorganisms that were spiked into feed water.

Experiment Number	*E. coli* B cfu/mL	MS2 pfu/mL	PRD-1 pfu/mL	echovirus 12 pfu/mL
2	200	100	100	–
3	200	1000	1000	600
4	400	–	–	1200
5	800	100	–	1000

E. coli, MS2, PRD-1 and Echovirus type 12

A pure culture of *E. coli* strain B (ATCC No. 11303) was grown to log phase in shaker culture flasks of tryptic soy broth at 36°C, was cooled to approximately 4°C, and stored for up to 7 days at 4°C and maintained stable concentrations of viable *E. coli*. Cultures were prepared weekly as needed during the four dosing experiments. The culture was serially diluted into lake water immediately prior to seeding and this stock was dosed into feed water to achieve the desired *E. coli* concentration for each daily charge to the filter.

Pure stocks of bacteriophages MS2 and PRD-1 were grown up in the *E. coli* strain F-amp and *E. coli* strain CN13, respectively, by a modification of the double agar layer procedure EPA Method 1602 (EPA, 2001); these were stored at −80°C. Aliquots of each stock were thawed each week, serially diluted into phosphate buffered saline, and stored at 4°C for up to 7 days. Aliquots of this dilution were then dosed into feed water to achieve desired concentration for each daily charge.

A pure stock of echovirus 12 was propagated in monolayers of Frhk-4 cells with maintenance medium at 37°C. The stock was freeze-thawed and extracted. Extracted stock was stored at −80°C until needed. Aliquots of enumerated stocked were thawed weekly, serially diluted in phosphate buffered saline, stored at 4°C and added into feed water to achieve the desired echovirus 12 concentration for each daily charge.

On the day before each microbial assay, a control sample of the spiked charge was stored at room temperature to measure the die-off of microbes with time, independent of the activity of the filter.

Water analysis methods

The assay frequency was approximately once per week while the filters received a charge of feed water with the challenge microorganisms each day. Three types of samples were collected for assay: (1) samples of feed water from the previous and current charge; (2) grab samples of the filtered water taken throughout a daily filtration cycle; (3) a composite sample of the filtered water to represent the daily average concentration of microbes.

E. coli concentrations in water were quantified via membrane filtration on MI agar BBL^{TM}(Becton-Dickinson, Franklin Lakes, New Jersey) using EPA Method 1604 (EPA, 2002). MS2 and PRD-1 concentrations were assayed using the single agar layer method (EPA Method 1602, EPA, 2001). Echovirus 12 samples were inoculated onto confluent monolayer of Frhk-4 grown in 60 mm plastic dishes (Manufacturer, City, State) using the plaque forming assay. Turbidity and pH were measured using a turbidimeter (Model 2100N, Hach, Loveland, CO.) and pH meter (Model 215, Denver Instruments, Denver, CO).

Log_{10} removal of microorganisms was calculated by subtracting the log_{10} filtered water concentration from the feed water concentration. The log_{10} removals from each assay during a portion of a filter run were averaged and converted to percent removal which represents the geometric mean removal for each time period.

The residence time distribution (RTD) of water within the BSF was measured by a step input tracer test. The pores of a clean, unripened plastic BSF (BioSand Water Filter^{TM}, Davnor, Inc., Calgary, Alberta, Canada), were filled with deionized water (DI) at the start of the tracer test. The tracer test began by rapidly replacing the DI water above the filter with a 200 mg/L of NaCl. The head above the filter was maintained by two peristaltic pumps (Cole Parmer Cat. No. 7553-0 and 7545-0) introduced at a constant rate and conductivity of the filtered water was measured. The NaCl concentration was measured by conductivity (Fisherbrand Traceable^{TM} Conductivity, Resistivity, and TDS Meter (Cat. No. 09-326-2). When conductivity in the filtered water reached the feed water value, the NaCl solution above the filter was removed and replaced by DI water. This provided another tracer test in which the subsequent decrease of conductivity in the filtered water was measured with time. The effect of head above the filter on RTD was determined by repeating the NaCl tracer tests at three head conditions: 5 cm ($\pm$1 cm); 13 cm ($\pm$1 cm); and head allowed to decrease with time from 17.3 to 2.0 cm.

Results and discussion

The results of the step input tracer test are presented in Figure 1.2. The rapid rise of conductivity to the feed value and the subsequent rapid decline of conductivity to the baseline value are indicative of plug flow. The head condition did not appear to affect the

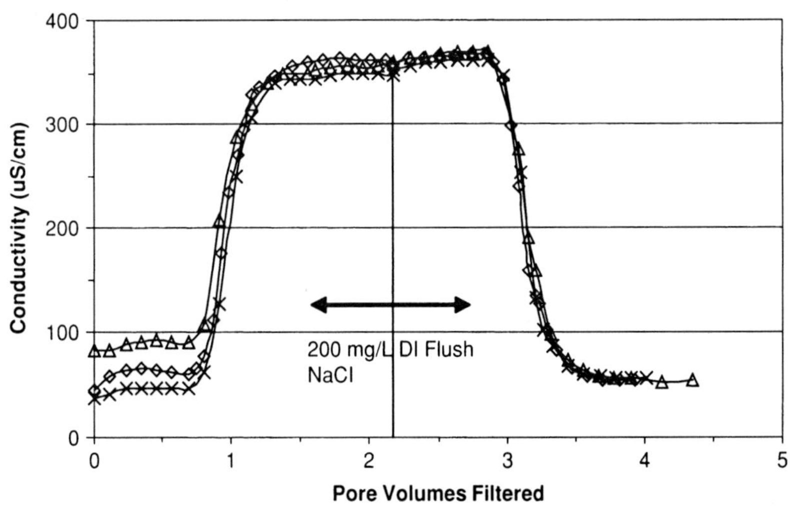

Figure 1.2 Results of three step-input tracer tests conducted wih the plastic BSF.

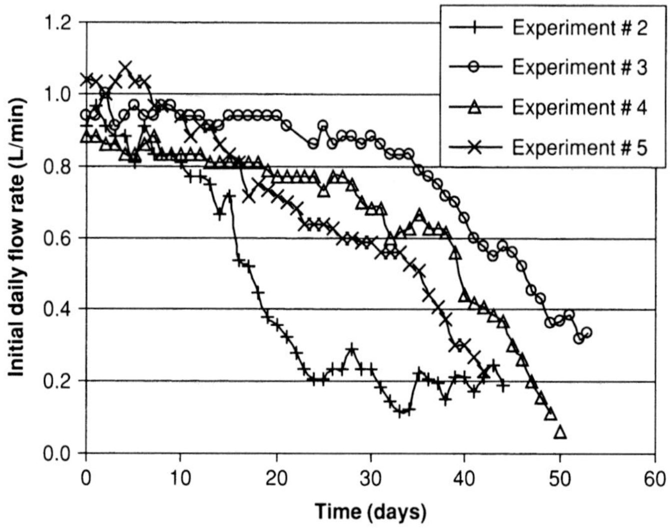

Figure 1.3 Daily flowrate after a 20 L charge for four filter experiments. Expts #2-4 conducted with plastic BSFs; Expt #5 with a concrete BSF.

result. The volume of water passed through the filter before conductivity rises to the feed value is also a measure of pore volume.

The media pore volume was determined empirically during the loading of three plastic filters and was found to be 18.3 (±0.1) L and 13.3 (±0.1) L for concrete BSFs. From the total volume of crushed granite gravel media in the filter, the porosity is 47%; this is somewhat high for filter media but not atypical of angular sands.

As shown in Figure 1.3, each of the four filter experiments demonstrated that filter flow rate decreased with filter maturation (or ripening) as has been widely reported for traditional SSF. During the ripening period, microbial reduction remained relatively low, often for as long as 30 days, and then increased. For experiments with the plastic BSF that used a daily charge of 40 L, the time-averaged geometric mean (GM) reduction of *E. coli* was only 73.6% during the first 30 days of filter operation but improved to 97.5% after day 31. For experiments with plastic BSF in which the daily charge was 20- instead of 40 L, the time-averaged GM reduction of *E. coli* was much higher (95%) during the first month but only slightly higher (99.5%) subsequently. Somewhat lower reductions of *E. coli* were found for the concrete BSF when operating with a charge of 20 L each day, with time-averaged *E. coli* reductions of 89.2% for the first 30 days and 95.6% on subsequent days.

Less reduction of *E. coli* in the plastic BSF with a daily charge of 40 L than with a daily charge of 20 L may be explained by the difference in the fraction of the charge that remains within the filter during the idle period between charges. Given plug flow behavior and a plastic BSF pore volume of 18 L, almost all of the filtered water that is produced (20 L) is the previous day's feed water that had resided in the filter during the overnight idle period. By contrast, the filtered water produced from a charge of 40 L is only about 50% composed of water (18 L) from the idle period. Therefore, a composite measurement of *E. coli* reduction

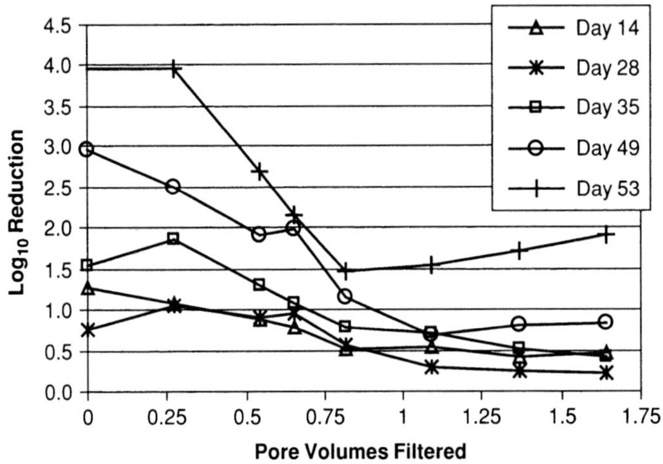

Figure 1.4 Run #3 *E. coli* reductions with pore volume. Results from Days 7 and 21 excluded for clarity (overlay Days 14 and 28).

from a 40 L charge could be less than from a 20 L charge because a smaller fraction of the filtered water had resided in the filter for the idle period. The reduction in *E. coli* is greater in water that remains overnight in an idle filter overnight than in water that passes through the filter bed without overnight idle, as indicated by the results described below.

The importance of idle time on *E. coli* reduction was explored further by measurements of *E. coli* concentrations in grab samples of filtered water taken at increments of time that correspond to cumulative pore volumes displaced during the days operation. The results for a series of 40 L charges (Experiment #3) are shown in Figure 1.4. These data indicate that: (1) microbial reductions improve with repeated daily charges of feed water due to filter maturation and (2) microbial reductions tend to be greater at the beginning each daily filtration cycle, probably due to processes by which microbes entering the filter on the day before are attenuated (or inactivated) within the filter during the overnight idle period.

The same general conclusions regarding *E. coli* reductions were reached for virus reductions, albeit these reductions were less than those for *E. coli* for some of the viruses tested. Only 69% reduction of coliphages MS2 and PRD-1 was obtained after filter maturation with a daily charge of 40 L; however, 90% reduction was reached at the beginning of daily filtration cycles in a mature filter. The removal of echovirus 12 was much higher than those for either coliphage, with an average exceeding 95%.

Switching from a plastic to a concrete BSF in Experiment #5 gave greater reduction of MS2 and echovirus 12 (often >99% and >99.9%, respectively). One possible explanation is a higher pH of the filtered water from the concrete (up to pH 9.3) than plastic BSF (pH < 8.2) due to leaching of lime from the concrete. The pH was notably higher at the start of the filtration cycle but decreased over time.

Summarizing all of the data thus far collected, bacteria and virus reductions by the BSF tend to be lower than those demonstrated for traditional SSF (Fox et al, 1994). Whether these microbial reductions are considered sufficient to provide microbiologically safe drinking water without addition of a disinfectant remains to be shown by epidemiological study of heath risks from BSF-filtered waters and by comparison of the observed reductions to those required by drinking water regulations in specific countries or regions.

Conclusions

The design and operating conditions of the BSF vary from those of continuously operated, traditional slow sand filters, resulting in different performance characteristics. Bacteria and virus reductions by the BSF tend to be lower than those demonstrated for traditional SSF but microbial reductions increase substantially with increasing retention time in a mature filter.

The intermittent operation of the BSF invites research on the mechanisms by which filter idle time attenuates microbes. The design implication is that better microbial reductions would be obtained by reducing the daily charge of feed water to that of the filter bed pore volume and/or providing a longer interval between introduction of each charge. The effects of filter maturation and the volume, frequency and duration of each water charge on microbial reductions need systematic study. With more detailed engineering analysis, operational practices that yield optimum performance and are acceptable to the user can be put forward. Further research is also needed to explain differences in extent of removals for different microbes, particularly for viruses.

Finally, the research question of most practical importance is whether microbial reductions observed in this study are typical of BSFs in household use globally. The ultimate test is whether microbial reductions achieved by the BSF correlate with reductions in household waterborne illnesses, such as diarrhea.

Acknowledgments

This research was carried out with the financial support of the United States Environmental Protection Agency (People, Prosperity and the Planet Student Sustainability Award), Centre for Affordable Water and Sanitation Technology (CAWST), Samaritan's Purse USA and Samaritan's Purse Canada; the W.K. Kellogg Foundation, and the Tinker Foundation, Inc. The primary author's graduate study is supported by a USEPA STAR Fellowship. Thanks also to Douglas Wait and other staff and students of Mark Sobsey's lab at UNC.

References

Buzunis, B.J. (1995) Intermittently Operated Slow Sand Filtration: A New Water Treatment Process, Master's Thesis, University of Calgary, Department of Civil Engineering. Calgary, Alberta.

Fox, K.R., Graham, N.J.D., and Collins, M.R. (1994), Slow Sand Filtration Today: An Introductory Review. In Slow Sand Filtration. Edited by M.R. Collins and N.J.D Graham. American Water Works Association. Denver, CO. Pp. 1–8.

Kaiser, N., Liang, K, Maertens, M. and Snyder, R. Biosand Household Water Filter Evaluation 2001. (2002) Samaritan's Purse Canada. Calgary, AB.

Murcott, S. Nepal Water Project 2001–2002. (2002) Massachusetts Institute of Technology, Department of Civil Engineering. Cambridge, MA (2002).

Palmateer, G., D. Manz, and A. Jurkovic. Toxicant and Parasite Challenge of Manz Intermittent Slow Sand Filter. *Environmental Toxicology*. **14**:217–255 (1999).

Pyper, G.R. and G.S. Logsdon. Slow sand filter design. In *Slow Sand Filtration*. Edited by G. Logsdon. American Society of Civil Engineers. Reston, VA. Pp. 122–148.

Sobsey, M.D. (2002), Managing water in the home: accelerated health gains from improved water supply, (WHO/SDE/WSH/02.07) WHO, Geneva.

U.S. Environmental Protection Agency (2001), Method 1602: Male-specific (F+) and somatic coliphage in water by single agar layer (SAL) procedure, Washington D.C., EPA 821-R-01-029.

U.S. Environmental Protection Agency (2002), Method 1604: Total Coliforms and Escherichia coli in Water by Membrane Filtration Using a Simultaneous Detection Technique (MI Medium), Washington D.C., EPA 821-R-02-024.

World Health Organization (WHO). Water Sanitation and Hygiene Links to Health – FACTS AND FIGURES. 2004.

33 Intermittent slow sand filters for household use – A field study in Haiti

D. L. Baker*, W. F. Duke**

*Centre for Affordable Water and Sanitation Technology, Canada, dbaker@cawst.org
**University of Victoria, Victoria, BC, Canada, bsduke@telus.net

Abstract An intermittent slow sand filter for household use, commonly referred to as the Biosand Filter (BSF), was developed by Dr. David Manz at the University of Calgary, Canada, in the 1990's. The version most widely in use is constructed of concrete, stands 95 cm in height and 36 cm in width. The flow rate is 20–40 litres per hour. The cost to produce a BSF is approximately 15 to 30 Euros (20–40 USD). Over 80,000 Biosand Filters are in use in 36 countries. Modifications to conventional slow sand filtration include; 1) reducing the height of the supernatant to 5 cm to provide the schmutzdecke or 'biofilm' with sufficient oxygen to allow intermittent use, and 2) eliminating the need for sand bed removal through a 'clean in place' technique which simplifies maintenance. These adaptations provide an affordable option for household application of slow sand filter technology to treat drinking water in rural and peri-urban communities of developing countries. In early 2005, a field study of the Biosand Filter was undertaken in Haiti by the Centre for Affordable Water and Sanitation Technology (CAWST), a Canadian NGO, and the University of Victoria, Canada. The effectiveness of the BSF in removing E. coli and decreasing turbidity was evaluated for 107 filters which had been in use for an average of 2.5 years (range 1 to 5 years). The sustained use and user satisfaction were assessed through interviews in each of the 107 households. Durability of the filter, maintenance requirements, and affordability were appraised by recording specific observations during the visits.

Keywords Intermittent slow sand filter, household water treatment, biosand filter effectiveness

Introduction

Over 1 billion people in the world lack access to safe water – a major feature of extreme poverty. Nearly all of these people live in developing countries, especially in rapidly expanding urban fringes, poor rural areas, and indigenous communities (Bartram, 2004). Community or municipal water treatment systems are frequently impractical and often unaffordable in these settings. At the present time, household water treatment provides the only reasonable alternative for many of these people.

 Recent Progress in Slow Sand and Alternative Biofiltration Processes edited by Rolf Gimbel, Nigel J.D. Graham and M. Robin Collins. ISBN: 9781843391203. Published by IWA Publishing, London, UK.

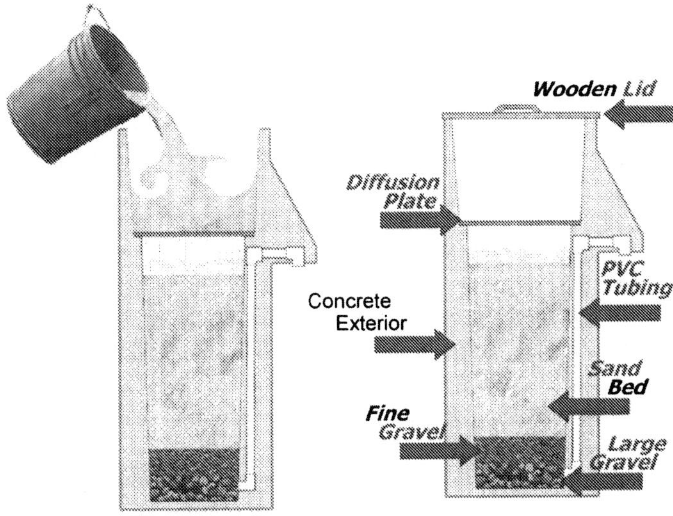

Figure 1.1 Biosand Filter

In order for a household water treatment technology to achieve widespread sustainable use among the poor, it must meet the 'criteria of the poor':

- effective in cleaning the water and improving its taste, smell and appearance
- easy to operate and maintain
- affordable and durable, with little or no recurring costs
- manufactured using local skills and materials
- does not use chemicals or energy

The Biosand Filter, an intermittent slow sand filter developed by Dr. David Manz, has become one of the most widely-distributed devices for household water treatment in developing countries around the world. This filtration device provides a critical component of a household water treatment (HWT) system involving; sedimentation, filtration and disinfection. A recent survey found that, as of June 2005, over 80,000 Biosand Filters have been installed, impacting over 500,000 people in 36 countries (CAWST, 2005).

Unpublished laboratory studies have shown the biosand filter to be effective in significantly reducing the number of parasites and bacteria in the source water (Palmateer, 1999; Bruzunis, 1993). There is, however, very little data available, published or unpublished, regarding the filter's effectiveness and sustained use over time within households in developing countries. A field study was conducted in Haiti, West Indies, during the first half of 2005 to help fill this gap in knowledge. Included in the objectives of this study were to:

(1) collect data about the filter's effectiveness in removing bacteria and reducing turbidity under typical working conditions in a developing country,
(2) record the perceptions of household users with regard to water quality from the filter, ease of use, and level of satisfaction with the filter to assess sustainability,
(3) make observations about the durability of the filter after prolonged use, and
(4) measure the levels of *E. coli* in household water from source to point-of-use.

Methods

In the Artibonite Valley of Haiti, approximately 2000 Biosand Filters had been installed in households, between 1999 and 2004, by the Community Development Division of Hospital Albert Schweitzer. The research team visited 110 households to request consent of the householders to participate in the study. This resulted in finding 107 households with working filters.

Interviews, observations, measurements and water samples were carried out by two teams of Haitian enumerators, each consisting of a nurse and a filter technician. Water analyses were performed by Haitian lab technicians using the membrane filtration method to determine *E. coli* colony counts as the indicator bacteria for faecal contamination. Training of the research team and supervision of the visits and lab analyses were carried out by the research investigators and CAWST staff.

Water samples were taken at the water source for each household when possible. The other relevant sample points were;

(1) transfer bucket used to carry water from the source; the **Influent** to the filter,
(2) the spout of filter where the water flows out; the **Effluent** of the filter
(3) the storage container in the household; the **Point-of-Use**.

All samples were tested for *E. coli* running 10% duplicate and 10% blank tests. Turbidity was also measured using the Hach 2100P Turbidimeter. The water in the storage container was tested for pH and free chlorine, immediately after taking the sample at the time of the household visit, to determine if chemical disinfection was being used.

Interviews, using prepared questionnaires, were carried out after consent was received from each householder. Translation was available to the research investigator during the interviews. Measurements of the filter flow rate and height of supernatant (height of water above the sand during the pause period) were made for each filter. Observations of the filter's physical condition including filter body, diffuser plate, lid, and overall hygiene were recorded.

Results and discussion

Shallow, hand-dug wells, were the only source of water for 61% of the households. The water from these wells contained an average of 234 *E. coli* cfu/100mL (median 53 cfu/100mL). Water piped from springs or deep wells, provided water for 26% of the households (average 195, median 27, *E. coli* cfu/100mL). Thirteen percent had access to both types of water sources and just 3% had plumbing in their home.

The Biosand Filters effectiveness in removing microbial pathogens from the water was based on the *E. coli* colony counts from the effluent. Eighty percent of the water out of the BSF had zero colonies in the sample, 17% tested between 1 and 10 colonies, while 3% were over 10 (cfu/100mL). Bacterial removal in the filter averaged 98.5% overall (n = 92).

Recontamination of the drinking water occurred post-treatment. While only 3% of samples from the filter effluent tested *E. coli* > 10 cfu/100mL, the samples taken from the storage containers at point of use tested 23% > 10, including 6% > 100. Fourteen of the 107 households reported that they had had problems with slow flow rates from the filters, and 6% said that they had to obtain assistance from Community Development to correct the

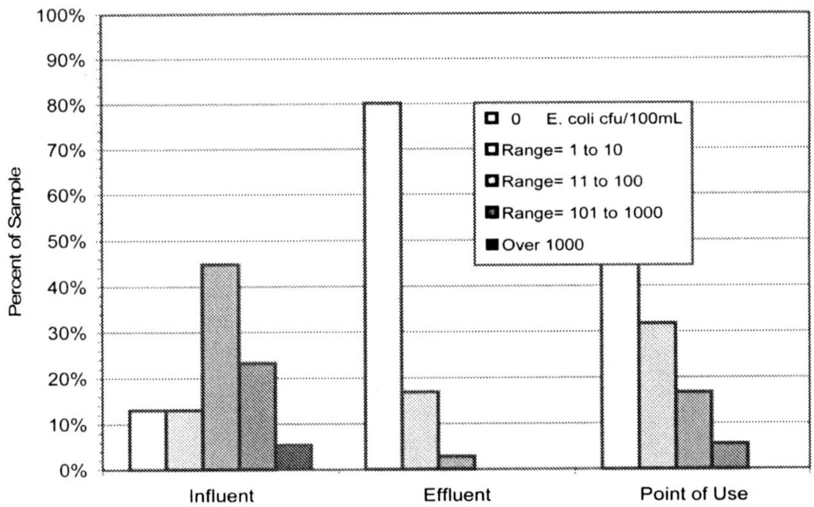

Figure 2.1 Effectiveness of Biosand Filter and Recontamination at Point of Use

problem. In all of these cases, the problem was blocking of the filter due to impurities and suspended solids in the water. All were corrected by the 'swirl-and-dump' (clean in place) procedure. The average flow rate for the filters was 35.4 L per hour, ranging from 11 to 95 L per hour.

In 92% of the cases, the filters were found to be well-maintained. Only minor problems such as cracked lids or diffuser plates were observed. One of the filters was found to have a crack near the lip of the filter above the spout, but it was still being used. At the time of the unannounced first visit, 97% were found to be functioning after an average 2.5 years in use.

In 91% of the households, the filtered water was used only for drinking. None of the households treated the water with chlorine after filtering. Eighty-five percent poured the source water directly into the filter, not allowing time for sedimentation or settling. One hundred percent reported that the filter was easy to use, even for older children. Ninety-nine percent said that the filter produced enough water for the entire household.

One hundred percent of the households reported that they "liked" their filters, citing better quality water (49%), health protection (22%), and "because it works well" (7%) as reasons. In 99% of the interviews, the participants answered that the filtered water appeared cleaner, tasted better, and smelled better than the source water. Ninety-five percent felt that their families' health had improved since they began using the filter, while 5% had not noticed any change. Ninety-five percent responded that they would recommend the filter to others.

Conclusions

Perceptions of Householders:

High level of satisfaction was reported by the 107 families using the Biosand Filter in terms of: (a) the taste, smell and appearance of the filtered water, (b) ease of use of the filter, and (c) sufficient quantity of water for the entire family.

Observations:

(1) Overall; filters were durable, well-maintained, with 97% functioning after avg. 2.5 yr
(2) Major user problem: plugging of the filter due to suspended solids in the influent
(3) Lack of knowledge regarding:
- Maintaining the filter to remove plugging material and restore flow rate
- Disinfecting the water post-filtering, safe water storage practices and containers

Water Analyses:

(1) Good turbidity removal; average: influent 6.2 NTU → effluent 0.9 NTU
(2) Significant bacterial removal; average 98.5% removal of *E. coli* overall
(3) Substantial recontamination of the stored water occurred post-treatment.

References

Bartram, J. Howard, G. (2004) Domestic water quality, service level and health. *Bulletin World Health Organization,* 82(3):

Centre for Affordable Water and Sanitation Technology (2005) Results: Key Performance Indicators. *CAWST 2005 Annual Report*, www.cawst.org , p. 10

Palmateer, G. Manz, D. (1999) Toxicant and parasite challenge of the Manz intermittent slow sand filter. *Environ.Tox*. 14:217–225.

Bruzunis, B.J. (1993) Laboratory report: conformational testing of an intermittently operated slow sand filter. *Department of Civil Engineering, University of Calgary, Calgary, Alberta, Canada.*

34 Appropriate technology for the treatment of drinking water in Roche, Tanzania

S. I. Pumphrey, D. W. Divelbiss and D. B. Oerther

University of Cincinnati, Cincinnati, OH, USA

Abstract In October 2005, three biosand filters were installed at a local school in the village of Roche, Tanzania in collaboration with the Village Life Outreach Project. In addition, sanitation conditions in the region were assessed. Based upon continued monitoring in the field by indigenous personnel, two of the three filters remained functional and in use four months after their initial installation. The results of this study are encouraging for continued efforts to improve water quality and access to safe drinking water in developing countries.

Keywords Appropriate technology, drinking water, point-of-use sand filtration, Africa, waterborne disease

Introduction

Situation – The village of Roche is located in the Tarime District of northern Tanzania, Africa just south of Lake Victoria (**Figure 1**) [1]. There is very limited infrastructure in Roche; it is without a sanitation system for wastewater, without a drinking water purification system, and without access to electricity. Travel throughout the area is primarily by foot or bicycle along packed dirt roads. Villagers are primarily herders of livestock and subsistence farmers who live in huts formed of mud, dung or brick. The school buildings are constructed of wood and concrete with corrugated metal roofing, and they border three sides of a common green area, which is the center of community activity in Roche.

The surrounding landscape is semi-arid, but there are two man-made ponds in the village, one for watering livestock and the other for human use. During the dry season, which lasts most of the year, the water in the human use pond becomes depleted, and people therefore take their water from the other pond. Additionally, most livestock is free-range, and consumes water from both ponds. The ponds are saturated with clay and are highly contaminated with fecal pollution (**Figure 2**), both directly from the livestock who defecate near or in the water and from human fecal matter which is flushed by surface water into the pond during the rainy season.

Each student brings a jug of water to school each day in the same manner they would bring water to their homes (**Figure 3**). Villagers transport water from its source to their homes or to the school in plastic, narrow-mouthed containers, usually with handles on the

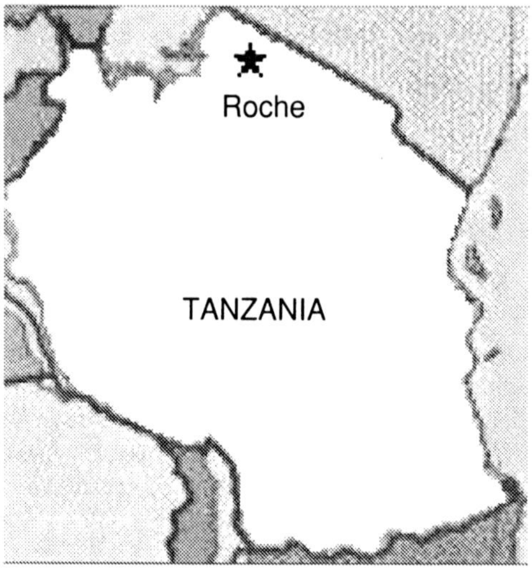

Figure 1 Location of Roche, Tanzania

top and sometimes with screw-on lids (**Figures 4**). The jugs must be mostly submerged to fill them at the pond. When asked, most villagers say that they know to boil the pond water before drinking it; however they use the water directly for washing and cooking purposes. Additionally, young children, the age range most susceptible to waterborne disease fatality, do not boil their own water when they are unaccompanied outside of the home – which is a regular occurrence in the local culture.

There is one bore-hole hand pump in Roche, from which most villagers collect their drinking water. The drinking water is collected and transported in the same containers as those used to gather water from the pond, without rinsing, washing or drying between switching sources, thus contaminating the bore-hole water with pollutants from the pond. Additionally, the ponds are located closer to many homes than the bore-hole is, making the ponds a more convenient source of water to many people.

Figure 2 Dung in the "livestock" pond

Figure 3 Students gathering water for school

Figure 4 Typical water jug

Table 1 Cases seen by VLOP Medical Team 20–25 October 2005 in Roche, Tanzania [2]

Probable Cause	Contaminated Bathing Water	Standing Water / Mosquito Breeding	Parasites, Flukes and Worms / Fecal-Oral Contamination / Poor Hygiene Practices			
Disease	UTI (urinary tract infection)	Malaria	Amebiasis	Helminthiasis	Schistosomiasis	Taeniasis
Number of Cases	17	91	6	26	10	5

Waterborne diseases-health of the people from Roche, Tanzania

The Village Life Outreach Project (VLOP) visits Roche annually with an objective of improving the overall well-being (life, health, and education) of the community in Roche. In October of 2005, for the second year in a row, a medical team composed of dedicated doctors and nurses from the United States of America saw and treated patients in Roche. During their five days in Roche, doctors treated 155 patients who were infected with diseases that could have been prevented with proper water treatment, storage, and distribution (details in **Table 1**) [2]. The medical team also treated many infected wounds which would not have healed (due to the contaminated water) if not for the medical treatment provided.

The VLOP leadership recognized the need to treat the source of these waterborne problems, and not just the symptoms, and requested an evaluation from an environmental engineering perspective. Recognizing the importance of education, the engineer prepared informational pamphlets to leave in Roche as a source of information about proper water storage when outside help, such as VLOP, is not available. Additionally, while in the village, several village members were selected (based on willingness to participate and effective communication skills) to learn along side of the water team so they could repeat the process on their own and maintain the biofilters established in Roche.

Appropriate methods of water treatment

Based upon a literature review, five major technologies were identified for drinking water treatment in developing countries, namely: 1) Sand Filtration, 2) Porous Clay Filters, 3) UV/Sunlight Disinfection, 4) Chlorination, and 5) Distillation. Based on available resources and considering issues of sustainable design, Sand Filtration or Porous Clay Filters were evaluated as the most appropriate methods for treating drinking water in Roche. These two methods were reviewed in depth prior to the trip (including constructing filters in the laboratory in the USA), and it was found that there is a high rate of failure with the Porous Clay Filters during their production and they require the use of colloidal silver, which is unavailable in Roche. The packing material for a sand filter is readily available in Roche, however, prior to the trip information about available building materials was inconclusive, so all material except the packing material was purchased in the USA and transported

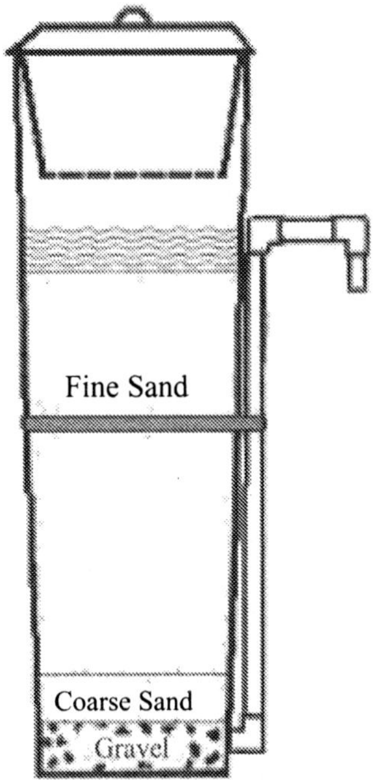

Figure 5 One Unit Slow Sand Filter

by VLOP to compensate for the short period of time available to construct the systems. Students from Hughes High School in Cincinnati, OH USA constructed hand-held sifting screens which were transported to Roche with VLOP. Team members of VLOP as well as several villagers collected unsorted rock, sand and gravel from a dried river basin near Roche.

Three slow sand filtration units were built in the following method: The unsorted earthen material was separated by size using the sifting screens. The material was washed first with wind and then with water from the bore-hole to remove the fine particles and organics. Holes to create a drainage pipe were melted into PVC piping using hot metal rods. Pipes were secured into the housing to form a free-flow (siphon) system. Material was packed into the housing as shown in **Figure 5**, with three inches of gravel on the bottom, two inches of course sand above that, and approximately 16 inches of fine sand above that [3]. The sand was hand packed when the housing already contained water, so as not to allow bubbles to decrease the volume of the filter bed.

Holes were nailed into the bottom of wash basins, which were set atop the filters to act as diffusers for the incoming water. The systems were designed to have lids, but did not have them at the VLOP time of departure, so the engineers informed village leaders about the missing lids and the responsibility to acquire them was designated to the leaders. Once assembled, villagers flushed pond water through the sand filters continually during daylight hours to simulate the anticipated usage and establish the schmutzdecke.

Results and discussion

Although faced with many challenges the village members and water team from VLOP fully assembled three one-meter-tall sand filters at the school in Roche. There are only three rooms in the school with secure locks, and since the water filters were new and considered a valuable commodity the school masters wanted the filters to all be in locked rooms. One of the rooms was used to store dry goods, and not an ideal place for a water filter that requires continual use. Therefore, one filter (the first one that was constructed) was placed in the school's main office, and the other two were in the resource room. The construction of the first two filters depleted the supply of high quality gravel that had been prepared, leaving only poor quality gravel for the third filter. As anticipated by the engineering team, the filter with the poor quality gravel clogged within the first four months of operation and is no longer functioning. Pond water was tested with the HACH PathoScreen Field Test Kit both before and after filtration. Prior to filtration, the sample turned black, indicating presence of harmful bacteria. After filtration through the remaining two filters the sample produced negative results, meaning the sand filters removed harmful bacteria from the water. These results indicate that the sand filters have improved the drinking water conditions for school children in Roche who previously consumed water directly from the pond.

Conclusions

The main complaint of the villagers was that the materials used in the construction of the sand filters were too expensive. The VLOP envisioned the sand filters as community efforts to clean the water, and therefore affordable for groups of people to construct together. However, documented evidence from the Center for Disease Control (CDC) indicates that community water supplies, however safe they may be, without a secure distribution system lead to minimal improvement in the overall rate of water born diseases in the community [4]. Since the area has very limited resources and infrastructure, providing Roche with a distribution system at this point in time would be very costly and inappropriate because of the lack of available maintenance systems. Therefore, education combined with point-of-use filtration is the best way to ensure that high quality drinking water is available to community members, especially young children. With the materials being as inaccessible as they are, individual families cannot afford to install full-sized sand filters into their homes; therefore the engineering team suggests altering the design of the sand filter. An option being considered is the **dual bucket slow filter,** which has been implemented in Guatemala by Margarito Vasquez and emulated in the laboratory in the United States [5]. The system stacks two equivalent sized (aprox. 19 L) buckets connected by a perforated, sand-filled segment of pipe which has been wrapped with rope around the sections of pipe with holes. The pipe can be lengthened by inserting lifts between the buckets, as in **Figure 6**, to achieve an optimal length of slow sand filtration. Dirty water goes in the top bucket and must pass through the rope, the holes in the pipe, the sand, and back out the holes and rope at the bottom of the pipe to flow into the bottom bucket. The bottom bucket remains closed except for a nozzle access point, thus providing water storage with the unit. The rope in the top bucket also provides a surface upon which the schmutzdecke can grow (**Figure 7**). This method was described to the people in Roche, but the duration of the trip was too short to

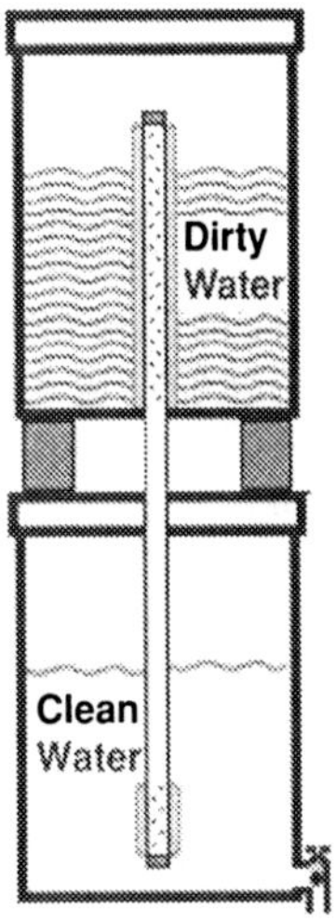

Figure 6 Duel Bucket Sand Filter Cross-section

Figure 7 Functioning Duel Bucket Sand Filter in Santa Maria de Jesus, Guatemala June 2005

implement this plan, and this can be further explored in efforts by VLOP in their 2006 trip, to make a significant impact in the villagers' health due to water consumption.

Acknowledgements

The author would like to thank Anna Hoessle, Tom Mboya Okelo, Dishon Boy, the Village Life Outreach Project, Adam Parrillo, Daniel Divelbiss, Linda Fowler, Donnie Bailey, Ashish Gandhi, Berdell Knowles Jr., Marc Philibert, Sharon Bachman and the Hughes High School students who participated in this project.

References

[1] Parrillo, A. (2005) Village Life Outreach Project. May–October 2005 Cincinnati, Ohio, USA.
[2] Wilson, A. (2005) Village Life Outreach Project Brigade. 20–26 October 2005 Roche, Tanzania.
[3] Bolton, M. (2005) Shoulder to Shoulder. 24 September 2005 Cincinnati, Ohio, USA.
[4] Lantagne, D. (2005) Center for Disease Control. 27 July 2005 Atlanta, Georgia, USA.
[5] Vasquez, M. (2005) Living Water International. 24–29 June, 2005 Santa Maria de Jesus, Guatemala.

Lantagne, Quick, and Mintz (2005) Household water treatment and safe storage options in developing countries: a review of current implementation practices. (*forthcoming*).

Lantagne, D. S. (2001) Investigation of the Potters for Peace colloidal silver-impregnated ceramic filter: intrinsic effectiveness and filed performance in rural Nicaragua. Alethia Environmental, Allston, MA 02134, USA.

Part V

ALTERNATIVE BIOFILTRATION – PROCESS BEHAVIOUR

35 Biomass development in biological activated carbon filters

L.T.J. van der Aa[1,2,*], A. Magic-Knezev[3,4], L.C. Rietveld[2], J.C. van Dijk[2,5]

[1]Waternet, P.O. Box 8169, 1005 Amsterdam, the Netherlands
[*]corresponding author: rene.van.der.aa@waternet.nl
[2]Delft University of Technology, P.O. Box 5048, 2600 GA Delft, the Netherlands
[3]Het Waterlaboratorium, J.W. Lucasweg 2, 21031 BE Haarlem, the Netherlands
[4]Wageningen University and Research, Department of Microbiology, Hesselink van Suchtelenweg 4 6703 CT Wageningen, the Netherlands
[5]Kiwa Water Research, P.O. Box 1072, 3430 BB Nieuwegein, the Netherlands

Abstract Pilot plant studies on surface water were performed to gain knowledge on biomass development in biological activated carbon (BAC) filters. Biomass activity was measured by ATP on carbon. Biomass took 8 months to fully develop; this is a long period compared to 2 months for start up of rapid and slow sand filters. Biomass developed in the entire BAC filter, although concentrations in the top layer were higher than in subsequent layers. A model was developed for simultaneous adsorption and biodegradation. The biodegradation was modelled by a uniform biomass on the outer surface of the carbon grain. Calibration parameters were taken from laboratory experiments. Model predictions for adsorption of dissolved organic carbon (DOC) describe pilot plant data well. Model predictions for biodegradation and biomass development are in the right order of magnitude, but not predicting pilot plant data successfully.

Keywords Adsorption, biodegradation, biological activated carbon filtration, biomass, drinking water, model.

Nomenclature

α_T = temperature correction factor [-]
β_p = pore diffusion coefficient [$m \cdot s^{-1}$]
β_s = surface diffusion coefficient [g carbon$\cdot m^{-2} \cdot s^{-1}$]
b = death rate biomass [s^{-1}]
c = concentration substrate in bulk liquid [g $C \cdot m^{-3}$]
c_s = concentration substrate in surface layer [g $C \cdot m^{-3}$]
d_{part} = particle diameter carbon grains [m]
D_f = diffusion coefficient laminair liquid film [$m^2 \cdot s^{-1}$]
D_p = pore diffusion coefficient [$m^2 \cdot s^{-1}$]

D_s = surface diffusion coefficient [$m^2 \cdot s^{-1}$]
ε = filter bed porosity, the ratio of interparticle pore volume to total volume [-]
δ_f = thickness laminar liquid film [m]
δ_s = thickness surface layer [m]
G_{min} = minimum germination time [s]
K_F = Freundlich constant K [g $C \cdot (g\ carbon)^{-1} \cdot (m^3 \cdot (g\ C)^{-1})^n$]
$K_{M,c}$ = Monod constant for substrate c [g $C \cdot m^{-3}$]
μ_{spec} = specific growth rate biomass [s^{-1}]
$\mu_{max,15°C}$ = maximum growth rate biomass at 15°C [s^{-1}]
n_F = Freundlich constant n [-]
q = substrate loading on carbon particles [g $C \cdot (g\ carbon)^{-1}$]
ρ_{filt} = filter density activated carbon [g $carbon \cdot m^{-3}$ filter]
t = time [s]
v = flow velocity, surface load [$m \cdot s^{-1}$]
δx = height filter layer [m]
X_s = concentration fixed biomass in surface layer [ng ATP$\cdot (g\ carbon)^{-1}$]
Y_{max} = maximum yield [ng ATP$\cdot (g\ substrate)^{-1}$]

Introduction

Pre-oxidation prior to granular activated carbon (GAC) filtration enhances biological activity resulting in biological activated carbon (BAC) filtration. Biodegradation in BAC filters removes part of the biodegradable dissolved organic matter (BDOC) resulting in a decrease in the DOC concentration and in a reduction of DOC loading on the carbon. The reduced DOC loading results in less competition: more adsorption sites remain available for adsorption of micro pollutants. BAC filter run times for both DOC and micro pollutants are longer than GAC filter run times (Orlandini, 1999, Aa et al., 2003). In this point of view biodegradation in BAC should be maximized. However, too much increase of biodegradability by pre-oxidation may result in incomplete removal of BDOC and biofilm formation in the distribution system. Also excessive biomass production might contribute to unacceptable high backwash rates.

The current article describes the development of biomass in BAC filters, and the corresponding BDOC removal. A BAC model was developed for simultaneous simulation of adsorption of DOC, biomass development and biodegradation of DOC.

Biodegradation in drinking water filters

Bacteria fix on the outer surface of the filter material, for example activated carbon grains. Substrate is taken up directly from the water phase and used for growth according to Monod kinetics and for maintenance. The maximum growth rate is a function of temperature and pH. The specific growth rate is a function of maximum growth rate and Monod kinetics depending on the local substrate concentrations in water. Biomass maintenance is linear to the amount of fixed biomass (Schlegel, 1992). Biomass also may release soluble microbial products, which may be a source for biomass development in subsequent layers (Billen et al., 1992, Rittmann et al., 2002).

More than 95% of the internal surface area of the activated carbon is in the micro pores. The micro pores of activated carbon are too small for bacteria to enter (Billen et al., 1992), bacteria may live in the macro pores (personal observation). Therefore direct substrate uptake by bacteria of adsorbed substrate—grazing—is neglected. When biomass on the outer surface of the carbon grains reduces the concentration of substrate below equilibrium concentrations or influent concentration is low, desorption may occur. Through counter diffusion the substrate could be transported to the outer surface of the carbon grains, where the substrate could be consumed by the biomass. Bioregeneration as a result of counter diffusion is reported for small compounds—e.g. glucose (Sontheimer et al., 1988). However, large DOC molecules are not likely to desorb (Kraaij, 2001). Therefore bioregeneration through counter diffusion for large NOM compounds is neglected (Kooij, 1983).

Biomass increases due to growth and attachment of biomass from the water phase. Especially at low substrate concentrations found in drinking water no true biofilm will develop (Rittmann et al., 1981) and attachment/detachment processes become relevant (Uhl, 2000).

From different BAC filters used for drinking water production bacteria species have been isolated and studied by Magic-Knezev. Different *Polaromonas* types were determined as the dominant cultivable species (Magic-Knezev et al., 2006).

Methods

Pilot plant study

A pilot plant was operated from September 2002 until June 2004, treating seepage water from the Bethune polder with a high concentration of humic substances. The pilot plant has the following characteristics: column diameter 0.745 m, bed height 2.15 m, flow 1.6 $m^3 \cdot hr^{-1}$, surface load 3.7 $m \cdot hr^{-1}$, EBCT 35 min, carbon type GAC830, ρ_{filt} 500 $kg \cdot m^{-3}$. The influent water was pre-treated with ozone and has the following characteristics: temp. 3–24 °C, pH 7.6–7.9, DOC 4.5–6 g $C \cdot m^{-3}$, UV 254 nm 8–11 m^{-1}, AOC 0.050–0.090 g $C \cdot m^{-3}$, ozone dosage 1.5–2.0 g $O_3 \cdot m^{-3}$. The pilot plant was monitored on water quality influent, effluent and over the bed height, once/twice per month. Important measured parameters were DOC, UV-extinction 254 nm, oxygen, phosphate, water temperature (all standard methods) and ATP on carbon measured according the method described by Magic-Knezev as a measurement of biomass activity (Magic-Knezev et al., 2004). BDOC in ozonated water was 1.2 mg C/l (1 observation), while the DOC concentration at that moment was 6.2 mg C/l.

Model description

A dynamic BAC model was developed to simulate simultaneous adsorption and biodegradation of NOM, and adsorption of organic micro pollutants. The model is implemented in

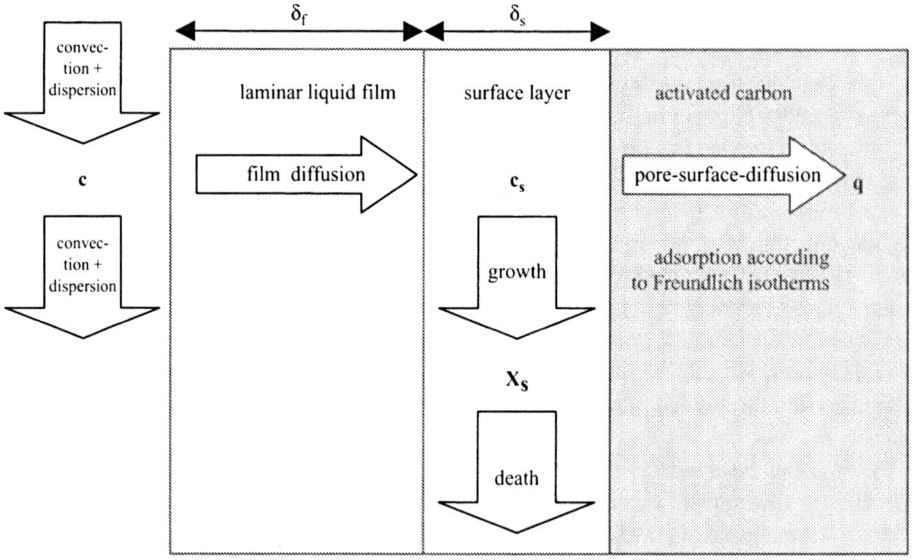

Figure 1 Model scheme of simultaneous adsorption and biodegradation

'Stimela', a Matlab-Simulink tool for modeling drinking water processes (Helm et al., 2002; Rietveld, 2005). In the model the BAC filter is divided into a number of completely stirred tank reactors (CSTR's) in series. The model scheme for each CSTR is shown in figure 1.

Raw water including substrates – biodegradable and non biodegradable dissolved organic carbon (BDOC and NBDOC), oxygen and phosphate—is led into the BAC filter. Through convection and dispersion the substrates are transported through the pores of the filter bed. The flow in the pores is turbulent. The carbon grains are surrounded by a laminar liquid film. Diffusion causes mass transfer through this film from the bulk liquid towards the surface of the carbon grain (Sontheimer et al., 1988).

Biomass is located on the outer surface of the carbon grain. BDOC, oxygen and phosphate are biodegraded according to Monod kinetics based on the 'visible' substrate concentration for the biomass. Suspended biomass, attachment and detachment of bacteria are neglected. Biomass death rate is linear to the amount of fixed biomass. Dead biomass is assumed to be washed out during backwashing. Biomass maintenance is neglected; all substrate taken up by biomass is converted into biomass.

NBDOC and BDOC that is not biodegraded diffuse into the pores of the carbon grain via pore and surface diffusion. Inside the pores of the carbon non reversible adsorption of NBDOC and BDOC takes place according to Freundlich adsorption isotherms (Sontheimer et al., 1988).

Essential in the model is the calculation of the concentration of the substrates at the outer surface of the carbon grain. This is the location of the biomass, and the substrate concentration in the surface layer is the 'visible' concentration for the biomass. Also the substrate concentration in the surface layer is the driving force for mass transport processes into the carbon grain, resulting in adsorption in the micro pores. The dynamically modelled variables are:

$$\frac{\partial c}{\partial t} = -\frac{v}{\varepsilon}\frac{\partial c}{\partial x} - \frac{1-\varepsilon}{\varepsilon}\frac{6}{d_{part}}\frac{D_f}{\delta_f}(c - c_s) \tag{1}$$

$$\frac{dc_s}{dt} = \frac{1}{\delta_s}\frac{D_f}{\delta_f}(c - c_s) - \frac{\beta_s}{\delta_s}\left(K_F c_s^{n_F} - \overline{q}\right) + \frac{\beta_p}{\delta_s}\left(c_s - \sqrt[n_F]{\frac{\overline{q}}{K_F}}\right) - \frac{\mu_{spec}}{Y_{max}} \cdot X_s \cdot \frac{\rho_{filt} \cdot d_{part}}{6 \cdot \delta_s (1-\varepsilon)} \tag{2}$$

$$\frac{d\overline{q}}{dt} = \frac{6(1-\varepsilon)}{d_{part}}\frac{\beta_s}{\rho_{filt}}\left(K_F c_s^{n_F} - \overline{q}\right) + \frac{6(1-\varepsilon)}{d_{part}}\frac{\beta_p}{\rho_{filt}}\left(c_s - \sqrt[n_F]{\frac{\overline{q}}{K_F}}\right) \tag{3}$$

$$\frac{dX_s}{dt} = (\mu_{spec} - b) \cdot X_s \tag{4}$$

$$\mu_{spec} = \min\left(\left(\frac{c_{s,1}}{K_{M,c_1} + c_{s,1}} \cdot \alpha_T \cdot \mu_{max,15^\circ C}\right), \left(\frac{c_{s,2}}{K_{M,c_2} + c_{s,2}} \cdot \alpha_T \cdot \mu_{max,15^\circ C}\right), etc.\right) \tag{5}$$

These equations are combined with the simplified competitive adsorption model (Sontheimer et al., 1988):

$$c_1 = \sqrt[n_{F,1}]{\frac{q_1^{\overline{n_F}}(q_1 + q_2)^{1-\overline{n_F}}}{K_1}} \tag{6}$$

$$q_1 = \sqrt[\overline{n_F}]{K_1 c_1^{n_{F,1}}}\left(\sqrt[\overline{n_F}]{K_1 c_1^{n_{F,1}}} + \sqrt[\overline{n_F}]{K_2 c_2^{n_{F,2}}}\right)^{\overline{n_F}-1} \tag{7}$$

Model parameters

DOC isotherms for ozonated BAC influent were determined in batch experiments using entire carbon grains (Aa et al., 2004), resulting in: $K_{F,DOC}$ 15.6 g C·(g carbon)$^{-1}$·(m^3·(g C)$^{-1}$)^n and $n_{F,DOC}$ 0.78. Both DOC-fractions are supposed to have the same diffusion and adsorption characteristics. They only differ in biodegradability.

At t = 0 days the substrate loading on the activated carbon is zero. Diffusion parameters were determined by fitting on initial DOC breakthrough data in the pilot plant, resulting in: β_p 1.4·10^{-6} m·s^{-1}, β_s 2.4·10^{-10} g carbon·m^{-2}·s^{-1} , D_f 4.9·10^{-10} m^2·s^{-1}. With a grain radius of 1 mm this gives: D_p 1.4·10^{-9} m^2·s^{-1} and D_s2.4·10^{-13} m^2·s^{-1}.

Biomass characteristics were taken from measurements on *Polaromonas* by Magic-Knezev: $\mu_{max,15^\circ}$ 19·10^{-6} s^{-1} (corresponds with G_{min} 10 hrs), $K_{M,BDOC}$ 1.5·10^{-3} g C·m^{-3}, Y_{BDOC} 0.8·10^6 ng ATP·(g C)$^{-1}$(Magic-Knezev et al., 2006) and b 1.6·10^{-6} s^{-1} (personal communication Magic-Knezev). Other biomass parameters were estimated at: $K_{M,O2}$ 2 g O_2·m^{-3}, $K_{M,PO4\text{-}P}$ 0.0005 g PO_4^{3-}-P·m^{-3}, α_T 1.07 (this equals a reduction of growth rate by 50% at every 10°C temperature drop),Y_{O2} 0.8·10^7 ng ATP·(g O_2)$^{-1}$, Y_{PO4-P} 0.8·10^8 ng ATP·(g PO_4^{3-}-P)$^{-1}$ and $X_{s,0}$ 0.02 ng ATP·(g carbon)$^{-1}$. BDOC concentrations were estimated at 25% of influent DOC (Billen et al., 1992), corresponding fairly well with measured value in the pilot plant (20%, 1 observation only).

Table 1 Model parameter values

	Applied value	Reference values	Reference
D_f [$m^2 \cdot s^{-1}$]	$4.9 \cdot 10^{-10}$	$4 \cdot 10^{-10}$–$10 \cdot 10^{-10}$	Lier, 1989
		$12 \cdot 10^{-10}$	Hozalski, 1996
K_F [g C·(g carbon)$^{-1}$·(m^3·(g C)$^{-1}$)^n]	15.6	0–80	Sontheimer et al., 1988
n_F [-]	0.78	0.10–0.80	Sontheimer et al., 1988
D_p [$m^2 \cdot s^{-1}$]	$14 \cdot 10^{-10}$	$0.56 \cdot 10^{-10}$–$100 \cdot 10^{-10}$	Sontheimer et al., 1988
		$0.06 \cdot 10^{-10}$–$510 \cdot 10^{-10}$	Lier, 1989
D_s [$m^2 \cdot s^{-1}$]	$24 \cdot 10^{-14}$	$0.34 \cdot 10^{-14}$–$29 \cdot 10^{-14}$	Sontheimer et al., 1988
		$0.08 \cdot 10^{-14}$–$117 \cdot 10^{-14}$	Lier, 1989
$\mu_{max.20\ C}$ [s^{-1}]	—	$8 \cdot 10^{-5}$	Billen, 1992
$\mu_{max.15\ C}$ [s^{-1}]	$1.9 \cdot 10^{-5}$	$1.9 \cdot 10^{-5}$	Magic-Knezev, 2006
	—	$0.13 \cdot 10^{-5}$–$7.9 \cdot 10^{-5}$	Kooij et al., 1984
$K_{M.BDOC}$ [g C·m^{-3}]	$1.5 \cdot 10^{-3}$	$1.5 \cdot 10^{-3}$	Magic-Knezev, 2006
		$50 \cdot 10^{-3}$	Billen, 1992
b[s^{-1}]	$0.16 \cdot 10^{-5}$	$2.9 \cdot 10^{-5}$	Hozalski, 1996
		$0.28 \cdot 10^{-5}$–$2.2 \cdot 10^{-5}$	Billen, 1992
α_T [-]	1.07	—	—
Y_{BDOC} [ng ATP·(g C)$^{-1}$]	$0.8 \cdot 10^{-6}$	—	Magic-Knezev, 2006

Results and discussion

Pilot plant results

Figure 2 shows DOC and biomass concentrations from the pilot plant. At day 0 initial DOC breakthrough is 0.3 g C·m^{-3}. The measured amount of biomass is 0.2 ng ATP·(g carbon)$^{-1}$. Between day 0 and day 200, DOC concentrations gradually increase. Adsorption is the main DOC removal process. Biomass concentration and corresponding biodegradation gradually increase. After day 200 a quasi-steady state is reached. The activated carbon is completely saturated and biodegradation is the main DOC removing process. About 1 g C·m^{-3} is removed in this phase. Biomass slowly develops until biomass concentration reaches a maximum of 532 ng ATP·(g carbon)$^{-1}$ in spring. At that time the filter has been in operation for 8 months. In summer the biomass concentration reduces again. Biomass concentrations in the top of the filter bed (EBCT 7 min) are approximately 50% higher than in the middle of the filter bed (EBCT 23 min). The maximum average biomass concentration for the complete filter is estimated at 300 ng ATP·(g carbon)$^{-1}$; the total biomass in the filter reaches values up to ($300 \cdot 10^{-9} \cdot 0.436 \cdot 2.15 \cdot 500 \cdot 10^3$ =) 0.14 g ATP. Assuming a ATP concentration of $2.5 \cdot 10^{-8}$ ng·cell^{-1} (Magic-Knezev et al., 2004) and a volume of $0.315 \cdot 10^{-18}$ m^3·cell^{-1}(Uhl, 2000), the maximum biomass volume per volume of filter material is ($532 \cdot 10^{-9} \cdot 500 \cdot 10^3 / 2.5 \cdot 10^{-17} \cdot 0.315 \cdot 10^{-18}$=) 0.0032 $m^3 \cdot m^{-3}$. This is a

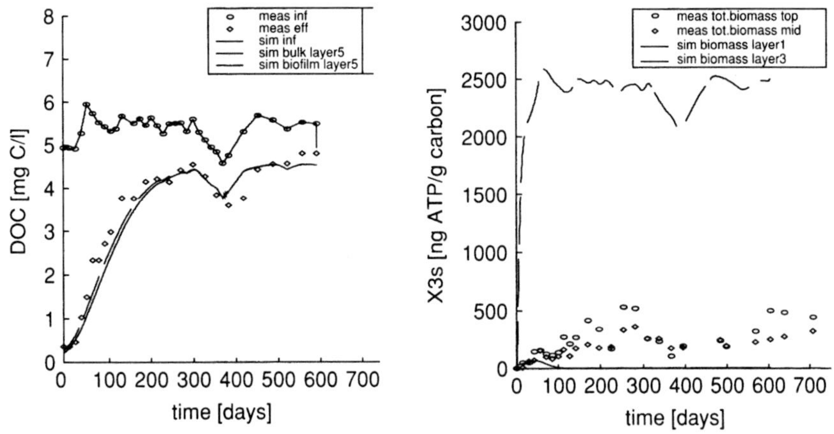

Figure 2 Pilot plant results and modeling results

small volume compared to other substances (suspended solids, algae, etc) causing filter bed resistance.

Model calibration results

Model results are also shown in figure 2. Initial DOC breakthrough was used to fit diffusion parameters, and therefore model results at filter start up correspond with the pilot plant data. Between day 0 and day 200 adsorption is the main DOC removing process. The applied Freundlich values from the laboratory tests describe the DOC breakthrough curve fairly good, although between day 100 and 200 the model predicts more (B)DOC removal than observed in the pilot filter. In the model biomass in the top layer develops until concentrations of approximately 2500 ng ATP·(g carbon)$^{-1}$. In the model biomass only develops in the top layer, in contrast with the observations in the pilot plant, where significant biomass concentrations were measured in the middle of the filter. In the model the total amount of biomass is estimated at ($2500 \cdot 10^{-9} \cdot 0.436 \cdot 2.15 \cdot 500 \cdot 10^3/5=$) 0.23 g ATP. This is 67% higher than the (estimated) maximum biomass amount in the pilot plant. In the model the biomass reaches a maximum concentration within two weeks. This is much earlier than experienced in the pilot plant, explaining the excess of (B)DOC removal in the model predictions from day 100 to 200.

Differences between model results and observed pilot plant data

In the quasi steady state in the model more biomass is produced than in the pilot plant, on the same amount of biodegraded NOM. Suggested is: (1) assumed biomass death rate is too small. The assumed biomass death rate is measured in laboratory experiments in absence of substrate. However, biomass death rate is larger due to predation; (2) assumed biomass yield is too large; (3) a significant amount of BDOC is used for cell maintenance.

In the model biomass develops much faster than in the pilot plant. Apparently the difference between the specific growth rate and the death rate is too large: (1) assumed maximum growth rate is too large; (2) assumed K_M-values are too small, substrate is not limiting growth; (3) assumed temperature correction is not correct; (4) assumed biomass death rate is too small.

In the model biomass only develops in the top layer, in contrast with pilot plant observations. Since space is not a limiting factor for biomass development, no true biofilm will develop (Rittmann et al., 1981). Therefore it is expected that the biomass in the top layer will develop until a limiting factor appears. If the limiting factor is substrate, no biomass will develop in subsequent layers. However, pilot plant shows that biomass does develop in subsequent layers. In the pilot filter either an other factor is limiting biomass development in the top layer, or some factor is stimulating biomass development in deeper filter bed layers. Hypotheses are: (1) biomass death rate is not constant in the filter bed, at high biomass concentrations more predation occurs and biomass death rate is larger; (2) biomass in the top layer releases soluble microbial products (Rittman, 2002), which are used as substrate for biomass development in deeper layers; (3) detachment of biomass from the top layer, and subsequent attachment in deeper layers, result in significant biomass development in deeper layers; (4) space/diffusion is limiting biomass development in the top layer; (5) transport of biomass to deeper layers as a result of mixing of filter material during backwashing.

In the pilot plant biomass concentrations is maximal in spring, and decreases in summer: (1) substrate concentrations (BDOC and/or phosphate) are not constant, but larger in spring, this may be caused by algae blooms (Billen et al, 1992); (2) predation is seasonal depending, resulting in variable biomass death rate.

When neglecting maintenance, in the quasi steady state the specific growth rate approximately equals the death rate. Increase of biomass death rate from $1.6 \cdot 10^{-6}\ s^{-1}$ to $2.9 \cdot 10^{-6}\ s^{-1}$ resulted in a total amount of biomass of 0.14 g ATP (results not shown), corresponding with observed pilot plant data. So far, adjustments in Monod-constants, maximum growth rate and temperature correction have not been successful in improving modeling results.

Conclusions

Biomass development in BAC filter is slow. Based on experiences in rapid and slow sand filters a start up time of approximately 2 months is expected. Pilot plant measurements show that development of maximum biomass activity in BAC filters may take 8 months. Biomass develops in the entire BAC filter. The biomass activity in the top layer (EBCT 7 min) is approximately 50% higher than in the middle of the filter (EBCT 23 min). Biomass activity peaks in spring, most probable due to higher substrate concentrations (BDOC and phosphate) due to algae blooms in the raw water. The volume of the maximum measured biomass concentration is estimated at $0.0032\ m^3 \cdot m^{-3}$. This is a relatively small volume. Therefore filter bed resistance due to living biomass is not significant.

Model predictions for adsorption of DOC – based on DOC Freundlich isotherms measured in batch experiments and on initial DOC breakthrough – describe pilot plant data well. Model predictions for biodegradation of substrates and biomass development – based on measured growth kinetics of *Polaromonas* species – are in the right order of magnitude, but not predicting pilot plant data successfully. Especially the development of biomass in deeper filter layers needs attention. Topics in study are: predation, maintenance, attachment/detachment/transport of bacteria, soluble microbial products, and seasonal variation of substrates (BDOC and phosphate).

Acknowledgments

This research is the result of cooperation between Waternet, Hydron, Kiwa Water Research, Norit, Delft University of Technology and Wageningen University and Research. The research has been co-funded by the Dutch Department of Economics, through Senter.

References

Aa, L.T.J. van der, Kolpa, R.J., Magic-Knezev, A., Rietveld, L.C. and Dijk, J.C. van (2003) Biological activated carbon filtration: pilot experiments in the Netherlands. In *Proc. of the Water Quality and Technology Conference*, Philadelphia, 2–6 November.

Aa, L.T.J. van der, Achari, V.S., Rietveld, L.C., Siegers, W.G. and Dijk, J.C. van (2004) Modeling biological activated carbon filtration: determination adsorption isotherms of organic compounds. In *Proc. of the WISA Conference*, Cape Town, 2–6 May.

Billen, G., Servais, P., Bouillot, P. and Ventresque, C. (1992) Functioning of biological filters used in drinking-water treatment – the Chabrol model. *JWSRT - Aqua* **41**(4), 231–241.

Helm, A.W.C. van der and Rietveld, L.C. (2002) Modelling of drinking water treatment processes within the Stimela environment. *Water Supply* ***2***(1), 87–93.

Hozalski, R.M. (1996) Removal of biodegradable organic matter in drinking water biofilters: experimental studies and model development. Ph.D. thesis, Johns Hopkins University, Baltimore, Maryland, United States of America.

Kooij, D. van der (1983) Activated carbon in drinking water technology. Kiwa & AWWA, Nieuwegein, the Netherlands.

Kraai, R. (2001) Sequestration and bioavailability of hydrophobic chemicals in sediments. Ph.D. Thesis, Utrecht University, Utrecht, the Netherlands.

Lier, W.C. van (1989) Mass transfer to activated carbon in aqueous solutions. Ph.D. thesis, Delft University of Technology, Delft, the Netherlands.

Magic-Knezev, A. and Kooij, D. van der (2004) Optimisation and significance of ATP analysis for measuring active biomass in granular activated carbon filters used in water treatment. *Wat. Res.* ***38***(18), 3971–3979.

Magic-Knezev, A. and Kooij, D. van der (2006, submitted) Nutritional versatility of two *Polaromonas* related bacteria isolated from biological granulated activated carbon filters. In *Proc. of the 4th Int. Slow Sand and Alternative biological Filtration Conference*, Mülheim an der Ruhr, 3–5 May.

Orlandini, E. (1999) Pesticide removal by combined ozonation and granular activated carbon filtration. Ph.D. thesis, Wageningen University and International Institute for Infrastructural, Hydraulic and Environmental Engineering, Delft, the Netherlands.

Rietveld, L.C. (2005) Improving operation of drinking water treatment through modeling. Ph.D. Thesis, Delft University of Technology, Delft, the Netherlands.

Rittmann, B.E. and McCarty, P.L. (1981) Substrate flux into biofilms of any thickness. *Journal of the Environmental Engineering Division-ASCE* ***107***(EE4), 831.

Rittmann, B.E. and Stilwell, D. (2002) Modelling biological processes in water treatment: the integrated biofilm model. *JWSRT - Aqua* **51**(1), 1–14.

Schleger, H.G. (1992) General Microbiology – 7th edition. Stuttgart, Germany.

Sontheimer, H., Crittenden, J.C. and Summers, R.S. (1988) Activated carbon for water treatment. AWWA – DVGW Forschungsstelle Engler Bunte Institut, Karlsruhe, Germany.
Uhl, W. (2000) Einfluss von Schüttungsmaterial und Prozessparametern auf die Leistung von Bioreaktoren bei der Trinkwasseraufbereitung; Praxisorientierte Untersuchungen, prozesskinetische Analyse und reaktionstechnische Modelliering. Ph.D. thesis, Rheinisch-Westfälisches Institut für Wasserforschung gemeinnützige Gmbh, Institut an der Gerhard Mercator Universität, Duisburg, Germany.

36 Nutritional versatility of two *Polaromonas* related bacteria isolated from biological granular activated carbon filters

Aleksandra Magic-Knezev[1,2,*] and Dick van der Kooij [3,4]

[1] Het Waterlaboratorium, J.W. Lucasweg 2, 2031 BE Haarlem, The Netherlands,
*corresponding author: aleksandra.magic@hetwaterlaboratorium.nl
[2] Waternet, Vogelenzangseweg 21, 2114 BA Vogelenzang, The Netherlands;
[3] Kiwa Water Research, P.O. BOX 1072, 3430 BB Nieuwegein, The Netherlands
[4] Wageningen University and Research, Department of Microbiology, Hesselink van Suchtelenweg 4, 63007 CT Wageningen, The Netherlands

Abstract Physiological properties of two bacterial strains isolated from GAC filters and identified as members of the genus *Polaromonas* were investigated in order to characterize the biodegradation processes in GAC filters used in water treatment. The results of batch culture growth experiments with mixed and individual substrates at concentrations typical for drinking water indicate that the two examined *Polaromonas* strains have similar physiological properties. Favourite substrates for growth are carboxylic acids and aromatic acids, while only a few amino acids and carbohydrates are utilized for growth. The maximum growth rate (0.09–0.11 h^{-1}) was reached at low substrate concentrations (10 μg C/l) of acetate. The saturation constant for growth of strain P-315 on acetate is 0.015μM and 0.039 μM for the growth on benzoate indicating a higher affinity of this strain for acetate. These results indicate that biodegradation of low molecular weight organic compounds in GAC filters is possible at very low concentration of substrate. The obtained results can be used to model the biodegradation in GAC filters for examining the effects of operational parameters on the efficiency of the removal of organic compounds by the combined adsorption and biodegradation in GAC filters.

Keywords Water treatment, GAC, biodegradation, *Polaromonas*, saturation constant, growth rate

Introduction

Biodegradation processes in granular activated carbon (GAC) filters are beneficial for the removal of organic compounds in the process of water treatment. The extent of biodegradation in GAC filters is determined by the concentration of active biomass and by the metabolic properties of this biomass for available substrates. For the optimal exploitation of biodegradation in GAC filtration detailed information is needed about quantitative and qualitative properties of biomass in the GAC filters. Several methods are available for the

 Recent Progress in Slow Sand and Alternative Biofiltration Processes edited by Rolf Gimbel, Nigel J.D. Graham and M. Robin Collins. ISBN: 9781843391203. Published by IWA Publishing, London, UK.

assessment of biomass concentration associated with GAC (Magic-Knezev and Van der Kooij 2004, onas strains isolated from biological granular activated carbon filters rioritering van DZH meneenen. Urfer and Huck 2000, Wang *et al.* 1995) but information about metabolic properties of bacteria from GAC is scarce. According to the ATP-based method (Magic-Knezev and Van der Kooij, 2004), 6–125 kg of active biomass can be present in a filter bed of 100 m^3 GAC. Further characterisation and optimisation of biodegradation in GAC filters requires information about the identity and physiological properties of predominant bacteria in GAC filters. In a study that was based on cultivation-dependant techniques, dominant cultivable bacteria in GAC filters used in drinking water treatment were identified as members of the genus *Polaromonas* of the *Comamonadaceae* family (Magic-Knezev *et al* 2003). Very little information about the physiological properties of species belonging to this genus is available. Therefore, the aim of this research was to determine growth characteristics of two strains belonging to the genus *Polaromonas* that were isolated from GAC filters. Experiments were designed to obtain the information about:

1. nature of favourite substrates at concentrations typical for water treatment and
2. kinetic parameters of growth on selected substrates at concentrations typical for water treatment.

The information about the nature of favourite substrates and kinetic parameters of growth for these two strains, together with the information about the concentration of biomass in a GAC filter contributes to the understanding of the removal of organic matter by the biodegradation processes in a GAC filter.

Methods

Bacterial strains

Strain P-315 was isolated from a GAC filter treating surface water that was pre-treated by coagulation and sedimentation followed by rapid sand filtration, softening and ozonation. Strain P-305 was isolated from a GAC filter treating ground water pre-treated with rapid sand filtration and softening.

According to 16S rDNA sequence analysis both strains were identified as members of the genus *Polaromonas*. Closest cultured relative with both strains is *Polaromonas vacuolata* with a sequence similarity >97% (NCBI accession number UAU14585). Although a high level of similarity was observed for 16S rDNA sequence, analysis of genomic DNA with AFPLP fingerprinting (not published) indicated that these two strains belonged to different genotypes.

Inoculum cultures to be used in the growth experiments were prepared by inoculating a 100-ml infusion flask containing basal medium supplemented with 1mg acetate-C /l with a suspension of a fresh culture grown on an R2A agar. Flasks were incubated at 15°C until the maximum colony counts were obtained (4–8 10^6 cfu/ml). These cultures are kept at 4°C.

Growth on mixed substrates

The growth experiments were performed in 1 l glass-stoppered Erlenmeyer flask treated with 550°C for 4 hours to eliminate traces of carbon. The pipettes were cleaned with a 10%

solution of $K_2Cr_2O_7$ in concentrated H_2SO_4, followed by 4 hour rinsing with tap water and heating overnight at 250°C. A total of four different types or organic compounds were used as a substrate for growth: I, 20 amino acids; II, 11 aromatic compounds; III, 17 carboxylic acids and IV, 20 carbohydrates. Solutions were prepared in 600 ml of dune-infiltrated river water treated by powdered activated carbon, followed by softening, rapid and slow sand filtration (assimilable carbon concentration (AOC) <5 μg acetate C/l). Indigenous bacteria were eliminated by pasteurisation of water at 60°C for 30 minutes. Mixtures of substrates were prepared by adding individual substrates from stock solutions, which had been prepared as described previously (Van der Kooij and Hijnen, 1985). The final concentration of each individual substrate in the mixture was 1 μg C/l. Phosphorus (Na_2HPO_4) and nitrogen were added (NH_4Cl) to avoid growth limitation other than by the carbon source. Flasks were inoculated with the pre-cultured suspension of the strains to reach a start concentration of 100–500 CFU/ml. The Erlenmeyer flasks were incubated without shaking in the dark at 15 ± 1°C and growth was monitored by periodic determination of viable cells counts on R2A medium (Oxoid) in triplicate after incubation at 25°C during 5–7 days. Parallel to the determination of viable cell counts, concentrations of ATP were measured. Each mixture or concentrations was tested in duplicate flasks.

Growth on individual substrates

For the preparation of 1 mg C/l solutions autoclaved tap water was supplemented with nitrogen, phosphorous and individual carbon source from the stock solution and inoculated with the 100–1000 CFU of the pre-cultured strain. Growth was carried out in 9-ml test tubes at 15 ± 1°C. The growth was monitored by measurement of the ATP concentration on days 7 and 14 after inoculation. Solutions containing 10 μg C/l of individual substrates were prepared in the same way as described for the solutions of mixed substrate except only one substrate was added per Erlenmeyer flask. Flasks were inoculated with the pre-cultured suspension of strain to reach a start concentration of 100 to 500 CFU/ml and monitoring of growth was identical as in the experiments with mixed substrates. For each substrate duplicate test was performed. The experiments conducted to determine kinetic parameters of growth (growth yield coefficient (Y), maximum growth rate ($1/G_{min}$) and substrate saturation constant K_s) were carried out in treated aerobic ground water containing a very low concentration of assimilable organic carbon. The test water was supplemented with nitrogen, phosphorous and the selected carbon source (acetate or benzoate) at concentrations ranging from 1 to 500 μg C/l. The flasks were incubated at 15 ± 1°C and growth was monitored by periodic determination of viable cell counts on R2A medium (Oxoid) in triplicate after incubation at 25°C during 5–7 days. Parallel to the determination of viable cell counts ATP concentration was measured.

ATP analysis

The measurement of ATP is based on the production of light in the luciferine-luciferase assay. ATP was released from suspended cells with nucleotide-releasing buffer (NRB, Celsis). The intensity of the emitted light was measured in a luminometer (Celsis Advance™) calibrated with solutions of free ATP (Celsis) in autoclaved tap water following the procedure as given by the manufacturer. The detection limit was 1 ng ATP L^{-1} of sample.

Growth kinetics

The growth yield Y was calculated from the maximum colony counts (N_{max}, CFU/ml) obtained for the applied substrate concentration (dS).

$$Y = dN_{max}/dS \qquad \text{(eq. 1)}$$

The doubling time (G in hours) of the cultures was calculated with the equation:

$$G = \log2(t' - t)/(\log N_{t'} - \log N_t) \qquad \text{(eq. 2)}$$

where t'− t is the incubation time in number of viable cells (N) increased from N_t to $N_{t'}$.

The saturation constant K_s and minimum generation time G_{min} was calculated with Lineweaver-Burk equation:

$$G = G_{min} + G_{min} \cdot K_s/S \qquad \text{(eq. 3)}$$

Where S is the concentration of the added substrate (dS) and the natural substrate concentration (S_n) which was derived from the N_{max} values observed in water without substrate addition. (Van der Kooij and Hijnen, 1985).

Results and discussion

Growth on mixed substrates

The growth on mixed substrate at the concentration of individual substrate of 1 μg C/l is performed for four major groups of organic compounds, vz 20 AA, 11 AR; 17 CA and 11 CH. The growth response is compared to the growth in water without added substrate. The strains responded similarly to substrate mixtures. Both strains showed significant growth in the mixtures of the AA, AR and CA compounds and a low response was observed with CH (Table 1). Best response was observed for the AR mixture indicating a high affinity of the strains for these types of organic compounds.

Growth on individual substrates

The growth of strains on individual carbon sources was tested for two concentrations: 1 mg C/l and 10 μg C/l. The results of the growth experiments on a wide range of individual compounds at a concentration of 1 mg C/l are presented in Table 2.

Table 1 Growth response as the ratio of the maximum concentration of two strains (C_m) in the suspensions of mixed substrates and in blank water(C_b) (R_{CFU} or R_{ATP} = C_m/C_b). AA = amino acids; CA = carboxylic acids; AR = aromatic compounds; CH = carbohydrates .

	Strain P-305		Strain P-315	
Mixture of substrates	R_{CFU}	R_{ATP}	R_{CFU}	R_{ATP}
Amino acids (AA)	8.1	6.6	6.8	9.1
Carboxylic acids (CA)	3.8	4.2	6.4	5.7
Aromatic acids (AR)	21.3	7.0	13.9	6.8
Carbohydrates (CH)	1.5	1.9	2.3	2.0

Table 2 Utilisation of individual compounds in drinking water by two *Polaromonas* strains at a concentration of 1 mg/l.

		Number (%) of compounds utilized	
Substrate type	***Number of compounds***	**Strain P-315**	**Strain P-305**
Amino acids (AA)	20	6 (30)	3 (15)
Aromatic compounds (AR)	11	6 (55)	6 (55)
Carboxylic acid (CA)	17	17 (100)	15 (88)
Carbohydrates (CH)	21	3 (14)	5 (24)

The strains showed some differences in response to the substrates from four major groups but both strains grew well on the majority of tested carboxylic acids and aromatic compounds. The utilisation of amino acids and especially grow on carbohydrates was limited to a number of compounds from this two groups. The strain isolated from the filter treating ground water (P-305) had a slightly lower response in terms of yield to individual substrates than the strain P-315 that was isolated from the filter treating surface water. Furthermore, strain P305 grew slower in the water without substrates added.

A number of compounds was selected for further growth experiments at the concentration of 10 μg C/l. Best response was observed for acetic acid, tartaric acid and benzoic acid (data not shown).

Growth kinetics of the *Polaromonas* strains for acetic and benzoic acid

Acetic acid and benzoic acid were selected as substrates for the determination of the kinetic parameters of growth. Acetic acid is a product of ozonation and forms an important part of assimilable organic carbon in drinking water. Benzoic acid was selected to obtain information about the degradation of aromatic compounds which are present in drinking water as a part of natural organic matter (humic and fulvic acid), or as a xenobiotic compounds. Furthermore, kinetic parameters of growth on acetate for *Spirillum sp.* (strain NOX) and *Pseudomonas fluorescens* (strain P17) which are widely applied for the analysis of AOC concentration in treated water, were determined earlier (Van der Kooij et al 1990).

The kinetic parameters of growth were determined for the concentration range 1–500 μg C/l. The growth curves observed in water supplied with substrates demonstrated that N_{max} values and the growth rate depend on the substrate concentration. Both substrates promoted the growth rate and increased the colony counts et al concentrations tested (Figure 1.).

The growth yield coefficients were calculated with maximum values of the ATP concentration and the plate count (CFU) for the applied substrate concentrations. These yield coefficient values differ slightly for the examined substrates and were used to calculate the natural indigenous substrate concentration (S_n) in the blank water (Table 3). The S_n concentration calculated from the yield coefficient derived from the ATP measurement were slightly higher than the concentrations calculated with the yield coefficient derived from the plate counts. The average value of the derived indigenous substrate concentration was used for the correction of S ($S_n + dS$) in the calculation of G_{min} and K_s with the Lineweaver-Burk equation (eq. 3). A plot of the generation time (G) of strain P-315 versus 1/S gave linear

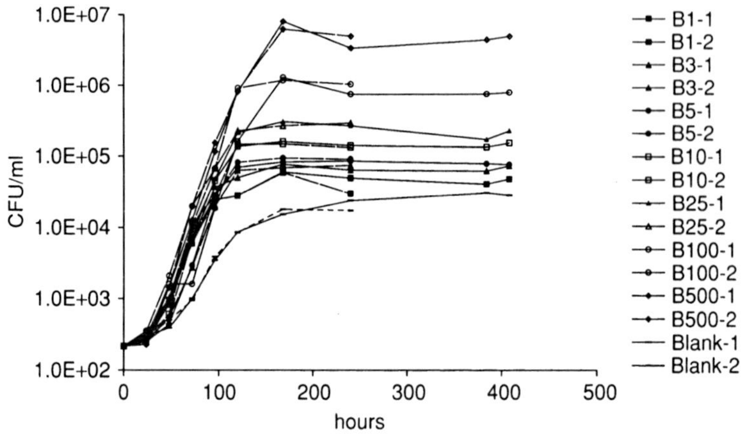

Figure 1 Growth of strain P-315 with different concentrations of benzoate at 15°C.

function for the growth on acetate and a non-linear function for the growth on benzoate (Figure 2).

These results reveal that S_n and added acetate were utilized simultaneously, while added benzoate is utilized after the depletion of S_n. Due to the preferential utilisation of S_n in the solutions with added benzoate, kinetic parameters for growth on benzoate could be determined for the concentrations above 10 μg C/l. The minimum generation time for both substrates was for strain P-315 between 9 and 11 h (0.09-0.11 h^{-1}) but the K_s value for acetate was lower than the K_s value for benzoate (table 4) indicating higher affinity of strain P-315 for acetate than for benzoate.

The observed K_s values are low and similar values have been reported for the growth of *Aeromonas hydrophyla* strain M800 on oleate (Van der Kooij and Hijnen 1988) and for the growth of *Flavobacterium sp.* strain S12 on oligosaccharides (Van der Kooij and Hijnen 1985). Furthermore, a similar saturation constant is reported for growth of a *Mycobacterium sp.* on phenantrene (table 4). Information about the physiological properties of bacteria belonging to the genus *Polaromonas* is scarce. There are only few species described in this genus and one of the species is isolated from a GAC filter treating polluted ground water (Coleman *et al.* 2002). The results of our research indicate that *Polaromonas* strains

Table 3 Growth yield on acetate (Y_{ac}) and benzoate (Y_{ar}) acid and the concentration of indigenous substrate (S_n) derived from the growth yield and the maximum growth in the blank.

Strain	Substrate	Y_{CFU} (CFU/μg C)	$S_{n(CFU)}$ (μg C/l)	Y_{ATP} (ng ATP/μg C)	$S_{n(ATP)}$ (μg C/l)
P- 305	Acetate	2.1 10^7	1.9	0.8	2.1
	Benzoate	1.7 10^7	2.4	0.8	2.1
P- 315	Acetate	1.4 10^7	1.7	0.8	2.4
	Benzoate	1.4 10^7	1.7	1.0	1.9

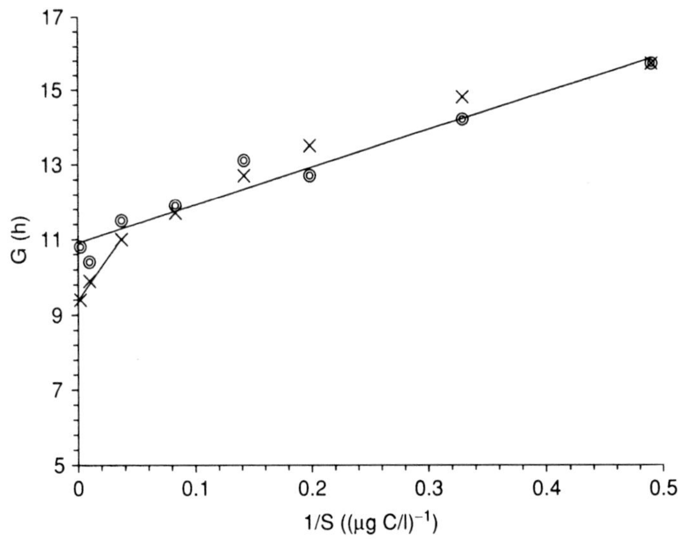

Figure 2 Lineweaver-Burk plot of strain P-315 for growth on acetate (○) and benzoate (×).

isolated from GAC filters are able to grow on very low concentration of carboxylic acids and aromatic compounds (1 μg C/l).

Carboxylic acids, which are the favourite substrates of *Polaromonas* strains, are formed by the oxidation of natural organic matter and are common constituents of organic matter at the end of the treatment train. Their presence in drinking water may contribute significantly to the biofilm formation in distribution systems. Due to their polar nature, carboxylic acids can not be removed by the adsorption on GAC, and their removal by biodegradation results in the improvement of biological stability of drinking water.

Table 4 Saturation constants for growth of some aquatic bacteria on different substrates.

strain	Substrate	K_s (μM)	G_{min} (h)	Reference
Polaromonas P-315	Acetate	0.015	10.9	This research
	Benzoate	0.039	9.4	
Polaromonas	cis-dichloroethene	1.6		Coleman *et al.* (2002)
Mycobacterium sp.	Phenantrene	0.026		Miyata *et al.* (2004)
Aeromonas hydrophyla	Oleate	0.01	4.3	Van der Kooij
Strain M800	Acetate	0.46	6.7	and Hijnen (1988)
Flavobacterium sp.	Maltotetraose	0.019	2.3	Van der Kooij
	Maltotetraose	0.016	2.2	and Hijnen (1985)
	Maltotetraose	0.015	2.2	
Spirillum sp.	Oxalate	0.14	4.1	Van der Kooij (1990)
Pseudomonas fluorescens	Acetate	0.17	5.5	
Pseudomonas spp. GF1TP	Toluene	2.1	2.6	Massol-Deya *et al.* (1997)
Pseudomonas spp. GF161	Toluene	2.7	3.0	
Ralstonia taiwanensis	Phenol	5.5		Chen *et al.* (2004)

The removal of organic compounds in GAC filters is a result of adsorption and biodegradation processes. The removal by the adsorption is determined by the adsorption capacity of GAC and the adsorption kinetic of organic compounds in the water, while the removal by the biodegradation is determined by the concentration of biomass and its metabolic potential for the organic compounds that are available as a substrate. The adsorption capacity of GAC for aromatic compounds can differ significantly, but aromatic compounds are more adsorbable and will therefore occupy the adsorption sites on GAC. When these compounds are biodegraded, adsorption capacity of GAC will be used more optimally for the target compounds that are non-biodegradable and adsorbable resulting in improved performance of a filter and improved water quality. The information about the biodegradation kinetics for carboxylic acids by bacteria from GAC will be used for modeling the removal of these compounds in GAC filters, aiming at optimizing the use of this treatment process.

Conclusions

Two strains of the genus *Polaromonas*, isolated as predominant cultivable bacteria from GAC filters used in water treatment, have a similar metabolic potential and the following physiological properties of these strains were observed:

1. Preferable substrates for growth are carboxylic acids and aromatic compounds while both strains grow only on a limited number of amino acids and carbohydrates;
2. Growth on benzoate and acetate occurs at very low concentration of these substrates (1 μg/l) with relatively low growth rate ($0.09\,h^{-1}$);
3. Higher Ks value for growth on benzoate (0.039 μM) than for acetate (0.015) indicate higher affinity of strain P-315 for acetate than for benzoate;
4. The ATP concentration is suited parameter for determination of bacterial growth yield and for determination of kinetic growth parameters at concentrations $>$10 μg C/l.

Hence, bacteria in GAC filters used in water treatment are able to utilise biodegradable low molecular weight compounds at very low concentrations, thus contributing to the production of biologically stable drinking water.

Acknowledgements

This investigation was part of a joint research project conducted at Kiwa Water Research and Wageningen University Research Centre and the Technical University of Delft. The financial support of Amsterdam Water Supply, Hydron Midden-Nederland and the Ministry of Economic Affairs (Senter) is gratefully acknowledged.

References

Chen, W.M., Chang, J.S., Wu, C.H., Chang, S.C. (2004) Characterization of phenol and trichloroethene degradation by the rhizobium Ralstonia taiwanensis. Res Microbiol. 155(8):672–680.

Coleman, N.V., T.E. Mattes, J.M., Gossett, and J.C. Spain. (2002) Biodegradation of cis-dichloroethene as the sole carbon source by β-proteobacterium. Appl. Environ. Microbiol. 68(6) 2726:2730.

Magic-Knezev, A., and D. van der Kooij (2004) Optimisation and significance of ATP analysis for measuring active biomass in granular activated carbon filters used in water treatment. Wat. Res. 38(18):3971–3979

Magic-Knezev A., Wullings, B., van der Kooij, D. (2003) Identification of cultivable bacteria from activated carbon filters used for water treatment 1st FEMS Congress of European Microbiologists, 2003 June 29 - July 3, Ljubljana, SLOVENIA.

Massol-Deya, A., R. Weller, L.Rios-Hernandes, J.Z. Zhou, R.F. Hickey, J.M. Tiedje. (1997) Succession and Convergence of Biofilm Communities in Fixed-Film Reactors Treating Aromatic Hydrocarbons in Groundwater. Appl. environ. Microb. 63(1) p. 270–276

Miyata, N., Iwahori, K., Foght, J.M., Gray, M.R. (2004) Saturable, energy-dependent uptake of phenanthrene in aqueous phase by Mycobacterium sp. strain RJGII-135: Appl. Environ. Microbiol. 70(1):363–9

Van der Kooij, D. and W.A.M. Hijnen (1985) Determination of the concentration of maltose- and starch-like compounds in drinking water by growth measurements with a well-defined strain of a *Flavobacterium* species. Appl. Environ. Microbiol. 49(4):765–771

Van der Kooij, D., and W.A.M., Hijnen (1988) Nutritional versatility and growth kinetics of an Aeromonas hydrophyla strain isolated from drinking water Appl. Environ. Microbiol. 54(11):2842–2851

Van der Kooij, D. (1990) Assimilable organic carbon (AOC) in drinking water. In (G.A. McFeters eds) Drinking water microbiology; progress and recent developments, pp 61–87. Springer-Verlag New York Berlin.

Urfer, D., and Huck, P.M. (2000) Measurement of biomass activity in drinking water biofilters using a respirometric method. Water Res. 35(6), 1469–1477

Wang, J.Z., Summers, R.S., and Miltner, R.J. (1995) Biofiltration performance: part 1, relationship to biomass. Filtration: Jour. Am. Water Works Assoc. 87: 55–63.

37 Biological fouling of structures in roughing filters used prior to slow sand filtration

M. J. Chipps, R. G. W. Bayley, M. E. Steele, R. White, A. Mikol, E. Fricker, and C. S. B. Fitzpatrick*

Thames Water R&D, Spencer House, Manor Farm Road, Reading, RG2 OJN, UK
*Dept. of Civil & Environmental Engineering, UCL, Gower Street, London, WC1E 6BT, UK

Abstract Populations of reservoir algae, zebra mussels in stored water tunnels, fungi and bacteria on filter media, and bryozoans, sponges and hydroids on filter nozzles present a series of challenges to the hydraulic and water quality performance of rapid gravity filters upstream of slow sand filters. Whilst they still protect slow sand filters and allow relatively high water throughputs, roughing filters are also identified as a potential constraint. This necessitates diligent monitoring and the future potential for improved designs for new and refurbished plants, including chemical cleaning.

Keywords Biological filtration, rapid gravity filter, nozzles, backwashing, starting head loss, biofouling, algae, data management, *Cordylophora, Spongilla, Plumatella, Dreissena.*

Introduction

Thames Water treats about 2000 Ml.d^{-1} of drinking water in London by slow sand filtration through over 100 slow sand filters (SSFs). The majority of the SSFs receive water that has been pre-treated by roughing filtration. Rapid gravity filters (RGFs) typically remove 30–80% of the lacustrine algae that proliferate in the eutrophic storage reservoirs, the removal depending partly on algal morphology. This enables the SSFs to be operated up to 0.5 m.h^{-1} filtration rate (Rachwal *et al.*, 1996). Cleasby *et al.* (1984) suggested a water quality of ≤ 5 NTU and 5 μg.l^{-1} chlorophyll-*a* for water reaching SSFs. Woodward *et al.* (1994) gave mean and peak RGF chlorophyll-*a* targets of <5 μg.l^{-1} and <15 μg.l^{-1} for London's SSFs to be operated at >0.3 m.h^{-1}.

In London, 118 RGFs serve 5 of the 6 SSF works, with 6 new RGFs being built at the remaining works. The RGFs were built between 1927 and 1995; refurbishments in the 1980s have improved older designs. With the exception of one works the filters do not use coagulants or upstream clarification processes. The stored water is not chlorinated, but ozone is used ahead of the dual media RGFs built in the 1990s.

This paper presents examples of biological fouling observed in London's RGFs. It uses head loss data to show the vulnerability to diverse biological organisms and describes some investigations that Thames Water has undertaken. Until recently such data were captured

 Recent Progress in Slow Sand and Alternative Biofiltration Processes edited by Rolf Gimbel, Nigel J.D. Graham and M. Robin Collins. ISBN: 9781843391203. Published by IWA Publishing, London, UK.

from SCADA systems by hand. A data processing tool named "RagMon" has been developed by Thames Water to present information from existing data sources in an easily understandable way for operators and designers.

Methodology

Thames Water's RGFs are monitored and controlled using SCADA systems and flow and differential pressure instruments. Turbidity meters are not used on individual roughing filters where there is no coagulant breakthrough to control. Turbidity measurement of the stored water and a combined filtrate is typical. The RagMon programme treats individual filter runs and backwashes as separate batch processes, generating performance indicators to describe filter and backwash operation. These indicators are: filter run length (hours); filter productivity ($m.run^{-1}$ i.e. run length × mean flow rate); flow and temperature normalised starting and terminal head loss (m); air scour flow rate; and low and high backwash water flow rates. The method for calculating the flow and temperature normalised starting head loss (NSHL) is derived from the Kozeny-Carman equation, and relies on other factors such as media size, shape and bed depth remaining constant (Logsdon *et al.*, 2002). The normalisation enables all data to be comparable, irrespective of filter throughput or seasonal changes in water viscosity. The formula is given in equation 1.

$$\mathrm{NSHL} = \mathrm{h} \cdot (10/Q) \cdot (1.139/\mu) \qquad \text{Equation 1}$$

Where: h = measured head loss (m) averaged over 15 minutes after the start, or preceding the termination, of the filter run

Q = measured flow rate ($m.h^{-1}$) (10 $m.h^{-1}$ used as standard)

μ = water absolute viscosity ($Ns.m^{-2}$) × 10^3 at measured water temperature (1.139×10^{-3} at 15°C used as standard)

Pilot trials are underway to compare biological colonisation on different nozzle and RGF underdrain materials. A test tank has been built in two sections, and each half fitted with identical floor types and samples of filter floor materials ("tokens"). Both sections are fed with RGF filtrate, but one side is also flushed daily for 50 mins with chloraminated water. Tokens are sampled monthly. Head loss and flow rate across each floor type are monitored.

Results and discussion

Effect of algae on filter productivity

The effect of algae clogging RGFs is relatively well known. Filamentous genera such as *Melosira, Tribonema* and *Aphanizomenon* cause surface blinding of the sand used in roughing filters (0.6–1.18 mm, effective size (E.S.) 0.7 mm). Dual media designs overlay this with anthracite (1.7–2.5 mm, E.S. 1.8 mm) reducing the impact of filter clogging algae. In filters exposed to daylight long filamentous algae, such as *Cladophora*, and higher plants can grow in the detritus that may coat the walls and wash troughs. These can break off and become embedded in the media, leading to the additional costs of loss of production and staff time for manual cleaning.

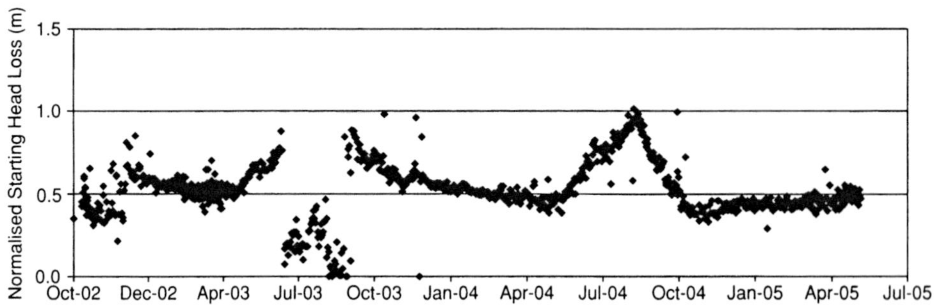

Figure 1 Flow and temperature NSHL data from works A. The typical rise and fall in 2004 is in contrast to the fall in values in summer 2003 caused by an influx of zebra mussel shells.

In London, RGFs are backwashed when a head loss trigger is reached or, during good water quality, after a preset run length. The latter spreads backwashes out throughout the day, avoiding backwash queues if reservoir water quality suddenly deteriorates. On sites with a sub-optimal backwash, biological material accumulates on the media in summer. The media is partially cleaned by extra backwashing in winter (figure 1).

By showing the head loss immediately preceding each backwash RagMon provides filter operators with an indication of the stored water quality in terms of the clogging effect of algae. The filter run length and filter productivity data also demonstrate the effect of algae or other blocking factors (e.g. zooplankton) in the reservoir water. A rising trend in terminal head loss could be used to show an algal bloom is developing, in which case a change in source water could be effected at certain treatment works. The data can also be used to show whether the filter design is satisfactory or whether future investment is required.

Backwashing effectiveness

Older designs of RGF were historically 'cleaned' using separate air scour and water phases. This proved ineffective at cleaning the sand so Thames Water modified its RGFs in the 1980s to use a combined air and water wash, although it was not possible to achieve the optimum collapse-pulse conditions defined by Amirtharajah *et al.* (1991). These filters tend to clog rapidly and accumulate large quantities of organic and inorganic material over time. For example, the NSHL data from works A show how the value rises in the summer and falls in the winter (figure 1, data for 2004), suggesting that the wash is less effective in the summer months compared to the winter, when the media is cleaned to some extent. The rise in the starting head loss as the water warms up is thought to be due to bacterial and fungal growth, with evidence from scanning electron micrographs (SEM) of filter sand samples (Fitzpatrick *et al.*, 2004). Other contributory factors during the summer may be 24 hour filter runs during good water periods and the influence lower water viscosity has on the backwash effectiveness.

Chipps *et al.* (1995) used NSHL data to demonstrate the advantage of collapse-pulse washing over separate air and water on pilot plant RGFs. The dual media filters were maintained in a relatively clean state, which enabled filtrate quality to be improved by using

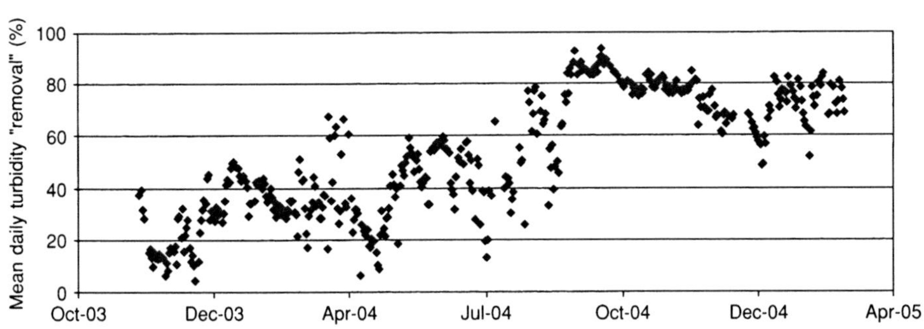

Figure 2 Apparent turbidity "removal" by a 0.7 mm E.S. pilot sand roughing filter with a sub-optimal backwash, showing the effect of biological "maturation" over time.

ozone and iron (III) sulphate contact filtration (Chipps, 1998). Subsequently dual media RGFs were constructed at works B and C, using air scour and optimum low and high wash water rates to achieve collapse-pulsing and reclassify the media layers respectively.

Zebra mussels

The zebra mussel, *Dreissena polymorpha*, has a long history of reducing hydraulic capacity in stored water tunnels and adverse effects on the operability of valves and pumps. Sites with pre-ozonation have reported zebra mussel shells accumulating in the bottom of contactors; likewise, mussel accumulations on filter walls require periodic removal. There is potential to cause undesirable taste and odours from decaying organisms (Hall *et al.*, 2004). The zebra mussel had not been thought likely to threaten SSF operation. However, a sudden reduction in NSHL on one RGF (figure 1, 2003 data) led to the discovery of an influx of dead shells into the filter (Elliott *et al.*, 2005). This indicated the flow was probably passing through a layer of shells rather than the full depth of sand, increasing the particle loading rate of the SSFs. It is therefore advised that filter plant designs should accommodate easy removal of mussel shells from pipes, RGFs and ozone contactors and incorporate settlement traps.

Media maturation

Some biological growth on RGF media is beneficial for water quality and, if managed effectively, there is little hydraulic penalty. Such filter beds can be termed "mature". Figure 2 shows how apparent "removal" of turbidity improved over 15 months on a sand pilot filter (600 mm depth, 0.7 mm E.S.) with a sub-optimal backwash, mimicking the operation of works A. If the biological coating on the media becomes excessive filter run lengths can be unacceptably short (Fitzpatrick *et al.*, 2004). Eventually mud-balls and "cracks" develop and filtrate quality deteriorates. The accelerated growth of biofilm on the media grains in summer indicates that any total bed media replacement should take place in the spring, to allow early media maturation and maintain filtrate quality.

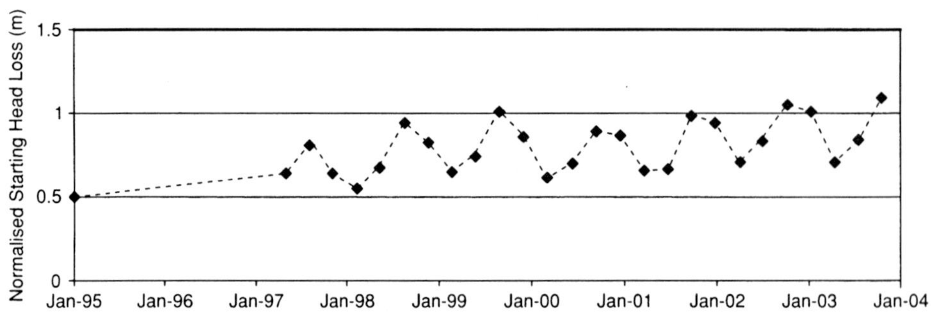

Figure 3 Mean normalised starting head loss for January, April, July and October for one filter at Works B.

Biological fouling of filter nozzles

At works B, observations and NSHL information showed that issues were arising with the dual media filters after three years of operation, despite using the collapse-pulse backwash. The mean monthly NSHL data for January, April, July and October 1997–2003 show a seasonal rise and fall (figure 3), similar to the observations at works A (figure 1). However, in this case there was also an underlying rising trend from the design NSHL of 0.5 m.

Investigations showed the following: media was found in some filter plenum chambers; air scour distribution at start-up was uneven with large 'jets' of air-water-media mix being ejected about 1 m above the water; a backwash over-pressure relief system allowed a proportion of the backwash flow to run to waste. On some filters the wash water reached the discharge trough on one half of the filter several seconds before the other half.

As a result the nozzles and plenum chambers of the most affected filters were examined. In 2002 a 2m^2 area of one filter was excavated. Twelve white nozzle caps were removed and replaced with new black ones.

It was found that the inside surfaces of the nozzle cap were partially clogged by a range of biological taxa ranging from brown filaments, white colonies and brown-black spheroids (figure 4a). The nozzle stems were also coated with biofilm, in some cases almost completely fouling the nozzle stem air slots. An initial microscopic examination was inconclusive because the organisms had degraded over time.

Independent expert help was sought with identification. The spheroidal material consisted of bryozoan statoblasts ("pipe-moss eggs") probably *Plumatella casmiana.* The fibrous material on freshly sampled nozzle caps was identified as a brown hydroid of the genus *Cordylophora.* This was also found on zebra mussel monitoring plates inserted into the stored water feed channels at works A. The biofilm on the nozzle stems was white-clear in colour and 'jelly-like' in consistency, but included *Cordylophora.* Microscopical examination indicated a community of microorganisms, and samples of the white material showed the spicules of a freshwater sponge, probably *Spongilla* (Aldridge, pers. comm., 2005, Folino-Rorem, pers. comm., 2005).

In 2004 four RGFs were given a full nozzle, tailpipe and media refurbishment. Each filter was examined and one was found to contain mud-balls. It took several days to empty

Figure 4a) Nozzle caps with *Cordylophora* and *Spongilla* (left) **4b)** Stem with *Cordylophora.*

the filter before the nozzles could be removed for inspection. After 2 years' use, the 12 black nozzles showed similar fouling to the 10 year old white caps (figure 4a).

The current hypothesis for the sequence of events over 10 years is: initial biofilm growth on tailpipes and nozzles, leading to attachment of higher organisms. The latter then obstructed the distribution of air and backwash water, resulting in dirty media, mud-ball formation and the high starting head loss values.

A number of measures were instigated to understand and mitigate future biofouling. Air flow monitoring was installed, together with additional head loss measurements to check changes across the nozzle floor. One filter was returned to clean condition to monitor changes closely using RagMon. On another filter iron (III) sulphate was dosed, based on pilot plant studies (Chipps, 1998), to reduce the particles available for filter feeders such as *Cordylophora.* An opaque cover was installed on another filter to prevent *in situ* algal growth and thus limit algal exudates as a food source for biofilm (Mann, 1988). The remaining filter was fitted with new nozzles with 36 slots of 0.5 mm width instead of the original 72 slots of 0.25 mm width. It remains to be seen if these become blocked by *Cordylophora.*

Chemical treatment options

Due to the potentially large number of species present and the decaying state of some of the organisms in nature, a test tank was constructed to identify the succession of species on different filter materials and determine the possibility of control by chloramination. These tests have run since June 2005. Results are preliminary, with no single material identifying itself as being more prone to biological colonisation. Head loss differences across each floor type are slight and SEM analysis reveals some biological growth on all the materials tested. There is currently no sign of any growth of *Cordylophora*, bryozoan or *Spongilla* species. The pilot trial is likely to run for at least 2 years.

Future work may examine the potential of chemical cleaning of media and nozzles *in situ*, or chlorinated backwashing as a control mechanism. It is noted that the RGF pilot plant study described by Chipps *et al.* (1995) used chloraminated backwash water. The contribution this played to preventing biofouling of the pilot plant nozzles is unknown. A survey of sites with chlorinated and non-chlorinated backwash water is required to accelerate understanding of this issue.

Discussion

Obtaining fresh and uncontaminated floor samples is difficult due to the inaccessibility of the nozzles and stems. One opportunity arose to examine the nozzle stems at works C. These were 12-year-old filters of similar design to works B. Observations made from the plenum chamber hatch showed a glutinous mass of biofilm, and something resembling *Cordylophora*, proliferating around the nozzle stem (figure 4b). This did not appear to be coating any other surface in the plenum chamber. At that time it was not possible to examine the nozzle caps.

A sample was obtained of underdrain nozzles and PVC laterals from works D. These were installed in the mid 1980s. The laterals contained dried remnants resembling *Spongilla* and *Cordylophora*, although precise identification was not possible.

Determination of the microorganisms present on dried, old samples of nozzles and stems was attempted. A large number of heterotrophic bacteria (mean of 2.3×10^{10}cfu/nozzle) and coliform bacteria (mean of 3.5×10^{8}/nozzle) were present. The characterisation and identification of a number of the organisms which grew under the test conditions showed them to be pseudomonads, which are commonly found in water and known to colonise plastic surfaces. The plant-growth promoting aspect of *Pseudomonas fluorescens* may suggest support for the presence and growth of the hydroid *Cordylophora*. Not all microbial taxa would have survived storage and/or desiccation and the list of organisms found must be incomplete.

Why *Cordylophora* and the other biofouling organisms should be causing a problem now and not, apparently, in the past is unknown. The past occurrence of *Cordylophora* is not recorded and this may be explained either by its complete absence hitherto or that it was present only as small colonies and thus went unnoticed. Statoblasts have certainly been observed although rarely appear to have caused significant problems with older designs of filter. One contributory factor may be the modern polymers used for nozzles as opposed to, for example, perforated steel pipes. Alternatively, narrower nozzle slits and stem slots may be more prone to fouling than older designs. Another factor may be aspects of water quality and changes in ionic strength. Chloride concentration in the early 1970s was approximately 35 $mg.l^{-1}$ whilst it can now be as high as 70 $mg.l^{-1}$. Such an increase might have enabled *Cordylophora* to exploit the nozzle/lateral/plenum habitats in recent years. Fulton (1962) demonstrated a rise in *Cordylophora* growth rates with increasing chloride concentrations.

Why bryozoan statoblasts were present, but not adults, is also unexplained. The development of adults would be extremely problematic in filter floors.

Discussions with colleagues in other water companies suggest that all these organisms comprise an emerging national problem. Again the reasons for this are unknown. The authors suspect that the reasons for fouling are likely to be complex and involve a number of interacting factors. In order to control or prevent organisms growing in filter underdrains it is important to understand more about their biology and ecology.

Conclusions

A wide range of organisms has been shown to impact on water quality and volumes produced by biological RGFs. Organisms identified include algae, *Dreissena, Cordylophora, Plumatella* and *Spongilla*. Once biological growth has blocked critical structures such as nozzles, the backwash is impaired leading to non-optimal rates of both air and water being applied to the filters. This, in turn, results in the filter media becoming dirty, shortening the filter run and, potentially, adversely affecting the filtrate quality. Poorer RGF filtrate quality or quantity have the potential to reduce SSF productivity, but not filtrate quality.

The factors affecting the biofouling of nozzle caps and stems are unknown, but likely to be complex and involve a number of interacting factors, some of which may be subtle. Without knowledge of the causal factors, remedial measures may prove only partially successful. Current replacement strategies or chemical cleaning of dirty components will not stop the organisms from growing. It might be possible to target the primary colonisers or juvenile stages of the invertebrates by the use of a free chlorine or monochloramine residual applied during backwashing. The adult stages are likely to resist realistic chlorine doses. As yet such strategies remain unproven: an industry wide survey could be helpful.

This study has produced the following recommendations based on current understanding:

- Treatment works designs should accommodate easy removal of zebra mussel shells from pre-ozone contactors or purpose built settlement traps to avoid risk of compromising RGF water quality performance.
- The enhanced growth of media biofilm in the summer suggests that for optimum water quality any media replacement programme should ideally take place in the spring, to allow early media maturation.
- In biological roughing filters it should be assumed that biological fouling of visible and hidden structures is possible. Therefore structures with small apertures should not be used.
- Accurate monitoring of head loss across media and floors, and accurate recording of backwash flows would be beneficial. In many cases instruments already exist, but the way data are captured and presented makes them difficult to interpret. RagMon is one means of attempting to correct this situation.

Acknowledgements

The authors thank their many colleagues for their assistance, and Thames Water for permission to publish this paper. The views expressed are those of the authors, and not necessarily those of Thames Water or UCL. A.Mikol is an EPSRC funded EngD student at UCL.

References

Aldridge, D. (2005). Personal Communication. Department of Zoology, University of Cambridge.

Amirtharajah, A., McNelly, N., Page, G. and McLeod, J. (1991). *Optimum Backwash of Dual Media Filters and GAC Filter-Adsorbers with Air Scour*. American Water Works Association Research Foundation, Denver, Colorado.

Chipps, M.J., Bauer, M.J. and Bayley, R.G. (1995). Achieving enhanced filter backwashing with combined air scour and sub-fluidising water at pilot and operational scale. *Filtration and Separation,* **32** (1), 55–62.

Chipps, M.J. (1998). *An Experimental Investigation into Filter Ripening: Contact Filtration of Lowland Reservoir Water*. Ph.D. Thesis, University of London.

Cleasby, J.L., Hilmoe, D.J. and Dimitracopoulos, C.J. (1984). Slow sand and direct in-line filtration of a surface water. *J. Amer. Wat. Wks. Assoc.*, **76** (12), 44–55.

Elliott P, Aldridge, D.C., Moggridge, G.D. and Chipps, M. (2005). The increasing effects of zebra mussels on water installations in England. *J. Chartered Institution of Water and Environmental Management* **19** (4), 367–375.

Fitzpatrick, C. S. B., Chipps, M. and Palupi, I. (2004). Optimising backwash of a fouled filtration medium. *Proc Am. Wat. Wks Assoc. Water Quality Technol. Conf.* San Antonio, Texas, USA Nov 14–18.

Folino-Rorem, N.C. (2005). Personal Communication, Faculty of Biology, Wheaton College, USA.

Fulton, C. (1962). Environmental Factors Influencing Growth of *Cordylophora*. *J. Exp. Zool.* **151,** 61–78.

Hall, T., Shepherd, D., Dillon, G., Hearn, K., Aldridge D. and Elliott, P. (2004). *Preventing Problems with Zebra Mussels*. WRc Confidential Portfolio Project Report, WRc, Swindon.

Logsdon, G.S., Hess, A.F., Chipps, M. J. and Rachwal, A.J. (2002). *Filter Maintenance and Operation Guidance Manual*. American Water Works Association Research Foundation, Denver, Colorado.

Mann, K.H. (1988). Production and use of detritus in various fresh-water, estuarine, and coastal marine ecosystems. *Limnology and Oceanography*, **33** (4), 910–930.

Rachwal, A.J., Bauer, M.J., Chipps, M.J., Colbourne, J.S. and Foster, D.M. (1996). Comparisons between slow sand and high rate biofiltration, In Graham, N.J.D & Collins, R., (Eds) *Advances in Slow Sand and Alternative Biological Filtration*. John Wiley & Sons, Chichester.

Woodward, C.A., Bauer, M.J. and Chipps, M.J. (1994). Pre-treatment of lowland river waters. In Collins, M.R. and Graham, N.G.D. (Eds.) *Slow Sand Filtration: An International Compilation of Recent Scientific and Operational Developments*. Americas Water Works Association, Denver, Colorado.

38 Influence of hydraulic retention time on the treatment efficiency of a biological aerated filter with shale gravel media

L. Qiu*, J. Ma and L. Zhang***

*School of Civil Engineering, Jinan University, No. 106 Jiwei Road, Jinan 250022, China;
**School of Municipal & Environmental Engineering, Harbin Institute of Technology, No. 202 Haihe Road, Harbin 150090, China

Abstract A bench-scale investigation was conducted to evaluate the influence of hydraulic retention time (HRT) on the efficiency and operational characteristics of biological aerated filter (BAF) by using a synthetic domestic wastewater as feed water and shale gravel as filter media. The variations of operation characteristics and treatment efficiency of the bioreactor resulted from the changing of hydraulic retention time were analysed. The experimental results demonstrated that it was effective to remove the organic matters and turbidity when the hydraulic retention time was longer than 0.6 h. However, the removal efficiency of organic matters and turbidity decreased remarkably when the hydraulic retention time was 0.4 h. The nitrification and denitrification ability of BAF declined sharply with decreasing the hydraulic retention time. The ammonia and total nitrogen removal efficiency were more than 80% and 50% respectively when hydraulic retention time was longer than 2.5 h. In addition, it was observed that nitrite accumulation could be accelerated when the hydraulic retention time was shortened, and that a linear relationship between hydraulic retention time and filtration time was also determined in this study.

Keywords biological aerated filter (BAF); shale gravel media; hydraulic retention time (HRT); treatment efficiency; nitrite accumulation; filtration time

Introduction

Many different types of filter and their associated media are in use today for potable and wastewater treatment. One such filter is the biological aerated filter (BAF), which has the excellent characteristics such as high treatment efficiency, low area occupancy, low construction and operation costs, convenient management and high impact load resisting ability (Amar and Partos, 1986; Canler and Perret, 1994; Pujol *et al*, 1994). Biological Aerated Filter (BAF) can be used to remove suspended solid (SS) and organic matters, nitrification, denitrification and phosphate removal in the secondary or tertiary treatment process. As a novel biological treatment technology, Biological Aerated Filter (BAF) has extensive potential in the small or middle point polluted source control and the advanced purification of secondary effluent (Ma and Qiu, 2002).

The media bed, which combined biological oxidation with suspended capture, played an important role in the biological filtration process. The media material represents a significant proportion of the initial capital outlay, but it can greatly affect day-to-day running costs such as energy costs in backwashing (Letterman, 1980). For more excellent biofiltration performance and economic running cost, it is important to determine the media's suitability for use and to help in the choice of alternative materials with appropriate characteristics such as density, fluidisation rates, size, porosity and attrition resistance (Kent *et al*, 1996).

So far the researches on BAF are mainly concentrated in synthetic materials such as polystyrene, lightweight floating Biostyr, and so on (Sagberg *et al*, 1992), but less in granular in nature and traditionally consisted of natural materials such as sand, shale and others expanded clays. It was considered that the production of synthetic material is usually costly though and media may not be significantly more effective than natural materials (Kent *et al*, 1996). In developing countries for example, there may be natural local materials that could provide cheaper alternatives to commercial filter media (Ogedengbe and Olawae, 1983). Shale gravel, an abundant mineral in northeast China, is a cheap material with great special surface area and favourable surface characteristics, having a promising advantage of initial capital cost. With the developing of BAF in China, to determine the media's suitability for use and to help in the choice of alternative materials become more and more significant (Ma and Qiu, 2002). The objective of this study is to get detail investigation on the performance of the shale media biofilter for advanced theoretical studies and engineering applications of BAF. So that the relations and regularity between variations of hydraulic retention time (HRT) and the treatment efficiency of COD, SS, ammonia and total nitrogen removal have been primarily investigated in this paper through an experiment of half synthetic domestic wastewater treatment by using shale media biological aerated filter (BAF).

Materials and methods

The filter used for carrying out the experimental works is made on Plexiglas having 250 cm of inside diameter and 2.5 meter height. The shale gravel particles of 3–5 mm of averaged diameter were filled in the filter and the media layer height is 1.6 meter. Throughout all the experimental period the reactor was fed with half synthetic domestic wastewater that was a mixed solution of starch, peptone, beef extract, NH_4Cl, KH_2PO_4, $CaCl_2 \cdot H_2O$, $MgSO_4 \cdot 7H_2O$, $FeSO_4 \cdot 7H_2O$ and raw domestic wastewater. The composition of synthetic liquid was determined and presented in Table 1. The feed wastewater flowed into the filter from the bottom making an up-flow mode in the reactor.

The reactor was initially inoculated with sludge previously grown in a SBR reactor for domestic wastewater treatment. After 20 days of continuous operation, 75% of COD and

Table 1 Composition of half synthetic domestic wastewater

Items	Data	Items	Data
COD (mg/L)	199.6~326.7	$N\text{-}NO_2$ (mg/L)	0~0.01
Ammonia (mg/L)	42.3~46.3	$N\text{-}NO_3$ (mg/L)	1.8~5.2
TN (mg/L)	48.6~51.2	TP (mg/L)	5~8
pH	5.8~7.7	—	—

60% of ammonia nitrogen had been removed in BAF, which could be considered as the initial stage of steady state in the reactor. And then the hydraulic retention time (HRT) was adjusted to 0.4, 0.6, 0.8, 1.25 and 2.5 hours (h) successively during the filter operation and steady performances were maintained in each period. The air/water ratio of superficial velocities was maintained at 5:1 throughout all the experiments, and the water temperature in the reactor was in the range of 20.5–24.5°C. Backwashing procedure was followed that gas 2 min firstly, then gas and water 3 min in the same time, and water 7–8 min in the end with backwashing rate 10 L/(m^2·s) both gas and water.

For all samples of COD, BOD_5, dissolved oxygen (DO), pH, total nitrogen, total phosphate, nitrite, nitrate, and ammonium nitrogen in the experiments were measured according to standard methods (SEPA of China, 2002).

Results and discussion

Influence of HRT on COD removal

Organic matters are the dominating constituents in domestic wastewater and of concern in wastewater treatment. Owing to the unevenness of water quality and quantity of middle and small point wastewater source, the HRT would changed in the filter operation that may be affected the performance of the biofilter. So that it is significant to investigate the influence of HRT on COD removal.

Figure 1 clear showed that the mean COD removal efficiency increased and the COD concentration in filter effluent descended gradually with the increasing in HRT. The mean COD removal efficiency increased from 85.9% to 93.7% and the COD mean concentration in filter effluent descended form 42.1 mg/L to 20.3 mg/L respectively when HRT increased from 0.6 h to 2.5 h, that indicated a promising filter performance in this operation condition. But while HRT shortened to 0.4 h, the mean COD removal efficiency decreased to 67.2% and the COD mean concentration in effluent increased to 102.6 mg/L respectively, illustrated evident decreasing in treatment ability. A black media layer near the filter bottom was observed frequently in the end of every operation circle that indicated dissolved oxygen deficiency in this area. It is considered that a great deal of solid matter in the influent were retained in the media layer near the filter bottom that resulted in anaerobic biological

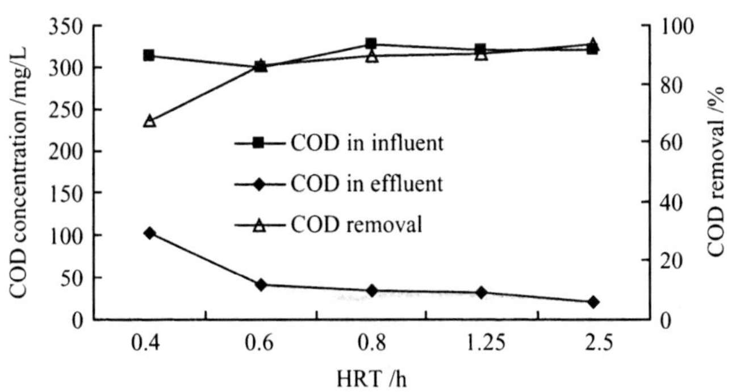

Figure 1 COD removal at different HRT

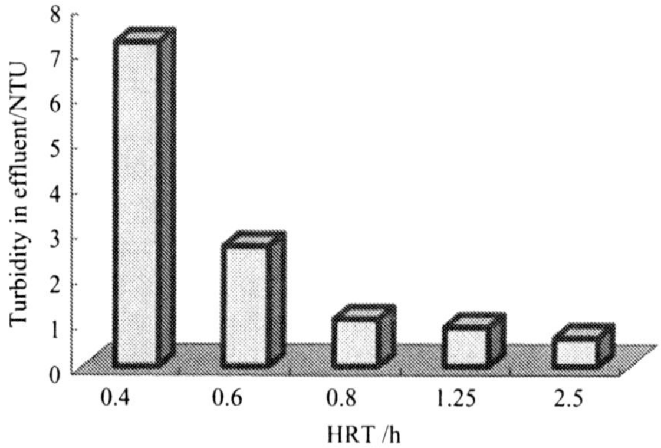

Figure 2 Turbidity in effluent with different HRT

oxidation because of plentiful oxygen consummation by prolific microorganisms. The above mentioned phenomena indicate the shale media bed has a promising performance to retain solid matter and to resist the hydraulic shock loading when HRT changed from 0.6 h to 2.5 h.

Influence of HRT on SS removal

It was difficult to exact determine the low SS value of half synthetic feed liquid in this study, therefore the turbidity of feed liquid was select as the constituent of suspended solid to consider the influence of HRT on SS removal. The result showed in Figure 2 reveals the variation of turbidity in effluent with different HRT. It can be observed in Figure 2 that the shale media biological aerated filter had a promising performance of suspended solid removal when HRT was longer than 8 h, and the mean turbidity in effluent decreased less than 1NTU under this operation condition. When HRT was shortened to 0.4 h, however, the mean turbidity in effluent increased to 7.18NTU suddenly. The results indicated that the HRT had no influence on SS removal in shale media BAF when HRT was more than 0.8 h but increased turbidity in effluent when HRT was less than 0.6 h within the tested range.

Influence of HRT on nitrification and denitrification

Many wastewater treatment plants are being required to meet increasingly stringent effluent standard, especially concerning nitrogen pollution, so that lots of commercial BAF systems utilize BAF unit in series for organic carbon and ammonia nitrogen removal respectively. But in some area, the combined organic and nitrogen removal in a single bioreactor is required because of land or money restrained, especially in small town in developing countries. Therefore, the influence of HRT on nitrification and denitrification in the shale media BAF was investigated in this research for detail consideration of this novel biofilter.

Figure 3 shows the ammonium and total nitrogen removal efficiency in the shale media BAF. The results reveal that the average removal efficiency of ammonium and total nitrogen increased from 21.7% to 89.6% and 1% to 60.1% respectively when the HRT was increased from 0.4 h to 2.5 h.

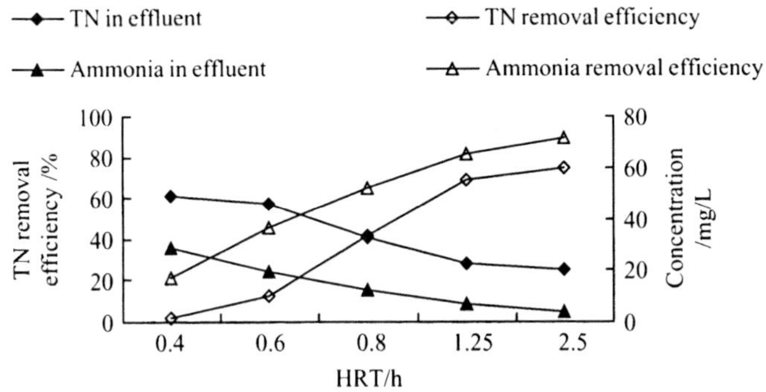

Figure 3 Ammonia and nitrogen removal with different HRT

Pujol *et al* (1998) considered that the decrease in HRT would improve filter's nitrification performance in a nitrification BAF. Whereas in a single biofilter for organic removal and nitrification simultaneously, such as the shale BAF in our research, increase in hydraulic loading will augment the organic loading in influent that will affect the ammonium nitrogen removal accordingly. It was considered that the ammonia removal rate in a bioreactor rested with the organic matter removal, which affected the activity and distributing of autotrophic microorganism. Because of the different competition ability to DO between the heterotrophic and autotrophic species, the heterotrophic bacteria became the preponderant population in the area where organic substrate was provided abundantly. At the same time, the growth of autotrophic microorganism was restrained because of lack of DO that caused by heterotrophic predominance in this domain, and the autotrophic microorganism would be washed out from the bioreactor in the backwash process. So, the organic loading, which will be increased with the increase in HRT, has a crucial influence on ammonia and total nitrogen removal in the shale media BAF.

The proportion of nitrite in the total nitrite and nitrate with different HRT were determined in the operation (Figure 4). A decrease in proportion of nitrite in total nitrite and

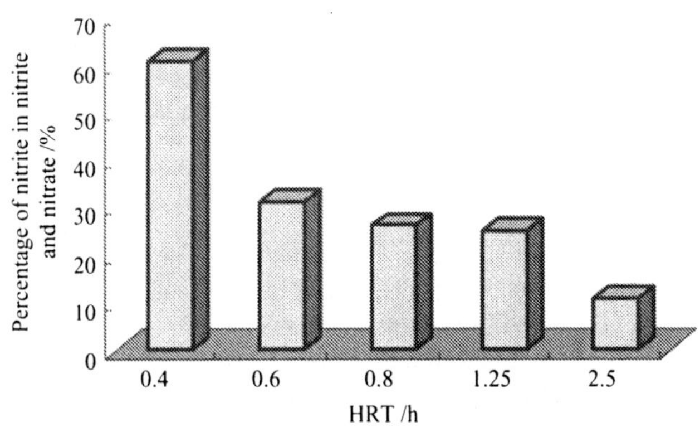

Figure 4 The proportion of nitrite in the total nitrite and nitrate with different HRT

nitrate with increase in HRT was observed in Figure 4. When the HRT declined from 2.5 h to 0.4 h, the proportion of nitrite in the total nitrite and nitrate increased from 10.6% to 60.7%. This result indicated that increase in hydraulic loading would improve the nitrite accumulation in the shale media BAF.

Nitrite accumulation, which has received considerable interest recently, would be accelerated by inhibiting the activity of nitrite oxidizer. It was reported that several chemical agents would inhibit the activity of nitrite oxidizers. Among these species, pH (Ford,1980), organic compounds (Sharma and Alert, 1977), heavy metals (Huang and Hopson, 1982), dissolved oxygen (Quinlan,1986; Hanaki *et al*, 1990), and especially free ammonia (FA) (Murray *et al*, 1975; Ford, 1980;Turk and Mavinic,1989; Abeling and Seyfried, 1992; Fdz-Planco *et al*, 1994; Fdz-Planco *et al*, 1996) was commonly reported as inhibitors of nitrite oxidizers. In this study, the temperature, pH and dissolved oxygen were maintained steady. Therefore, the free ammonia was also stable (Ford,1980). So that it was conferred that the primary cause of nitrite accumulation in the shale media BAF was the increasing in organic loading resulted from the shortened HRT. Sharma and Alert (1977) considered that increasing in organic loading would inhibit the activity of nitrite oxidizer. Owing to difference of special multiplication rate and oxygen saturation constant, the ammonia oxidizers and nitrite oxidizers occupy different ecological niche in the biofilm and have different metabolic characteristics with the variation of filter operational condition. In fixed-film processes, however, mass transfer resistance associated with both liquid phase and the biofilm could result in significant concentration gradients from bulk liquid to biofilm reaction sites, and may affect system performance. The concentration gradients from top to bottom of filter and from bulk to biofilm reaction sites provide different conditions of oxygen, substrates and other agents for heterotrophs and autotrophy species. The physiological differences between nitrifying autotrophy and heterotrophs aerobes lies in specific growth rates of up to 0.76 and 0.84 day^{-1} for ammonia and nitrite oxidizers respectively, and 4.8 day^{-1} for heterotrophs aerobes, have been reported by several authors (Rittmann and Snoeyink, 1984; Watanabe *et al*, 1992). The substrate and oxygen competition results in the formation of nitrifying population mostly positioned in a middle layer and heterotrophs dominate the outer layer. The ammonia oxidizers occupied advantageous position in competing with nitrite oxidizers for oxygen (Laanbroek and Gerards, 1993) and substrate, in combined with period perturbations caused by backwash, nitrite accumulation in the bioreactor will be accelerated by the variation of HRT which result in the frequency of backwash. This study only identified the nitrite accumulation and investigated the influences of operation variables on the performances of shale media BAF. Further research on nitrite accumulation in BAF should be developed.

The relationship of HRT and filtration time

With the hydraulic loss reached 120cm as the terminal of filtration time, the relationship of HRT and filtration time of shale media BAF was investigated in this paper (Figure 5). It can be observed in Figure 5 that the relationship between HRT and filtration time is linear. When the HRT decreased, the filtration time declined gradually. As HRT was shortened to 0.4 h, the hydraulic loss reached 120 cm in 18 h and the decrease in effluent quality was also determined. Obviously, the suitable HRT in shale media BAF for domestic wastewater treatment is not less than 0.8 h. The influence of the decrease in HRT on filtration time can

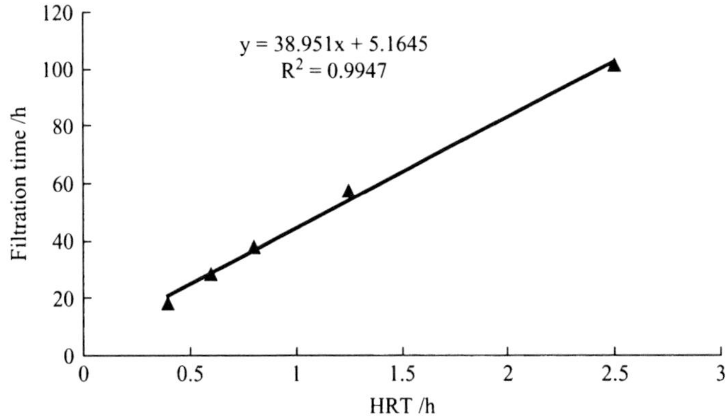

Figure 5 Relationship of filtration time and HRT

be considered in two ways: first, to shorten HRT will increase organic and SS load that will increase solid accumulation in filter bed; second, to shorten HRT will increase substrate supply and improve mass transfer velocity that will accelerate biomass augmenting in the media bed. The accumulation of solid and biomass will mushroom the hydraulic loss in the filter that will shorten the filtration time of shale media biological aerated filter.

Conclusions

HRT has great influence on the performance of shale media biological aerated filter whereas increase in HRT will improve the treatment performance effectively. The experimental results demonstrated that it was effective to remove the organic matters and turbidity when the hydraulic retention time was longer than 0.6 h. However, the removal efficiency of organic matters and turbidity decreased remarkably when the hydraulic retention time was 0.4 h. The ammonia and total nitrogen removal efficiency were more than 80% and 50% respectively when hydraulic retention time was longer than 2.5 h.

Nitrite accumulation in the shale media filter was determined and that could be accelerated when the hydraulic retention time was shortened

A linear relationship between hydraulic retention time and operation time was considered in this study. To shorten HRT will accelerate the increase in hydraulic loss and the decrease in the filtration time. The correlation equation is $H = 38.951HRT + 5.1645$, and $R^2 = 0.9947$.

Acknowledgements

This work was financially supported by Shandong Youth and Middle-aged Scientist Prize Fund (2005BS08008), People's Republic of China.

References

Abeling U. and Seyfried C.F. (1992) Anaerobic-aerobic treatment of high-strength ammonium wastewater: nitrogen removal via nitrite. *Wat. Sci. Technol.*, **26**(5–6), 1007–1015

Amar D. and Partos J. (1986) The use of an upflow fixed bed reactor for treatment of a primary settled domestic sewage. *Wat. Res.*, **20**(1), 9–14

Canler J. P. and Perret J. M. (1994) Biological aerated filters: assessment based on 12 sewage treatment plants. *Wat. Sci. Technol.*, **29**(10–11), 13–22

Ford D. L. (1980) Comprehensive analysis of nitrification of chemical processing wastewaters. *J. W .P. C. F.*, **52**(11), 2726–2746

Pujol R., Hamon M. and Kandel X. (1994) Biofilters: flexible, reliable biological reactors. *Wat. Sci. Technol.*, **29**(10–11), 33–38

Fdz-Polanco F., Garcia P. and Villaverde S. (1994) Temperature effect over nitrifying bacteria activity in biofilters: activation and free ammonia inhibition. *Wat. Sci. Technol.*, **30**(11), 121–130

Fdz-Polanco F., Villaverde S and Garcia P. (1996) Effects causing nitrite build-up in biofliter. *Wat. Sci. Technol.*, **34**(3), 371–378

Hanaki K., Wantawin C. and Ohgaki S. (1990) Nitrification at low levels of dissolved oxygen with and without organic loading in a suspended-growth reactor. *Wat. Res.*, **24**(3): 297–302

Huang C. S. and Hopson N. E., (1982) Metal inhibition of nitrification. In: *Proc. 37th. Ind. Waste Conf.* Purdue University, Indiana, pp. 85–98

Kent T. D., Fitzpatrick C. S. B. and Williams S. c. (1996) Testing of biological aerated filter (BAF) media. *Wat. Sci. Technol.*, **34**(3–4), 363–370

Laanbroek H. J. and Gerards S. (1993) Competition for limiting amounts of oxygen between Nitrosomanas europaea and Nitrobacteria winogradskyi grown in mixed continuous cultures. *Arch. Microbiology*, **159**(6), 453–459

Letterman R. D. (1980) Economic analysis of granular bed filtration. *J. Env. Eng. (ASCE),* **106**(2), 279–291

Ma J. and Qiu L. (2002) Biological aerated filter and its research progress. *Environmental Engineering* (in Chinese), **20**(3), 7–11

Murray I., Parsons J., and Robinson K. (1975) Inter-relationship between nitrogen balance, pH and dissolved oxygen in an oxidation ditch treating farm animal waste. *Wat. Res.*, **9**(1), 25–30

Ogedengbe O. and Olawale O. (1983) palm kernel shells as filter media. *Filtration and Separation.* **20**(2), 138–140

Pujol R., Lemmel H and Gousailles M. A(1998) Keypoint of nitrification in an upflow biofiltration reactor. *Wat. Sci. Technol.*, **38**(3): 43–49

Quinlan A. V. (1986) Optimum temperature shift for nitrobacter *winogradskyi* effect of dissolved oxygen and nitrite concentration. *Wat. Res.*, **20**(5), 611–617

Rittmann B. E. and Snoeyink V. L. (1984) Achieving biological stable drinking water. *J. Amer. Wat. Works Assoc.*, **76**(10), 106–114

Sagberg p., Dauthville p. and Hamon M. (1992) Biofilm reactors: A compact solution for the upgrading of wastewater treatment plants. *Wat. Sci. Technol.*, **26**(3–4), 733–742

SEPA of China (2002) *Methods for the examination of water and wastewater*. China Environmental Science Press, Beijing

Sharma B. and Alert R. C. (1977) Nitrification and nitrogen removal. *Wat. Res.*. **11**(10), 897–925

Turk O. and Mavinic D. S., (1989) Maintaining nitrite build-up in a system acclimated to free ammonia. *Wat. Res.*, **23**(11), 1383–1388

Watanabe Y., Masuda S. and Ishiguro M. (1992) Simultaneous nitrification and denitrification in microaerobic biofilms. *Wat. Sci. Technol.*, **26**(3–4), 511–522

39 Influence of applied loading on the competition between nitrifiers and heterotrophs in a two-stage submerged biofilter

Y. Z. Peng, H. D. Wang, S. Y. Wang, S. J. Zhang

College of Energy and Environmental Engineering, Beijing University of Technology, 100 Pingleyuan, Chaoyang District, Beijing P.R. China 100022

Abstract Study of the factors affecting the competition between nitrifiers and heterotrophs is imperative for optimizing bioreactors for enhanced nitrogen removal. This paper studied the influence of applied loading, temperature and biofilm culturing on the competition. The experimental results showed that the temperature affected nitrification more greatly than COD removal in the biofilm culturing phase. The nitrification efficiency increased obviously in the second filter indicated the advantage of the filter colonized by nitrifying microorganisms on ammonia oxidation. The activity of ammonia oxidizers and nitrite oxidizers, which increased gradually along the filter height at different loadings, was higher at high loading. At the higher loading, ammonia oxidizers showed higher activities along the filter than nitrite oxidizers. The profile of activities showed the heterotrophic bacteria, being much faster growers than the nitrifiers, dominate the inlet of the reactor. Biomass at different loadings decreased along the filter height, but the biomass at high loading was higher.

Keywords Nitrifiers, heterotrophs, submerged biofilter, loading, activity of microorganism, competition.

Introduction

Nitrification and carbonaceous removal are important reactions in the biological treatment of wastewater and slightly polluted raw water. Nevertheless, nitrification is unstable frequently due to the low maximum specific growth rate of nitrifiers and competition with heterotrophs for space and oxygen (Fdz-Polanco *et al.*, 2000). The understanding of the factors that affect the competition and the growth of nitrifiers is imperative for optimizing bioreactors for enhanced nitrogen removal.

Biological Aerated Filters (BAF) are fixed-film reactors developed recently (Pujol *et al.*, 1994). Carbon removal and nitrification can be carried out in two units operated in series or in a single unit (Pujol *et al.*, 1998; Chudoba *et al.*, 1998). However the BAF process presents some limitations for nitrification of wastewater with organic carbon (Neufeld *et al.*, 1986; Gilmore *et al.*, 1999). Some studies have showed the influence of substrate carbon and

 Recent Progress in Slow Sand and Alternative Biofiltration Processes edited by Rolf Gimbel, Nigel J.D. Graham and M. Robin Collins. ISBN: 9781843391203. Published by IWA Publishing, London, UK.

nitrogen ratio on the spatial competition and growth dynamics of heterotrophs and nitrifiers (Akiyoshi *et al.*, 1995; Fdz-Polanco *et al.*, 2000). The main goal of this paper is to study the influence of applied loading, temperature and biofilm culturing on the competition between nitrifiers and heterotrophs, and the influence of the specific activity of ammonia and nitrite oxidizers and heterotrophs on the biochemical reactions of these populations.

Methods

The study was made in a two-stage upflow biological aerated filter (UBAF) filled with expanded clay granular media approximately 2 to 3 mm in diameter. The two units were operated in series. The first stage, which was used to remove organic matters, received the domestic sewage (C/N ratio of 3) collected from Beijing University of Technology. The second stage was used as a nitrification filter. The characteristics of influent, the experimental conditions and the analysis methods of the experiment were described in Peng *et al.* (2005).

Results and discussion

It has made two start-up tests at temperature 14 and 27°C to study the performance of COD removal and nitrification (figure 1). Removal efficiency of ammonia was 60% after 15 days from the start-up at 14°C, after day 21, up to 90% of ammonia was oxidized. While at 27°C, just 8 days after the start-up the NH_4^+-N removal rate reached 97%. With respect to COD removal, though 60% of COD was removed a little early at temperature 27°C than at 14°C, it took 21 days to achieve the steady removal rate of 70% at both temperature conditions. These results showed the temperature affected nitrification more greatly than COD removal in the biofilm culturing phase.

The applied COD loading in the first filter could be up to 9.4 kgCOD/m^3·d. Under this COD loading condition, the effluent COD concentration was less than 60 mg/l and 76% of COD was removed, while the ammonia removal efficiency was 25%; nitrification efficiency was consistently at or near 100% at applied loading 1.9 kgCOD/m^3·d. As to the second filter,

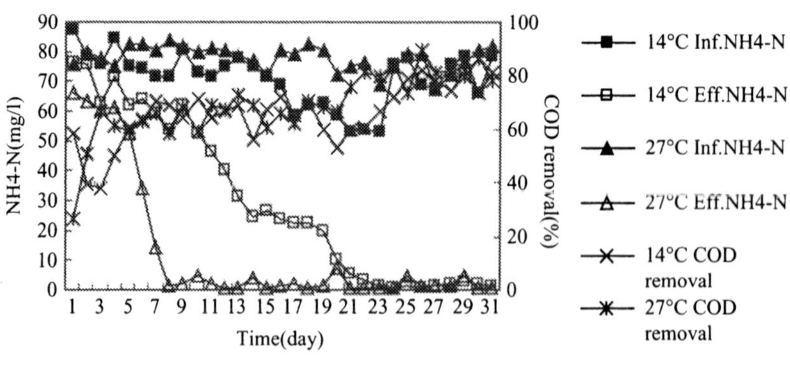

Figure 1 Time changes of NH_4^+-N and COD removal rates for two start-up phases (14°C and 27°C)

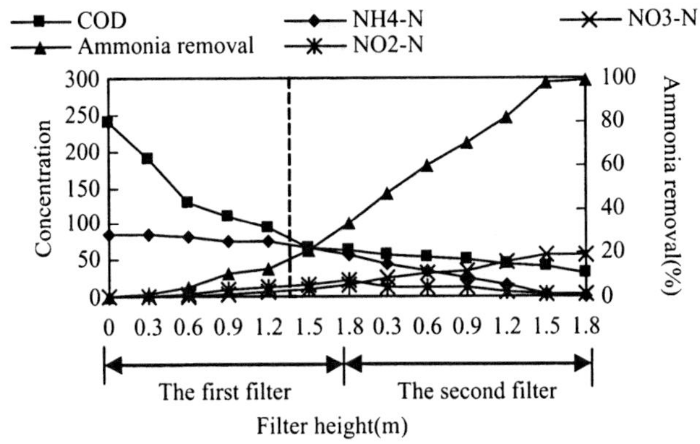

Figure 2 Profiles of COD and nitrogen concentration thorugh the two filters. Data were collected when the COD applied load was 3.1 kgCOD/m^3·d.

its nitrification efficiency related to the applied loading on the first filter closely. Figures 2 and 3 show the effects of COD removal and nitrification along the flow path of two filters, with COD applied loading of 3.1 and 11.3 kgCOD/m^3·d respectively.

Due to the increased COD load applied to the first filter, the maximum nitrification rate region was moved up along the filter height. The characteristic of COD removal changed along the filter height. That is, COD removal occurred mainly below the middle of the reactor with higher removal rate at low loading, while with the increase of COD loading, heterotrophs dominated the higher area of the reactor, which made the changes of COD removal rate along the filter height become not obvious. Nitrification point was relegated to the upper of the filter due to the higher influent COD loading, while the higher residual COD bleeding from the first filter resulted in the decrease of nitrification efficiency in the second filter.

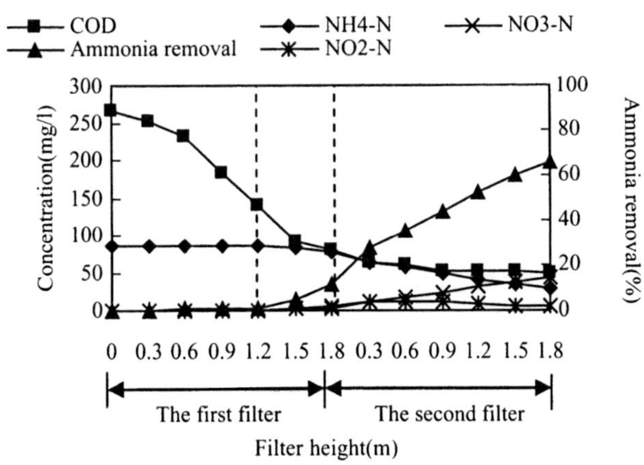

Figure 3 Profiles of COD and nitrogen concentration thorugh the two filters. Data were collected when the COD applied load was 11.3 kgCOD/m^3·d.

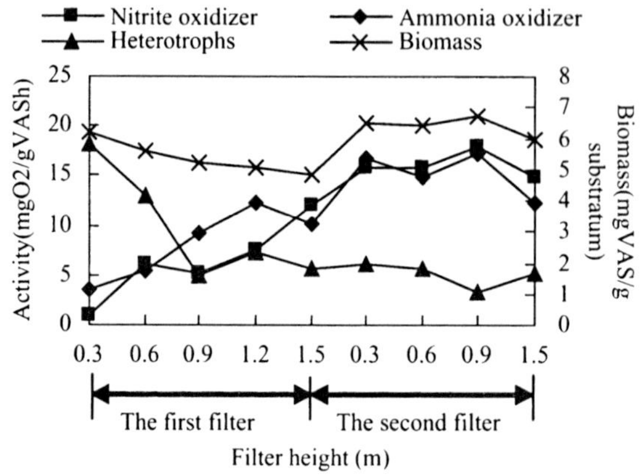

Figure 4 Profiles of the activities of ammonia oxidizers, nitrite oxidizers and heterotrophs with distance through the two filters. Data were collected when the COD applied load was 3.1 kgCOD/m^3·d.

The nitrification efficiency increased obviously in the second filter indicated the advantage of the filter colonized by nitrifying microorganisms on ammonia oxidation. In order to reduce the competition between different bacterial species and enable the microorganisms to grow under optimal conditions, operating the two stages of carbon oxidation and nitrification in separate reactors will be beneficial.

Figures 4 and 5 showed the tendency that the activity of ammonia oxidizers and nitrite oxidizers increased gradually along the filter height. It can be seen that the activity was higher at high loading, and reached the highest value when 70% ammonia was oxidized. At the higher loading, ammonia oxidizers showed higher activities along the filter than nitrite

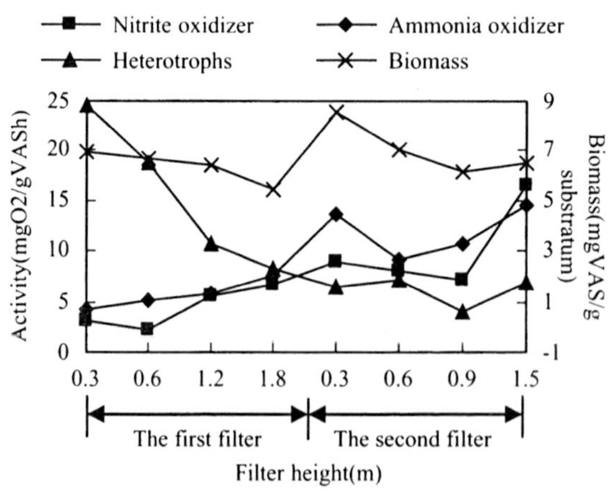

Figure 5 Profiles of the activities of ammonia oxidizers, nitrite oxidizers and heterotrophs with distance through the two filters. Data were collected when the COD applied load was 11.3 kgCOD/m^3·d.

oxidizers. The profile of activities showed the heterotrophic bacteria, being much faster growers than the nitrifiers, dominate the inlet of the reactor. Their activities increased with the increasing loading, and the decrease rate of the activity along the filter height at high loading was lower than that at low loading. Biomass at different loadings decreased along the filter height, but the biomass at high loading was higher.

Acknowledgements

This work was financially supported by the Key Projects under the National High Technology Research and Development Program (863 Program) of China (2003AA601010).

References

Chudoba, P. and Pujol, R. (1998). A three-stage biofiltration process: Performances of a pilot plant. *Wat. Sci. Tech.*, 38(8-9), 257–265.

Fdz-Polanco, F., Mendez, E., Uruena, I. A., Villaverde, S. and Garcia, P. A., (2000). Spatial distribution of heterotrophs and nitrifiers in a submerged bilfilter for nitrification. *Wat. Res.*, 34(16), 4081–4089.

Gilmore, K. R., Husovitz, K. J., Holst, T. and Love, N. G. (1999). Influence of organic and ammonia loading on nitrifier activity and nitrification performance for a two-stage biological aerated filter system. *Wat. Sci. Tech.*, 39(7), 227–234.

Neufeld, R., Greenfield, J. and Rieder, B. (1986). Temperature, cyanide and phenolic nitrification inhibition. *Wat. Res.*, 20, 633–642.

Ohashi, A., Viraj de Silva, D. G., Mobarry, B., Manem, J. A., Stahl, D. A. and Rittmann, B. E. (1995). Influence of Substrate C/N Ratio on the Structure of Multi-species Biofilms Consisting of Nitrifiers and Heterotrophs. *Wat. Sci. Tech.*, 32(8), 75–84.

Peng, Y. Z., Wang, H. D., Wang, S. Y., Zhang, S. J. (2005). Enhanced nitrogen and phosphorus removal with a two-stage submerged biofilter for reclaimed water with the stringent standards for scenic environment use. *Proceedings of the International Conference "Water Conservation and Management in Coastal Area"*. Qingdao, China.

Pujol, R., Hamon., Kandel, X. and Lemmel, H. (1994). Biofilters: Flexible reliable biological reactors. *Wat. Sci. Tech.*, 29(10-11), 33–38.

Pujol, R., Lemmel, H. and Gousailles, M. (1998). A keypoint of nitrification in an upflow biofiltration reactor. *Wat. Sci. Tech.*, 38(3), 43–49.

Part VI

ALTERNATIVE BIOFILTRATION – TREATMENT PERFORMANCE

40 The removal of green fluorescent labelled *Escherichia Coli* by pilot scale drinking water biofilters

M. Silva*,, S. McLellan** and J. Li****

*Department of Civil Engineering and Mechanics, University of Wisconsin-Milwaukee, Milwaukee, Wisconsin 53201, USA.
**Great Lakes WATER Institute, University of Wisconsin-Milwaukee, 600E Greenfield Avenue, Milwaukee, Wisconsin 53204, USA.

Abstract Biological filters combining microbial activity and rapid sand filtration are used in drinking water treatment plants for enhanced biodegradable organic matters (BOM) removal. Biofilms formed on filter media are responsible for the adsorption of BOM and attachment of planktonic microorganisms. This study investigated the removal of *E. coli* cells injected into pilot scale biofilters and the role of biofilm in retaining the injected *E. coli*. Green fluorescent protein (GFP) was used as a specific marker to detect and quantify *E. coli* in biofilms. Two sets of experiments were set up to compare the filter performance using different feed composition and microbial community. The first set of comparative tests consisted of feed solutions containing high and low nutrients. The second set consisted of an enriched and non-enriched biofilm comparative test. Filter effluent samples were collected at timed intervals and GFP *E. coli* were enumerated using real-time PCR (qPCR) and fluorescence in situ hybridization (FISH). Injected cells were unevenly distributed inside the filter and more than 60% of the cells were detected by qPCR at the upper part of the column, except in the non-enriched biofilm experiment. FISH results were lower than qPCR results probably due to the increase of cell clumping with higher biofilm density. The low nutrient biofilm as well as the enriched biofilm appeared to have a higher efficiency to retain *E. coli* in the extracellular polymeric substance (EPS).

Keywords Biofilm; *E. coli*; green fluorescent protein (GFP); filtration; drinking water purification; FISH.

Introduction

Natural organic matter (NOM) in drinking water contributes to the regrowth of microorganisms in distribution systems (Rittman, 1990, Huck, 1990, Moll *et al.*, 1998) and reacts with disinfection by-products (DBPs), which is of health concern (Moll *et al.*, 1998). In recent years, the use of ozonation followed by biological filtration for DBPs reduction and biological stabilization in drinking water has gained considerable attention in the United States (Huck, 1990, Moll *et al.*, 1998). In addition to NOM and particles, biofilms are able to capture pathogenic microorganisms or indicator bacteria such as *Escherichia coli* (Langmark, 2005). Biofilm research has become fundamental to discover the formation and properties of biofilms at the molecular level in the past three decades. The main purpose of this study was to determine fate of Green Fluorescent Protein (GFP) labelled *E. coli* injected to biofilters, by comparing the detection of GFP *E. coli* by real-time PCR and fluorescence in situ hybridization (FISH). Due to its great sensitivity, real-time PCR has been used to rapidly detect and quantify genes and microorganisms in complex environmental systems, including biofilms, soils and sediments (Stults et al., 2001; Wang et al., 2004). In our study, qPCR was used to quantify GFP *E. coli* whereas FISH was used mainly to visualize the location of GFP *E. coli* in the EPS matrix. Additionally, the properties of biofilm were investigated by varying operating conditions: high and low nutrient, and enriched and non-enriched biofilm to determine their impact on the *E. coli* removal efficiency.

Methods

The biofilm inocula used to establish initial microbial communities in the bioreactor were collected from the anthracite coal-sand filters at the Milwaukee Water Works (MWW) – Linwood treatment plant, Milwaukee, Wisconsin. Biofilm samples were collected from the anthracite coal-sand filters in December 2004 and in April 2005.

Filter set up and operation

Biofilm communities were developed in two polycarbonate plastic bioreactors. The column dimension is 2.45 cm ID × 30.5 cm. Both columns were packed with 0.55 mm spherical glass beads to a depth of 25 cm and had a porosity of 40 %.

After initial inoculation, the reactors were operated with 100% recirculation for a few days to ensure biofilm establishment on filter media. For the high and low nutrient experiment, feed solution was replaced with 50% fresh feed once daily whereas for the enriched and non-enriched experiment, fresh feed solution was flashed throughout the system three to five hours once daily. Heterotrophic plate counts were performed on the reactor effluent daily using TSA agar plates. Bacteria population reached steady state remaining constant at 10^6 CFU/mL after a few weeks of operation.

After the biofilm became mature on the filter media, 500 mL feed solution containing GFP labelled *E. coli* were injected into each column. The filter effluent was distributed into a fraction collection device and *E. coli* cells were quantitatively assessed by real time PCR. After the number of *E. coli* in filter effluent became very small and constant, glass beads were collected from different locations inside the filter and the biofilms were detached for bacteriological analysis. Plate counts with LB/ampicillin agar plates, real time PCR, and fluorescence in situ hybridization (FISH) were used to detect and count *E. coli* inside the biofilm.

Two comparative tests of experiments were set up. The first set consisted of a high and low nutrient comparative test. The second set consisted of an enriched and non-enriched biofilm comparative test. For the high and low nutrient experiment, one column was fed with a feed solution consisted of sodium acetate (20 mg/L), $NaNO_3$ (11.85 mg/L), and K_2HPO_4 (2.19 mg/L). The other column was fed with a feed solution ten times less concentrated. In both columns the nutrients were dissolved in tap water and the flow rate was 6.3 mL/min.

Biofilm was enriched at 37°C and resuspended in 500 mL of feed solution described in the next comparative test. Once it reached steady state, feed solution was changed for the high and low nutrient solutions.

For the enriched and non-enriched biofilm experiment, both columns were fed with feed solution that consisted of sodium acetate (23 mg/L) dissolved in phosphate buffered mineral media (21.8 mg/L; K_2HPO_4; 8.5 mg/L KH_2PO_4; 17.7 mg/L Na_2HPO_4; 2.7 mg/L $NaNO_3$; 22.5 mg/L $MgSO_4$; 0.25 mg/L $FeCl_3$; 36.4 mg/L $CaCl_2$), as the substrate for bacterial growth. In both columns the nutrients were dissolved in Mill-Q water and the flow rate was 3.6 mL/min. In one experiment, biofilm was enriched at 37°C incubator shaking at 160 rpm for 24 h by adding some flocs of biofilm into 5 mL TYE media one day prior to inoculating the column. In the other experiment, some flocs of indigenous biofilm were added into 5 mL of water from original environment and kept on the bench for 24 hrs one day prior to inoculating the column.

GFP *E. coli* solution preparation for injection

A single colony of GFP *E. coli* was taken into 5 mL of LB/ampicillin media. *E. coli* inoculum was shaken in incubator at 37°C overnight. One mililiter of *E. coli* culture was transferred into 25 mL LB/ampicillin media and it was shaken again at 37°C overnight. *E. coli* inoculum was centrifuged and resuspended twice. Finally, the pellets were resuspended in 500 mL feed solution and injected into the system.

Bacteria enumeration

TSA agar plates were used to count culturable heterotrophic bacteria. LB/ampicillin agar was used to cultivate the GFP gene cloned *E. coli.* Additionally, total bacteria was determined by filtering samples onto black polycarbonate filters, staining samples with the nucleic acid stain DAPI (4', 6'-diamidine-2'-phenylindole dihydrochloride) at a concentration of 0.1 mg/mL DAPI, and checked in the epifluorescent microscope by using DAPI filter.

GFP *E. coli* detection and enumeration

DNA extraction for qPCR and generation of standard curve

For the high and low nutrient, biomass was detached from beads by stirring in 0.1% $Na_4P_2O_7$ solution. Stirred suspension was analyzed by qPCR and FISH. For qPCR, DNA was extracted using fast DNA spin soil kit (Qbiogene, Irvine, CA) and resuspended into 75 μL solution. For the enriched and non-enriched biofilm experiment, 500 mg of beads from the top of the columns were collected and DNA was extracted. Additionally, total beads from the top, middle, and bottom sections were washed twice in 0.1% $Na_4P_2O_7$ solution in a 50 mL tube. Samples for FISH were collected and tubes were spin for 10 min at 4°C, and 10,000 rpm. Pellets were resuspended into 35 mL 0.1% $Na_4P_2O_7$ and vortexed. Tubes were spin again and pellets were resuspended in Tris.Cl EDTA and transferred to 1.5 mL tube. Bacteria from this saturated liquid culture were lysed and proteins removed by digestion with proteinase K and lysozyme. DNA was extracted with phenol/chloroform/isoamyl alcohol.

All qPCR was carried out on a Cepheid Smart Cycler (Sunnyvale, CA). The 25 μl PCR reactions consisted of *Taq* PCR buffer (Invitrogen, Carlsbad, CA), 2.5U/reaction *Taq*; 2.5 mM MgCl2, 200 μM of each dNTP, primer was present at a final concentration of 0.35 μM, and probes were present at a final concentration of 0.2 μM. Reaction conditions consisted of two-step cycles of 45 sec at 95°C and 60 sec at 65°C, for a total of 35 cycles.

To generate the qPCR standard curve, a culture of GFP cloned *E. coli* was incubated at 35°C for 24 hrs. Saturated cultures of *E. coli*, containing 4.73×10^8 cells/mL, were serially diluted 1:10 once and then 1:2 for a total of ten dilutions in duplicate. Dilution series were then plated on ampicillin added LB plates for plate count and used to obtain cycle threshold numbers for GFP in qPCR. The standard curve was determined to be $y = -0.2463x + 9.9583$ with $R^2 = 0.9706$.

Fluorescence in situ hybridization (FISH)

Samples were preserved with 10% paraformaldehyde (1:1 ratio) before refrigeration at 5°C until analyzed. Briefly, 1mL of preserved samples was filtered through a 0.2 μm polycarbonate filter and rinsed with 1 mL sterile deionized water. Filters were then placed into Petri dishes, treated with 40 μL Texas red labelled probe and placed in zip-loc bags with paper towels moistened with water and then incubated at 45°C for 4 hrs. After the 4 hr incubation, the filters were washed twice with 400 μL washing buffer (0.9 M NaCl, 0.02 M Tris-pH 7, 0.1% SDS) and placed in clean plates with 80 μL of buffer followed by 10 minutes incubation at the hybridization temperature. The filters were then rinsed twice with 400 μL of sterile deionized water before they were mounted on slides with immersion oil. The probe signal could be seen through a fluorescent microscope using Texas red filter.

Results and discussion

Establishment of biofilm

To monitor the biofilm formation in our experiments, we collected effluent samples throughout the time. The system reached steady state when bacteria population stabilized. Fig. 1 and Fig. 2 show our results of quantitative assessment of indigenous biofilm population. Bacteria population reached steady state remaining constant at 10^6 CFU/mL after 32 days of operation for the high and low nutrient experiment and 14 days for the enriched and non-enriched experiment, including the 100% biofilm recycling time. Our experiments are in agreement with other studies reporting the accumulation of steady-state biofilm between

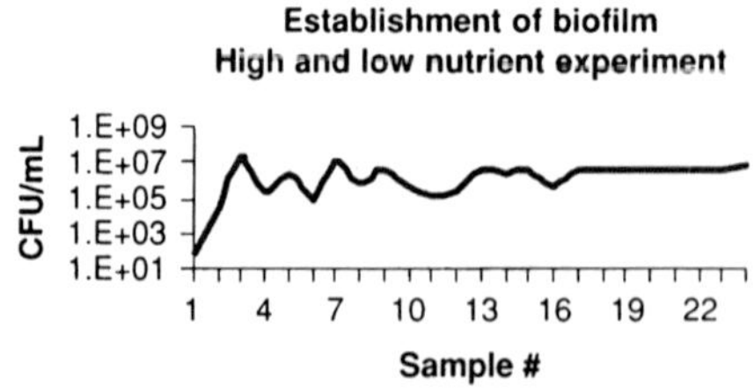

Figure 1 Establishment of biofilm for the high and low nutrient experiment.

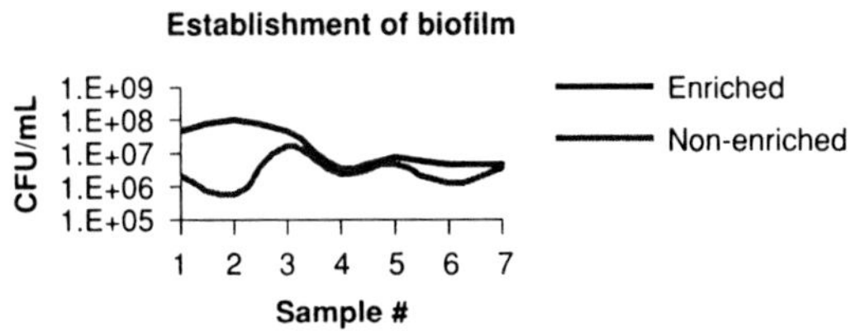

Figure 2 Establishment of biofilm for the enriched and non-enriched experiment.

6 days and 4 months, depending on experimental conditions (Block et al., 1993, Camper, 1995, Gauthier et al., 1996, Rompre et al., 1995, Volk and LeChevallier, 1999, Dai and Hozalski, 2002). As it was expected, biofilm reached steady state faster when 100% fresh feed solution was flashed throughout the system instead of 50%. In addition, enriched and non-enriched biofilm reached steady state simultaneously, however the bacteria population was 0.14-log unit lower for the non-enriched biofilm, considering the average bacteria count of the last three days. Another important observation is that lower flow rate, which was reduced from 6.3 mL/min (first set of experiments) to 3.6 mL/min (second set of experiments), might have allowed easier mass transport and utilization of the nutrients within the biofilm, therefore the steady state condition was achieved within a shorter period of time (Rittman, 1990, Escher and Characklis, 1990).

Assessment of *E coli* in the filter effluent

The total cells injected into the system were 7.8×10^9 CFU for the high and low nutrient experiments and 5.9×10^9 CFU for the enriched and non-enriched biofilm experiments. Fig. 3 shows that 64%, 61%, 47%, and 20% of the cells came out in the effluent for the high nutrient, low nutrient, enriched biofilm, and non-enriched biofilm experiments, respectively. The other 0.07% to 2.4% of the cells came out at later time points.

Survival of *E. coli* cultures exposed to low-nutrient conditions has been reported previously (Camper et al., 1991, Fass et al., 1996, Lisle et al., 1998). In addition, Banning

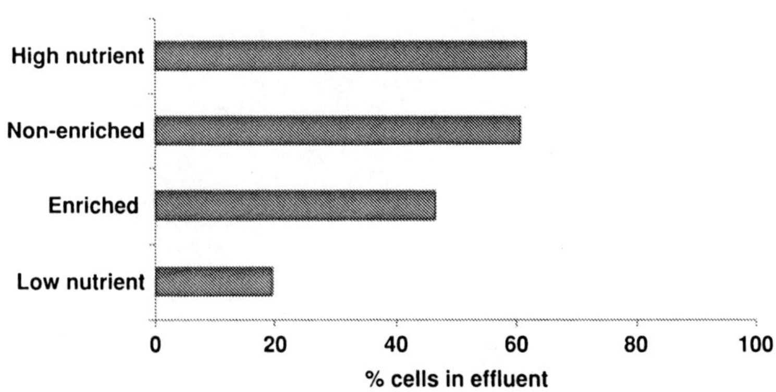

Figure 3 Cumulative % of GFP *E. coli* population present in the filter effluent after cell injection (determined by real time PCR).

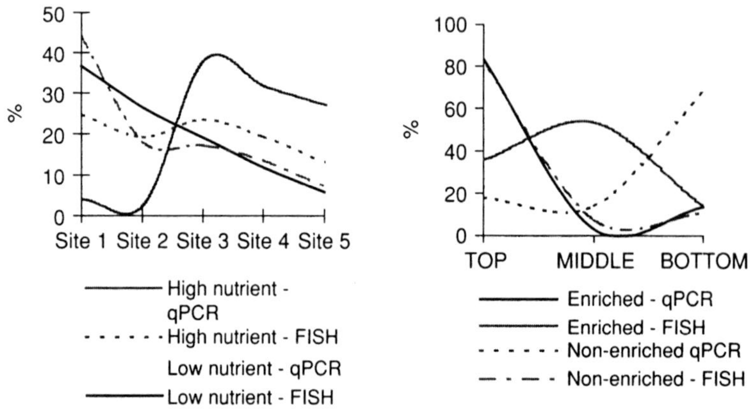

Figure 4 Comparison of cumulative % of GFP *E. coli* population present in biofilm matrix by qPCR and FISH analysis (data given in terms of percentage of cells detected in matrix).

et al. (2003) suggested that an increase in available nutrients may reduce E. *coli* survival potentially due to either enhanced competition for nutrients or enhanced antagonism by the indigenous microbial population. In addition, enriched biofilm might have contributed to the attachment of microorganisms due to a higher production of EPS. This hypothesis is supported by the fact that the majority of total bacteria population measured by DAPI for the high nutrient experiment were found in sites 2 and 3 (85%), exactly where 46% of the E. coli cells were found by qPCR.

E. coli biofilm matrix results

For the high and low nutrient experiment, the distribution of the *E. coli* population inside the column was examined by vertically dividing the filter into five sections and obtaining about 50 g of glass beads from each section. For the enriched and non-enriched biofilm experiment,

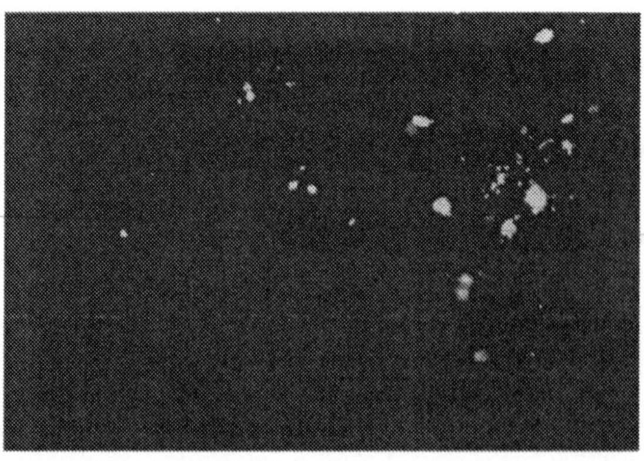

Figure 5 Fluorescence in situ hybridization (FISH) image of E. *coli* cells trapped in biofilm from the high nutrient experiment, site 4.

the columns were divided into three sections: top, middle, and bottom, obtaining 90 g beads per section. A quantitative assessment of the distribution of *E. coli* cells was performed using real time PCR and FISH. The majority of the cells detected by qPCR were found in upper part of column (63% in sections 1 and 2 for high nutrient experiment; 62% in sections 1 and 2 for the low nutrient experiment, and 83% on the top of the column for the enriched biofilm experiment). However, for the non-enriched experiment, the majority of the cells were detected on the bottom part of column by qPCR (68%). It might be due to a lower EPS production since the natural biofilm has a lower bacterial density. Therefore, this was the only experiment in which *E. coli* could travel more inside of the filter until eventually settling in the lower part of the column. Majority of the cells detected by FISH were found in lower parts of the column (56% in sections 3 to 5 for the high nutrient experiment; 57% in sections 4 and 5 for the low nutrient experiment, and 66% on the middle and bottom sections for the enriched biofilm experiment). However, for the non-enriched experiment, majority of the cells were detected on the top by FISH (82%). The discrepancy between qPCR and FISH results might be due to bacteria cell clumping when direct imaging using florescent probes was performed. Although *E. coli* cells were detected by both microscopy and qPCR, the challenges were how to accurately quantify GFP *E. coli* due to cell clumping, unequal distribution inside the column, and difficulty of harvesting cells from the biofilm. Our results indicate that qPCR is might be a better quantitative tool to assess *E. coli* in biofilm system and FISH could be used for visualization of the cell location but less accurate when quantitative analysis is desired.

Conclusions

Based on the experimental results obtained from this research, we conclude:

(1) Biofilters with low nutrient biofilm and enriched biofilm are more efficient in retaining GFP E. *coli* in the biofilm matrix compared to biofilters with high nutrient biofilm and non-enriched biofilm.

(2) E. *coli* cells were able to survive inside the filter biofilm matrix for a prolonged period after the contamination event. Many E. *coli* cells are clumped together and intertwined with other biofilm components.

(3) There was a discrepancy between the quantitative results of E. *coli* cells retained in biofilm matrix performed by FISH and real time PCR. Our study suggests that real time PCR is a probably more accurate than FISH in quantitative assessment in high density biofilms.

References

Banning, N., Toze, S. and Mee, B. (2003) Persistence of biofilm-associated *Escherichia coli* and *Pseudomonas aeruginosa* in groundwater and treated effluent in a laboratory model system. *Microbiology* **149**:47–55.

Block J.C., Haudiddier, K., Paquim, J.L., Miazga, J. and Levi, Y. (1993) Biofilm accumulation in drinking water distribution systems. *Biofouling* **6**:333–334.

Camper, A.K. 1995. Factors influencing biofilm growth in drinking water distribution systems. Ph.D. thesis. Montana State University, Bozeman.

Camper, A.K., Jones, W.L. and Hayes, J.T. (1996) Effect of growth conditions and substratum composition on the persistence of coliforms in mixed-population biofilms. *Appl. Environ. Microbiol.* **62**(11): 4014–8.

Camper, A. K., McFeters, G.A., Characklis, W.G. and Jones, W. L. (1991) Growth kinetics of coliform bacteria under conditions relevant to drinking water distribution systems. *Appl. Environ. Microbiol.* **57**:2233–2239.

Dai, X. and Hozalski, R.M. (2002) Effect of NOM and biofilm on the removal of Cryptosporidium parvum oocysts in rapid filters. *Water Research* **36**(14): 3523–32.

Escher, A. and Characklis, W.G. (1990) Modeling the initial events in biofilm accumulation. In *Biofilms*. Edited by Characklis, W.G. and Marshall, K.C. (Wiley, New York) p. 445–86.

Fass, M.D., Reasoner, D.J., Gatel, D. and Block, J.C. (1996) Fate of *Escherichia coli* experimentally injected in a drinking water distribution pilot system. *Water Research* **30**(9): 2215–2221.

Gauthier, C., Prevost, M., Mailly, J., and Rompre, A. (1996) Impact de la teneur en acides amines sur la stabilite microbiologique de l'eau potable: essais em reacteurs annulaires, p. 21-1–21-6. *In Proceedings of ISWA Workshop on Influence of Natural Organic Matter Characteristics on Drinking Water Treatment and Quality.*

Gomez-Suarez, C., Pasma, J., Van der Borden. A. J., Wingender, J., Flemming, H., Busscher, H. J. and Van der Mei, H. C. (2003) Influence of extracellular polymeric substances on deposition and redeposition of *Pseudomonas aeruginosa* to surfaces. *Microbiology* **148**:1161–1169.

Huck, P.M. (1990) Measurement of biodegradable organic matter and bacterial growth potential in drinking water. *J. Amer. Water Works Assn.* **82**(7): 78–86.

Langmark, J., Storey, M.V., Ashbolt, N.J. and Stenstrom, T.A. (2005) Accumulation and fate of microorganisms and microspheres in biofilms formed in a pilot-scale water distribution system. *Appl. Environ. Microbiol.* **71**(2): 706–12.

Lisle, J.T., Broadway, S.C., Prescott, A. M., Pyle, B. H., Fricker, C. and McFeters, G. A. (1998) Effects of starvation on physiological activity and chlorine disinfection resistance in *Escherichia coli* O157:H7. *Appl. Environ. Microbiol.* **64**:4658–4662.

Rittmann, B. E. (1990) Analyzing biofilm processes used in biological filtration. *J. Amer. Water Works Assn.* **82**(12): 62–66.

Moll, D., Summers, R. S. and Breen, A. (1998) Microbial Characterization of biological filters used for drinking water treatment. *Appl. Environ. Microbiol.* **64**:2755–2759.

Robinson, P.J., Walker, J.T., Keevil, C.W. and Cole, J. (1995) Reporter genes and fluorescent probes for studying the colonisation of biofilms in a drinking water supply line by enteric bacteria. *FEMS Microbiology Letters* **129**(2–3): 183–8.

Rompre, A., Prevost, M., Coallier, J.,Gauthier, C., Serkedjieva, R., Lafrance, P. and Laurent, P. (1995) Impact of biodegradable organic matter on fixed and suspended biomass: results from full scale distribution systems and annular reactor study, p. 2157–2174. *In* Proceedings of AWWA Water Quality Technology Conference.

Stults, J.R., Snoeyenbos-West, O., Methe, B., Lovley, D.R. and Chandler, D.P. (2001) Application of the 5′ fluorogenic exonuclease assay (TaqMan) for quantitative ribosomal DNA and rRNA analysis in sediments. *Appl. Environ. Microbiol.* **67**(6): 2781–9.

Volk, C.J. and LeChevallier, M. (1999) Impacts of the Reduction of Nutrient levels on bacterial water quality in distribution systems. *Appl. Environ. Microbiol.* **65**(11): 4957–4966.

Wang, G., Gentry, T.J., Grass, G., Josephson, K., Rensing, C. and Pepper, I. L. (2004) Real-time PCR quantification of a green flurorescent protein labeled, genetically engineered Pseudomonas putida strain during 2-chlorobenzene degradation in soil. FEMS Microbiology Letters **233**: 307–314.

Williams, M.M. and Braun-Howland, E. B. (2003) Growth of *Escherichia coli* in Model Distribution System biofilms exposed to hypochlorous acid or monochloramine. *Appl. Environ. Microbiol.* **69**(9):5463–5271.

41 Removal and inactivation of waterborne viruses using zero-valent iron

Youwen You[1†], Jie Han[1], Liping Zhang[1], Yan Jin[1], Pei C. Chiu[2*]

[1]Department of Plant and Soil Sciences, University of Delaware, Newark, DE 19716, USA
[2]Department of Civil and Environmental Engineering, University of Delaware, Newark, DE 19716, USA
[†]Current affiliation: School of Veterinary Medicine, University of Pennsylvania, Kennett Square, PA 19348, USA
[*]Corresponding author, Tel: 302-831-3104; Fax: 302-831-3640; E-mail: pei@ce.udel.edu

Abstract The effectiveness of zero-valent iron (ZVI) to remove two viruses, φX174 and MS-2, from water was evaluated. In a control column filled with oxide-free sand, removal of these viruses by filtration was limited. In contrast, in a column containing 35% ZVI, the removal for both viruses was $>$4-log in the first pulse test and $>$5-log in all three subsequent pulse tests. The contact time with iron was 20 min for all tests. We propose the improved efficiency was due to continuous formation of new iron (hydr)oxides which served as virus adsorption sites. The results suggest that ZVI may be useful for disinfecting drinking water and wastewater and may help to reduce chlorine use and minimize disinfection by-product formation.

Keywords Zero-valent iron; virus; bacteriophage; water disinfection

Introduction

Zero-valent iron (ZVI) has been used in permeable reactive barriers (PRBs) for groundwater remediation for the past decade. Since the installation of the first full-scale PRB in Sunnyvale, California in 1995 (NRC, 1999), over 120 full- and pilot-scale PRBs have been installed, mostly in the U.S. and Europe (http://www.eti.ca). Most of the PRBs contain ZVI as a reactive material, typically in the form of cast iron granules. The iron PRB process is continuous and passive and requires minimal maintenance. ZVI filings in existing PRBs have been shown to remain reactive in degrading pollutants in groundwater for years without rejuvenation (US EPA 2002a,b). While initially proposed to treat chlorinated solvents in contaminated aquifers, ZVI has since been used to remove a broad range of organic and inorganic contaminants including metals, radionuclides, pesticides, hydrocarbons, Freons, and nutrients (http://www.rtdf.org).

More recent studies further suggest that ZVI may also be applied to treat wastewaters containing refractory compounds such as azo- and nitro-aromatics and explosives (Perey et al., 2002; Oh et al., 2004, Oh et al., 2005a). Reductive treatment with ZVI can rapidly transform

 Recent Progress in Slow Sand and Alternative Biofiltration Processes edited by Rolf Gimbel, Nigel J.D. Graham and M. Robin Collins. ISBN: 9781843391203. Published by IWA Publishing, London, UK.

these refractory pollutants to products that are either innocuous or more degradable in subsequent chemical and biological oxidation (Oh et al., 2003; Oh et al., 2005b). Furthermore, ZVI was also proposed to remove arsenic in drinking water (Melitas et al., 2002).

However, most studies on ZVI to date have focused on chemical pollutants. ZVI was not shown to remove *biological agents* from water until a recent study from our laboratory (You et al., 2005). Results of that study, which involved a batch experiment and a short-term column experiment, suggest that ZVI may be an effective and inexpensive option to remove viruses and potentially other microbial pathogens from drinking water and wastewater.

This study was conducted to validate our previous results and to evaluate the performance of ZVI over an extended time. Specifically, the objective of this study was to determine whether commercial granular iron can consistently remove model viruses (bacteriophages) from water at high efficiencies for a prolonged period of time without rejuvenation, a characteristic that is of critical importance for large-scale applications.

Methods

Two bacteriophages, MS2 and ϕX174, were used in this study. MS2 and ϕX174 were chosen because of their structural resemblance to many human enteric viruses and because they had been used as surrogates for human enteric viruses in many studies. MS2 is an icosahedral single-stranded RNA phage with a diameter of 26 nm (Vanduin, 1988) and an isoelectric point (pH_{isp}) of 3.9 (Zerda, 1982). It was grown on bacterial lawns of Escherichia coli from the American Type Culture Collection (ATCC 15597). ϕX174 is a single-stranded DNA bacteriophage, approximately 23 nm in diameter with a pH_{isp} of 6.6 (Ackerman and Dubow, 1987). ϕX174 was grown on its host E. coli ATCC 13706. Both MS2 and ϕX174 were assayed by the double-agar-overlay method described by Adams (1959) with their corresponding hosts. One mL of log-phase E. coli host and either 1 or 0.5 mL of aqueous sample (original or diluted) were mixed in a tube of molten tryptic soy agar (TSA, Difco Laboratories, Detroit, MI) and poured into TSA plates. The plates were incubated at 37°C overnight for MS2 and 5 hours for ϕX174. Samples were plated in duplicates and only dilutions that resulted in 10 to 300 plagues per plate were considered countable. These phages differ in size and surface properties (e.g., hydrophobicity, charge) and thus may help to identify characteristics that control virus removal by ZVI.

A saturated column system had been developed and applied successfully in previous studies to investigate the reactions and transport of viruses in porous media (Jin, 1997, 2000; Chu et al., 2001, 2003). Well-characterized Accusand sand (Unimin Corp, Le Sueur, MN) was pre-treated using the standard methods described in Kunze and Dixon (1986) and Amonette and Zelazny (1994) to remove impurities (mainly metal oxides). It was shown in our previous studies, and confirmed in this study (see **Results and Discussion**), that treated sand is inert and does not interact with viruses (Chu et al., 2001 and 2003; You et al., 2005). Therefore, the sand column serves as a control against which virus removal by ZVI can be quantified. Columns were packed wet with vacuum-degassed artificial groundwater (AGW), which contained 0.075 mM of $CaCl_2$, 0.082 mM of $MgCl_2$, 0.051 mM of KCl, and 1.5 mM of $NaHCO_3$ (ionic strength $\approx$ 2 mM). Vacuum degassing reduces the pressure of total dissolved gases to less than 1 atm and prevents bubble formation in the column during experiment. After autoclaving and degassing, the pH of AGW was adjusted to 7.5 using NaOH or HCl prior to use.

Figure 1 Column experiment setup. The influent entered the columns in an upward direction.

The experiment was conducted at a constant temperature of 5 ± 1°C in a large refrigerator. Two identical acrylic columns with a 3.8-cm i.d. and a 10-cm length were set up. The control (sand) column was wet-packed with cleaned sand in AGW. The ZVI column was packed in a similar fashion with 3 cm of cleaned sand and 7 cm of 1:1 (v/v) mix of cleaned sand and Peerless™ iron granules. The surface area of Peerless iron measured in our laboratory using the BET method with nitrogen was 1.3 m^2/g, similar to that of other commercial iron granules (Oh et al., 2002; Alowitz and Scherer, 2002). The pore volumes (PVs) of the sand and ZVI columns were approximately 42 and 56 mL, as determined by bromide tracer tests. Prior to introducing viruses, each column was flushed with 10 PVs of degassed AGW to establish a steady flow condition.

For each removal efficiency test, influent solution containing approximately 1×10^8 pfu/L of the two phages was introduced as a step input of 5 PVs using a peristaltic pump at an approximate flow rate of 1.0 mL/min. This gives an effective ZVI contact time of about 20 min. Four separate tests were conducted at different times: immediately following column packing (i.e., time zero) and after 1, 2, and 5 weeks. The columns were operated for 35 days, or approximately 1,200 PVs. Between any two tests, the influent was switched to virus-free AGW (degassed, sterilized, and pH-adjusted) and effluent samples were monitored for another five PVs. Effluent samples were collected using a fraction collector and analyzed for viruses using the procedures described above. For each test, a breakthrough curve was obtained for each phage and each column, resulting in a total of 16 breakthrough curves. The experimental setup for the column experiment is shown in Figure 1.

Results and discussion

The breakthrough curves of MS2 and ϕX174 for the sand and ZVI columns from the four virus removal tests are shown in Figure 2(a)–(d).

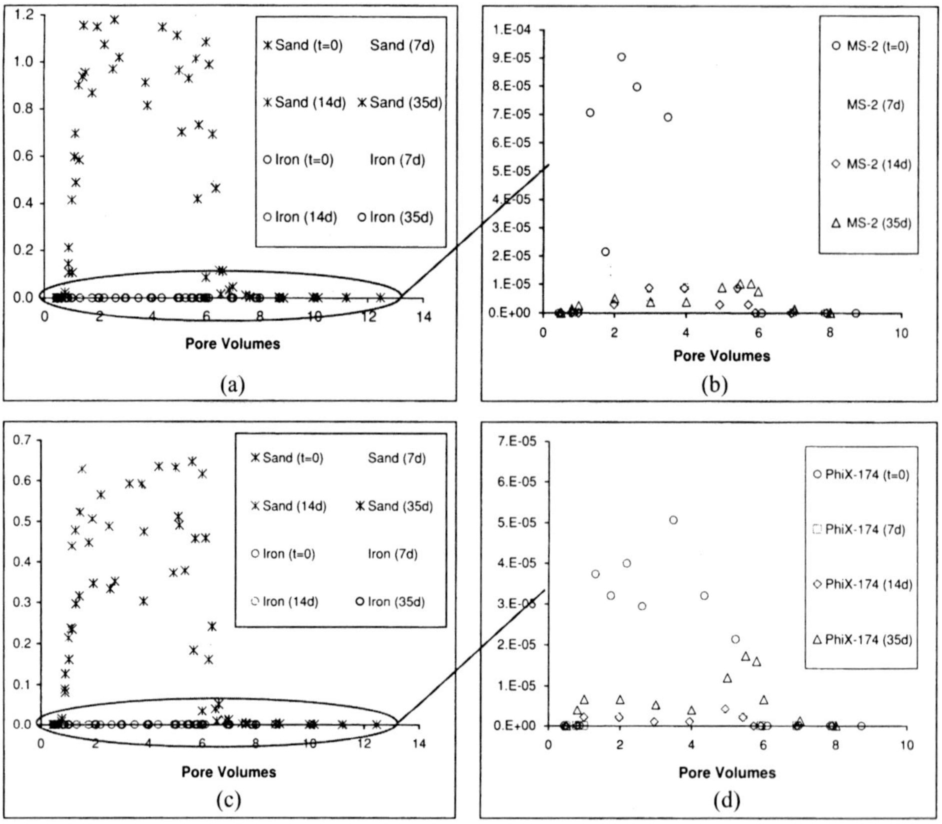

Figure 2 MS2 and ϕX174 concentrations in sand and iron column effluents from the four tests.

It is evident from the data in Figure 2(a) that the bacteriophage MS2 broke through completely from the sand column in all four tests. The breakthrough concentration of MS2 was essentially identical to the influent MS2 concentration in each test, indicating that no removal occurred in the sand column over the 35 days. In sharp contrast, very few viable MS2 particles were eluted from the iron column during the experiment (Figure 2(a)). While the effluent concentrations of MS2 from the iron column appear to be zero in all four tests, their breakthrough curves are not identical, as illustrated in Figure 2(b). In the first test, conducted shortly after column packing, the effluent MS2 concentration was approximately 8E-5, corresponding to a removal efficiency of 0.99992 (99.992%). After flushing the column for 1 week, however, the MS2 concentration in the effluent decreased further to approximately 1×10^{-5} (i.e., ~99.999% or 5-log removal). This improved efficiency may be due to formation of new iron (hydr)oxides during the first week as a result of iron corrosion by dissolved oxygen and proton in the influent. This 5-log removal efficiency was sustained (actually, improved slightly) for the rest of the experiment, as shown by the 2- and 5-week results.

The ϕX174 result is similar to that of MS2, as shown in Figure 2(c). A noted difference between ϕX174 and MS2 results was that about 50% of ϕX174 in the influent was removed by the sand column. This removal was observed in all four tests and did not appear to vary with time. The mechanism responsible for the removal is not yet clear but may be related to the higher pH_{isp} of ϕX174 than that of MS2. Nonetheless, compared to the sand column result, the breakthrough concentration of ϕX174 from the iron column in the first test was far lower (~4E-5, Figure 2(d)), corresponding to ~99.992% removal. The removal efficiency increased further after one week to approximately 99.999% and remained constant for the remaining of the experiment. This again may be due to continuous iron corrosion and formation of new iron oxyhydroxides, which are known to serve as virus adsorption sites (Ryan et al., 2002).

The data in Figure 2 clearly illustrate that zero-valent iron, as an active component in a simple sand filtration column, can consistently achieve >4- to 5-log removal of the bacteriophages MS2 and ϕX174 in 20 minutes of effective contact time with iron. The result also suggests that the effectiveness of ZVI to remove viruses may be long-lasting and sustainable due to continuous corrosion and (oxyhydr)oxide formation. Although the columns were operated for only 1,200 PVs, the capacity of ZVI to remove viruses is probably significantly greater, as suggested by the long service life of groundwater PRBs. This capacity will depend on the formation rate of new adsorption sites, which is controlled largely by water chemistry including dissolved oxygen, pH, buffering capacity (acidity), and specific ions that promote corrosion (e.g., chloride) or passivate surface through formation of complexes or precipitates (e.g., carbonate, phosphate, and humic materials). Further research is necessary to better understand how virus removal efficiency and ZVI capacity are controlled by solution parameters and whether periodic rejuvenation or medium replacement would be necessary to sustain system performance.

Conclusions and implications

Results of this study support our previous finding that ZVI can consistently remove viruses from water over weeks of operation (> 1,000 PVs). These results have potentially important implications for drinking water and food safety and public health.

There are approximately 165,000 public drinking water systems in the U.S., treating 34 billion gallons of water daily, and most people (268 million) obtain drinking water from public water systems. Chlorine is by far the most common disinfectant used in the U.S. to control microbial pathogens, but it is not always effective against viruses and protozoa. Disinfection with chlorine has serious drawbacks, including formation of toxic/carcinogenic DBPs via reaction of chlorine with natural organic matter (NOM). In addition, accidental and deliberate release of chlorine gas can have catastrophic consequences and has been listed by the Homeland Security Council as one of the National Planning Scenarios (HSC, 2004). Moreover, some chlorine-manufacturing facilities still use mercury cell electrolysis, a process that can release massive amounts of mercury into the environment. Therefore, it is desirable to minimize chlorine usage and storage in water and wastewater treatment plants and other facilities. The greatest challenge facing the U.S. water industry and regulatory agencies today is, arguably, *how to simultaneously control microbial pathogens, disinfection by-products (DBPs), and residual disinfectant in drinking water, and to do so at a cost that is acceptable to society.*

For the aforementioned reasons, and to meet the ever more stringent regulations and growing demand for potable water, alternative disinfection technologies that are effective, reliable, environmentally friendly, and inexpensive are urgently needed. ZVI may be able to help disinfect water and wastewater and thus reduce or eliminate chlorine use. For example, ZVI may be incorporated into a sand filter that precedes coagulation. Such filtration systems can pre-disinfect source water and may replace pre-chlorination, which is practiced by many water treatment plants. The corrosion products formed, namely ferrous and ferric species, are common coagulants and can improve efficiency of coagulation. Pretreatment of source water with ZVI may enable water treatment plants to meet the disinfection goals by using less chlorine or replacing chlorine with other disinfectants such as chloramines. For treatment systems using membrane processes, elemental iron can help to ensure adequate removal of viruses, which are more difficult to filter out than bacteria and protozoa. Finally, a significant added advantage of ZVI is co-removal of many chemical pollutants in source water, as described earlier, including metals, arsenic, and most importantly, NOM, which is the precursor of many toxic DBPs. Results of this study suggests that further research is warranted and necessary to assess the effectiveness of ZVI against other microbes and to examine the impact of water chemistry on ZVI performance.

References

Ackerman, H. W. and Dubow, M. S. (1987) Viruses of Prokaryotes. In: *CRC Press*, Boca Raton, FL., **2**, 190–191.

Adams, M.H. (1959) *Bacteriophages*, Interscience Publishers, New York. p. 1–4.

Alowitz, M. J. and Scherer, M. M. (2002) Kinetics of nitrate, nitrite, and Cr(VI) reduction by iron metal. *Environ. Sci. Technol.* **36**, 299–306.

Amonette, J. C. and Zelazny, L. W. (1994) *Quantitative methods in soil mineralogy*, Soil Sci. Soc. Am.Misc. Publ., Soil Sci' Soc. Am., Madison, WI.

Chu, Y., Jin, Y. and Yates, M. V. (2000) Virus transport through saturated sand columns as affected by different buffer solutions. *J. Environ. Qual.* **29**, 1103–1110.

Chu, Y., Jin, Y., Flury, M. and Yates, M. V. (2001) Mechanisms of virus removal during transport in unsaturated porous media. *Water Resour. Res.* **37**, 253–263.

Homeland Security Council, Planning Scenarios Executive Summary, version 2.0, July, 2004.

Jin, Y., Yates, M.V., Thompson, S. S. and Jury, W. A. (1997) Sorption of viruses during flow through saturated sand columns. *Environ. Sci. Technol.* **31**, 548–555.

Jin,Y., Chu, Y. and Li, Y. (2000) Virus removal and transport in saturated and unsaturated sand columns. *J. Contam. Hydrol.* **43**, 111–128.

Jin, Y. and Flury, M. (2002) Fate and transport of viruses in porous media. *Adv. Agron.* **77**, 39–102.

Kunze, G. W. and Dixon, J. B. (1986) Pretreatment of mineralogical analysis. In *Klute, A (ed.), Methods of Soil Analysis. Part 1. Physical and Mineralogical Analysis*, pp. 91–100, Soil Sci. Soc. Am., Madison, WI.

Melitas, N., Conklin, M. and Ferrell, J. (2002) Electrochemical Study of Arsenate and Water Reduction on Iron media Used for Arsenic Removal from Potable Water. *Environ. Sci. Technol.* **36**, 3188–3193.

National Research Council. (1999) *Groundwater and Soil Cleanup: Improving Management of Persistent Contaminants*, pp. 304, The National Academies Press, Washington, DC.

Oh, S. Y., Cha, D. K. and Chiu, P. C. (2002) Graphite-Mediated Reduction of 2,4-Dinitrotoluene with Elemental Iron. *Environ. Sci. Tehcnol.* **36**, 2178–2184.

Oh, S. Y., Cha, D. K., Kim, B. J. and Chiu*, P. C. (2004) Reduction of Nitroglycerin with Cast Iron: Pathway, Kinetics, and Mechanisms. *Environ. Sci. Technol.* **38**, 3723–3730.

Oh, S. Y., Chiu, P. C., Kim, B. J. and Cha, D. K. (2003) Enhancing Fenton Oxidation of TNT and RDX through Pretreatment with Zero-Valent Iron. *Water Research.* **37**, 4275–4283.

Oh, S.-Y., Cha, D. K., Kim, B. J. and Chiu*, P. C. (2005a) Transformation of Hexahydro-1,3,5-trinitro-1,3,5-triazine (RDX), Octahydro-1,3,5,7-tetranitro-1,3,5,7-tetrazocine (HMX), and Methylenedinitramine with Elemental Iron. *Environ. Toxicol. Chem.* **24**, 2812–2819.

Oh, S. Y., Chiu, P. C., Kim, B. J. and Cha, D. K. (2005b) Zero-Valent Iron Pretreatment for Enhancing the Biodegradability of RDX. *Water Res.* **39**, 5027–5032.

Perey, J. R., Chiu, P. C., Huang, C. P. and Cha, D. K. (2002) Zero-valent iron pretreatment for enhancing the biodegradability of azo dyes. *Water Environ. Res.* **74**, 221–225.

Ryan, J. N., Harvey, R. W., Metge, D., Elimelech, M., Navigato, T., Pieper, A. P. (2002) Field and laboratory investigations of inactivation of viruses (PRD1 and MS-2) attached to iron oxide-coated quartz sand. *Environ. Sci. Technol.* **36**, 2403–2413.

Vanduin, J. (1988) The single-stranded RNA bacteriophages. In *The Bacteriophages*, R. Calendar ed., 1, pp.117–167, Plenum, New York.

U.S. Environmental Protection Agency. (2002a) *Field Applications of In Situ Remediation Technologies: Permeable Reactive Barriers*, Washington, DC, January

U.S. Environmental Protection Agency. (2002b) *Research Brief: Long-term Performance of Permeable Reactive Barriers Using Zero-Valent Iron: An Evaluation at Two Sites*, EPA/600/S-02/001, Cincinnati, OH, March.

You, Y., Han, J., Chiu, P. C. and Jin, Y. (2005) Removal and Inactivation of Waterborne Viruses Using Zero-Valent Iron. *Environ. Sci. Technol.* **39**, 9263–9269.

Zerda, K.S. (1982) Adsorption of viruses to charge-modified silica. Ph.D. diss., Baylor Coll. of Medicine, Houston, TX.

42 Removal of MIB and geosmin by full-scale biological sand filters

D.H. Metz*, R. C. Pohlman*, J. Vogt* and R. S. Summers**

*Greater Cincinnati Water Works, 5651 Kellogg Avenue, Cincinnati, OH 45228, USA
**University of Colorado Boulder, CO, USA

Abstract Taste and odour control is extremely important to water utilities around the world. Customer confidence depends on an aesthetically pleasing product, and many utilities are willing to make large expenditures to insure aesthetic quality. The microbial by-products, 2-methylisoborneol (MIB) and/or trans-1,10-dimethyl-trans-9-decalol (geosmin) are often times the cause of musty water problems. These compounds can be controlled by biofiltration. Greater Cincinnnati Water Works (GCWW) of Cincinnati, Ohio (USA) has had intermittent problems with these compounds in the settled water. However, sustained biological removal of MIB (80 percent) and geosmin (50 percent) was found over a six-year period in the GCWW's sand filters. During this period, the filter influent concentration of TOC, MIB and geosmin varied greatly. It should be noted that no disinfectant was added before the filters and the water was not ozonated. Additionally, the filter sand media was completely replaced and the hydraulic loading increased 20 percent. However, consistently high removals of MIB and geosmin were observed throughout the monitoring period. The microbiological acclimation time or modest change in empty bed contact time did not present a problem relative to taste and odour goals. The filter effluent water seldom had perceptible musty or earthy odours as a result of MIB or geosmin concentrations.

Keywords Biofiltration, MIB, geosmin, full-scale removal, taste and odor

Introduction

Drinking water customers often judge the potability of their drinking water by taste and odour (T&O). Water with an objectionable taste or odour is perceived to be of poor quality, and customers complain to the water utility or to public officials. A survey conducted of more than 800 water utilities in United States and Canada found that 16 percent of utilities experience serious T&O problems, and that utilities spend an average of about 4.5 percent of their total budget on T&O control, (Westerhoff et al., 2005). Taste and odour compounds are often caused by the microbial metabolic by-products, 2- methylisoborneol (MIB) and/or trans-1,10-dimethyl-trans-9-decalol (geosmin), present in the surface water. While biological filtration has been practiced for many years in Europe, only recently has it gained popularity in the United States. Its efficacy for taste and odour control has not

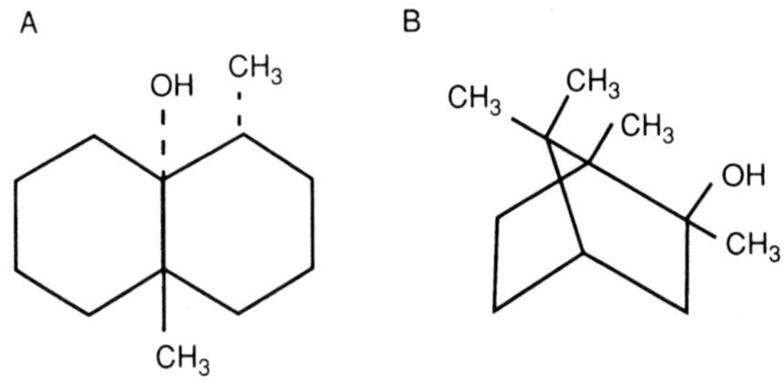

Figure 1 Chemical Structure of Geosmin (A) and MIB (B)

been sufficiently studied. Few studies quantify the extent of MIB and geosmin removal in well acclimated and controlled biofilters. The goal of this paper is to examine biofilter performance for MIB and geosmin control in a full-scale treatment plant under variable influent concentrations, acclimation times and flow conditions.

The chemical structures of geosmin and MIB are shown in Figure 1. MIB and geosmin are uncharged organic molecules, moderately water soluble, volatile and relatively hydrophobic. Based on these properties, treatment is usually chemical oxidation by ozone, adsorption with deep-bed granular activated carbon (GAC) contactors or powdered activated carbon (PAC) addition. These compounds do not biodegrade rapidly, but will biodegrade over extended time periods, (Meyer et al. 2005).

MIB and geosmin are metabolic products of prokaryotic organisms, cyanobacteria (blue-green algae) and actinomycetes, that occur naturally in many surface waters. While MIB and geosmin are not generally considered a health risk, they do create an objectionable "earthy" or "musty" odour. The concentration at which odours are discernible is called the odour threshold concentration (OTC). The OTC for MIB and geosmin is between 2 and 10 ng/L. Therefore, the treatment technique employed must be very efficient at low contaminant concentrations in order to produce drinking water consistently below the OTC.

Biofiltration

Biofiltration is a water treatment process that is commonly practiced in Europe and is starting to gain popularity in the United States (Urfer et al., 1997). In the United States, conventional drinking water treatment includes single media (sand or occasionally anthracite) or multi-media (sand, anthracite or occasionally GAC) filters, but normally a disinfectant is added before filtration. The disinfectant residual normally precludes biological growth on the non-GAC filters and biodegradation does not occur. In the U.S., an oxidant such as ozone sometimes is used prior to biological filtration. The oxidant breaks the natural organic matter (NOM) into smaller molecular weight fractions, rendering it more biodegradable. As a result of this process, biofiltration can effectively reduce NOM.

Because of the reliance on living organisms various operational factors can impact the performance of biofilters. Media type, temperature, empty bed contact time (EBCT), presence/absence of disinfectant in backwash water, and acclimation time are a few of the factors that have been studied. All of these factors have been investigated with respect to NOM removal. However, less research has been conducted on the impact of these factors on biofilter performance for taste and odour control. Studies show varying effects for the impact of temperature on biofilter performance. Elhadi et al. (2003) reported a significant temperature effect when operating biofilters for MIB and geosmin removal.

EBCT has an impact on the level of removal for many compounds of interest. Swertfeger (1996) found that aldehyde and ketoacids were degraded with the least contact time followed by assimable organic carbon (AOC), biodegradable organic carbon (BDOC), chlorine demand and dissolved organic carbon (DOC). The appropriate design EBCT depends on the compound of interest and the ultimate water quality objective. More biorefractory compounds require a longer EBCT.

Biofilters perform best after sufficient acclimation time, and it is important to note that biofilters can be acclimated with respect to one parameter and not another. Swertfeger (1996) reported no detectable increase in bacteria biomass beyond initial sampling, but a minimum of one month was required for sand filters to achieve steady-state removal of AOC. A minimum of three months was required for the adsorptive media filters to reach steady-state AOC removal. Dugan (1998) observed that steady-state had been achieved with respect to DOC removal before biomass had reached steady-state conditions. This work demonstrates that there may be a minimum amount of biomass required to degrade DOC, above which increased biomass does not yield increased DOC removal.

Research suggests that biofilters removing MIB and geosmin may require additional acclimation time. Namkung and Rittmann (1987) found that steady-state conditions for MIB and geosmin removal were achieved after 290 days of acclimation. However, Meyer et al. (2005) observed that once biomass acclimation to MIB and geosmin was developed, it was possible to achieve significant percent removals and effluent concentrations below the OTC. Bench-scale experiments performed at the University Colorado using GCWW acclimated media indicated that ozone was not required upstream of the biofilters in order to form adequate biomass required for excellent removal of MIB and geosmin, Meyer et al. (2005). Acclimation time is a concern since MIB and geosmin episodes can be unpredictable in occurrence, magnitude and duration.

Methods

Plant Operation

The Greater Cincinnati Water Works (GCWW) of Cincinnati, Ohio USA, operates a 832 m^3/d (220 MGD) conventional surface water treatment plant. In October 1992, GCWW initiated a new treatment process – adsorption with post filtration GAC contactors. When the GAC process was initiated, pre-chlorination before rapid sand filtration was discontinued, and no other disinfectant was added before filtration. Hence, the rapid sand filters became biologically active as was evidenced by the reduction of total organic carbon (TOC) through the filters and the relatively high biomass levels on the media. Phospholipid biomass measurements (2004), indicated levels of 150 to 200 nmol phosphate/gram of dry media. After the treatment change, unbeknownst to the utility, levels of MIB and geosmin began increasing in the settled reservoir water. In the winter of 1999 when the GAC was more spent than usual (average 200 days in service vs. a usual average 100 days in service), a rash of customer complaints occurred reporting an objectionable musty taste and odour. The

Table 1 Filter Influent TOC, MIB and Geosmin Concentrations 2000–2005

	TOC mg/L	MIB ng/L	Geosmin ng/L
Minimum	1.11	<0.5	<0.5
Maximum	3.16	398	17
Median	2.05	2.0	2.0
Average of detectable values	1.99	42	4.4

cause was elucidated by through-the-plant monitoring. MIB was found in the settled water at concentrations as high as 400 ng/L and in the finished water at levels as high as 15 ng/L. GCWW staff was surprised to observe prior to the GAC adsorption, 80 to 90 percent removal of MIB and geosmin was achieved through the rapid sand filters. The sand filters had a depth that ranged between 0.61m and 0.76 m and were operated at a average hydraulic loading rate 6.13 m/hr. This yielded an EBCT of between 6.0 min and 7.4 min. Often the concentrations in the filter effluent were below the detection limit. In these cases the analytical limit of detection was used as the effluent value in determining the percent removal. Encouraged by these results, GCWW initiated a more intense long-range monitoring to determine the plant's ability to remove MIB and geosmin through biofiltration.

Analytical Methods

Initially, the MIB and geosmin were analyzed by Standard Methods #6040B with closed loop stripping concentration which had a detection limit of 1.3 ng/L. The more recent samples were analyzed by Standard Methods 6040D concentration with solid phase micro-extraction (SPME). The concentrated samples were run by ion trap GCMS using a Varian Model 2000 GCMS. The detection limit was either 0.5 ng/L or 2.0 ng/L. The total organic carbon (TOC) analyses were run by method #5310B of Standard Methods for the Examination of Water and Wastewater (2005) using a Tekmar 800 TOC analyzer.

Results and discussion

GCWW's evaluation of MIB and geosmin monitoring data during the 1999 customer complaint incident concluded that excellent removal efficiencies through biofiltration were not limited to the high concentration event, but could be consistently observed. The water quality of the sand filter influent during a subsequent heightened monitoring period (2000 to 2005) is shown in Table 1. The average influent MIB concentration was about an order of magnitude higher than that of geosmin.

The monthly percent TOC reduction through the biofilters ranged between 7 and 20 percent. The type of NOM comprising the TOC can change seasonally and with coagulation practices. Temperature can also impact the TOC removal. Figure 2 represents the monthly average TOC removal and the average monthly temperature during the period 1998 to 2005. Greater removal of TOC occurred during the warmer weather months when there was more bioactivity. The maximum average TOC removal occurred in the month of July (20 percent), the minimum occurred in February (7 percent). The seasonal variations

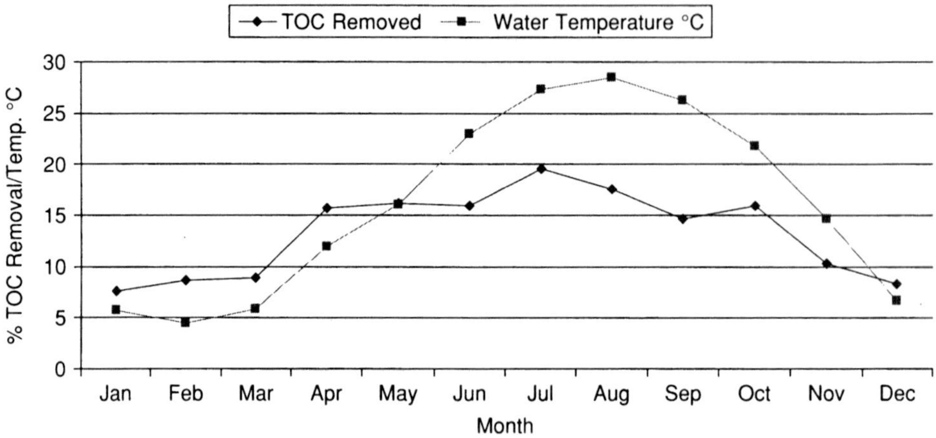

Figure 2 Percent TOC Removal Through Biological Sand Filtration and Temperature °C Monthly Averages 1998 – 2005

in removal efficiencies observed can be correlated to the seasonal variations in temperature (averaging 4 to 28 °C).

Figure 3 depicts the annual averages for TOC removal. The maximum average TOC occurred in 2000 (17 percent), while the minimum TOC average removal occurred in 2003 (8.3 percent). The lowest TOC removal for the eight year period was 3.2 percent, while the greatest removal was 37 percent. Potential reasons for these annual differences will be discussed subsequently.

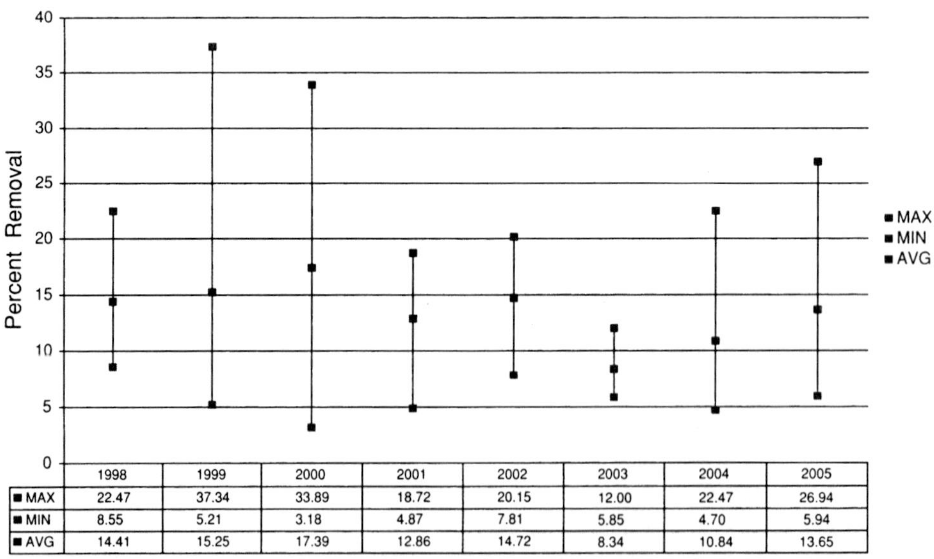

	1998	1999	2000	2001	2002	2003	2004	2005
MAX	22.47	37.34	33.89	18.72	20.15	12.00	22.47	26.94
MIN	8.55	5.21	3.18	4.87	7.81	5.85	4.70	5.94
AVG	14.41	15.25	17.39	12.86	14.72	8.34	10.84	13.65

Figure 3 Annual Average Percent TOC Removed Through Sand Filters

Table 2 Average Percent Removal By Biological Filtration 2000–2005

Excluding BDL Data			Using BDL as Value in Effluent		
TOC	MIB	Geosmin	TOC	MIB	Geosmin
12.2	80.4	52.5	12.2	>73.8	>66.2
n = 208	n = 40	n = 17	n = 208	n = 126	n = 111

From 2000 to 2005 TOC removal through biological filtration averaged 12.2 percent, Table 2. During this same time period MIB removal averaged 80.4 percent in samples that had detectable concentrations in the filter influent and effluent, and >73.8 percent removal when the analytical detection limit was used as a value in the effluent, Table 2. Geosmin removal averaged 52.5 percent in samples that had detectable concentrations in the filter influent and effluent, and >66.2 percent removal when the analytical detection limit was used as a value in the effluent, Table 2.

Because GCWW's data is representative of a full-scale operating water treatment plant rather than a controlled study, some unexpected variables came into play. In fact, over the course of the heightened monitoring period all of the sand was replaced in the rapid sand filters and the hydraulic loading was increased twenty percent. In December, 2000 GCWW began replacing all of the sand in the plant's 47 filters. By the summer of 2003 this project was complete. There was good removal of MIB and geosmin before the filter media was replaced and during the time period when the sand media was being replaced, as shown in Figures 5 and 6. However, on some occasions, effluent detections above the OTC (5 ng/L) were noted (28 of 229 MIB samples and 1 of 220 geosmin samples) before and during the media change out. After the sand media was changed, no effluent detections of MIB or geosmin above the OTC were observed. However, the influent concentration of the MIB has

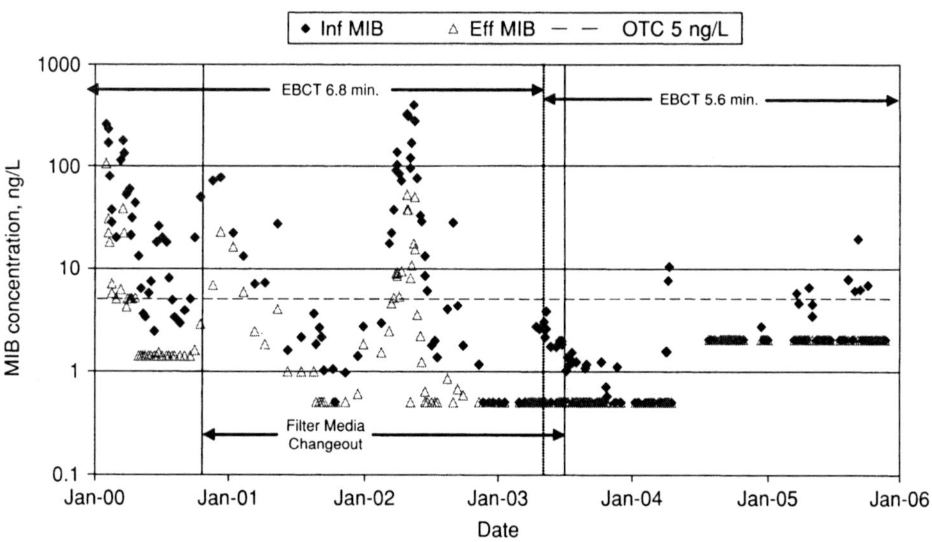

Figure 5 MIB Concentrations in GCWW Filter Influent and Effluent

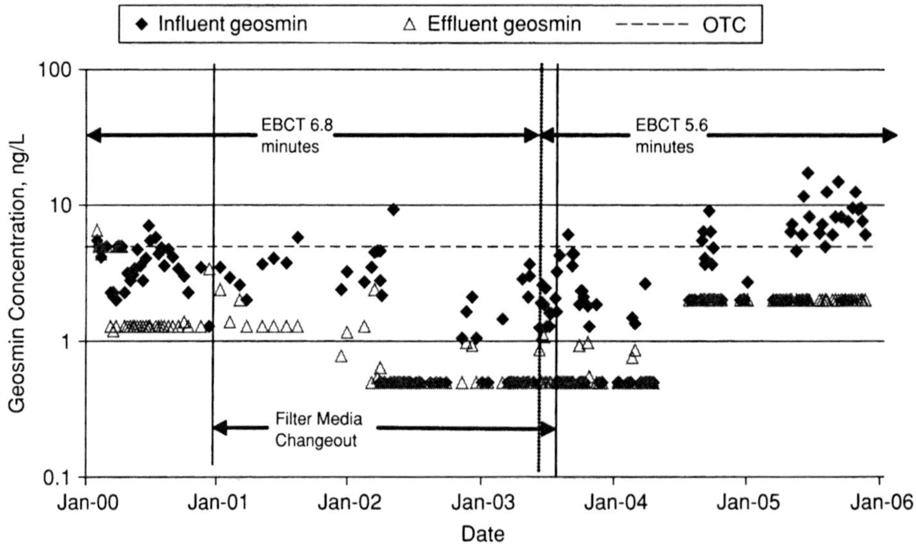

Figure 6 Geosmin Concentrations in GCWW Filter Influent and Effluent

been significantly lower in recent years; 50 ng/L before media replacement, 35 ng/L during media replacement and 2.1 ng/L after media replacement.

A slight reduction in TOC removal occurred during the sand replacement period and after replacement. TOC removal was 16.0 percent before replacement, 13.1 percent during sand replacement and 11.3 percent after replacement. The influent TOC was similar 2.21 mg/L before replacement, 1.98 mg/L during sand replacement and 1.93 mg/L after replacement.

Similarly, since the change in EBCT (6.8 min. to 5.6 min.) that occurred in May of 2003, GCWW has observed no MIB or geosmin effluent concentrations above the OTC, but lower MIB influent concentrations have precluded a direct comparison, Figures 5.and 6. A slight reduction in TOC removal was observed after the EBCT was decreased, 13.7 percent vs. 11.1 percent. However, the influent TOC concentration was similar before and after the change 2.03 mg/L and 1.95 mg/L, respectively. Note also that the lowest annual TOC reduction was 8.3 percent in 2003, Figure 3. In this year a number of newly sanded filters were put into service and the EBCT was reduced.

Conclusions

Sustained biological removal of MIB (80 percent) and geosmin (50 percent) was found over a six-year period in the GCWW's sand filters. During this period, the filter influent concentration of TOC, MIB and geosmin varied greatly. Additionally, the filter sand media was completely replaced and the hydraulic loading increased 20 percent. However, consistently high removals of MIB and geosmin were observed throughout the monitoring period. The microbiological acclimation time or modest change in empty bed contact time did not present a problem relative to taste and odour goals. The filter effluent water seldom had perceptible musty or earthy odours as a result of MIB or geosmin concentrations.

Acknowledgements

The authors appreciate the help of Niranjan Selar, GCWW chemist, for performing the MIB and geosmin analyses and the mentoring of Jack DeMarco, GCWW Superintendent of Water Quality and Research.

Support for Prof. Summers was provided through a grant to the University of Colorado from Korea Institute of Water and Environment, Korea Water Resources Corporation (KIWE-KOWACO).

References

Dugan, N.R. (1998) Determination of Biofiltration Performance, Estimation of Biofiltration Model Parameters and Validation of Model Performance at Pilot-and Full-Scale, and with Respect to pH, in Drinking Water Biofilters. M.S. thesis. Department of Civil and Environmental Engineering, University of Cincinnati, U.S.

Elhadi, S.L.N., Huck, P.M. and Slawson, R.M. (2003) Removal of Earthy/Musty Odours from Drinking Water by Biological Filtration: Temperature and Media Effects. In *Proc. AWWA Water Quality Technology Conference (WQTC)*, Philadelphia, Pennsylvania, U.S.

Meyer, K.J., Summers, R.S., Westerhoff, P. and Metz, D.H. (2005) Biofiltration for Geosmin and MIB Removal. In *Proc. American Water Works Annual Conference*, San Francisco, CA., U.S.

Namkung, E., and Rittmann, B. (1987) Removal of Taste- and Odor-Causing Compounds by Biofilms Grown on Humic Substances. *Jour. Amer. Water Works Assoc.*, **79**(7),107–112.

Swertfeger, J.W. (1996) Biological Degradation of Ozone Byproducts During Drinking Water Treatment. M.S. thesis, Department of Civil and Environmental Engineering, University of Cincinnati, U.S.

Standard Methods for the Examination of Water and Wastewater, 21st ed. (2005) Clesceri, L.S., Greenberg, A.E., Eaton, A.D., eds., American Public Health Association, American Water Works Association and Water Environment Federation, Washington, D.C.

Urfer, D., Huck P.M., Booth, S.D.J. and Coffey, B.M. (1997). Biological Filtration for BOM and Particle Removal: A Critical Review. *Jour. Amer. Water Works Assoc.*, **89**(12), 83–98.

Westerhoff, P., Summers, R.S. Chowdhury, Z. and Kommineni, S. (2005) *Ozone-Enhanced Biofiltration for Geosmin and MIB Removal.* Report 91075, Amer. Water Works Assoc Research Foundation, Denver, CO.

43 Removal of geosmin and MIB in biofilters - on the role of biodegradation and adsorption

W. Uhl*, F. Persson, G. Heinicke***, M. Hermansson**, T. Hedberg*****

*Institute for Urban Water Management, Water Supply Engineering Group, Technische Universität Dresden, Dresden, Germany. email: wolfgang.uhl@tu-dresden.de
**Cell and Molecular Biology, Microbiology, Göteborg University, Sweden.
***Water Environment Transport, Chalmers University of Technology, Göteborg, Sweden.

Abstract Geosmin and 2-methylisoborneol (MIB) are two substances causing earthy/musty odours that are difficult to remove by conventional drinking water treatment. In this study removal of geosmin and MIB by biofiltration of untreated surface water was investigated using granular activated carbon (GAC) and crushed expanded clay (EC) as filter media. Biofiltration through both GAC and EC removed geosmin and MIB present at low (20 ng l^{-1}) concentrations by at least 97% at an empty bed contact time of 30 minutes and a temperature of 15°C. At lower temperature (6–12°C) and simultaneously lower biomass concentrations, removal efficiency was similar in the GAC but considerably lower in the EC biofilter, pointing to a second mechanism different from biodegradation. Consequently, microbial activity was suppressed with azide to enable discrimination between biodegradation and adsorption. During azide dosage, the GAC biofilters still removed geosmin and MIB nearly unaffectedly. In the EC biofilter, however, removal of both odorants ceased completely. Methylene blue adsorption confirmed that the GAC, even after almost four years of operation receiving surface water, had capacity to remove geosmin and MIB by adsorption. Since odour episodes commonly occur during the warm season when microbiological activity is high, EC constitutes a viable option as carrier medium for direct biological filtration of surface water. The additional GAC adsorption capacity however adds robustness to the removal process.

Keywords Biofiltration, drinking water, geosmin, MIB, adsorption, biodegradation

Introduction

The seasonal occurrence of biogenic odorants in surface waters is a common problem for water suppliers. Dissolved substances with odour threshold concentrations in the ng l^{-1} range, such as trans-1-10-dimethyl-trans-9-decalol (geosmin) and 2-methylisoborneol (MIB) are mostly unaffected by conventional chemical water treatment processes [1, 2] and the resulting taste and odour episodes could undermine consumer trust in tap water.

Due to their hydrophobic nature [3], geosmin and MIB are readily adsorbed onto granular activated carbon (GAC). They have however to compete with natural organic matter (NOM)

for adsorption sites [4]. In particular, NOM fractions of similar molecular weight and chemical characteristics [5, 6] may impair odorant removal by GAC by orders of magnitude [7, 8].

Filtration processes with long retention times such as slow sand filtration [9], trickling filtration [10], bank filtration [11], and artificial groundwater recharge [2] have been shown to achieve high removal of substances causing earthy-musty odour. Also at shorter retention times, biofilters have demonstrated substantial removals of both geosmin and MIB [12, 13]. Removal rates have been observed to depend on temperature, as well as initial concentration [14] and biofilter biomass [15].

As shown in the literature, biodegradation and adsorption of odorous compounds have been investigated under various operational conditions. Biodegradation and adsorption could however occur simultaneously in a biofilter with an adsorptive filter medium, as shown by Yagi *et al.* [16] for high concentrations (mg l^{-1}) of geosmin and MIB in a laboratory scale study using GAC with and without inoculation with *Bacillus subtilis*. In order to optimise biofiltration with respect to removal of odorous compounds, it is thus essential to understand and assess both biodegradation and adsorption, under different realistic operational conditions.

Materials and methods

Experimental setup

The study was conducted at Lackarebäck waterworks, Göteborg, Sweden. The waterworks is treating surface water with concentrations of geosmin and MIB reaching 10–20 ng l^{-1} and 5 ng l^{-1} respectively, during late summer. Even though the observed concentrations of geosmin ($\leq$ g ng l^{-1}) and MIB ($\leq$1.4 ng l^{-1}) in drinking water are in the lower range of published odour threshold values [17], their occurrences correlate with odour-related consumer complaints. Geosmin and MIB have also been proven responsible for the perceived earthy/musty odour, by means of the column-sniffing methodology [2]. The surface water is soft and moderately humic with low turbidity, quite typical for Swedish surface waters. For details of surface water quality, see [18].

Two parallel sets of 50 mm inner diameter glass columns with a total of 1 m of filter bed received untreated surface water. Each set consisted of three consecutive columns with 20 cm, 30 cm and 50 cm of filter media to allow sampling at several bed depths, (Figure 1) and was kept in the dark. The influent to each column was below the supernatant surface. The flow rate was 2 m h^{-1}, corresponding to a total empty bed contact time (EBCT) of 30 minutes. One set of columns was filled with GAC (Filtrasorb 300, Calgon Carbon Corporation, USA, effective size 0.8–1.0 mm) and the other with crushed expanded clay (EC, Filtralite NC 0.8–1.6 mm, Filtralite, Norway, effective size 0.85–0.95 mm). The surface area of the media grains including pores with a diameter larger than 0.6 μm was 1272 cm^2 and 505 cm^2 per cm^3 of GAC and EC respectively, measured by mercury porosimetry (surface angle 130° and surface tension 0.45 N m^{-1}).

Standards of geosmin and MIB (Wako, Japan) were diluted in autoclaved MilliQ water (Millipore) in an amber glass bottle and this stock solution was continuously fed to the surface water influent to give final concentrations of 20 ng l^{-1}.

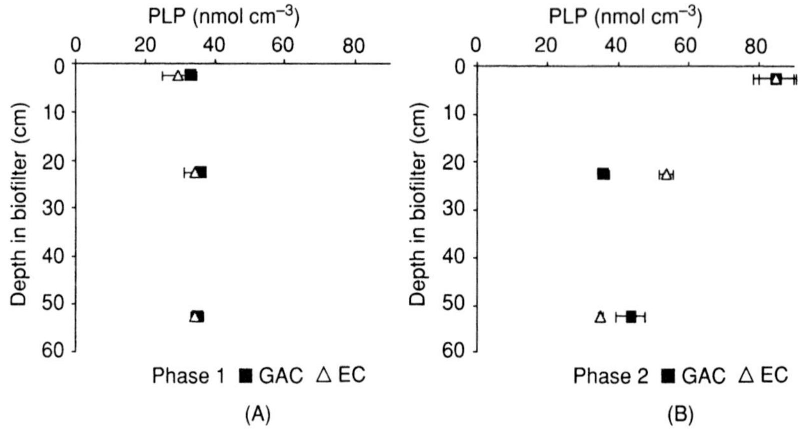

Figure 1 Biomass concentration (as phospholipid phosphate) per volume of biofilter bed as function of bed depth during phase 1 (A) and phase 2 (B). Error bars show one standard deviation (n = 3).

During phase 1 (Nov. 2002) the GAC and EC biofilter media had been receiving untreated surface water for 21 months in larger biofilters with the same flow rate as the biofiltration equipment used in this study. Biofilter media were transferred from mid bed depth of the larger biofilters to the biofiltration equipment 4 weeks before start up of the measurements.

In phase 2 (Oct. 2004) the biofilter media had been receiving untreated surface water for 44 months. Biofilter media were transferred to the biofiltration equipment from mid bed depth of the larger biofilters 10 months prior to the study and were receiving surface water with a set temperature of 15°C during that period, using a heater. Samples of the influent and effluent concentrations of geosmin and MIB were taken 6 hours after the start of the addition, and after 7 and 14 days. The columns were then fed with surface water amended with 400 mg l^{-1} of sodium azide during continuing odorant addition.

Analytical methods

Samples of geosmin and MIB were taken in glass bottles and were refrigerated until analysis by closed-loop air stripping and GC-MS. Odorants were stripped from 1 l sample bottles for 2 hours, and adsorbed onto an activated carbon pellet (1.5 mg). Stripping was performed at 55°C with deuterated geosmin as internal standard and CS_2 as eluent for all samples, except those taken directly before and during addition of azide.

Sampling of biofilter media

Biofilter media were sampled after draining the columns. Cores (5 cm) were sampled from the top of every column using a 10 mm inner diameter plastic pipe, corresponding to bed depths of 0–5, 20–25 and 50–55 cm. The bed density of about 50 ml of dried (60°C) GAC and EC was measured to be able to relate data of biomass and activity to volume of the filter beds.

Viable biomass

Viable biomass in the biofilters was measured as phospholipid phosphate (PLP) [20] according to a modified set up [21], with the exception that extracted lipids were digested at 95°C for 48 h. After lipid extraction, the biofilter media were dried at 60°C and weighed.

For determination of oxygen consumption rate (OCR), approximately 3 cm^3 of biofilter media were added to 35 ml glass tubes with glass caps, which were then filled with surface water from the biofiltration equipment, *i.e.* with or without addition of azide. The oxygen concentration was measured with an oxygen meter with a DO probe.

Results and discussion

Biomass

In order to evaluate biodegradation in the biofilters the distribution of viable biomass, analysed as phospholipid phosphate (PLP), was measured (Figure 1). For the phase 1 experiments, filter media had been transferred from mid bed depth of larger biofilters 4 weeks prior to the measurements. This explains the small differences in biomass between bed depths in phase 1 (Figure 1a). The biomass levels were higher and more stratified in both biofilters during phase 2 (Figure 1b). This was due to the fact that the filter media were transferred from the larger biofilters 10 months prior to the measurements and had since then been receiving surface water of at least 15°C, enabling biomass development during a prolonged period of time.

To convert biomass levels to numbers of bacteria, a conversion factor of 1.5×10^{-8} nmol PLP per bacterium was used [18], a factor that corresponds well to earlier obtained factors [20, 22]. The biomass levels corresponded to a number of $2.0 - 5.7 \times 10^9$ bacteria per cm^3 of biofilter media. The surface area of 1272 cm^2 per cm^3 of GAC and 505 cm^2 per cm^3 of EC, including pores with a diameter larger than 0.6 μm, may well represent the surface area available of bacterial attachment and growth. Relating the bacterial numbers to this surface area would result in one bacterium per 22–65 μm^2 of GAC surface area and 9–26 μm^2 of EC surface area. Since bacterial size generally is 0.5–2 μm this indicates that both biofilter media were far from covered by bacteria.

Removal of geosmin and MIB

Both geosmin and MIB were removed to at least 96 % in the GAC biofilters at 30 min EBCT in both measurement phases. In the EC biofilter removal of both geosmin and MIB was considerably lower than in the GAC biofilter during the first measurement phase, but was similar to the GAC biofilter during the second phase (Figure 2). At 15 min EBCT removal of geosmin and MIB was about 90 % in the GAC biofilter during both measurements phases, while for the EC biofilter removal of geosmin and MIB was 69 % and 55 % in phase 1 and 90% for both odorants in phase 2 (data not shown). It is worth to point out that in phase 2 the GAC and EC biofilters removed both odorants efficiently already 6 hours after the start of odorant addition. This indicates that no acclimatisation to the elevated concentrations of odorants was needed, which is valuable since odour episodes often are transient.

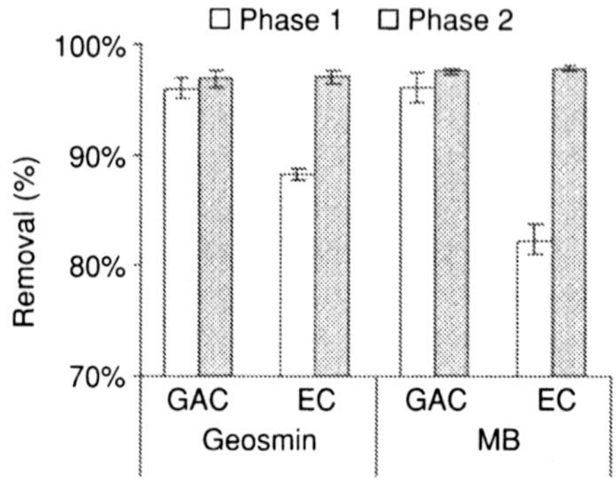

Figure 2 Removal of Geosmin and MIB in the GAC and EC biofilters during phase 1 and phase 2, measured at 100 cm bed depth corresponding to 30 min EBCT. Error bars show one standard deviation (n = 2).

Since geosmin and MIB are assumed to be degraded as secondary substrates, their biodegradation depends on the metabolic rate of the primary substrates, *i.e.*NOM [24]. NOM removal and biomass development in biofilters have been shown to be temperature dependent [25, 26] and a similar temperature dependence has recently been demonstrated for the removal of the secondary substrates geosmin and MIB [15].

Temperature and total biomass in the biofilters were higher in phase 2 than in phase 1, which consequently resulted in considerably higher removal efficiency in the EC biofilter in phase 2 than in phase 1, while nearly complete removal was achieved in the GAC biofilter in both phases. This could have been an effect of much higher biomass concentrations in the GAC filter than in the EC filter. However, comparable biomass concentrations were found. Thus, it can be concluded that in the GAC, which had been in operation for several years, there must have been another mechanism for odorant removal besides biodegradation, which was likely adsorption, especially considering the high methylene blue adsorption values. In order to experimentally verify this, bioactivity was suppressed with sodium azide to differentiate between adsorption and biodegradation. The azide anion blocks electron transfer and thereby inhibits energy production in the cell, which will cease energy dependent intracellular activities.

Suppression of microbial activity with azide

As a function of bed depth, concentrations of geosmin and MIB normalised to the influent concentration (C/C_0), during normal operation and azide dosage, are shown in Figure 3. The removal of geosmin and MIB ceased completely in the EC biofilter during dosages of azide, while removal in the GAC biofilter was practically unaltered for geosmin and somewhat reduced for MIB.

A complete inactivation of microbial DOC degradation applying the same azide concentration and exposure time as in this study has previously been demonstrated [27]. To however ensure that microbiological activity in the biofilters was really suppressed by azide, oxygen consumption rate (OCR) was measured and counts of CTC positive bacteria was

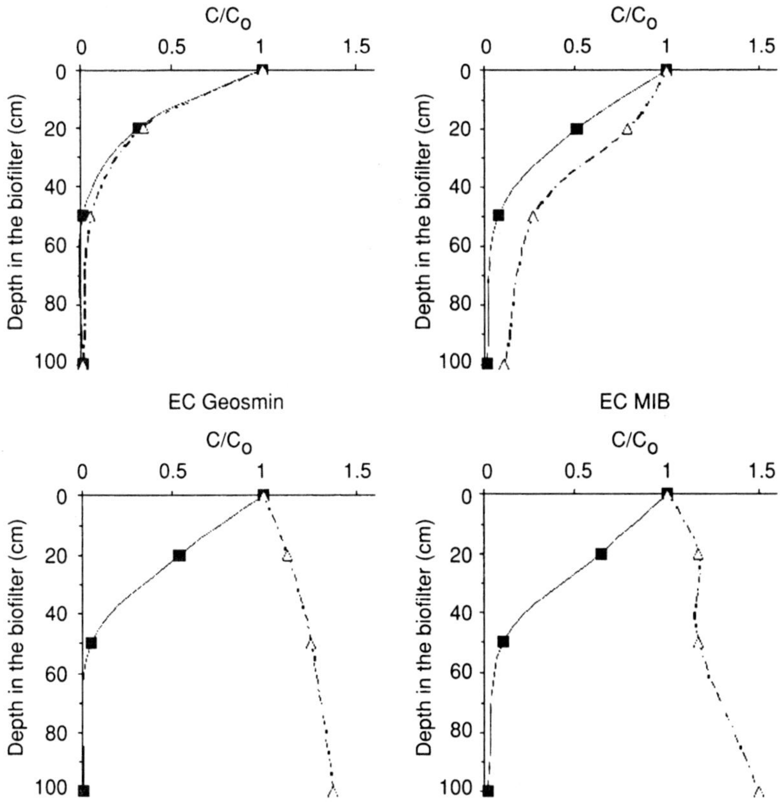

Figure 3 Normalised concentrations (C/C_0) of geosmin and MIB during normal operation —■— and azide dosage - -△- - of the EC and GAC biofilters. Bed depths of 20, 50 and 100 cm corresponded to EBCT of 6, 15 and 30 min.

performed before and during azide dosage. The CTC assay indicated that the respiratory activity was effectively inhibited by azide. Besides the percentage of numbers of respiring bacteria to total numbers of bacteria, oxygen consumption rate (OCR) is a measure of microbial metabolic activity. OCR was also reduced at a high degree by the addition of azide.

Since no removal of odorants was observed in the EC biofilter after suppression of the metabolic activity with azide, we conclude that the EC removal of geosmin and MIB was entirely due to biodegradation, and not due to adsorption to the EC filter media or the microbial biofilm to any appreciable extent. This was also in accordance with the low methylene blue adsorption values (see above). In fact, the concentrations of geosmin and MIB increased over the filter bed (Figure 3). We do not believe that these observations were due to cell disruption since measurements have shown a complete recovery of azide-suppressed bacteria 4 hours after azide was removed, as measured by CTC (data not shown). Sodium azide has also been used in numerous studies at similar or even higher concentrations, seemingly without affecting cell integrity, see *e.g.* [28]. The concentration of both odorants in the feed water samples for the azide amended EC biofilters was however found to be 26% lower than during normal operation. Since the samples were taken sequentially starting at 100 cm bed depth and finishing with the feed water, the temporary under-dosage of odorants could explain most of the observed negative removal of geosmin and MIB in the azide amended EC biofilter.

In the GAC filter, MIB removal was reduced to some extent by the addition of azide while geosmin removal remained practically unaltered, indicating a substantial influence of adsorption. It has previously been shown that GAC adsorbs MIB to a lesser extent than it adsorbs geosmin [29, 30]. The observation that the azide-suppressed GAC biofilter and the EC biofilter without azide removed geosmin and MIB similarly, suggests that the removal by adsorption and biodegradation were of a similar rate. However, the contribution of biodegradation and adsorption in the unaffected GAC biofilter cannot be derived directly from our results. The observation that only a minor part of the GAC surface was covered by microbial biomass (see above) indicates the possibilities of simultaneous biodegradation and adsorption to the GAC surface. Since both biofilters carried a similar biomass with a similar OCR, originating from the same surface water, it is reasonable to assume that biodegradative properties of the EC biofilm community for geosmin and MIB were present also in the GAC biofilter and that biodegradation and adsorption were competing processes in the GAC biofilter.

The operation time during which GAC maintain satisfactory adsorption capacity for geosmin and MIB depends on process parameters such as EBCT and feed water NOM [31], but is commonly a few months to 2 years [29, 32]. The GAC medium in this study was evidently not exhausted for geosmin and MIB adsorption despite an operation time of 44 months receiving surface water with an average total organic carbon (TOC) of 4.7 mg l^{-1}. Freundlich coefficients from the literature were applied to estimate adsorption capacity. With competition by NOM and 20 ng l^{-1} of geosmin and MIB in the water phase, the theoretical equilibrium solid phase concentration on the GAC surface equals 143–186 μg of geosmin and 10–220 μg of MIB per gram of GAC, using coefficients from the literature [5, 7, 8, 33]. The estimated load of 0.40 μg of geosmin and 0.30 μg of MIB per gram of GAC during the 10 months in the second phase was therefore far below any of the equilibrium concentrations. Earlier loading with naturally occurring geosmin and MIB was considered negligible, since the filter media originated from 50 cm bed depth in larger biofilters where they should not receive any appreciable concentrations.

Conclusions

To conclude, with 30 min EBCT GAC supported a higher removal efficiency of geosmin and MIB (at least 96 %) compared to EC (88% and 82% respectively) at low temperature and when the filter media carried a low overall biomass. Removal at higher temperature and simultaneously higher biomass concentrations was similar for the two media (at least 97 %). After suppressing the microbial activity with azide, GAC still removed geosmin and MIB substantially, while no removal on the EC was observed. This shows that adsorption played a significant role for the removal on GAC, even after receiving trace concentrations of odorants for 10 months and surface water NOM for almost 4 years, while removal on EC was entirely due to biodegradation.

Since earthy-musty odour episodes commonly occur during the warm season, supporting a high microbial activity, EC constitutes a viable option as carrier medium for direct biological filtration of surface water. The GAC adsorption capacity however contributes to additional robustness, since even for GAC in operation for several years, odorant removal was not solely dependent on microbial activity as it was for the EC medium.

References

[1] Bruce, D., Westerhoff, P. and Brawley-Chesworth, A., Removal of 2-methylisoborneol and geosmin in surface water treatment plants in Arizona. *J. Water S. R. T. – Aqua*, **51**, 183–197 (2002).

[2] Sävenhed, R., Borén, H., Grimvall, A., Lundgren, B. V., Balmér, P. and Hedberg, T., Removal of individual off-flavour compounds in water during artificial groundwater recharge and during treatment by alum coagulation sand filtration. *Water Res.*, **21**, 277–283 (1987).

[3] Pirbazari, M., Borow, H. S., Craig, S., Ravindran, V. and McGuire, M. J., Physical-chemical characterization of 5 earthy-musty-smelling compounds. *Water Sci. Technol.*, **25**, 81–88 (1992).

[4] Herzing, D. R., Snoeyink, V. L. and Wood, N. F., Activated carbon adsorption of odorous compounds 2-methylisoborneol and geosmin. *J. Am. Water Works Assoc.*, **69**, 223–228 (1977).

[5] Newcombe, G., Drikas, M. and Hayes, R., Influence of characterised natural organic material on activated carbon adsorption. 2. Effect on pore volume distribution and adsorption of 2-methylisoborneol. *Water Res.*, **31**, 1065–1073 (1997).

[6] Newcombe, G., Morrison, J., Hepplewhite, C. and Knappe, D. R. U., Simultaneous adsorption of MIB and NOM onto activated carbon - II. Competitive effects. *Carbon*, **40**, 2147–2156 (2002).

[7] Pirbazari, M., Ravindran, V., Badriyha, B. N., Craig, S. and McGuire, M. J., GAC adsorber design protocol for the removal of off-flavors. *Water Res.*, **27**, 1153–1166 (1993).

[8] Chen, G., Dussert, B. and Suffet, I., Evaluation of granular activated carbons for removal of methylisoborneol to below odor threshold concentration in drinking water. *Water Res.*, **31**, 1155–1163 (1997).

[9] Yagi, M., Kajino, M., Matsuo, U., Ashitani, K., Kita, T. and Nakamura, T., Odor problems in lake Biwa. *Water Sci. Technol.*, **15**, 311–321 (1983).

[10] Lundgren, B. V., Grimvall, A. and Savenhed, R., Formation and removal of off-flavour compounds during ozonation and filtration through biologically-active sand filters. *Water Sci. Technol.*, **20**, 245–253 (1988).

[11] Chorus, I., Klein, G., Fastner, J. and Rotard, W., Off-flavors in surface waters - how efficient is bank filtration for their abatement in drinking-water. *Water Sci. Technol.*, **25**, 251–258 (1992).

[12] Sumitomo, H., Biodegradation of 2-methylisoborneol by gravel sand filtration. *Water Sci. Technol.*, **25**, 191–198 (1992).

[13] Terauchi, N., Ohtani, T., Yamanaka, K., Tsuji, T., Sudou, T. and Ito, K., Studies on a biological filter for musty odor removal in drinking water treatment processes. *Water Sci. Technol.*, **31**, 229–235 (1995).

[14] Egashira, K., Ito, K. and Yoshiy, Y., Removal of musty odor compound in drinking-water by biological filter. *Water Sci. Technol.*, **25**, 307–314 (1992).

[15] Elhadi, S. L. N., Huck, P. M. and Slawson, R. M., Impact of biomass concentrations on the removal of earthy/musty odors from drinking water by biological filters. In: *Proc. of the Am. Water Works Assoc. Water Quality Technology Conference*, San Antonio, Texas, USA (2004).

[16] Yagi, M., Nakashima, S. and Muramoto, S., Biological degradation of musty odor compounds, 2-methylisoborneol and geosmin, in a bio-activated carbon filter. *Water Sci. Technol.*, **20**, 255–260 (1988).

[17] Young, W. F., Horth, H., Crane, R., Ogden, T. and Arnott, M., Taste and odour threshold concentrations of potential potable water contaminants. *Water Res.*, **30**, 331–340 (1996).
[18] Heinicke, G., Biological pre-filtration and surface water treatment. Microbial barrier function and removal of natural organic and inorganic compounds. PhD thesis. WET, Chalmers University of Technology, Sweden (2005).
[19] Elhadi, S. L. N., Huck, P. M. and Slawson, R. M., Determination of system losses of geosmin and MIB in bench-scale filtration apparatus. *Water Qual. Res. J. Canada*, **39**, 207–212 (2004).
[20] Findlay, R. H., King, G. K. and Watling, L., Efficacy of phospholipid analysis in determining microbial biomass in sediments. *Appl. Environ. Microbiol.*, **55**, 2888–2893 (1989).
[21] Wang, J. Z., Summers, R. S. and Miltner, R. J., Biofiltration performance: part 1, relationship to biomass. *J. Am. Water Works Assoc.*, **87**, 55–63 (1995).
[22] Balkwill, D. L., Leach, F. R., Wilson, J. T., McNabb, J. F. and White, D. C., Equivalence of microbial biomass measures based on membrane lipid and cell-wall components, adenosine-triphosphate, and direct counts in subsurface aquifer sediments. *Microb. Ecol.*, **16**, 73–84 (1988).
[23] Graham, D., Characterization of physical adsorption systems. 3. The separate effects of pore size and surface acidity upon the adsorbent capacities of activated carbons. *J. Phys. Chem.*, **59**, 896–900 (1955).
[24] Namkung, E. and Rittmann, B. E., Removal of taste-causing and odor-causing compounds by biofilms grown on humic substances. *J. Am. Water Works Assoc.*, **79**, 107–112 (1987).
[25] Moll, D. M., Summers, R. S., Fonseca, A. C. and Matheis, W., Impact of temperature on drinking water biofilter performance and microbial community structure. *Environ. Sci. Technol.*, **33**, 2377–2382 (1999).
[26] Uhl, W., Einfluss von Schüttungsmaterialien und Prozessparametern auf die Leistung von Bioreaktoren in der Trinkwasseraufbereitung (In German). PhD Thesis. Gerhard Mercator Universität Duisburg, Germany (2000).
[27] Ribas, F., Frias, J., Ventura, F., Mohedano, L. and Lucena, F., Assessment of biological-activity and fate of organic-compounds in a reactor for the measurement of biodegradable organic-carbon in water. *J. Appl. Bacteriol.*, **79**, 558–568 (1995).
[28] Bugg, T., Foght, J. M., Pickard, M. A. and Gray, M. R., Uptake and active efflux of polycyclic aromatic hydrocarbons by Pseudomonas fluorescens LP6a. *Appl. Environ. Microbiol.*, **66**, 5387–5392 (2000).
[29] Ridal, J., Brownlee, B., McKenna, G. and Levac, N., Removal of taste and odour compounds by conventional granular activated carbon filtration. *Water Qual. Res. J. Canada*, **36**, 43–54 (2001).
[30] Jung, S. W., Baek, K. H. and Yu, M. J., Treatment of taste and odor material by oxidation and adsorption. *Water Sci. Technol.*, **49**, 289–295 (2004).
[31] Gillogly, T. E. T., Snoeyink, V. L., Vogel, J. C., Wilson, C. M. and Royal, E. P., Determining GAC bed life. *J. Am. Water Works Assoc.*, **91**, 98–110 (1999).
[32] Hattori, K., Water-treatment systems and technology for the removal of odor compounds. *Water Sci. Technol.*, **20**, 237–244 (1988).
[33] Lalezary, S., Pirbazari, M. and McGuire, M. J., Evaluating activated carbons for removing low concentrations of taste-producing and odor-producing organics. *J. Am. Water Works Assoc.*, **78**, 76–82 (1986).

44 Biodegradation of MIB and geosmin in biological sand and BAC filters: acclimation, steady-state and varying influent

R.S. Summers*, S. Chae*, **, S.M. Kim*, ** and H.W. Ahn**

*University of Colorado, Boulder, CO, USA
**Korean Water Resources Company (KOWACO), Korea

Abstract This study evaluated the removal of the odour causing compounds 2-methylisoborneol (MIB) and geosmin by biofiltration with sand and granular activated carbon (GAC) media. After four months of operation about 50 % of the MIB was removed by the biologically active sand filters at lab temperature, 20 C, and no removal was seen at 4 C. The biologically active filters with GAC displayed MIB removals of about 65%

Keywords Biotreatment, biofilters, BAC, MIB, geosmin

Introduction

Biological filtration is becoming more commonly used in drinking water treatment. However, its ability to control taste and odor (T&O) is not well understood. Few studies are available that directly quantify the extent of taste and odor removal in well acclimated and controlled biofilters. The microbial metabolic by-products, 2- methylisoborneol (MIB) and trans-1,10-dimethyl-trans-9-decalol (geosmin), present in many surface waters, are thought to be the most common source of T&O. MIB and geosmin are uncharged organic molecules, moderately water soluble, volatile and relatively hydrophobic. Based on these properties, treatment is usually chemical oxidation or activated carbon adsorption. However, these processes are relatively expensive. While, MIB and geosmin have not been shown to biodegrade rapidly, biodegradation over extended time periods has been demonstrated (Meyer et al., 2005; Elhadi, et al., 2003; Namkung, E., and Rittmann, B., 1987). The goals of this project are to assess the biodegradation of MIB and geosmin by biologically active sand (BAS) and biologically active activated carbon (BAC). In addition to media type, the impact of acclimation and temperature will be discussed in this paper. Other variables being currently investigated are the impact of variable odour concentration, background natural organic matter (NOM) concentration, and preozonation.

 Recent Progress in Slow Sand and Alternative Biofiltration Processes edited by Rolf Gimbel, Nigel J.D. Graham and M. Robin Collins. ISBN: 9781843391203. Published by IWA Publishing, London, UK.

Table 1 Biological Filters and Performance Data.

	BAS 1	**BAS 2**	**BAS 3**	**BAS 4**	**BAS 5**		**BAC 1**	**BAC 2**
	SAND	**SAND**	**SAND**	**SAND**	**SAND**		**GAC**	**GAC**
Media	**Acclim-ated**	**Acclim-ated**	**Acclim-ated**	**Acclim-ated**	**Un-acclimated**		**Acclim-ated & Exhausted**	**Acclim-ated**
Temperature C	20	20	20	5	20		20	20
TOC					<60 days	>60 days		
Removal (%)	12	14	11	4	3	8	19	15
Std Dev	8	6	13	11	5	7	4	11
Count	27	27	27	27	7	20	11	8
MIB >120 days								
Removal (%)	48	46	42	1	7		66	63
Std Dev	4	4	6	3	4		13	10
Count	8	8	8	8	8		11	11

Methods

Pilot Plant Operation

Table 1 shows the operation conditions of the seven parallel filters run in Phase I of this study. The acclimated media, both sand and granular activated carbon (GAC), were from the Greater Cincinnati Water Works (GCWW) of Cincinnati, Ohio USA, surface water treatment plant. The GAC had been in use for about 10 months and the sand had been in use for at least 2 years. The 1 cm columns were operated at a hydraulic loading rate of 5 m/hr and an empty bed contact time of 7 min. The biological active sand (BAS) filter 4 was temperature controlled at 5 C. The BAS in column 5 was chlorinated to inactivate the biomass prior to the run. The GAC in the BAC 1 filter had been exposed to additional amounts of MIB and geosmin in batch reactors prior to the run to exhaust its remaining adsorption capacity. Biomass in the filter was measured using the phospholipids extraction method developed by Findlay et al. (1989), as modified by Wang (1995) (see Westerhoff et al., 2005). The phospholipid biomass measurements prior to the start of the filter columns yielded levels of about 60 nmol phosphate/gram of dry media for sand and 40 nmol phosphate/gram of dry media for the GAC. The influent water was dechlorinated City of Boulder tap water. The influent water quality is shown on Table 2. Geosmin and MIB were spiked in at a target concentration of 100 ng/L. Aqueous based geosmin and MIB was used for the first 3 months of operation, after which radiolabeled MIB was used.

Results and discussion

The results after 4 months of operation are shown in Table 1. TOC removal in the BAS 1 through 3 showed about 12 % removal at the lab temperature, while only 4 % removal at

Table 2 Biological Filter Influent Water Quality.

	pH (-)	Alkalinity (mg/L as CaCO3)	Temp (C)	Temp - BAS 4 (C)	TOC (mg/L)	UVA (/m)	MIB (ng/L)
Mean	7.54	43	21.0	5.7	1.02	1.35	108
Median	7.58	44	21.0	5.5	1.01	1.30	107
Max	7.68	47	22.3	7.5	1.40	2.30	128
Min	7.40	40	19.1	3.8	0.85	0.90	98
Std Dev	0.10	3	1.0	1.1	0.14	0.30	7.4
Count	7	7	10	26	28	28	11

5 C with the pre-acclimated BAS. For the first two months the un-acclimated sand in BAS 5 averaged 3 % removal, while after two months the average removal in creased to 8 %. The TOC removal in the BAC columns was 20 to 40 % higher than that with BAS. After four months of operation, the MIB removal in the pre-acclimated BAS filters 1 through 3 averaged 46 % removal at the lab temperature, while nearly no removal (1 %) was found at 5 C with the pre-acclimated BAS. After four months the un-acclimated sand in BAS 5 averaged 7 % removal. The MIB removal in the BAC columns was about 50 % higher than that with BAS at lab temperature. No difference in performance was found between the acclimated BAC and the acclimated and adsorption exhausted BAC for either TOC or MIB. This indicates that GAC, taken from the top of GCCW's post-filter adsorber, was exhausted with respect to TOC and MIB.

Conclusions

After four months of operation about 50 % of the MIB was removed by the biologically active sand filters at lab temperature and no removal was seen at 4 C. The biologically active filters with GAC displayed MIB removals of about 65 %.

Acknowledgements

This research was funded through a grant to the University of Colorado from Korean Institute of Water and Environment, Korean Water Resources Corporation (KIWE-KOWACO) as part of project, "Development of Decision Making Guideline for Taste and Odor Control in Drinking Water". The authors appreciate the assistance of Debbie Metz and others with the GCWW for providing the acclimated sand and GAC.

References

Elhadi, S.L.N., Huck, P.M. and Slawson, R.M. (2003) Removal of Earthy/Musty Odours from Drinking Water by Biological Filtration: Temperature and Media Effects. In *Proc. AWWA Water Quality Technology Conference (WQTC)*, Philadelphia, Pennsylvania, U.S.

Meyer, K.J., Summers, R.S., Westerhoff, P. and Metz, D.H. (2005) Biofiltration for Geosmin and MIB Removal. In *Proc. American Water Works Annual Conference*, San Francisco, CA., U.S.

Namkung, E., and Rittmann, B. (1987) Removal of Taste- and Odor-Causing Compounds by Biofilms Grown on Humic Substances. *Jour. Amer. Water Works Assoc.*, **79**(7),107–112.

Standard Methods for the Examination of Water and Wastewater, 21st ed. (2005) Clesceri, L.S., Greenberg, A.E., Eaton,A.D., eds., American Public Health Association, American Water Works Association and Water Environment Federation, Washington, D.C.

Westerhoff, P., Summers, R.S. Chowdhury, Z. and Kommineni, S. (2005) *Ozone-Enhanced Biofiltration for Geosmin and MIB Removal*. Report 91075, Amer. Water Works Assoc Research Foundation, Denver, CO.

45 Microcystin-LR removal by bench scale biologically-activated-carbon filters

E. Mesquita*, J. Menaia*, M. J. Rosa, V. Costa***

*Urban Water Division, National Civil Engineering Laboratory (LNEC) Av. do Brasil, 101–1700-066 Lisbon, Portugal
**Center of Marine and Environmental Research, Faculty of Marine and Environmental Sciences, University of Algarve, Campus de Gambelas, 8005-139 Faro, Portugal

Abstract The occurrence of cyanobacterial blooms in surface water sources for human consumption is a worldwide problem as they often comprise cell masses of toxin producing species. The removal of cyanobacterial cells is effectively accomplished by conventional drinking water treatment processes (*e.g.*, coagulation / flocculation / sedimentation, rapid sand-filtration). However, such processes do not remove cyanotoxins satisfactorily and may disrupt cyanobacterial cells with the consequent release of intracellular cyanotoxins to the water. Thus, the treatment of cyanotoxin contaminated water mainly rely on membrane filtration or activated carbon adsorption processes. Among these, granular activated carbon filters (GAC) efficiently remove cyanotoxins from water and, in addition, constitute a permanent barrier against toxin intrusion into drinking water distribution systems. However, GAC's lifetime is limited by their adsorption capacity, so the bed's carbon needs be regenerated periodically. In addition, during operation GACs' bed invariably becomes colonised by microorganisms which may degrade cyanotoxins, then synergistically adding to adsorption in their removal. In the herein reported studies, the removal of microcystin-LR (MC-LR) by colonised granular-activated carbon filters (BAC) was investigated at bench-scale. Four filters were used with a bed (1.2 cm Ø, 3.0 cm depth) of crushed (0.35–0.50 mm) NORIT 0.8 Supra activated carbon. The removal of MC-LR was investigated in two BAC, which had been previously loaded and colonised by feeding them with the laboratory tap water for six months. A BAC filter, which had its biological activity inhibited with sodium azide (6 mM) throughout the 325 days of experiments, served as control. A non-colonised and non-loaded filter (GAC) was used as an additional control. All the filters were continuously fed with cell extracts of *Microcystis aeruginosa* (with predetermined MC-LR concentrations) in dechlorinated water, at a empty bed contact time of 11 min. The effect of changing the cell extract content or of the presence of supplements (microbial growth substrates) was examined. In addition to the capability of BAC's microbiota to significantly degrade MC-LR, the release of microcystin through supplement driven desorption was observed.

 Recent Progress in Slow Sand and Alternative Biofiltration Processes edited by Rolf Gimbel, Nigel J.D. Graham and M. Robin Collins. ISBN: 9781843391203. Published by IWA Publishing, London, UK.

Keywords Biologically active activated-carbon (BAC), microcystin, cyanotoxin removal

Introduction

The occurrence of cyanobacterial blooms in surface water sources for human consumption is a worldwide problem, as they emerge in many climate sets and often comprise cell masses of toxic cyanotoxin producing species. Among these, *Microcystis aeruginosa* is common and brings concerns due to the very high toxicity of its cyanotoxins (Carmichael, 1988), particularly the hepatotoxic microcystin-LR (MC-LR), the most common of the microcystin variants (Svrcek and Smith, 2004). This hepatotoxic heptapeptide, which can be also produced by members of the genera *Anabaena*, *Planktothrix*, *Nostoc*, *Anabaenopsis*, *Gomphosphaeria*, *Coelosphaerium* and *Hapalosiphon* (Sivonen and Jones, 1999; Hitzfeld *et al.*, 2000), exhibits a significant stability, so it can persist active in non-chlorinated water for long periods (Watanabe *et al.*, 1992).

Ozonation and chlorination degrade cyanotoxins to some extent (Rositano, 2001; Nicholson, 1994). However, conventional drinking water treatment processes (*e.g.*, coagulation/flocculation/sedimentation, rapid sand-filtration) are efficient in the removal of cyanobacterial cells, but do not remove cyanotoxins satisfactorily and may disrupt cyanobacterial cells with the consequent release of intracellular cyanotoxins to the water (Newcombe and Nicholson, 2004).

Currently available technologies for the efficient removal of cyanotoxins from drinking water mainly rely on membrane filtration or activated carbon adsorption processes (Teixeira and Rosa, 2005; Svrcek and Smith, 2004). Sporadically, the addition of powdered activated carbon (PAC) to the coagulation tanks is effectively used for such purpose, provided that the correct type and doses of PAC are applied (Newcombe and Nicholson, 2004). However, in addition to not being cost-effective, PAC utilization does not provide a permanent barrier to cyanotoxin intrusion into the distribution networks. On the other hand, granular activated carbon filters (GAC) efficiently remove toxins from water (Lambert *et al.*, 1996; Falconer *et al.*, 1989). However, GAC's lifetime is limited by their adsorption capacity, so periodical regenerations of the filters' bed needs to be carried out (Newcombe and Nicholson, 2004). Nonetheless, during operation GACs' bed invariably becomes colonised by microorganisms which, as for other organic micropollutants, may degrade cyanotoxins, thus assisting adsorption in their removal. In fact, the capability of bacteria from diverse ecosystems to degrade MC-LR has been described (Jones *et al.*, 1994; Rapala *et al.*, 1994; Bourne *et al.*, 1996; Cousins *et al.*, 1998, Saitou *et al.*, 2002). In addition, evidence for the enhancement of microcystin removal by GAC due to the establishment of biological activity has been produced (Chorus and Bartram, 1999; Newcombe and Nicholson, 2004). Hence, synergistically to adsorption, biologically active GAC filters (BAC) may advantageously take profit of biodegradation and, thus, extend GAC's lifetime.

In this paper, studies on the removal of MC-LR by bench-scale BAC filters are reported. In addition to biodegradation, the adsorption and desorption contributions to cyanotoxin removal were assessed.

Methods

Filters set-up

The tested filters were borosilicate glass cylindrical columns with an internal diameter of 1.2 cm and a bed depth of 3.0 cm. The bed consisted of NORIT 0.8 Supra activated carbon, which was crushed and sieved to a granular size of 0.35–0.50 mm. The NORIT 0.8 Supra has a relatively good suitability for microcystin adsorption (Falconer *et al.*, 1989).

Of the four filters tested, three (BAC) were preloaded with water natural-organic-matter (NOM), while being colonised under identical conditions. For this purpose, the filters were run with laboratory tap-water with a total organic carbon content of ca. 2 mg L^{-1}, during 6 months at empty bed contact times (EBCT) of 8 min., till ca. 100 000 bed-volumes (BV) were accomplished. One filter (GAC) did not undergo preloading and its colonisation was prevented afterwards by adding sodium azide (6m M) to its influent. Likewise, the biological activity of one BAC filter (I-BAC) was inhibited throughout the MC-LR removal experiments. The BAC filters' biological activity was assessed by determining their O_2 uptake rates against GAC and I-BAC.

Following BAC preloading and colonisation, the study of MC-LR removal was done with tap water (dechlorinated by heating at 121°C for 40 min) amended with *Microcystis aeruginosa* (PCC 7813) cell extracts with predetermined MC-LR concentrations and a pH adjusted to 8.0 ± 0.1.

Then, four filters were assayed for MC-LR removal. These comprised two biologically active BAC filters (BAC-1, BAC-2), one control BAC filter with inhibited biological activity (I-BAC) and an initially unloaded control-GAC filter, which colonisation was further prevented with sodium azide throughout the whole experimental period.

Microcystis aeruginosa cell extracts

The cell extracts were obtained from *M. aeruginosa* (PCC 7813) mass cultures (4 L) in modified BG-11 media (Stanier *et al.*, 1971). The cultures were grown for 3–4 weeks at 21 ± 2°C, under a light regime of 12 hours light / 12 hours dark, while being sparged with sterile air. Cells were harvested by centrifugation (10 min., 5000× g, 20°C) or filtration through on microfiber glass filters (GF/C or GF/F, Whatman). To obtain *M. aeruginosa* extracts, cells were frozen, thawed and MC-LR was extracted overnight with 75% (v/v) methanol. Both the harvested cells and the methanolic extracts were stored at −18°C, the later for less than 5 weeks. Prior to use, the MC-LR concentration in the methanolic cell extracts was determined by HPLC analysis, the methanol was then evaporated in a rotavapor (R-200, Büchi) and the dried residue dissolved in the filters fed water.

Microcystin-LR analysis

Microcystin-LR sample preparation and quantification was based on methods described elsewhere (Meriluoto and Codd, 2005; Spoof, 2004). The HPLC system consisted of a low pressure gradient pump (P680A, Dionex), a reversed-phase C18 column, 3 μm particle size, 50 mm × 4.6 mm i.d. (Acclaim 120, Dionex) coupled to a photodiode array detector (PDA-100, Dionex). The column was protected by a C18 guard column, 5μm particle size, 10 mm × 4.3 mm i.d. (Acclaim 120, Dionex).

A linear gradient of HPLC grade trifluoroacetic acid and acetonitril, both at 0.05% (v/v), in Milli-Q water, was utilised as mobile phase. The flow rate, column temperature, and injection volume were 1 mL/min, 40°C and 10μL, respectively. The identification of MC-LR peaks was done with basis on standards retention time and UV spectrum. The standards were pure MC-LR (kindly supplied by Abo Akademi, Finland).

Prior to analysis, MC-LR in the filters influent and effluent samples (ca. 50–100 mL) was concentrated by solid phase extraction on C18 silica cartridges (1 g mass sorbent, 6 mL reservoir; Isolute), and eluted with methanol 75% (v/v). The eluate was evaporated at 50°C with air sparging, the obtained residue was re-dissolved in 0.5 or 1 mL of methanol 75% (v/v) (LiChrosolv, Merck) and analysed.

Filters operation

All the filters were identically fed with the same multi-channel peristaltic pump (Ismatec, BV-GES) for 325 days, in a down-flow mode, at 20 ± 2°C and at an empty bed contact time (ECBT) of ca. 11 min. In order to hasten the cyanotoxin breakthrough, to assess the influence of the cell extract concentration or the availability of microbial substrates and nutrients on MC-LR removal, the composition of the feeding water was changed within five sequential periods. Accordingly, during the first 10 days of operation the filters were fed with water amended with cell extracts producing influent MC-LR concentrations of ca. 50 μg L^{-1}, afterwards it ranged from 100 μg $L^{-1}$to 260 μg L^{-1} till 149 days of operation. From the 150th to the 302nd day the cell extract amendment was lowered to levels corresponding to MC-LR concentrations of 40–60 μg L^{-1} and no cell extract was added after this period. Except for BAC-2, during the last 42 days of experiments the filters were run with sterile water supplemented with sodium acetate (0.5 mM, Merck), glucose (0.5 mM, Merck), yeast extract (100 mg L^{-1}, Merck) and peptone (100 mg L^{-1}, Trypticase-BBL).

Results and discussion

Filters start-up and microcystin biodegradation

Microcystin-LR breakthrough was detected for the three BAC (including the I-BAC) by the 18th day of testing, following a raise of the added cell extract concentration (Fig. 1). In the GAC's effluent the microcystin was detectable on the 21st day, but at lower concentration than observed at the other filters. Afterwards, the MC-LR effluent concentration increased over time at a significantly slower pace than in BACs' effluents, what may be ascribed to the "virgin" state of the GAC's carbon.

After MC-LR breakthrough occurred in BAC and I-BAC, while the influent cyanotoxin concentration was the highest assayed (200–260 μgL^{-1}), the cyanotoxin concentration in I-BAC's effluent was similar or lower than in the BACs' ones. Hence, BAC MC-LR biodegradation was not apparent till the 50th day of experiments. Then and over 30 days, upon a decrease in the fed cell extract concentration to levels corresponding to 70–160 μgL^{-1} of

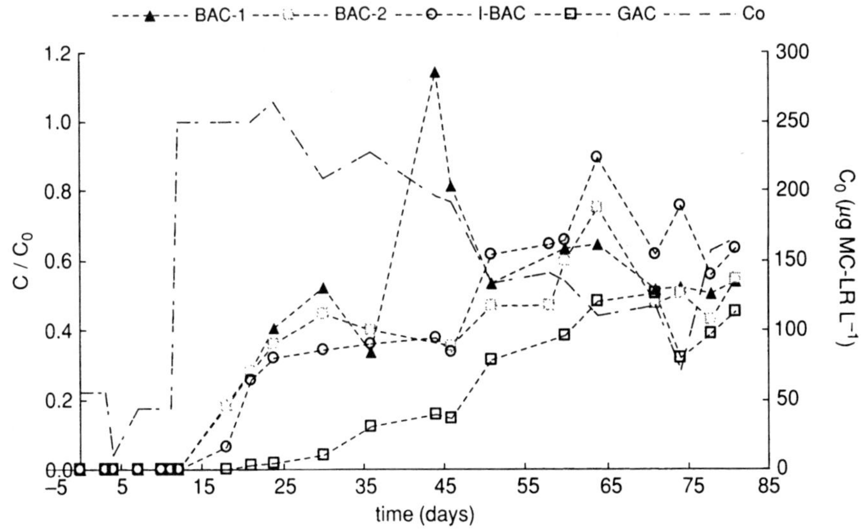

Figure 1 Relative microcystin-LR output concentration ratios during the first 80 days of filters' operation.

MC-LR, BAC cyanotoxin output concentrations were significantly lower than the I-BAC one (fig. 1). Within this period, both for BAC and I-BAC, the MC-LR effluent-influent concentration ratios (C/Co) trended to steadiness with a slight decrease and were significantly lower for BAC-1 (17%) and BAC-2 (21%) than for I-BAC. Correspondingly, BAC O2 uptake rates ($23 \pm 5\,\mu g\,h^{-1}\,cm^{-3}$) were ca. 15 times higher than the background one measured for I-BAC ($3 \pm 2\,\mu g\,h^{-1}\,cm^{-3}$), which were comparable to those obtained for GAC ($5 \pm 1\,\mu g\,h^{-1}\,cm^{-3}$). Hence, biodegradation contributed significantly (ca. 19%) for the overall MC-LR removal by BAC.

Over the 30 days period under consideration and for the conditions of the experiment (EBCT = 11 min.), MC-LR biodegradation rates were important ($0.078 \pm 0.016\mu g\,h^{-1}cm^{-3}$) and likely to exceed those which are generally required to treat cyanotoxin contaminated source-water. However, while these results are promising, the used experimental conditions are far from simulating industrial BAC systems, where the microbiota may behave differently due to the existing conditions (*e.g.*, water composition, colonisation intensity, microbial activity).

The capability of BAC's microbiota to degrade MC-LR was further assessed in batch cultures in medium with a composition identical to BAC's influent (cell extract at a concentration correspondent to ca. $85\,\mu g\,L^{-1}$ of MC-LR). The cultures (500 mL) were inoculated with 50 mL of BAC-1 effluent and incubated at 23°C with orbital shaking (100 R.P.M.). As depicted in fig. 2, a relatively fast decreasing in cyanotoxin concentration, which was concomitant to microbial growth, was observed in the cultures during the first 11 days of incubation. Then, after the consumption of ca. 20% of the initially available MC-LR, such decreasing was sharply attenuated and stopped.

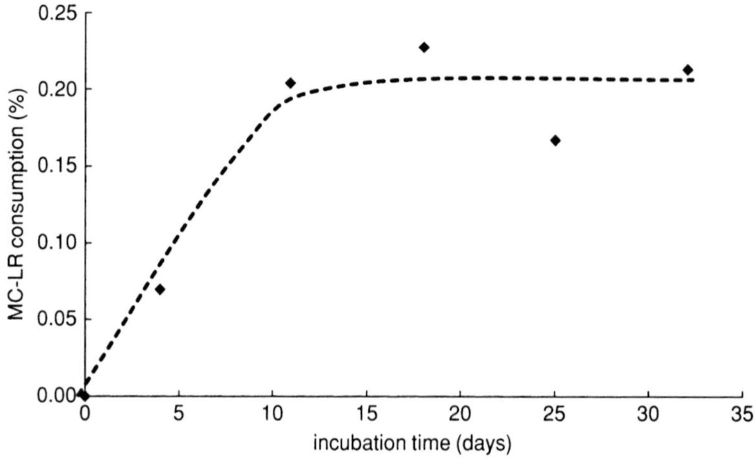

Figure 2 MC-LR biodegradation in batch cultures.

As it is unlikely that the observed microbial growth, which was coupled to MC-LR degradation, had been chiefly supported by the available cyanotoxin amount, the possibility of a co-metabolic MC-LR removal was hypothesised. Then, observed MC-LR biodegradation would have been done by microorganisms primarily growing on other cell extract constituents, whose depletion disallowed both microbial growth and cyanotoxin degradation.

Influence of cell extract concentrations on cyanotoxin biodegradation

To investigate a possible influence of the availability of *M. aeruginosa* cell constituents on MC-LR biodegradation in the BAC filters, from the 86th to the 144th day of filters running, the cell extract in the influent was raised to levels equivalent to MC-LR concentrations of 170–260 μgL^{-1}. Accordingly, cyanotoxin biodegradation by BAC's microorganisms would be promoted. However, during this period influent cell extract concentrations were not kept steady and I-BAC effluent MC-LR concentrations resulted mostly comparable or slightly lower than those from BAC (Fig. 3). Hence, as biodegradation by BAC filters was not perceptible under these conditions, on the 151th day of experiments the resuming of MC-LR biodegradation by BACs was attempted by lowering gradually the influent cell extract concentration to levels correspondent to 40–60 μg MC-LR L^{-1}.

Apparently, such conditions did not allow any consistent and significant recover of the BAC's cyanotoxin-biodegrading capabilities over a period of at least 64 days. Admissibly, MC-LR desorption may have overcome biodegradation, which therefore may have been undetectable.

However, on the 275th day of filters running and during the following 8 days I-BAC cyanotoxin outputs consistently exceeded the BAC's ones significantly (21% to 42%), what in average corresponded again to the biodegradation of 19% of the BAC's MC-LR input.

Nevertheless, inexplicably GAC effluent MC-LR concentrations did not follow the I-BAC ones. Moreover, unexpectedly a MC-LR complete-breakthrough ($C/Co = 1$) was

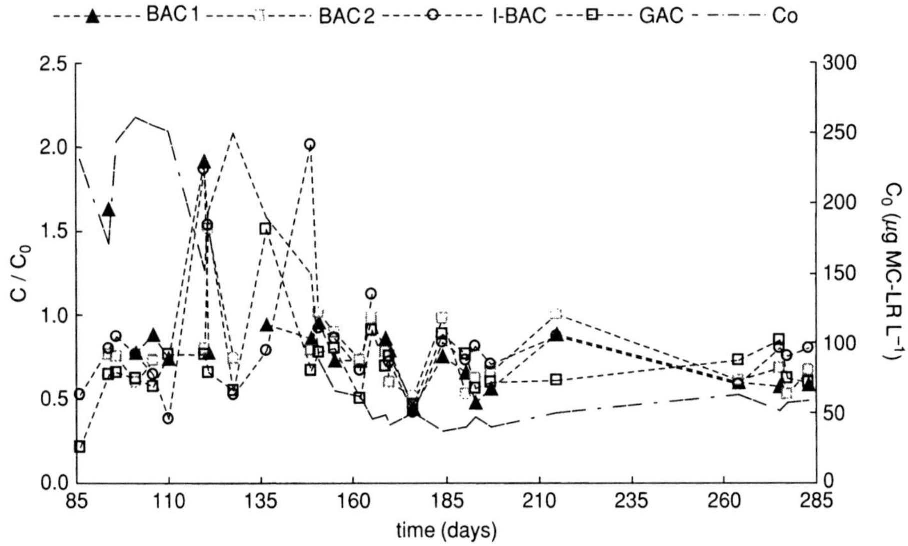

Figure 3 The influence of the of *Microcystis aeruginosa* cell extract concentration on microcystin removal.

neither observed for GAC nor for I-BAC. Even if the measured BAC O_2 consumption rates (ca. 25 μg h^{-1} cm^{-3}) were consistently greater than the observed background (ca. 1 μg h^{-1} cm^{-3}) in GAC and I-BAC filters, MC-LR anaerobic-biodegradation in the azide inhibited filters can not definitely be ruled out – acquired microbial-resistance to azide was observed in previous bench scale studies on MC-LR adsorption on NORIT 0.8 Supra columns, as O_2 consumption tests had shown.

Effects of amending the filters influent with microbial substrates and nutrients

As a further attempt to hypothetically enhance MC-LR biodegradation at BAC filters, the effect of supplementing the influents with compounds (glucose, acetate) and media constituents (yeast extract, peptone) which are generally able to support microorganisms' growth was studied. So, from the 284th to the 302nd day of filters running, the BAC-1, I-BAC and GAC influent was supplemented. The BAC-2 influent was only supplemented for three days (from the 284th to the 287th day).

Then, cyanotoxin concentrations in all filters effluent exhibited a sharp and significant increase that was concomitant to the beginning of supplement addition, which is mostly ascribable to MC-LR desorption (C/C0 > 1). Afterwards, I-BAC MC-LR output concentration did not greatly decline, whereas the opposite was observed for the other filters. While BAC-2 C/C0 values dropped and approached those observed prior to the supplements addition (fig. 4) as soon as the influent supplementation was discontinued, the C/Co decrease was slower for the supplemented BAC-1. In addition, contrarily to I-BAC behaviour, GAC C/Co values were lower and, after seven days, declined to the levels detected prior to supplementation.

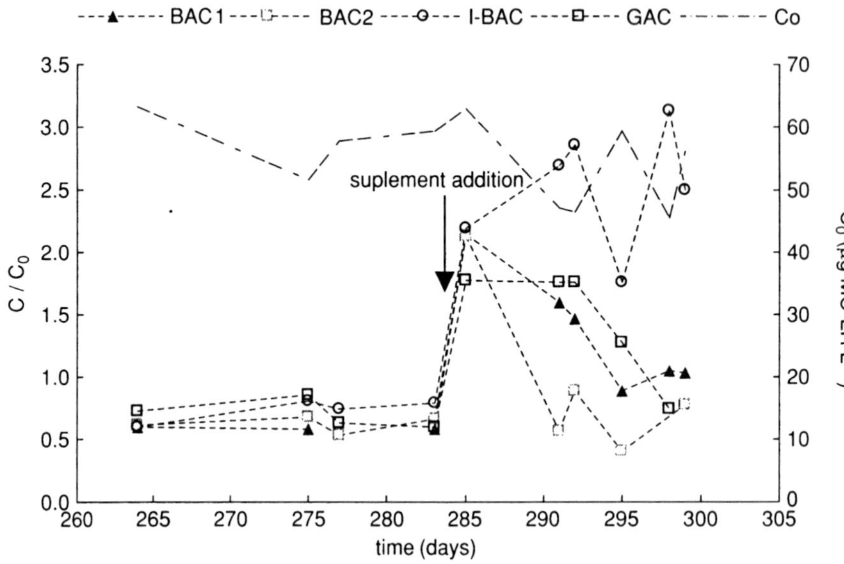

Figure 4 The influence of supplements amendment on microcystin-LR removal.

Assuming that the adsorptive properties of both filters were kept close to similar, the observed differences between I-BAC and BAC-1 MC-LR output concentrations might have approached MC-LR biodegradation in the non-inhibited filter. Even if such interpretation is limited by the impossibility of approaching the adsorption / desorption contributions to MC-LR net-outputs, the obtained results suggest that cyanotoxin biodegradation in BAC-1 was greatly promoted by the presence of growth substrates and nutrients. Therefore, it is likely that in such type of systems the cyanotoxin removal strongly depends on the water composition, namely its contents in organic compounds able to support microbial growth.

Microcystin-LR desorption

Mycrocistin-LR desorption was further appraised by discontinuing *M. aeruginosa* cell extract addition to all filters influent from the 310th day to the end of the experiment (325th day), while maintaining supplementation, except for BAC-2.

In samples taken three days after stopping cell extract addition, the I-BAC effluent MC-LR concentration was ca. 4 times and ca. 10 times higher than that found in the BAC-1 and BAC-2 effluents, respectively (fig. 5). Presumably, the supplements previously adsorbed on BAC-1's bed sustained MC-LR biodegradation. Again, the MC-LR output from I-BAC was not followed by cyanotoxin desorption from GAC. Assuming that till supplements' addition began (284th day, 36085 BV) GAC loading was qualitatively and quantitatively similar to that of I-BAC and estimating the overall amount of MC-LR released from these two filters, a superior capacity of I-BAC for MC-LR adsorption becomes apparent (fig. 5). In fact, over the period in which supplement's addition drove the desorption of MC-LR, the overall amount of microcystin released from BAC-1 was close to twice that released from GAC.

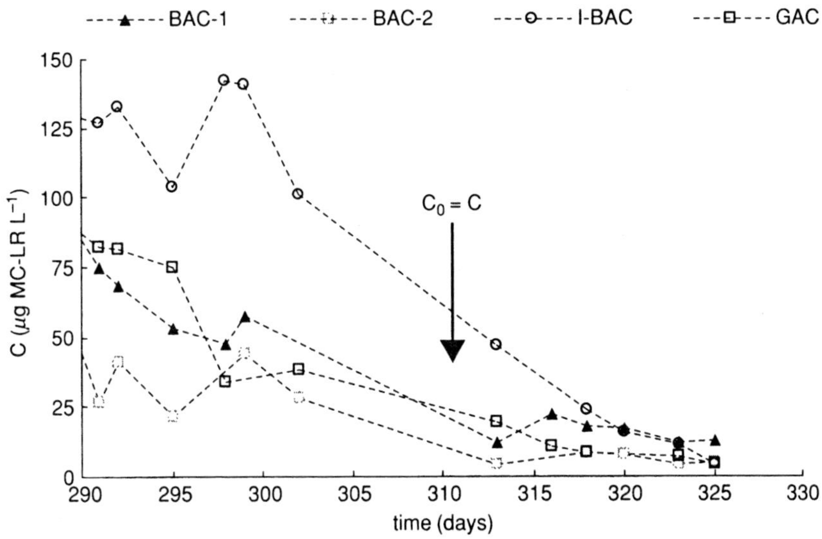

Figure 5 Microcystin-LR desorption trends.

Accordingly, a weaker desorption of MC-LR from GAC occurred, so that three days after discontinuing microcystin addition the MC-LR output from GAC was significantly lower than from I-BAC.

Afterwards MC-LR output concentrations became mostly similar for BAC-1and I-BAC, decreased slowly but persisted till the end of the experiments. The MC-LR concentration evolved likewise in the BAC-2 and GAC effluents, but in both the cyanotoxin was at a lower strength than in BAC-1and I-BAC effluents. The observed differences between BAC-1 and BAC-2 effluent MC-LR concentrations are likely to be due to the absence of supplement driven desorption in BAC-2.

Cyanotoxin desorption from all filters was observed over the last fifteen days of assay and would continue for an indeterminate time (fig. 5). Over this period the estimated overall-release of MC-LR was higher from the I-BAC (ca. 710 μg) than from the other filters. In addition, more MC-LR was released from the BAC-1 (ca. 573 μg) than from the non-supplemented BAC-2 (321 μg), what may corroborate that desorption was chiefly driven by the presence of supplements and that biodegradation had a minor role, if any, during this period. Comparatively, MC-LR release from GAC (ca. 278 μg) was less important, what may confirm the lower cyanotoxin adsorption capacity of the non-colonised carbon.

Due to the magnitude of the cyanotoxin concentrations utilised and to the presence of high levels of supplements in the filters influent, the experimental conditions were far from comparable to the ones occurring in real water treatment settings. However, the fact that desorption from the non-supplemented BAC-2 was significant for fifteen days after discontinuing the MC-LR input brings concerns of cyanotoxin release to drinking water following cyanobacterial bloom events. Unless microcystin biodegradation in treatment-plant colonised-GAC systems occurs faster than the cyanotoxin loading, such release may occur.

Conclusions

The investigation of organic micropollutant removal by biological filters is complex. This particularly applies to studies with filters having an activated carbon bed, as in such systems adsorption/desorption and biodegradation may occur simultaneously and are difficult to differentiate and follow. Moreover, the highly limited availability of pure MC-LR and the complexity of its analysis added difficulties to these studies. Even so, the obtained results provide insights into the physical-chemical and biological processes underlying BAC microcystin removal and support the following conclusions:

- the microbial capability for MC-LR biodegradation is likely to be widespread, as it occurs among a drinking water distribution system microorganisms;
- MC-LR biodegradation occurs in biologically activated carbon-filters and, hence, may possibly lessen BAC microcystin-loading during bloom events, therefore mitigating the potential for a later cyanotoxin release to the water through desorption;
- however, MC-LR biodegradation in BAC is probably co-metabolic and depends on the availability of other compounds that may serve as chief substrates for the filter microcystin degrading microorganisms;
- accordingly BAC's performance for microcystin biodegradation rely on the water "assimilable organic carbon" (AOC) content;
- at least for the carbon used in these studies, microbial colonization enhances the BAC's bed capacity for MC-LR adsorption;
- as expected, MC-LR desorption from BAC's bed is intensified by the presence of other soluble organic-compounds in the water, therefore, in addition to biodegradation, the water "dissolved organic carbon" concentration and composition affect the net release of the adsorbed microcystin;
- such release, which may persist while no cyanotoxin input occur, namely after the end of a bloom event, deserves attention owing to the associated public health hazard.

Nonetheless, further in-depth studies need to be carried out, desirably at a larger scale, in order to achieve the understanding of BAC's cyanotoxin removal mechanisms, which is still required for a full and secure exploitation of such systems in the treatment of cyanotoxin contaminated waters. A better knowledge of the microbial processes involved, including their controlling parameters, is of paramount importance for this purpose.

This work was supported by EU Project TOXIC - Barriers Against Cyanotoxins in Drinking Water - Contract N.° EVK1 – CT2002-00107.

The authors are grateful to Salmon & Cia (representating NORIT in Portugal) for supplying the activated carbon used in the experiments and to Vitor Napier for technical support.

References

Bourne, D.G., Jones, G.J., Blakeley, R.L., Jones, A., Negri, A.P., and Riddles, P. (1996) Enzymatic pathway for the bacterial degradation of the cyanobaterial cyclic peptide toxin microcystin-LR. *Appl. Environ. Microbiol.* **62**, 4086–4094.

Carmichael, W.W., Beasley, V.R., 125Bunner, D.L., Eloff, J.N., Falconer, L.R., Gorham, P., Harada, K.-I., Krishnamurthy, T., Yu, M.J., Moore, R.E., Rinehart, K., Runnegar, M., Skulberg, O.M., and Watanabe, M. (1988) Naming of cyclic heptapeptide toxins of cyanobacteria (blue-green algae). *Toxicon.* **26**, 971–973.

Chorus, I., Bartram, J. (Ed.) (1999) *Toxic Cyanobacteria in Water: a guide to their public health consequences, monitoring and management.* E & FN Spon, London.

Cousins, I.T., Bealing, D.J., James, H.A. and Sutton, A. (1998) Biodegradation of microcystin-LR by indigenous mixed bacterial populations. *Water Res.* **30,** 481–485.

Falconer, I., Runnegar, M., Buckley, T., Huyn, V. and Bradshaw, P. (1989) Using activated carbon to remove toxicity from drinking water containing cyanobacterial blooms. *Journal of AWWA* **81**, 102–105.

Hitzfeld, B.C., Hoeger, S.J. and Dietrich, D.R. (2000) Cyanobacterial toxins: removal during drinking water treatment, and human risk assessment. *Environ. Health Perspect.* **108**, 113–122.

Jones, G.J., Bourne, D.G., Blakeley, R.L. and Doelle, H. (1994) Degradation of the cyanobacterial hepatotoxin microcystin by aquatic bacteria. *Nat. Toxins* **2**, 228–235.

Lambert T.W., Holmes, C.F.B. and Hrudey, S.E. (1996) Adsorption of microcystin-LR by activated carbon and removal in full scale water treatment. *Water Res.* **30**, 1411–1422.

Newcombe, G. and Nicholson, B. (2004) Water treatment options for dissolved cyanotoxins. *J. Water Supply: Res. Technol. – Aqua* **53.4**, 227–239.

Nicholson, B.C., Rositano, J. and Burch, M.D. (1994) Destruction of cyanobacterial peptide hepatotoxins by chlorine and chloramine. *Water Res.* **28**, 1297–1303.

Rapala, J., Lahti, K., Sivonen, K. and Niemelä, S.I. (1994) Biodegradability and adsorptionon lake sediments of cyanobacterial hepatotoxins and anatoxin-a. *Lett. Appl. Microbiol.* **19**, 423–428.

Rositano, J., Newcombe, G., Nicholson, B. and Sztajnbok, P. (2001) Ozonation of NOM and algal toxins in four treatedwaters. *Water Res.* **35**, 23–32.

Saitou,T., Sugiura, N., Itayama, T., Inamori, Y. and Matsumura, M. (2002) Degradation of microcystin by biofilm in practical treatment facility *Wat. Sci. Technol.* **46** (11–12), 237–244.

Sivonen, K. and Jones, G. (1999) Cyanobacterial toxins. In *Toxic cyanobacteria in water: a guide to their public health consequences, monitoring, and management.* (Ed. I. Chorus and J. Bartram)E & FN Spon, London

Spoof, L. (2004) High-performance liquid chromatography of microcystins and nodularins, cyanobacterial peptide toxins. Ph thesis, Dept. Biochem. Pharm. Abo Akademi University, Turku-Abo, Finland.

Stanier, R. Y., Kunisawa, R., Mandel, M., Cohen-Bazire, G. (1971) Purification and properties of unicellular blue-green algae (order *Chronocalles*). *Bacteriol. Rev.* **35**, 171–205.

Svrcek, C. and Smith, D.W. (2004) Cyanobacteria toxins and the current state of knowledge on water treatment options: a review. *J. Environ. Eng. Sci.* **3**, 155–185.

Teixeira, M.R., and Rosa, M.J. (2005) Microcystins removal by nanofiltration membranes. *Sep. Purif. Technol.* **46**, 192–201.

Watanabe, M.F., Tsuji, K., Watanabe Y., Harada,K.-I. and Suzuki, M. (1992) Release of heptapeptide toxin (microcystin) during the decomposition process of Microcystis aeruginosa. *Nat.Toxins* **1**, 48–53.

WHO (1998) Cyanobacterial toxins: microcystin-LR. In *Guidelines for drinking-water Quality.* Addendum to vol. 2, 95–110, WHO, Geneva.

46 Removal of organic pollutants from micro-polluted source water by O_3-BAC process

Z.-Y. Wang, H.-J. Han, W.-C. Ma, M. Xue

(School of Municipal and Environment Engineering, Harbin Institute of Technology, Harbin, 150090, China) imy115@163.com

Abstract In this paper the character of the Micro-polluted Source Water and the mechanism of the ozone-biological activated carbon (BAC) were introduced. The condition parameters of O_3-BAC process were confirmed by the test, such as ozone dosage, reaction time and empty bed contact time of BAC (EBCT). Besides some routine tests, water quality was analyzed by Ames test and GC/MS test. The results indicated that the raw water from the river showed positive for TA98 strain and for TA100 strain while the effluent showed negative for TA98 strain but positive for TA100 strain. It was showed that 119 kinds of organic pollutants have been detected in the raw water and 6 kinds in the effluent. Therefore, the O3-BAC process can be an effective treatment for micro- organic pollutants and can offer a safe drinking water with high quality. Meanwhile, the influence of temperature was investigated. The test showed that the bio-activity of BAC was reduced with the decline of temperature, and biomembrane forming was influenced badly by the low temperature.

Keywords O_3-BAC process, GC/MS test, Ames test, low temperature.

Introduction

The drinking water source in China is mostly contaminated by organic contaminant and turns out to be micro-pollutants water source, it can not be removed effectively by using conventional water treatment process. To solve this problem, many techniques have been studied to improve the situation. Among those techniques, the ozone-biological activated carbon (O_3-BAC) process is distinguished by its high effect removal of dissolved organic pollutants and mutagenic materials, and its safety and high quality. The O_3-BAC process is the unite of physical chemistry adsorption of activated carbon, ozone chemistry oxidation, biology oxidize degradation and ozone disinfection.

Experimental device and the condition parameters

Experimental device

This experiment used the water of songhua river as inlet. The experimental device was showed as figure 1.

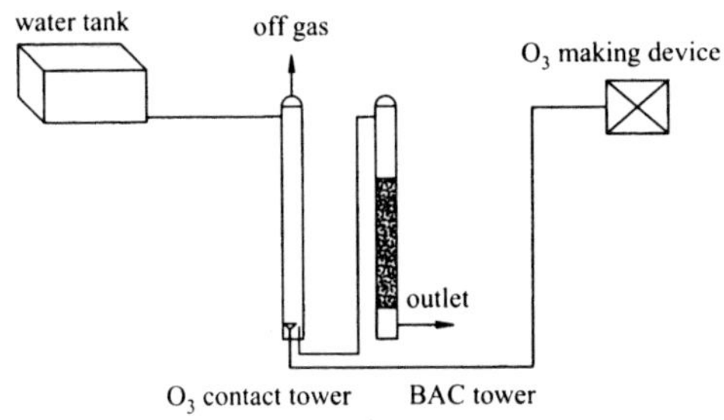

Figure 1 Flowsheet of the O3-BAC plant used in this study

The figure 1.1 showed that this device mainly contained the ozone contact tower (efficiency highness was 1.8m, diameter was 8cm, volume was 9.0L), ozone generation device and BAC contact tower (efficiency highness was 1.4m, diameter was 12cm, volume was 15.8L). Chose the granule activated carbon (grain diameter was 1mm, length was 2~3mm). The ozone generated by the ozone generation device entered the bottom of ozone contact tower, the gas and water were interflowed in order to fully mix and react, off-gas directly entered the environment via the pipe in the top. Outlet entered the top of BAC tower from the bottom of O_3 contact tower. Using gravitation flux mode, from top to bottom, the water was filtrated and adsorbed.

The ozone adding quantity and contact time

This experiment respectively chose 2, 3, 4, 5 mg/L ozone dosage, and 10, 20, 30 min reacting time to do the orthogonal test, and the results were showed as table 1.

On the condition of fixed reaction time, the COD removal increased as the ozone dosage increased. But there were few differences between ozone dosage of 4 mg/L concentration and 5 mg/L concentration. On the condition of fixed ozone dosage, the COD removal increased as the reaction time increased. But there were few differences between reaction time of 20 min and 30 min. Because of the influence of the ozone dosage and the reaction time on the cost of engineering, ozone dosage of 4 mg/L concentration and reaction time of 20 min were chosen in this study. On these parameters, the high molecular weight materials in the raw water were decomposed to the organic pollutants which could be adsorbed and

Table 1 The results of ozone dosage and reaction time

Ozone Dosage (mg/L)	**2**			**3**			**4**			**5**		
Reaction Time (min)	10	20	30	10	20	30	10	20	30	10	20	30
COD Removal of O_3 contact tower (%)	12	15	17	28	34	37	36	41	43	39	45	47
COD Removal of the system (%)	29	32	36	48	53	56	64	70	72	66	70	73

biodegraded by the BAC tower. Therefore, the adsorbability and the biodegradability of the BAC tower and the COD removal of the system were increased.

The empty bed contact time of BAC (EBCT)

The empty bed contact time of BAC (EBCT) was 30min at the start of the test. It was changed to 20min when the equipment ran for one month and 10min for two month. In these days, the removal efficiency did not obviously change as the EBCT changed.

The results and discussions

The results of routine tests

On the condition of the parameters above, the system ran for three month. The removal rate of COD_{Mn} could reach 70%. The removal rate of residual chlorine could reach 99%. Meanwhile, the removal rate of UV_{254} could reach 95%. The results indicated that the O3-BAC process could remove the organic pollutants which were toxic and deleterious in the raw water effectively.

The results of GC/MS test

The water quality of the raw water and the outlet from the BAC tower were analyzed by GC/MS test. The results of GC/MS test were shown in Table 2.

119 kinds of organic pollutants have been detected in Songhua River. From the classification of pollutant by USEPA, there were five kinds of those pollutants which did significant harm to human. It indicated that these pollutants could not be removed by conventional treatment process. By the GC/MS test, 119 kinds of organic pollutants which were non-toxic have been detected in the effluent. It indicated that because of the oxidation of ozone and the biodegradation of BAC, the O_3-BAC process had a great effect on the removal of unbiodegradable pollutants.

The results of Ames test

The results of Ames test were shown in Figure 2. As shown in Figure 2, the raw water showed positive for TA98 strain and for TA100 strain. The effluent of the O_3-BAC process showed negative for TA98. It indicated that the O_3-BAC process could make the mutagenic activity in the raw water decrease. But the effluent of the O_3-BAC process showed little positive for TA100 which had been introduced in some reports. The reason may be that

Table 2 The results of GC/MS test

Results	Kind of organic compounds	Pollutant which should be gotten rid of by USEPA
Raw water	119	5
outlet	6	0

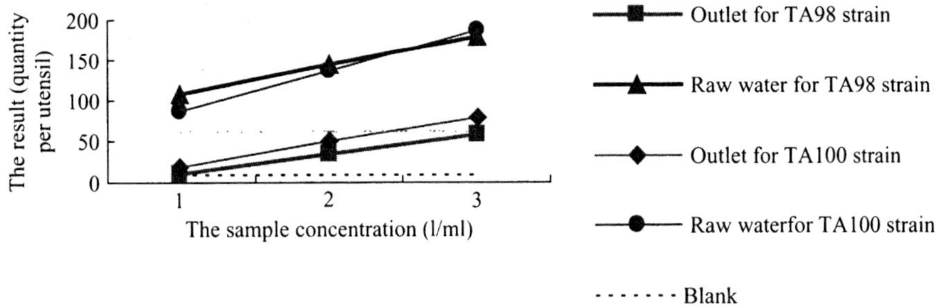

Figure 2 The results of Ames test

ozone oxidized the pollutants to the mutagenic materials which were not adsorbed by BAC, which made the mutagenic activity increased. But the quantity of these mutagenic materials was too little. Taking one with another, the O_3-BAC process could decrease the mutagenic materials in the raw water effectively.

The discussions at the low temperature

The test did an elementary study on the low temperature's influence on the removal efficiency of the O_3-BAC process. This study used the technology of high speed raising microbiofilm. Biomembrane forming beginning time was 5 days and about 10 days biomembrane forming would be completed at the normal temperature (20°∼25°). But at the low temperature, the speed for biomembrane forming was very slow. Biomembrane was very thin after about 20 days and after about one and a half month biomembrane forming would be completed, which indicated that temperature had great influence of the biomembrane forming speed.

In addition, the influence of the low temperature on the microbial activity was studied. The temperature for most bacteria growing was 20°∼40°. At the low temperature the activity of mesophilic bacteria was restrained badly, so the removal efficiency of the system was low. SOUR was used as the indicator to investigate the microbial activity. The results indicated that the microbial activity at 10° was 50% less than at 20°. When the temperature was 4°, SOUR was close to zero which indicated that the microbial activity was very low. How to improve the removal efficiency under low temperature needs further study.

Conclusions

Ozone dosage of 4 mg/L concentration and reaction time of 20 min were chosen in the test. The empty bed contact time of BAC (EBCT) was 30 min at the start of the test. It was changed to 20 min when the equipment ran for one month and 10 min for two month. On the condition, the removal rate of COD_{Mn} could reach 70% and the effluent could reach the drinking water standards.

119 kinds of organic pollutants have detected in the raw water and 6 kinds were tested in the effluent by GC/MS which indicated that the O_3-BAC process had a great efficiency

on the removal of unbiodegradable pollutants. And the results of Ames test indicated that the O_3-BAC process had a great effect on the removal of the mutagenic materials.

It was shown that one and a half month was needed on biomembrane forming at low temperature. In addition, the microbial activity badly decreased as the temperature decreased. How to improve the removal efficiency under low temperature needs further study.

Reference

Camel V., Bermond A. (1998) The use of ozone and associated oxidation processes in drinking water treatment. Wat Res. 32(11):3208–3222.

47 Iron and manganese removal by multi-stage filtration (MSF)

L.D. Sanchez*, L.M. Burbano*

*Universidad del Valle, Instituto Cinara, A.A 25157, Cali, Colombia

Abstract The presence of iron and manganese in drinking water causes serious problems for network infrastructure, valves and micro-meters, due to incrustations and the risk of biofilm forming inside pipes. Users tend to reject this water, particularly groundwater, because of its colour, the presence of suspended solids and its taste. These are considered critical factors as they are associated with aesthetic and operational problems. The aim of this study was to research iron and manganese removal by Multi-Stage Filtration (MSF) in a full scale plant, using a groundwater source. In this study the groundwater source had the following values: iron between 1.6 mg/L and 2.5 mg/L and manganese between 0.004 mg/L and 0.6 mg/L. In addition, the source showed a variation in turbidity levels between 14 and 20 NTU, in colour between 6 and 17 colour units, and in faecal coliforms between 0 and 10 CFU/100 ml. The effluent of the treatment plant recorded iron concentrations between 0.004 mg/L and 0.2 mg/L, with an average value of 0.16 mg/L. In relation to manganese, the system registered values between 0.004 mg/L and 0.1 mg/L, and an average concentration of 0.05 mg/L. The registered values for iron and manganese below the standards set by WHO guidelines in 2004 (0.3 mg/L and 0.1 mg/L, respectively). Total removal for the effluent of the treatment plant was 89% for manganese and 92% for iron. This technology is easy to operate and maintain, especially for small communities in rural areas.

Keywords Groundwater, iron, manganese, filtration rate, removal efficiency

Introduction

In 1899, for the first time, Harry Clark reported the presence of iron and manganese in groundwater, due to the percolation of water saturated with dissolved oxygen that was in contact with organic matter. It was only in 1960 that Griffin discovered that the presence of manganese was associated with bacterial activity. Groundwater contained greater concentrations of iron and manganese than surface water, due to its low pH, high concentration of CO_2 and low dissolved oxygen content. In contrast, surface water undergoes alterations in quality as it receives waste generated in population centres, industries, agricultural activities and run-off. This contamination limits the existing and potential uses of the resource, especially when the source is used in agricultural activities, which are affected by the presence of iron and manganese (Avendaño, 2002).

A lack of surface water resources of good quality in some zones, especially in the Valle del Cauca region in Colombia, has led the public and private sectors to improve their use of groundwater by building deep wells as a water supply alternative (Ascuntar, 1995). However, due to the presence of iron and manganese in ground and surface water and its effect on supply systems, these metals shall be removed. Oxidation – filtration are the most common treatment barriers (Sommerrfeld, 2000).

As groundwater contains very little or no dissolved oxygen, aeration facilitates the dragging or sweeping of volatile substances thanks to the turbulent mix, and thus the oxidation of metals and gases. Aeration provides the necessary dissolved oxygen to convert iron and manganese from its form in groundwater to its insoluble form through use of simple methods such as aerating cones, and aeration trays (Petrusevski, 2003). However, as the precipitates formed are not sufficiently heavy to be removed by simple sedimentation, filtration systems must be used as treatment barriers. Taking into account the wide use of MSF in the region, this research aimed to study this technology for iron and manganese removal in groundwater.

Methods

The study was carried out in a real-scale plant fed with a groundwater source with iron and manganese problems. The methodology used to develop the research is described below.

Experimental units

The experimental unit was made up of aeration trays (4) and a MSF plant, as described in Table 1, and a general diagram of the plant is given in Figure 1.

Sampling and parameter determination

The parameters monitored daily were iron, manganese, turbidity, colour and faecal coliforms, 100 ml plastic bottles were used for the iron and manganese parameter analysis. The bottles were washed with acids at 10% (HCl for iron and HNO3, for manganese). HNO3 and HCl at 96 and 98% respectively were used to fix the samples. Iron determination was carried out using the phenanthroline method. The manganese determination was carried out using ammonium persulphate as an oxidising to produce highly coloured permanganate ions which are measured with visual or photometric comparison (Romero, 2002).

Table 1 Experimental units

Component	Filtration rate (m/h)	Filter bed		
		Size (mm)		Total height (m)
Upflow roughing filter (URF)	0.48– 0.55	Gravel:	3.81- 0.32	1.2
Slow sand filters (SSF)	0.15	Gravel: Coarse sand: Sand:	12.7 2 - 4 0.15 – 0.35	1.05

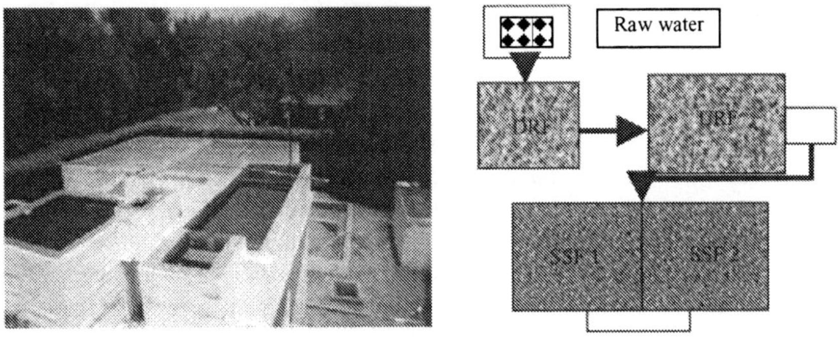

Figure 1 MSF plant for iron and manganese removal

Results and discussion

Affluent water quality

Iron and manganese behaviour were maintained between 1.6 mg/L and 2.5 mg/L and between 0.04 mg/L and 0.6 mg/L, respectively. In addition the source presented variations in turbidity between 14 and 20 NTU, colour between 17 and 6 colour units, and faecal coliforms between 0 and 10 CFU/100 ml. The concentrations exceed the limit recommended for drinking water (WHO Guidelines, 2004) and therefore, the source needs treatment to be used for human consumption and for protection of the distribution system.

Iron and manganese behaviour

Figure 2 shows the iron and manganese behaviour in the different treatment stages. The effluent from the URF had an iron concentration between 0.42 mg/L and 0.90 mg/L, with an average of 0.56 mg/L, Standard deviation (STD) 0.28. The effluent of the SSF recorded iron concentrations between 0.004 and 0.2 mg/L, with an average of 0.16 mg/L, STD 0.15 (according to WHO (2004) the reference value 0.3 mg/L). For manganese, the system was more efficient up to the URF stage, obtaining an effluent between 0.004 and 0.1 mg/L, and an average concentration of 0.05 mg/L, STD 0.05 (reference value 0.1 mg/L). In the case of oxidation with aeration and MSF, the literature reports the predominance of filtration

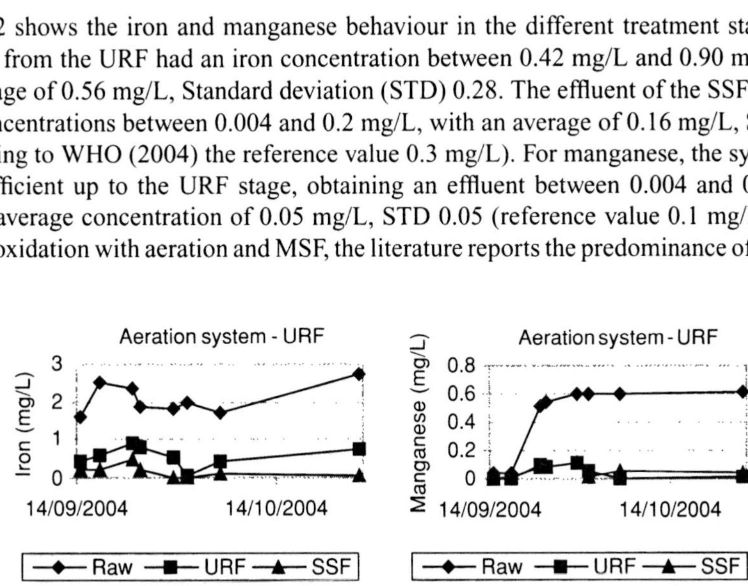

Figure 2 URF behaviour in iron and manganese removal

mechanisms and biological activity (Sanchez *et al.,* 1999). Therefore, in these filtering beds, there may also be bacteria able to contribute to oxidation of the iron and manganese present in the water and facilitating their removal through the filtration units.

Total removal for the effluent of the treatment plant (MSF) was 89% for manganese and 92% for iron. The URF removed 50% of the iron and 75% of the manganese while the SSF removed 71% of the iron and 20% of the manganese. These values were similar than those reported by Galvis et al., (1999), who showed typical reduction of 50% for URF and of between 30% and 90% for SSF, for the two parameters mentioned. However, surface water was examined and not groundwater, as in that study. This technology is very attractive for surface water with high iron and manganese content, given that it also removes other microbiological and physiochemical parameters, without the addition of chemical substances. Turbidity, colour and faecal coliforms were also removed at 97%, 58% and 100% respectively. In other words, at the end of the system, the water for consumption had turbidity between 0.2 and 0.5 NTU, colour between 2 and 10 colour units, and 0 faecal coliforms.

In addition, no obstruction problems were identified in the filtration media during the study. The duration of a filtration run for the URF was an average of 6 days and 120 for the slow sand filters, with a maximum headloss for the URF is 32 cm. In any case, hydraulic washes were done each 6 days to prevent sludge compacting in the filtering media and filter obstruction. In the SSF, the maximum headloss was of 70 cm, which equals an operation run of 120 days. The results obtained gave evidence that aeration and MSF technology has great potential for application in solving problems such as the presence of iron and manganese in groundwater, which can cause serious problems in water supply system infrastructure, particularly in pipes, valves and micrometers. In this study, good results were obtained with filtration rate between 0.48– 0.55 m/h for upflow roughing filters and 0.15 m/h for the slow filters. However, it is relevant to study the effect of maturation on the filters, because according to the studies made by Sanchez et al., (1996), as a filtering system matures, it becomes more efficient in removing contaminants.

Conclusions

The results obtained gave evidence that aeration and MSF technology has great potential for application in solving problems such as the presence of iron and manganese in groundwater, which can cause serious problems in water supply system infrastructure, particularly in pipes, valves and micrometers. Total removal for the effluent of the treatment plant (MSF) was 89% for manganese and 92% for iron; Turbidity, colour and faecal coliforms were also removed at 97%, 58% and 100% respectively.

References

Ascuntar, O. (1995) El Agua Subterránea una Alternativa de Abastecimiento Público. Cali. Colombia. 10–30.

Avendaño, N. (2002) Remoción de Hierro y Manganeso, Unidad de Vigilancia de la Salud y Ambiente. Mexico. 1–4.

Galvis, G.; Latorre, J. (1999) Filtración en Múltiples Etapas, Tecnología Innovativa para el Tratamiento de Agua, Cinara, Universidad del Valle, IRC. Cali. Colombia, 70–160.
Petrusevki, B. (2003) Groundwater Resources & Treatment, 5–20.
Romero, J. (2002) Calidad del agua, Escuela Colombiana de ingeniería, Bogotá. Colombia, 38–127.
Sanchez, L. (1996). Limpieza de la Biomembrana en un Filtro Lento en Arena y su Efecto al Inicio de la Carrera de Filtración., Universidad del Valle. Cali. Colombia, 25–40.
Sommerrfeld, E. (2000) Iron and Manganese Removal Handbook.
World Health Organization (2004) Guidelines for Drinking – Water Quality. Geneva. **1**. 3rd ed.

Part VII

ALTERNATIVE BIOFILTRATION – PROCESS DEVELOPMENTS

48 Ozonation/biofiltration for treatment of humic surface water

H. Ødegaard*, E. Melin, T. Leiknes***

*Department of Hydraulic and Environmental Engineering, Norwegian University of Science and Technology, N-7491 Trondheim, Norway
**SINTEF Water and Environment, N-7465 Trondheim, Norway

Abstract In the 90's our group started research on an alternative method for removal of humic substances from surface water based on ozonation and biofiltration in small water works. The method has now become quite popular in Norway. This paper presents an overview of the research and practical experiences gained with the method over the last 10 years. The main topics are the impact of ozonation on humic substances (colour removal), design of biofilters (choice of EBCT and biofilm carriers) and the use of separation filters for biomass removal. A novel OBM (oxidation/bio/membrane filtration) concept is also discussed.

Keywords Biofiltration; drinking water; membrane; ozonation; ozonation by-products

Introduction

In Norway, over 90% of the water treatment plants use surface water as the water source. One of the major problems of using surface water as the water source in northern climates is high content of humic substances, also referred to as natural organic matter (NOM). This results in high colour and total organic carbon (TOC) content.

The majority of the drinking water plants are small. The operator in these plants is often a part time employee with no educational background from sanitary engineering. In addition, these plants often take water from small reservoirs with large variations in water quality throughout the year making operation of the treatment process challenging. There was a need for a very simple treatment method that is easy to operate, do not require chemical supply, could be run with minimal attention and produces a minimum of sludge, and at the same time could meet the water quality standards on colour as well as giving safe pathogens control.

Ozonation and biofiltration was identified as an interesting treatment method for these small plants and in the 90's a research program was launched to test and study the process Norwegian conditions in mind. The use of activated carbon filters following ozonation acting as biofilters was already well known, especially from Germany, but our idea was to use biofilm carriers similar to those used in wastewater treatment plants and to use ozonation directly on the humic water without any coagulation or other methods for NOM removal.

 Recent Progress in Slow Sand and Alternative Biofiltration Processes edited by Rolf Gimbel, Nigel J.D. Graham and M. Robin Collins. ISBN: 9781843391203. Published by IWA Publishing, London, UK.

The research was started in the university environment mainly as student projects. The ideas behind the method and the initial research work were presented at the conference on Advances in slow sand and alternative biological filtration in London (Ødegaard, 1996). Since then several projects have been carried out in Norway and the method has become quite popular. This paper presents an overview of the research and practical experiences gained with the method over the 10 years that has passed since the method was proposed.

Impact of ozone on humic substances

Ozone is effective in colour removal. Figure 1 shows the results from batch tests carried out with different source waters. When the removal efficiency is plotted against colour specific ozone dose ($mgO_3/mgPt$), similar removal efficiencies are achieved regardless of water source and quality (Figure 1b). The following model was developed to estimate the colour removal (Ødegaard, 1996):

$$\frac{C}{C_0} = \frac{1}{1 + 20 \cdot D/C_0}$$

where C_0 is colour in the raw water (mgPt/l), C is colour after ozonation (mgPt/l) and D is the ozone dose (mgO_3/l). Figure 1b also illustrates that the practical removal efficiency for ozonation/biofiltration process is around 80%, which is achieved at doses around 0.15–0.20 $mgO_3/mgPt$. By increasing ozone dose above this, little additional removal can be expected. In our experiments, the ozonation stage is solely responsible for colour removal and no additional removal in biofilter has normally been observed (Melin and Ødegaard, 1999; Melin *et al.*, 2006).

The removal of colour is due to modification of the structure of NOM rather than complete oxidation of organic matter into CO_2. Typically, TOC removal in ozonation stage is under 10% with ozone doses required for about 80% colour removal (Melin and Ødegaard, 1999; Melin *et al.*, 2002) although with high dosages higher TOC removal has been observed in the ozonation stage. The reactions between ozone and NOM create a large number of small molecular weight compounds which are easily biodegradable. These compounds can

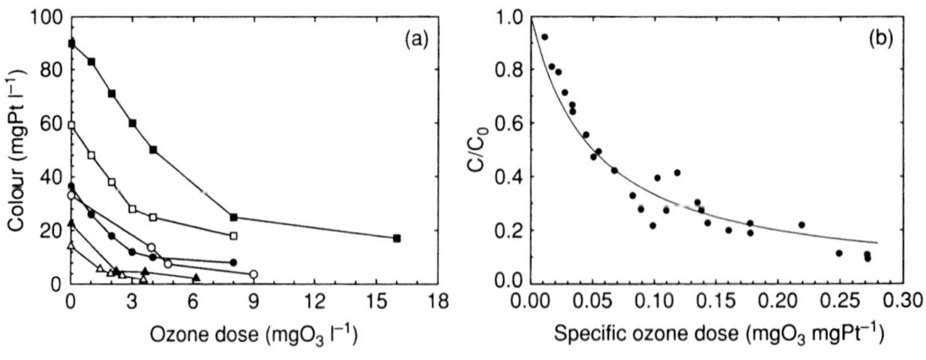

Figure 1 Colour removal in different source waters (a) and removal efficiency as a function of colour specific ozone dose (b). The solid line presents the model prediction. From Ødegaard (1996) with additional results.

cause excessive bacterial growth in the distribution net and biofiltration must be applied after ozonation. Due to removal of these compounds, the major TOC removal mechanism in the process is biodegradation in the biofilter.

One draw-back of using relatively high dosages of ozone is the formation of bromate. Norwegian surface waters have normally low alkalinity, low pH, low temperature and relatively high TOC – all factors that reduce bromate formation (Song *et al.*, 1996). No measurements carried out in either of the pilot- or full scale plants in operation have 2 10 µg BrO_3^-/l and in general they are below the very strict Norwegian standard of 5 µg BrO_3^-/l , even at relatively high bromide concentrations.

Biofilter performance

One of the design factors for biofilter is the choice of filter media. Granular activated carbon (GAC) is widely used as biofilter media and therefore our research has mostly studied possible alternatives for GAC. In the early phases, slow sand filtration was used (Ødegaard, 1996). This was replaced by Kaldnes biofilm carrier media (KMT). The KMT carrier elements are shaped like small cylinders (about 10 mm in diameter and 8 mm in height) with a cross inside the cylinder and longitudinal fins on the outside. The carrier elements are ordinarily made of polyethylene (density 0.95 g/cm^3) but the media made of PVC (density 1.45 g/cm^3) were used in most of the experiments to achieve a stationary bed. The media was originally developed for use in a moving bed biofilm reactor for wastewater treatment (Ødegaard *et al.*, 1994). One of the advantages with the media is an open structure and very low head loss. If additional removal of turbidity or bacteria is desired, an optional separation filter may be used after KMT filter (see section separation filters). Other tested biofilter media include expanded clay aggregates (marketed as Filtralite) and calcium carbonate.

Table 1 presents some results achieved with different biofilters in pilot- and full-scale plants. Although differences, for example, in water quality, type of NOM and ozone doses

Table 1 Average removal efficiencies with different biofilter configurations

Raw water TOC mg/l	Average O_3 dose ($gO_3/gTOC$)	Filter media	EBCT (min)	TOC removal (%)	Reference
5.0	1.4	KMT + sand filter	30	38	Ødegaard (1996)
4.0	1.5	KMT + sand filter	30	29	Ødegaard (1996)
5.0	1.1	Filtralite	16	23	Melin & Ødegaard (1999)
4.7	1.7	Filtralite	27	31	Melin & Ødegaard (1999)
3.6	1.7	KMT (low density)	27	38	Melin *et al.* (2000)
3.6	1.7	KMT (high density)	27	38	Melin *et al.* (2000)
3.6	1.7	$CaCO_3$	27	37	Melin *et al.* (2000)
3.6	1.7	Filtralite	27	38	Melin *et al.* (2000)
3.4	1.5	$CaCO_3$	35	35	Melin *et al.* (2006)
3.4	1.5	Anthracite/sand/$CaCO_3$	35	33	Melin *et al.* (2006)
1.0-6.4	1	KMT + sand filter*	60	21	Melin & Ødegaard (1998)
2.9-4.8	1.5	GAC/phonolith/$CaCO_3$*	28	23	Melin *et al.* (2002)
2.4-3.9	1.7	GAC*	25	23	Ongoing study

*) Full-scale plant

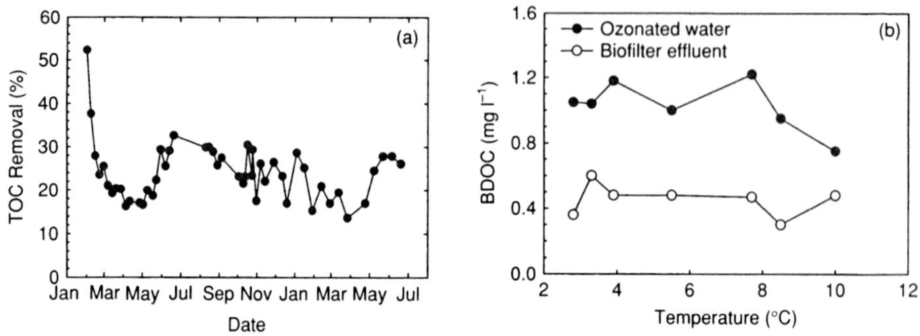

Figure 2 TOC removal in full-scale plant during a long-term testing (a) and BDOC removal at different raw water temperatures in the same plant (b) (Melin *et al.*, 2002).

in different studies make comparison of process performance difficult, it appears that none of the filter media stand out as clearly inferior to the others. It also appears that in full-scale plants lower average TOC removals have been observed. One of the reasons can be seasonal variation in TOC removal efficiency.

Figure 2a presents TOC removal in a small full-scale ozonation/biofiltration plant operated at Lake Leirsjøen in Trondheim for 1.5 years (Melin *et al.*, 2002). The biofilter was three-media filter with GAC as the first layer. Adsorption to GAC explains the high initial TOC removal. Most of the adsorption capacity was used after 1.5 months operation. The water temperature was between 2 and 14°C. Figure 2a shows a pattern where the lowest removal efficiency is observed in the end of the winter while the highest removal efficiency is achieved during summer. Although the removal efficiency increases in the beginning of the summer when temperature increases, it does not decrease as fast as temperature in fall and winter giving poor overall correlation between temperature and TOC removal. The BDOC (biodegradable dissolved organic carbon) removal efficiency did not show significant decreasing trend in samples taken at periods of different raw water temperatures (Fig. 2b), also indicating that biofilter performance was not impaired by very low water temperatures. Removal efficiency neither correlated with raw water TOC concentration, so one possible explanation is changes in the NOM characteristics during the year.

Figure 3 shows results from experiments carried out with biofilters having Filtralite as filter media. The experiments were carried out with two different ozone doses and different empty bed contact times (EBCT). Besides one experiment, the raw water had TOC around 5 mg/l and colour around 50 mgPt/l. As demonstrated in previous batch tests, increasing the ozone dose from 1.1 to 1.7 mgO_3/mgTOC had very little effect on colour removal and the removal with two different water qualities was about the same (Figure 3a). The ozone dose had much larger impact on the TOC removal in the process than biofilter EBCT in the tested range (Figure 3b). However, even with the highest dose the TOC removal was still quite modest compared to other treatment methods. Therefore, ozonation/biofiltration alone is most suitable in cases where significant reduction of TOC is not required.

The head loss is has typically been low in biofilters and the backwash intervals have been several weeks (Melin and Ødegaard, 1999; Melin *et al.*, 2006). In the Filtralite tests

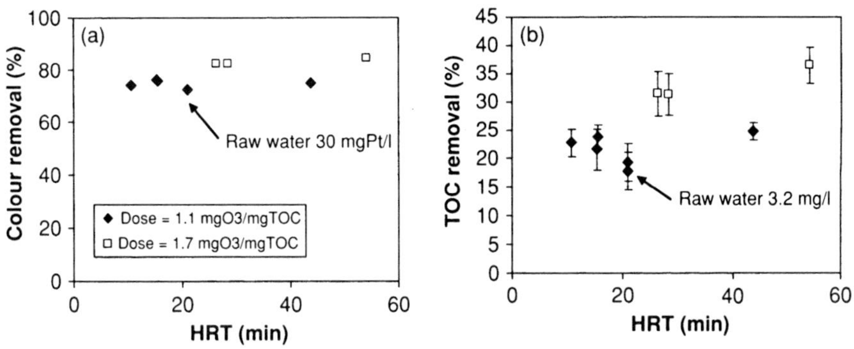

Figure 3 Removal of colour (a) and TOC (b) in ozonation/biofiltration process at different EBCTs. The raw water TOC and colour were around 5 mg/l and 50 mgPt/l, respectively, except where noted (Melin and Ødegaard, 1999).

(expanded clay combined biofilter and separation filter) with EBCTs of 11–21 min (filter velocities of 5–10 m/h), the terminal head loss was reached after 20–30 days of operation.

Although the TOC removal in the biofilter is not high, the major task of the biofilter is to remove all the easily biodegradable compounds to prevent regrowth problems in the distribution net and optimising the biofilter design is essential for the process. To investigate the effect of biofilter EBCT more accurately, it was decided to use gas chromatographic analyses of some of the known organic ozonation by-products (OBP) as an indicator for removal of the easily biodegradable compounds. The analysed OBPs were formaldehyde, acetaldehyde, glyoxal, methylglyoxal, acetone, glyoxylic acid, pyruvic acid and ketomalonic acids (Melin and Ødegaard, 1999; 2000).

A simple model was used predict the removal of OBPs in the biofilter:

$$S = (S_0 - S_{\min}) \cdot e^{-k\theta} + S_{\min}$$

where S is OBP concentration in the biofilter effluent, S_0 is the influent biofilter concentration, S_{min} is the minimum achievable substrate concentration, k is the first-order reaction constant and θ is the EBCT in the biofilter.

Figure 4a shows results for the sum of analysed OBPs (calculated as carbon) in the biofilter effluent. Different ozone doses (0.5–1.2 $mgO_3/mgTOC$) were used to achieve variety of influent OBP concentrations. Figure 4b shows OBP removal profile along the Filtralite biofilter column in a different experiment. The lines show the model fit. The OBP concentrations level off at higher EBCTs indicating that a maximum achievable removal efficiency has been reached (S_{min}). Apparently, this is affected by influent OBP levels (Figure 4a) but Figure 4b indicates that backwashing may also temporarily decrease the removal rates in biofilters by removing some of the biomass. However, for typical Norwegian surface water the optimal EBCT seems to be in a range of 15 to 20 min.

Temperature is not the only environmental factor in northern climates that may limit bacterial growth and TOC degradation in the biofilters. Nutrients, especially phosphorus, can be growth limiting factor. In a batch test with ozonated water from Lake Leirsjøen, maximum heterotrophic plate counts of 5.0×10^5 cfu/ml were measured while with the

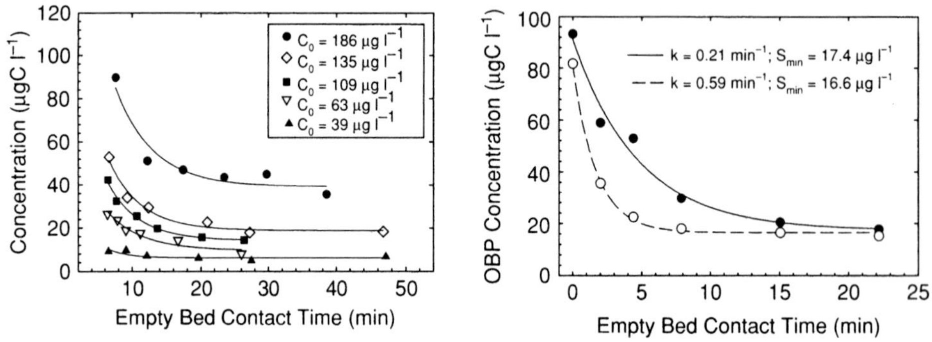

Figure 4 The effluent concentrations of OBPs at different EBCTs and influent concentrations (a) (Melin and Ødegaard, 2000) and OBP profiles along a biofilter column one (•) and 21 (○) days after backwash (b) (Melin and Ødegaard, 1999).

presence of excess phosphorus this increased an order of magnitude (to 50–53 × 10^5 cfu/ml) (Melin *et al.*, 2002). It was not studied whether this has any effect on biofilter performance although good removal of organic matter was observed in the plant at location (Figure 2).

Biomass separation filters

The effluent from the biofilter usually contains high numbers of heterotrophic bacteria that should be removed from the water. The levels can be in a range of 4.4 to 8.6 × 10^3 cfu/ml (Melin *et al.*, 2000). Different filter configurations were tested to see how they were able to remove the biomass. The filter configurations tested were monomedia (silica sand), dual-media (Filtralite, silica sand) and three-media (anthracite, silica sand, garnet sand) filters without chemical addition (Table 2).

Although stable removal of heterotrophic bacteria was achieved, the average removal efficiencies were around 84–88% which is not sufficient. The filter configuration, influent plate counts or filtration rate did not have significant effect on the removal efficiency. The

Table 2 Average removal of heterotrophic bacteria in separation filters (Melin *et al.*, 2006).

Filter configuration	Filter velocity m/h	Heterotrophic plate counts Influent 10^3 cfu/ml	Effluent 10^3 cfu/ml	Removal efficiency (%)	Head loss rate cm/d
1-media	2	7.8	1.1	85	3.8–4.6
2-media	5	6.2	0.73	88	2.8
2-media	7.5	5.4	0.63	88	5.9–7.6
3-media	2	6.5	0.83	87	0.5
3-media	2	16	2.3	85	0.3
3-media	5	13	1.5	88	0.95–1.1
3-media	5	16	2.3	84	2.0

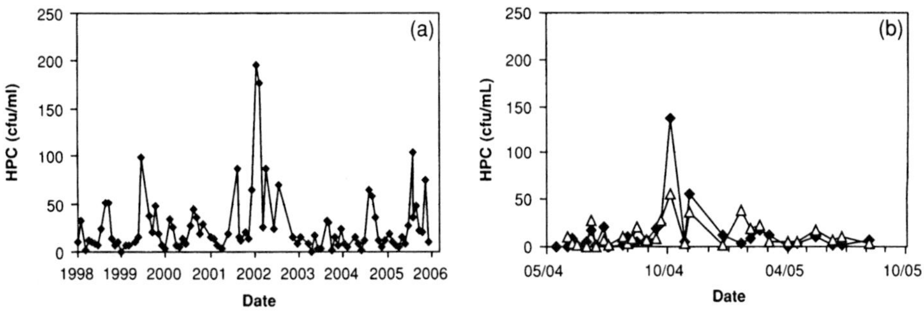

Figure 5 Levels of heterotrophic plate counts in distribution net in Nes water treatment plant (a) (updated from Melin and Ødegaard, 1998) and in Klungset treatment plant (b).

head loss rates in the filters were very low with the multi-media filters having the lowest head loss rates (see Table 2).

A small full-scale ozonation plant has been built based on KMT biofilter and sand separation filter without final disinfection (Ødegaard, 1996). After initially high levels of heterotrophic bacteria in samples taken from the distribution net, the levels have normally been well below the recommended 100 cfu/ml during the last seven years (Figure 5a).

Most of the full-scale plants based on ozonation/biofiltration in Norway has been built with a combined biofilter and separation filter directly after ozonation. They have normally final UV-disinfection in order to comply with the requirement of a double hygienic barrier. Figure 5b shows levels of HPC at two points in distribution net from such a plant. An alternative would be to use membrane filtration for biomass and pathogen separation.

Ozonation/biofiltration with membrane filtration

The first studies on the combination of ozonation/biofiltration with membrane filtration were based on using a rotating disk membrane (RDM) reactor after the bioreactor (Leiknes *et al.*, 2003; Ødegaard *et al.*, 2002). The RDM reactor applies polysulfone ultrafiltration (UF) membrane disks (MWCO of 750.000 Daltons) attached to rotary shafts. These are immersed in the reactor and operated with a low vacuum pressure where the permeate is extracted through the shafts. As in all membrane systems, fouling minimization and fouling control is a central issue in obtaining a sustainable operation. For the RDM unit, fouling can be controlled by varying the rotation rates to induce hydrodynamic shear forces, however, an alternative approach of applying mechanical cleaning with sponges in the reactor was also investigated. As anticipated, fouling rates were found to be a function of fluxes where a flux of around 20 l/m2·h was found to give an acceptable performance, i.e. low fouling rate. TOC concentrations were monitored throughout the process and TOC removal in the membrane reactor by the accumulated biomass was observed. The total TOC-removal was 25-30%.

The bioreactor was subsequently omitted and the effect of biofilm formation on the membrane disks, biodegradation and control of biofouling was investigated. Fouling control by rotation was not found to be sufficient, however, mechanical cleaning by sponges proved to be effective. TOC removal was comparable to initial results observed (Leiknes *et al.*, 2004). Fouling rates gave approximately one week of operation before cleaning was required, a task that took 15 minutes giving essentially a 100% water recovery.

These preliminary studies indicate that when sufficient biomass is maintained, either in the bioreactor or in the membrane reactor, TOC reduction by biodegradation is achieved. However, an underlying process requirement in using the membranes as a biofilm carrier is to have sufficient biomass to obtain the necessary biodegradation of the biodegradable NOM while preventing excessive biofouling which prohibits the membrane performance.

Further investigation of the treatment process has therefore focused on separating the treatment steps such that each stage can be designed and optimized for the respective goals. Initial studies using a bioreactor coupled with an immersed hollow fibre membrane reactor (Zenon ZeeWeed) have been conducted. TOC reduction was comparable to results from the studies described above.

The mode of operation was based on a cyclic phase of 30 min production which included a 30 second backwash using permeate. The backwash rate was set at 1.5 times the production rate (flux) and air scouring was introduced during the backwash period. Retentate was withdrawn from the reactor continuously at a rate giving an overall water recovery of 90%. Fluxes of 40, 60 and 75 $l/m^2 \cdot h$ were tested respectively and terminated when the TMP approached the set maximum value of 0.5 bar. Operating times varied from a little more than one week for the flux of 75 $l/m^2 \cdot h$ to an excess of 3 weeks with a flux of 40 $l/m^2 \cdot h$ (TMP well below set maximum value) before extensive cleaning was necessary. An estimated projection of operating time for the lower flux indicates that an expected operating time of 70–80 days may be achievable. These initial results indicate a potential for sustainable operation of the process. Fouling minimization and control through determining membrane reactor designs and operating modes will therefore be an important aspect of this process concept to achieve sustainable operation of the membrane filtration step.

The process combination of advanced oxidation, biofiltration and membrane filtration will be studied further and developed in the EU-project TECHNEAU, started in 2006. Here more focus will be given to the potential of the process to remove organic micropollutants through investigating alternative advanced oxidation processes (UV/TiO_2, O_3/H_2O_2 etc.) optimized biodegradation reactors combined with efficient particle separation with membranes.

Conclusions

It has been shown that ozonation is effective in colour removal from humic surface water. The dose required for 80% colour removal that has typically been achieved in pilot- and full-scale plants, is around 0.15–0.20 $mgO_3/mgPt$ which translates to 1.0–1.5 $mgO_3/mgTOC$.

Biofiltration is required to remove easily biodegradable organic ozonation by-products. The results show that for typical Norwegian water, the EBCT should be around 20 min. However, more detailed studies are required to establish more accurate correlation between biofilter loading and effluent regrowth potential. No significant difference between the performance of different biofilter media during long term biofilter operation has been experienced. The total TOC removal is relatively low, but the easily biodegradable ozonation products are efficiently removed in biofilters.

Separation of biomass may take place in the biofilter or in separate granular filters at very low head loss rates. However, to achieve two secure hygienic barriers especially against parasites, more efficient separation like membrane filtration should be used.

Combining ozonation/biofiltration with membrane separation is a promising process concept. This will be studied in detail in the EU-project TECHNEAU in the coming years.

References

Leiknes, T., Ødegaard, H., Ohme, H. and Lazarova, M. (2003) Ozonation/biofiltration for NOM-removal using rotating disk membranes. In *Proc. of the 5. Int. Membrane Science and Technology Conf.*, Sydney, 10–14 November.

Leiknes, T., Lazarova, M. and Ødegaard, H. (2004) Development of a hybrid ozonation biofilm-membrane filtration process for the production of drinking water, In *Proc. of the IWA Specialist conference Water Environment Membrane Technology (WEMT 2004)*, Seoul, 7–10 June.

Melin, E.S and Ødegaard, H. (1998) Ozonation/biofiltration process for small treatment plants - Case study in Nes community, Norway. Poster presented at *19'th Biennial conference of the International Association on Water Quality*, Vancouver, 21–26 June.

Melin, E. and Ødegaard, H.(1999) Biofiltration of ozonated humic water in expanded clay aggregate filters. *Wat. Sci. Tech.* **40**(9), 165–172.

Melin, E., Bohne, R., Sjøvold, F. and Ødegaard, H. (2000) Treatment of ozonated water in biofilters containing different media. *Wat. Sci. Tech.* **41**(4–5), 57–60.

Melin, E. and Ødegaard H. (2000) The effect of biofilter loading rate on the removal of organic ozonation by-products. *Wat. Res.* **34**(18), 4464–4476.

Melin, E., Eikebrokk, B., Brugger, M., Ødegaard, H. (2002) Treatment of Humic Surface Water at Cold Temperatures by Ozonation and Biofiltration. *Wat. Sci. Tech.: Water Supply*, **2**(5–6), 451–457.

Melin, E., Skog, R. and Ødegaard, H. (2006) Ozonation/biofiltration with calcium carbonate as biofilter media. In *Proc. of the 4'th Int. Conf. on Slow Sand and Alternative Biological Filtration Conference*, Mühlheim, 3–5 May.

Ødegaard, H., Rusten, B. and Westrum, T. (1994) A new moving bed biofilm reactor. *Wat. Sci. Tech.* **29**(10–11), 157–165.

Ødegaard, H. (1996) The development of an ozonation/biofiltration process for the removal of humic substances. In *Advances in slow sand and alternative biological filtration* (Graham and Collins, eds), John Wiley & Sons Ltd, Chichester.

Ødegaard, H., Melin, E. and Ohme, H. (2002) Ozonation/biofiltration for NOM-removal by using the rotating disc membrane filtration system. In *Proc. Japanese/Norwegian Workshop on Membrane Filtration Technology for Water Treatment*, Trondheim, 30 September.

Song, R., Donohoe, C., Minear, R., Westerhoff, P., Ozekin, K. and Amy G. (1996) Empirical modeling of bromate formation during ozonation of bromide-containing waters. *Wat. Res.*, **30**(5), 1161–1168.

49 Ozonation/biofiltration with calcium carbonate as biofilter media

E. Melin*, R. Skog, H. Ødegaard****

*SINTEF Water and Environment, N-7465 Trondheim, Norway
**Department of Hydraulic and Environmental Engineering, Norwegian University of Science and Technology, N-7491 Trondheim, Norway

Abstract Some water works in Norway do not have treatment process for removal of humic substances, but only chlorinate the water. However, since the surface water is often soft and corrosive, these plants have calcium carbonate filters to increase pH and alkalinity for corrosion control. For upgrading these treatment plants, ozonation/biofiltration is an attractive alternative. Ozone would provide additional disinfection and colour removal and the existing $CaCO_3$ filters could be used for both biofiltration and corrosion control. In this study, feasibility of using $CaCO_3$ as biofilter was studied. As comparison, a three-media filter (anthracite/sand/$CaCO_3$) was also tested. The treatment efficiencies with these biofilters were similar to biofilters tested in other studies. With ozone doses of 1.5 and 2.1 mgO_3/mgTOC, the average TOC removal was 34% and colour removal was 92%. The BDOC removal ranged from 47 to 61%. The two biofilters did not have differences in overall performance. The results show that $CaCO_3$ alone can be used as biofilter media in ozonation/biofiltration process.

Keywords Biofilter; calcium carbonate; drinking water; filter media; ozonation

Introduction

Several large Norwegian surface water treatment plants use water sources that have lower concentrations of organic matter and colour than required in drinking water standards (TOC $<$ 5 mg/l and colour $<$ 20 mgPt/l). Several of these waterworks do not employ treatment methods for organic matter removal, but chlorinate the water. However, since the surface waters typically are very soft and corrosive, many of these plants employ calcium carbonate filters (often together with CO_2 dosing) to increase pH and alkalinity for corrosion control. These water works may need to upgrade the treatment process in a near future. Recent studies have shown that the amount of organic matter and colour are increasing in surface waters (Eikebrokk *et al.*, 2004). If this trend continues, the raw water quality will exceed the maximum values set by drinking water regulations. In addition, the new Norwegian drinking water regulations require two independent barriers against microbial and chemical contamination. At the same time, parasites not removed by present chlorination practice are an increasing concern especially after recent major *Giardia* outbreak in city of Bergen.

 Recent Progress in Slow Sand and Alternative Biofiltration Processes edited by Rolf Gimbel, Nigel J.D. Graham and M. Robin Collins. ISBN: 9781843391203. Published by IWA Publishing, London, UK.

One interesting process alternative for these water works is ozonation and biofiltration. Ozone is a powerful disinfectant and also has capacity to break down many organic micropollutants. The existing filters for corrosion control could be used as biofilters minimising requirement for plant expansion. The ideal case would be to maintain calcium carbonate as the filter media so that the filters could be used both for corrosion control and removal of the easily biodegradable organic ozonation by-products. Calcium carbonate is not usually used as biofilter media. An earlier study indicated, however, that $CaCO_3$ may function well as such (Melin *et al.*, 2000). This project was carried out to study the applicability of calcium carbonate as biofilter media in ozonation process in more detail. The performance was compared to three-media filter consisting of anthracite, sand and $CaCO_3$. This type of filter is often used in coagulation/contact filtration plants in Norway since it gives good particle removal and corrosion control. In addition, it is similar in configuration to a commercially marketed ozonation/biofiltration system (Melin *et al.*, 2002).

Methods

In biofilter experiments, the raw water was tap water (colour ca 13 mgPt/l) to which natural humic substances were added in the form of regenerate solution from an ion exchange plant. The content of humic substances was increased to give colour of 25 mgPt/l and total organic carbon (TOC) around 3.4 mg/l. The water temperature in biofilter stage ranged from 15 to 19°C. The raw water pH was 8.1–8.6 (average 8.4) and it was reduced 0.3 units in the ozonation stage.

Figure 1 shows the schematic of the pilot plant. The ozonation stage consisted of two 4 m high ozonation columns in series with diameter of 15 cm and diffusers in the bottom.

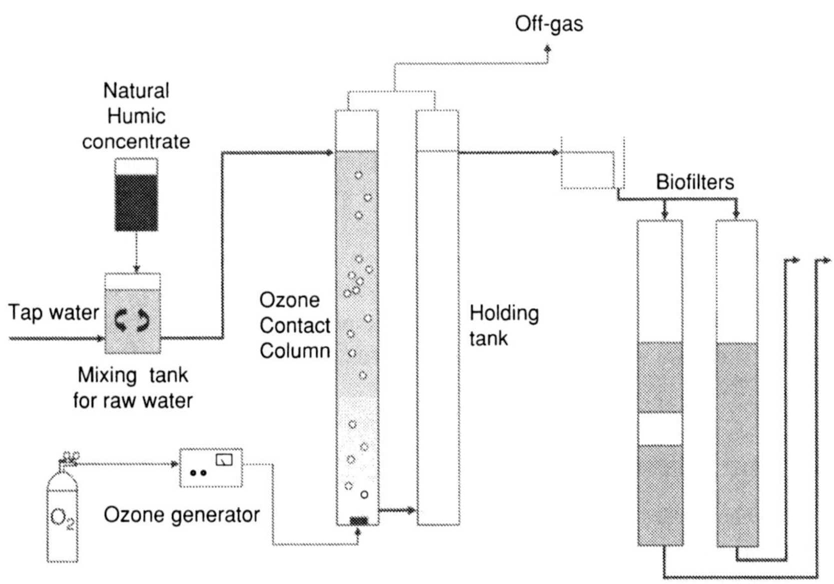

Figure 1 Schematic of the pilot plant.

Table 1 The biofilter configuration.

Filter media	Diameter (mm)	Density (kg/m^3)	Media height (m) Filter 1 (F1)	Filter 2 (F2)
Anthracite	0.8–1.6	1.4	0.5	-
Sand	0.4–0.6	2.65	0.35	-
$CaCO_3$	1–3	2.65	1.1	1.95
Gravel support	1–10		0.25	0.25

Table 2 Experimental program.

Experiment	Average ozone dose (mgO_3/l)	Average specific ozone dose ($mgO_3/mgTOC$)	3-media filter (F1) operation	$CaCO_3$ filter (F2) operation
1	1.2	0.34	downflow	downflow
2	1.9	0.57	downflow	upflow
3	5.0	1.5	downflow	downflow
4	7.0	2.1	downflow	upflow

Only the first column was used for ozonation and the second was used as a holding tank to remove residual ozone. Since the ozonation columns had originally been designed for higher flow rates than used in these experiments, the retention times in the columns were relatively long (45 min for each column). After the columns, the flow was equally divided into biofilters through overflow tank. The ozone was produced from technical grade oxygen by Erwin Sander 300.5 ozone generator.

The biofilters had height of 3.6 m to the inlet and had inner diameter of 12 cm. The filter media had height of 1.95 m exclusive gravel support in the bottom. Table 1 shows the media bed heights in the biofilters. The filters were designed for an empty bed contact time (EBCT) of 35 min, which is the same as used for $CaCO_3$ filters in Trondheim water treatment plant for corrosion control. This gives a filter velocity of 3.8 m/h, which is comparable (?) to that used in practice with similar filters.

Four different ozone doses were used in the biofilter experiments (Table 2). The filters were normally operated as downflow filters, but upflow mode was tested with the CaCO3 filter in two experiments. For clarity, the experimental periods are numbered by increasing ozone dose, but in practice, experiment number 2 was carried out last. The biofilters were operated for 1.5 months before the experiments were started. Biodegradable dissolved organic carbon (BDOC) was analysed with the method developed by Servais et al. (1989) and the other analyses were carried out using standard methods for water examination (*Melin* et al., 2002).

Results and discussion

Initial tests were carried out to study the effect of ozone dose on colour removal. One experiment was carried out with tap water (colour = 13 mgPt/l and TOC = 2.5 mg/l) which is the present water quality in Trondheim water treatment plant. The second experiment was carried out adding humic concentrate to increase colour (colour = 25 mgPt/l and

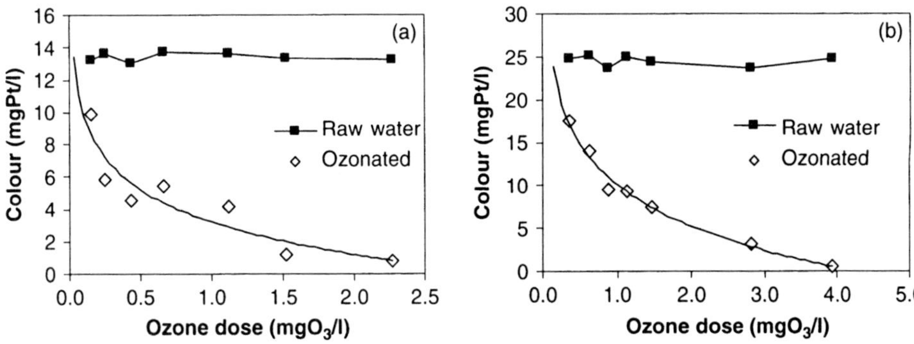

Figure 2 Effect of ozone dose on colour removal with raw water colour of 13 mgPt/l (a) and 25 mgPt/l (b).

TOC = 3.8 mg/l) to simulate a possible future situation. Good (80%) colour removal were achieved at relatively low ozone doses of 1.2 and 2.1 mgO_3/l with raw water colour of 13 and 25 mgPt/l, respectively (Figure 2). However, these doses do not yet give very high residual ozone concentrations after the ozonation column. Figure 3 shows residual ozone in water samples taken from the holding tank. The retention time in the holding tank to the sample point was about 5 min. The results show that even if good colour removal was achieved at low ozone doses, these still resulted in low residual ozone concentrations. Therefore, the biofilter experiments were carried out with four different ozone doses; two with low doses where the aim would primarily be colour removal with pathogen control and two with higher doses aiming at parasite removal.

The biofilter tests were carried out with raw water colour of 24 mgPt/l and TOC of 3.4 mg/l. Figure 4a shows average TOC in raw water, ozonated water and biofilter effluents. The results show that with low ozone doses (experiments 1 and 2), the oxidation of organic matter and removal of TOC in the biofilters were low. Ozonation stage removed on average 6 and 3% of TOC, respectively, in these experiments. After biofiltration, the total TOC

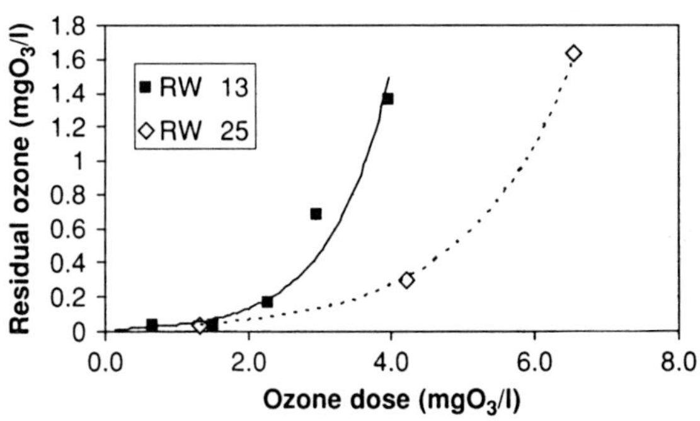

Figure 3 Effect of ozone dose on residual ozone after 5 min retention time in holding tank with raw water colours of 13 and 25 mgPt/l.

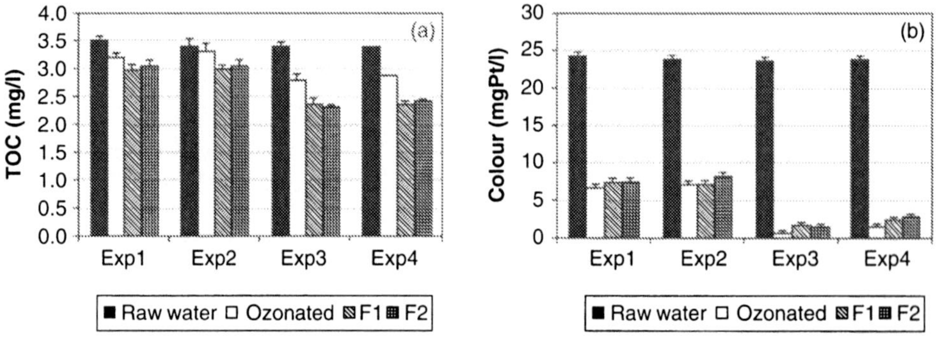

Figure 4 Average TOC concentrations (a) and colour (b) in different stages of the process during the experimental periods.

removals were around 13 and 11% without any significant differences between the biofilters. Higher ozone dose in experiments 3 and 4 increased TOC removals both in the ozonation stage (to 18 and 15% in experiments 3 and 4, respectively) and in the biofilters. The average TOC removals after biofilters were in the range of 32 to 35% with no consistent differences between the biofilters. These removal efficiencies are in the same range as observed in our other pilot studies (Ødegaard *et al.*, 2006).

Figure 4b shows colour in the different stages in the process. The colour is removed during in the ozonation stage with no additional removal in the biofilters. The average removals were 67–71% at lower doses and 89–94 % at higher doses. The colour removals were in the same range as observed in initial experiment, although in the biofilter test there is no difference between experiments 1 and 2 or 3 and 4. This may be due to analytical uncertainty at low levels of colour.

Removal of UV_{254} absorbance was similar to colour with no significant differences between ozonated water and biofilter effluents. The average raw water absorbance was 13.7 E/m and it was reduced to 8.0 E/m in experiments 1 and 2 and to 2.7 E/m in experiments 3 and 4. The average removals were 42 and 80%, respectively.

Table 3 shows the results from BDOC analyses. The raw water had a low BDOC which is expected since the raw water source is a large lake with no industrial discharges into it. The BDOC in ozonated water follows a typical pattern with increasing BDOC with increasing ozone dose until a plateau is observed at the highest doses (Carlson and Amy, 1997). The lowest BDOC removal efficiency was observed with the lowest ozone dose. In the other

Table 3 BDOC concentrations (mg/l) during different experimental periods.

Experiment	Ozone dose ($mgO_3/mgTOC$)	Raw water	Ozonated	F1	F2
1	0.30	0.21	0.30	0.21	0.23
2	0.47	0.02	0.53	0.23	0.21
3	1.4	0.18	1.01	0.49	0.54
4	2.0	0.22	1.02	0.51	0.40

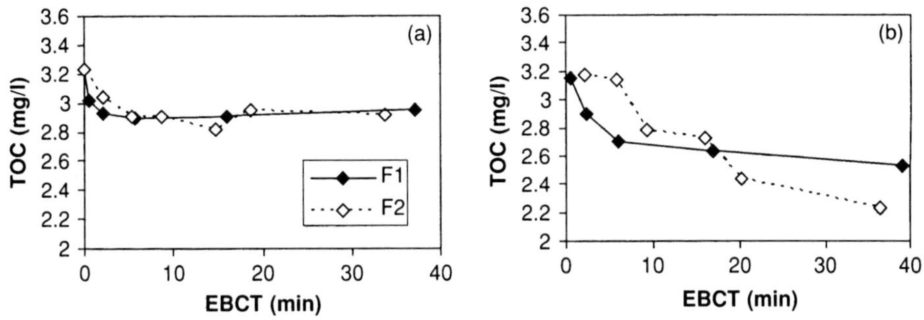

Figure 5 TOC removal profiles in biofilters with ozone dose of 0.36 $mgO_3/mgTOC$ (a) and 1.6 $mgO_3/mgTOC$ (b).

tests, removal efficiencies were 47–61% which is in the same range as analysed from a small full-scale plant (Melin et al., 2002). Neither of the biofilters had consistently better BDOC removal.

The biofilters had sampling ports along the filter column, and samples were taken to study the TOC degradation in the filters. Figure 5 shows TOC removal in the biofilter beds with low and high ozone doses. With low dose, little BDOC is produced and all the biodegradation seems to take place within 5.5 min EBCT (Figure 5a). In the case of high dose, more BDOC is formed and in this case the TOC removal profiles are different (Figure 5b). In the three-media filter, majority of the TOC is still removed within 6 min EBCT but there seems to be additional removal along the rest of the filter. Earlier studies have indicated that an EBCT of 20 min should be usually used (Ødegaard *et al.*, 2006) and this is reasonable for the three-media filter. Due to variation in TOC analyses, the $CaCO_3$ filter result is difficult to interpret. In addition, the effluent TOC is for some reason 0.3 mg/l lower in $CaCO_3$ filter than in three-media filter in this test. However, Figure 5b shows clearly that biodegration is slower in $CaCO_3$ filter. This may not be a critical factor, since most probably EBCT in a CaCO3 filter would be determined by corrosion control aspects rather than by biodegradation rates.

Table 4 shows turbidity in raw water and biofilter effluents. The raw water turbidity was very low, so the turbidity removals are modest. The results indicate that the three-media filter (F1) had better turbidity removal but differences are very small. Since the $CaCO_3$ filter (F2) was not backwashed before starting upflow experiments, some of the retained particles probably came out from the filter bed increasing turbidity in the upflow experiments.

Table 4 Average turbidity values (NTU) during different experiments.

	Raw water	F1	F2
Experiment 1	0.23	0.10	0.13
Experiment 2	0.22	0.12	0.26
Experiment 3	0.21	0.09	0.12
Experiment 4	0.22	0.09	0.16

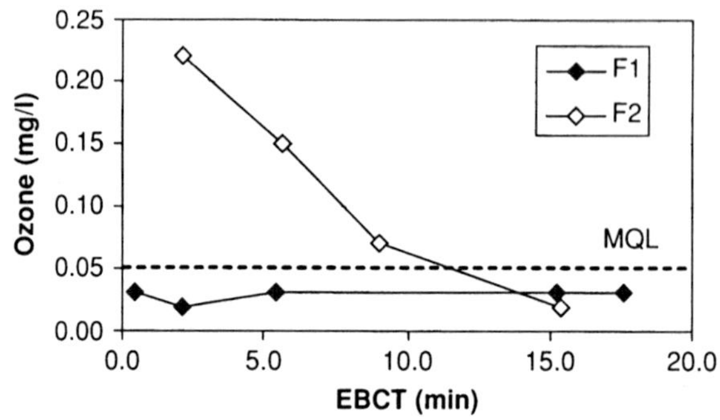

Figure 6 Removal of residual ozone in the biofilters. The water out from ozonation column had ozone concentration of 0.68 mg/l. The method quantification limit (MQL) was 0.05 mg/l.

The head loss was low during these experiments due to low raw water turbidity and apparently low biomass accumulation. During 2.5 months of operation, the $CaCO_3$ filter needed no backwash. The three-media filter was backwashed approximately once a month.

Heterotrophic plate counts (HPC) were measured from biofilter effluents once in each experimental period. The three-media filter had heterotrophic plate counts ranging from 27 to 723 cfu/ml and the $CaCO_3$ filter had range from 49 to 6020 cfu/ml. The lowest counts were analysed in the first experiment. The three-media filter effluent had average HPC of 350 cfu/ml and the $CaCO_3$ filter had 2000 cfu/ml. As with turbidity, the highest HPCs were measured from F2 during upflow mode. Therefore it is uncertain whether differences in plate counts are due to better detainment of bacteria in three-media filter or due to detachment of bacteria after changing F2 to upflow operation. With other parameters, there was no difference between the downflow or upflow operation (Figure 4).

During the biofiltration test, a holding tank was used after ozonation column to remove residual ozone before water entered the biofilters. A test was made to study the fate of ozone in the biofilters in case that the biofilter feed water contains dissolved ozone. In this test, the water for the biofilters was taken directly from the ozonation column. Figure 6 shows results from a test where the water out from the ozonation column contained 0.68 mg/l of ozone. The results show that anthracite removed efficiently ozone in the first few centimetres of the filter bed in the top of the filter. The ozone quantification limit in the analysis was 0.05 mg/l so the low levels of ozone in Figure 6 are probably only background noise. In case of $CaCO_3$ filter, the ozone penetrated almost 60 cm into the biofilter before concentration was reduced below 0.05 mg/l. Therefore, to prevent that a part of the $CaCO_3$ filter would become biologically inactive, the treatment process should be designed to remove residual ozone by a layer of anthracite or activated carbon in the filter, or by having a holding tank with long enough retention time.

Conclusions

The results show that calcium carbonate alone can be used as biofilter media in ozonation/biofiltration process. There were no significant differences between $CaCO_3$ filter and three-media (anthracite, sand, $CaCO_3$) filter. Both had treatment results that are comparable to those we have obtained with other biofilter media. It did not matter whether the $CaCO_3$ filter was operated in upflow or downflow mode. However, if the biofilter feed water contains residual ozone, it may penetrate into the $CaCO_3$ filter making part of it biologically inactive.

The ozone dose affected the treatment results with increased colour and TOC removal at higher ozone doses. With a dose of 1.5 mgO_3/mgTOC, 93% colour removal and 34% TOC removal was achieved. The results also showed that with low turbidity water, the filter run times are at least one month before backwash is needed.

Since the raw water used in this study was tap water, the effect of biofilm growth on pH and alkalinity control could not be investigated. This will be done in a pilot plant that is under construction in Trondheim water works. It will also be used to compare ozonation/biofiltration with coagulation and contact filtration.

References

Carlson, K. and Amy, G. (1997) The formation of filter-removable biodegradable organic matter during ozonation. *Ozone Sci. Eng.* **19**(2), 179–199.

Eikebrokk, B., Vogt, R.D. and Liltved, H. (2004) NOM increase in Northern European source waters: discussion of possible causes and impacts on coagulation/contact filtration processes. *Wat. Sci. Tech. - Water Supply* **4**(4), 47–54.

Melin, E., Bohne, R., Sjøvold, F. and Ødegaard, H. (2000) Treatment of ozonated water in biofilters containing different media. *Wat. Sci. Tech.* **41**(4-5), 57–60.

Melin, E., Eikebrokk, B., Brugger, M., Ødegaard, H. (2002) Treatment of Humic Surface Water at Cold Temperatures by Ozonation and Biofiltration. *Wat. Sci. Tech.: Water Supply*, **2**(5-6), 451–457.

Ødegaard, H., Melin, E. and Leiknes, T. (2006) Ozonation/biofiltration for treatment of humic surface water. In *Proc. of the 4'th Int. Conf. on Slow Sand and Alternative Biological Filtration Conference*, Mühlheim, 3–5 May.

Servais, P., Anzil, A. and Ventresque, C. (1989) Simple method for determination of biodegradable dissolved organic carbon in water. *Appl. Environ. Microbiol.* **55**(10), 2732–2734.

50 The effect of permanganate preoxidation on biomass nitrification

Z. Ren*, J. Ma**

* PO BOX 2627, Harbin Institute of Technology. 202 Haihe Road, Nangang District, Harbin 150090, P.R China. E-mail: renzhijun2003@126.com
** PO BOX 2627, Harbin Institute of Technology. 202 Haihe Road, Nangang District, Harbin 150090, P.R China. E-mail: majun@hit.edu.cn

Abstract A laboratory scale experiment was conducted to analyze the mechanism of enhanced nitrification of permanganate preoxidation. In this study, the population and species of nitrite bacteria and nitrate bacteria, the potential nitrosification and nitrification were tested for the analysis of nitrification. The results showed that the distribution and species of nitrobacteria were the same in both two BAC processes. *Nitrosospira*, *Nrtrosominas* and *Nitrobacter*, *Nitrospina* were the predominant bacteria in the biofilter. Along the direction of water flow, the population of nitrite bacteria reduced gradually. However, the population of nitrite bacteria increased firstly and then reduced and got mixed at the bottom of the biofilters. On the other hand, compared with BAC alone, there had been more population of nitrite bacteria and nitrate bacteria and better performance of nitrification and nitrosification in the BAC process with permanganate preoxidation, which showed that BAC followed with permanganate preoxidation had better removal performance for ammonia and nitrite.

Keywords Permanganate; BAC; nitrite bacteria; nitrate bacteria; nitrosification; nitrification

Introduction

Recent research showed that more and more water resources were polluted for the widely used nitrogen fertilizer in China, which may cause water quality problems during distribution, such as bacterial regrowth, taste and odor (M. Kajino et al, 1999), production of nitrites by incomplete oxidation of ammonia (O.Nowak et al, 1999)and during disinfection by chlorine (Daniel J et al, 1999). So ammonia removal is a major challenge in meeting with the increasingly stringent effluent standards. In drinking water production, biological processes, like biological filtration, seem to be the most promising techniques to remove ammonia. Biological nitrification is a two-step process with sequential oxidation of ammonia into nitrite, and then oxidation of nitrite into nitrate. It was admitted that *Nitrosomonas* and *Nitrobacter* were the most common genera of autotrophlic bacteria, known respectively as ammonia oxidizers and nitrite oxidizers.

The influencing factors on autotrophic bacteria, such as filtration material (A. Kihn et al, 2002)and temperature (Oleszkiewicz et al, 1988), have been clearly established. However, limited information is available related to the effect of oxidants pretreatment on nitrification. In previous pilot experiments, two processes, permanganate- coagulation- sedimentation - BAC and coagulation-sedimentation- BAC, had been conducted to evaluate the effect of permanganate preoxidation on the performance BAC process in Fuyang second water treatment plant, which is located in Anhui province. The results showed that the removal rate of organic maters, ammonia and nitrite in the BAC filter followed by permanganate preoxidation were 5.4%, 10.9% and 13% higher respectively than that without permanganate preoxidation (Majun, 2005). It is interesting to see that BAC process with permanganate preoxidation had a better performance for ammonia removal. However, the direct effect of permanganate preoxidation on autotrophic bacteria nitrification has not been clearly established because the nitrification efficiency were only investigated by measuring the ammonia concentration in the inlet and outlet waters of biofilters in pilot-scale study.

The purpose of this work is to analyze the mechanism of permanganate preoxidation on nitrification. In this study, biofilters with and without permanganate preoxidation were operated in the laboratory. Not only the variation of ammonia and nitrite of inlet and outlet of biofilters but also the population and species of nitrite bacteria and nitrate bacteria, the potential of nitrosification and nitrification were determined.

Methods

System configuration and operational conditions

In this study, a laboratory-scale experiment was conducted in Harbin Institute of Technology. The raw water was drawn from Songhua River and its characteristics were shown in Table 1.1.

The schematic diagram of lab-scale experiment is shown in Figure 1.1. The treatment includes coagulation with and without permanganate pretreatment, rapid sand filtration and biological activated carbon filtration. The treatment has an average volume loading of 20 liter per day. The sand filtration apparatus (34 mm in diameter) fills with 100cm sand (1 mm). The biofiltration column (34 cm in diameter and 2000 mm in length) is packed with 90 cm bed depth of GAC (2–3 mm) and 10 cm sand (1 mm). The characteristics of GAC used in the experiments are presented in Table1.2.

Raw water was firstly treated with and without permanganate preoxidation respectively. The dosage of permanganate was 1.0 $mg{\cdot}L^{-1}$ and the reaction time was controlled at 30 mins. At the end of this time, aluminum sulphate was used as a flocculant and the dosage was 20 $mg{\cdot}L^{-1}$. After fast stirring for 1mins at 100 rpm, the water was allowed to stand quiescently for one hour. Then the supernatant was taken into regulating tank by pump. After sand filtration, the effluent of sand filtration was pumped as the influent of BAC filtration by peristaltic bump. Peristaltic bump is used to maintain the appropriate flow rate through the filters.

Table 1.1 The characteristics of raw water

Parameters	NH_3-N /$mg{\cdot}L^{-1}$	NO_2-N/$mg{\cdot}L^{-1}$	NO_3-N/$mg{\cdot}L^{-1}$	COD_{Mn}/$mg{\cdot}L^{-1}$	Temperature/°
Concentration	2.88–4.44	0.14–0.46	0.93–1.35	4.41–10.7	18–24

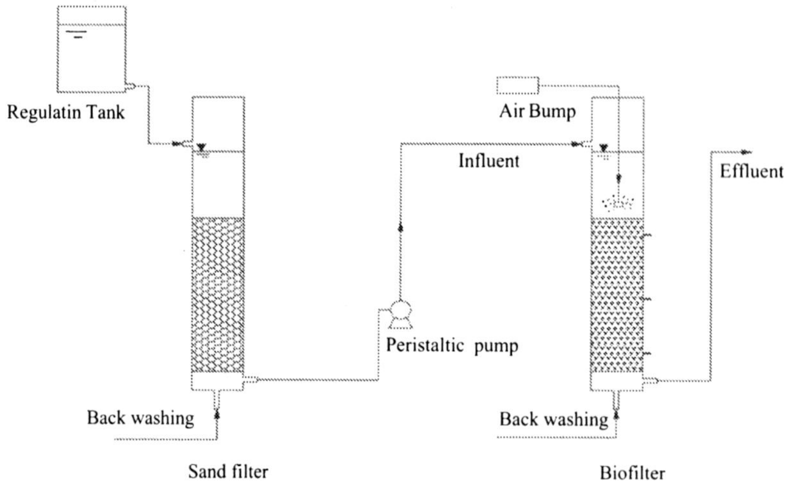

Figure 1.1 The schematic diagram of lab-scale experiment

Table 1.2 The characteristics of GAC

Index	Parameter
Specific surface/$m^2.g^{-1}$	≥900
pH value	6–8
Dry bulk density/$g.L^{-1}$	450–520
Granule size/Mesh	20–40
Ash content/%	≤10
Intensity/%	≥90

Backwashing periods were different between sand filters and biofilters. The backwashing of sand filter was performed in every five days, but backwashing period of biofilter was determined according to the increase of head losses. The backwashing conditions of sand filter and biofilter were similar: the clear water was introduced from the filter bottom with the expanding rate from 25% to 35% until the effluent turbidity of filter was lower than 3NTU.

Analytical methods

Ammonia, nitrite and nitrate

Ammonia, nitrite and nitrate were determined according to the standard methods (Ministry of Public Health of China, 2001).

Population of nitrite bacteria and nitrate bacteria

The population of nitrite bacteria and nitrate bacteria on both biofilters are determined by the most probable number (MPN) methods. About 1.0 g wet weight of the packing material was sampled and homogenized in 19 ml of deionized water at 500 rpm for 30 min. The culture medium for nitrite bacteria contains 2.0 g $(NH_4)_2SO_4$, 0.25 g NaH_2PO_4, 0.01 g $MnSO_4.H_2O$, 0.75 g K_2HPO_4, 0.03 g $MgSO_4.7H_2O$, 5.0 g $CaCO_3$ per liter at pH

7.2. The culture medium for nitrate bacteria contains 1.0 g Na_2NO_2, 1.0 g Na_2CO_3, 0.25 g NaH_2PO_4, 1.0 g $CaCO_3$, 0.75 g K_2HPO_4, 0.01 g $MnSO_4$ and 0.03 g $MgSO_4.4H_2O$ per liter at pH 7.2. The homogenized suspension was diluted with culture medium, then 1.0 ml of the suspended solution with different dilution ratios was transferred to 9.0 ml culture medium in test tubes, and incubated at 25° to 28° for 10 to 14 days. At the end of the incubation period, each tube for nitrite bacteria was scored by adding Griess reagent as an indictor to test the presence of nitrite. A red color reaction indicated that nitrite was formed, and the tube was scored positive. The absence of a red color was scored negative. Each tube for nitrate bacteria was scored by adding an indicator (diphenylamine and concentrated H_2SO_4) to test the presence of nitrate. A blue color reaction indicated that nitrate was formed, and the tube was scored positive. The absence of a blue color was scored negative. By referring to the MPN table, the positive number can be used to estimate the population of nitrite bacteria and nitrate bacteria.

The separation of nitrite and nitrate bacteria

The separation of nitrite and nitrate bacteria is performed according to MPN methods. About 1.0 g wet weight of BAC was sampled and incubated in 500mL liquid culture medium for nitrite and nitrate bacteria. After incubating at 25° to 28 ° for 10 to 14 days, the 1.0mL of suspended solution with different dilution ratio was diluted with 9mL deionized water in test tubes. Then 0.1mL different dilution ratio solution was smeared on the solid culture medium for nitrite and nitrate bacteria, and incubated at 25° to 28 ° for 10 to 14 days. At the end of incubation period, the single strain was separated.

The potential of nitrosification and nitrification

The applied method consisted of estimating the potential of nitrosification and nitrification, which is proportional to nitrite bacteria and nitrate bacteria. The parameter was determined by measuring the production rate of oxidizer nitrogen (NO_2 and NO_3).

The method to determine the potential of nitrosification was as follow: 1.0 g wet weight of packing material at the different height of biofilter was sampled and the sample was incubated in one liter distill water with potassium chlorate to prevent nitrite oxidizing into nitrate. After one and a half hour, 10 mg/L ammonia was added then the concentration of NO_2 was determined every one hour and the nitrite was analyzed according to the standard methods for examination of water and wastewater.

The method to determine the potential of nitrification was similar to the method of potential nitrosification. 1.0 g wet weight of packing material at different height of biofilter was sampled and the sample was incubated in one liter distill water. One and a half hour later, 5 mg/L nitrite was added then the concentration of NO_3 was determined every one hour. The nitrate was determined using ion chromatography (DIONEX model 4500i, USA).

Results and discussion

Biofilter start up

Before this study, the biofilter had stop running for about one year. So the dates in this study included two periods, BAC process start-up and steady running.

The way of biofiters start-up was operated in continuous flow-mode and the effluent of BAC was the settled water with and without permanganate pretreated, respectively. Two biofilters were operated in serious and the effect of oxygen and temperature were considered the same during nitrification. In order to meet the oxygen consumption, an air pump, as aeration device, was used to sparge the influent water to reach the desired DO concentration. The dissolved oxygen concentration was always over 2 mg/L in the effluent of bilfiters. The two equipments were in the laboratory and temperature was maintained from 18° to 24° throughout all the experimental period.

Ammonia, nitrite and nitrate removal

Table1.3 shows the variation of ammonia, nitrite and nitrate in coagulation, sand filtration and biofiltration processes with and without permanganate preoxidation. The results indicated that ammonia was not effectively removed in both coagulation and sand filtration processes, whilst, permanganate preoxidation is not effective in aiding the coagulation process and sand filtration to remove ammonia. The ammonia removal rates are about 10 percent in coagulation process and 25 percent in sand filtration respectively.

The biofilter is the main part to remove ammonia. With the growth of nitrite bacteria, the ammonia concentration in the effluent of biofilter decreased gradually. After 30 days, the ammonia removal got steady (the variation of ammonia in biofiltration did not show in this paper). During the whole process, the average ammonia and nitrite concentration of effluent biofilter was 1.11 mg/L and 0.152c in the biofilter without permanganate preoxidation. Compared with biofilter without permanganate preoxidation, the average ammonia and nitrite concentration of effluent biofilter was 0.90 mg/L and 0.094 mg/L resperctively.

The lower concentration of ammonia and nitrite in the effluent of BAC process suggested that nitrification was improved obviously in BAC process with permanganate preoxidation.

Table 1.3 Variation of NH_3-N, NO_2-N and NO_3-N in BAC and PPC-BAC process

			Coagulation		Sand filtration		BAC process	
Process		**Raw water**	**BAC**	**PPC-BAC**	**BAC**	**PPC-BAC**	**BAC**	**PPC-BAC**
Ammonia	Range	2.88-4.44	1.9-4.67	1.98-4.72	0.77-4.52	0.83-4.6	0-3.86	0-3.5
	Average	3.85	3.50	3.44	2.88	2.79	1.11	0.90
	Removal rate	—	9.04	10.61	25.25	27.43	71.09	76.70
Nitrite	Range	0.12-0.46	0.08-0.37	0.098-0.4	0.021-0.7	0.02-0.7	0.005-0.85	0.0045-0.79
	Average	0.332	0.183	0.226	0.194	0.135	0.152	0.094
	Removal rate	—	44.96	31.86	41.59	59.47	54.34	71.86
Nitrate	Range	0.93-1.35	0.98-1.85	0.98-1.96	0.99-2.68	0.98-2.75	1.94-3.96	2.39-4.3
	Average	1.11	1.47	1.49	1.86	1.92	3.33	3.62
	Removal rate	—	0.36	0.37	0.75	0.81	2.22	2.51

Table 1.4 The separation of nitrite bacteria

Name of nitrite bacteria	BAC Population	Percent	PPC-BAC Population	Percent
Nitrosospira	2	33.2%	2	50%
Nrtrosominas	1	16.6%	1	25%
Other	3	49.8%	1	25%

Table 1.5 The separation of nitrate bacteria

Name of nitrite bacteria	BAC Population	Percent	PPC-BAC Population	Percent
Nitrobacter	1	33%	1	25%
Nitrospina	2	66%	3	75%

Species of nitrite and nitrate bacteria

Nitrite bacteria and nitrate bacteria at the height of 20 cm in both biofilters with and without permanganate preoxidation were separated.

Table 1.4 shows the nitrite bacteria separation results. Four strains and six strains are separated respectively from biofiters with and without permanganate preoxidation. By testing the percent of nitrite and watching strain morphological property, the results showed that the species of nitrite bacteria in two biofilters were the same and the species of nitrite bacteria were *Nitrosospira* and *Nrtrosominas*. On the other hand, some other strains were found in both biofilters. After testing the percentage of nitrite, red color reaction was not found, which indicated that these strains were not nitrite bacterias. However, the effects of these strains are not clear on nitrification and the following study will continue.

Table 1.5 is the nitrate bacteria separation result. The nitrate bacteria separation results show that four strains and three strains are separated respectively from biofiters with and without permanganate preoxidation. The species of nitrate bacteria were same and the species of nitrate bacteria were *Nitrobacter* and *Nitrospina*.

The nitrite and nitrate bacteria separation indicates that permanganate preoxidation does not change the species of nitrite and nitrate bacteria.

Potential of nitrosification and nitrification

The potential of nitrosification and nitrification was significantly correlated to the biomass, which can be an index of the population of nitrite and nitrate bacteria. According to the potential, the nitrosification and nitrification rate at different depths of BAC process were determined and results were presented in Figure 1.2 and Figure 1.3.

At 10 cm, 50 cm and 80 cm of BAC process alone, the nitrosification rate and nitrification rate were 6.18×10^{-6} gNO_2-N·$h^{-1}g^{-1}$, 5.196×10^{-6} gNO_2-N·$h^{-1}g^{-1}$, 1.826×10^{-6} gNO_2-N·$h^{-1}g^{-1}$and 2.7918×10^{-5} gNO_2-N·$h^{-1}g^{-1}$, 7.1637×10^{-5} gNO_2-N·$h^{-1}g^{-1}$, 2.4399×10^{-5} gNO_2-N·$h^{-1}g^{-1}$, respectively. Compared with BAC alone, the nitrosification rate and nitrification rate were 7.584×10^{-6} gNO_2-N·$h^{-1}g^{-1}$, 6.18×10^{-6} gNO_2-N·$h^{-1}g^{-1}$, 1.966×10^{-6} gNO_2-N·$h^{-1}g^{-1}$ and 3.0422×10^{-5} gNO_2-N·$h^{-1}g^{-1}$, 7.2408×10^{-5} gNO_2-N·$h^{-1}g^{-1}$, 4.3886×10^{-5} gNO_2-N·$h^{-1}g^{-1}$respectively.

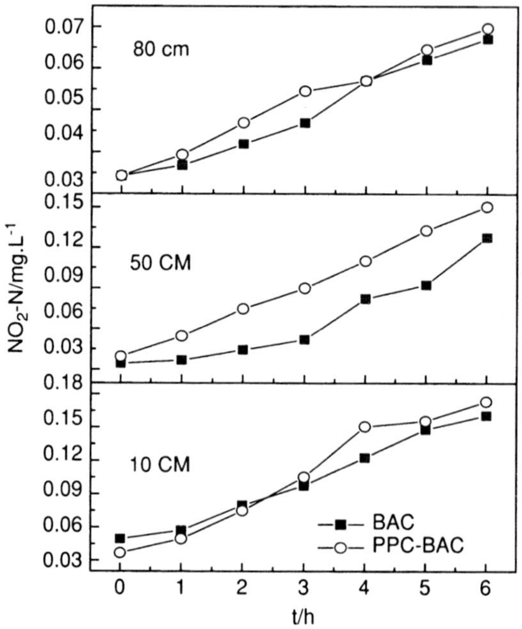

Figure 1.2 The effect of PPC preoxidation on nitrosification performance

The results show that the potential of nitrosification and nitrification is in accordance with the population of nitrite and nitrate bacteria. At different height of biofilter, biomass with permanganate preoxdiation shows that the higher nitrosification rate for ammonia oxidizing into nitrite and nitrification rate for nitrite oxidizing into nitrate.

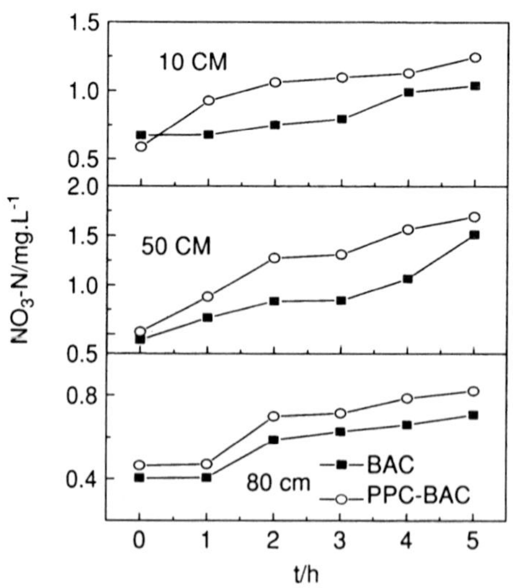

Figure 1.3 The effect of PPC preoxidation on nitrification performance

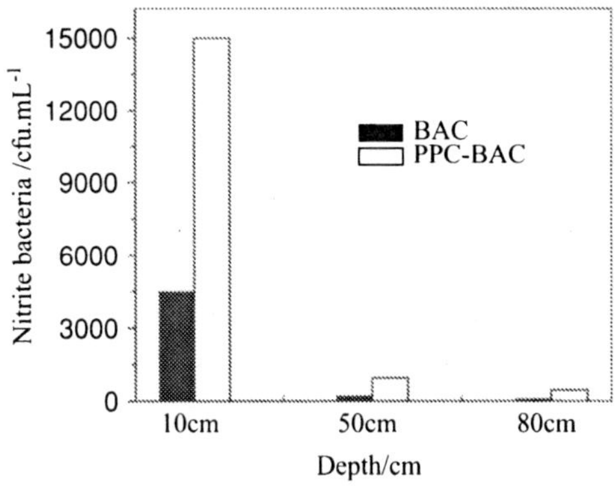

Figure 1.4 The effect of PPC preoxidation on nitrite bacteria distribution in BAC process

Population of nitrite bacteria and nitrate bacteria

Removal of nitrogen compounds from water can be accomplished by nitrification in water treatment. The population of nitrite bacteria and nitrate bacteria at the height of 10 cm, 50 cm and 80 cm in both biofilters with and without permanganate preoxidation were determined by MPN method. The results were shown in Figure1.4 and Figure 1.5.

The cell distribution in biofliter was clear and the law of cell distribution is the same in both biofilters whether with permanganate preoxidation or not. Along the direction of water flow, the population of nitrite bacteria decreased gradually. At the top of biofliter, the population of nitrite bacteria is about from 10^3 cfu.mL^{-1} to 10^4 cfu.mL^{-1} and at the bottom of biofilter, the population of nitrite bacteria is only about from 10^1 cfu.mL^{-1} to 10^2 cfu.mL^{-1}.

However, the population of nitrate bacteria showed the different distribution in the biofilters. Along direction of water flow, the population increased firstly and got most at the middle of biofilter, then the population decreased gradually and got mixed at the bottom of biofilter.

The same cell distribution in both biofilters can be explained by the facts that the dominant bacteria were most oligotrophic bacteria. For low nutrition concentration, the

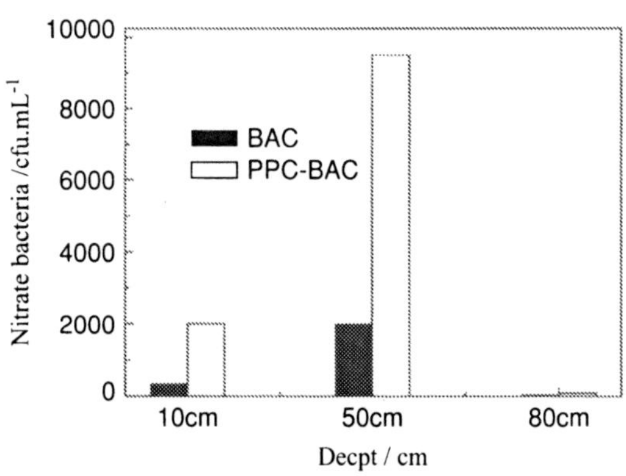

Figure 1.5 The effect of PPC preoxidation on nitrate bacteria distribution in BAC process

competition between heterotrophic microorganism and autotroph bacteria is not obvious. DO and nurture are the main factors for the nitrite bacteria and nitrate bacteria distribution.

Although the MPN method gives a statistical estimation of autotrophic nitrifiers, the difference in the cell numbers shown above in biofilters with and without permanganate preoxidation is significant. At 10 cm and 80 cm of biofilter with permanganate preoxidation, the population of nitrite bacteria are 1.5×10^4 cfu.mL^{-1} and 0.45×10^4 cfu.mL^{-1}, respectively. At the same depth of BAC alone, the population of nitrite bacteria are only 4.5×10^2 cfu.mL^{-1} and 0.9×10^2 cfu .mL^{-1} respectively.

At 10 cm, 50 cm and 80 cm of BAC, MPNs of nitrate bacteria in biofiter with permanganate preoxidation were 2×10^3 cfu $\cdot$mL^{-1}, 9.5×10^3 cfu $\cdot$mL^{-1} and 0.9×10^2 cfu $\cdot$ mL^{-1}respectively. However, at the same depth, the population of nitrate bacteria were 0.35×10^3 cfu $\cdot$mL^{-1}, 2.0×10^3 cfu $\cdot$mL^{-1} and 0.4×10^2 cfu $\cdot$mL^{-1} in the biofilter without permanganate preoxdation.

MPNs of nitrite and nitrate bacteria in BAC with permanganate peroxidation are higher than that without permanganate pretreatment. This results indicated that BAC with permanganate preoxidation provide a more suitable environment for nitrifying bacteria than BAC process alone, although the details of the interaction between biofilter and permanganate preoxidation are not clear.

Discussion

It is said that many factors can influence nitrification process, such as the concentration and the kinds of organic maters, temperature, dissolved oxygen, pH value and so on. In our research, two sets of experimental equipments were operated in parallel and the temperature and aeration condition were the same, so the variety of the concentration and kind of organic maters should be a main factor to influence nitrification. The concentration of organics, especially with large molecule weight, was reduced to a certain extent and the kinds of organics were changed by permanganate preoxidation, which might change kinds of bacteria in biological filter. The variation of heterotrophic bacteria helped to increase the quantity of autotrophic bacteria and improved nitrification.

It is known that the main byproducts of permanganate are manganese (II) and manganese (IV). Some researches showed manganese could influence the circulation of nitrogen. Wang Xiu-heng et al (2003) found that both the dosage and reaction period of manganese (Mn^{2+}) could influence the nitrification. A short reaction time or low dosage resulted in stimulating effect, while prolonged reaction time or raised dosage reversed the effect to an inhibiting effect. The best dosage and time of manganese ion was 5 mg/L and 24 h respectively and the minimum toxicity limit was 40 mg/L after 48 h.

On the other hand, enzyme is necessary in nitrification process. Tan Wen et al (1999) had found that oxide of manganese is a kind of chemical catalyzer, which can improve the nitrification process.

Conclusions

Laboratory experiments were set up to estimate the performance of biofilters with and without permanganate preoxidation. The performance of nitrification was investgated. Our researches suggested that biofilter with permanganate preoxidation had a higher ammonia and nitrite removal rate, 6.57% and 8.75% respectively than that without permanganate preoxidation. The performance of nitrification and species of nitrite and nitrate bacteria were determined. The results showed that nitrite bacteria and nitrate bacteria had the same distribution in both biofilters. Along the direction of water flow, the population of nitrite bacteria reduced gradually. However, the population of nitrite bacteria increased first and then reduced and got mixed at the bottom of the biofilters. In addition, permanganate preoxidation did not change the species of nitrite and nitrate bacteria. The main nitrite bacteria were *Nitrosospira* and *Nrtrosominas* and nitrate bacteria were *Nitrobacter* and *Nitrospina* in two biofilters.

Finally, the population of nitrite and nitrate were determined by MPNs. More population of nitrite bacteria and nitrate bacteria were obtained in the BAC with permanganate preoxidation, which was the main reason that lead to a better performance for removing the ammonia and nitrite.

Acknowledgement

The support by the Ministry of Education under the scheme of Cultivation Foundation for Key Research Project is greatly appreciated.

References

M.Kajino, K.Morizane, T.Umetani, K.Terashima. (1999) Odors arising from ammonia and amino acids with chlorine during water treatment. *Wat. Sci. Tech*, **40**(6), 107–114.

O.Nowak, G.Wandl, K.Svardal.(1999) Long-term experience with biological pretreatment of rendering plant effluent. *Wat, Sci, Tech,* **40**(1),37–44.

Fisher, Daniel J.;Burton, Dennis T. ;Yonkos, Lance T. ;Turley, Steven D. ;Ziegler, Gregory P. (1999) The relative acute toxicity of continuous and intermittent exposures of chlorine and bromine to aquatic organisms in the presence and absence of ammonia. *Wat. Res.,* **33**(3), 760–768.

A.Kihn, Andersson, A.; Laurent, P.; Servais, P.; Prevost, M (2002) Impact of filtration material on nitrification in biological filters used in drinking water production. *Water Supply: Research and Technology-AQUA, IWA*, 35–45.

Oleszkiewicz J.A. Berquist S.A. (1988) Low temperature nitrogen removal in sequencing batch reactors. *Wat. Res.* **22**(9), 1163–1171.

Majun, Renzhijun. (2005) Permanganate preoxidation /BAC process for treatment of polluted surface water. *China Water & Wastewater*, **21**(11), 59–61.

Wang Xiuheng, Ren Nanqi, Wang Aijie, Ma Fang. (2003) Effect of ferrous and manganese ion on nitrification. *JOURNAL OF HARBIN INSTITUTE OF TECHNOLOGY*, **35**(1), 122–125.

Tanwen, Jiangwenjun, Liujun Cao Lihua. (1999) The Nutritional Risistance of Tobacco Individual Development to Diseases. *Technology of Tobacco*, (1):46–48.

51 Integrated biological filtration and reverse osmosis treatment of cold poor quality groundwater on the North American prairies

H. Peterson(1, 2), R. Pratt(2, 3), R. Neapetung(2, 4), O. Sortehaug(5)

(1) WateResearch Corp, Saskatoon, Canada
(2) Safe Drinking Water Foundation, Saskatoon, Canada
(3) Gordon First Nation, Punnichy, Canada
(4) Yellow Quill First Nation, Yellow Quill, Canada
(5) Maxit Group, Madrid, Spain

Abstract Groundwater supplies on the Canadian prairies are often contaminated with arsenic, high levels of sulphate, ammonium or nitrate, iron and manganese as well as organic material. These typically brackish water sources (TDS levels >1000 mg/L) are supplying potable and non-potable water requirements of water-short communities. The high cost of supplying better quality water (pipelines) to remote native communities sparked the development of water treatment processes that can effectively deal with local water sources. During the past four years treatment processes for these poor quality groundwater sources have been developed. A community with a long-standing boil water advisory (total duration 9 years), Yellow Quill First Nation, was selected for the initial evaluation and development of different water treatment processes. During a 20 month on-site pilot conventional (including manganese greensand), advanced (including ozone), and emerging (including biological) water treatment processes were evaluated. The most promising processes were biological and the majority of our effort was expended developing a combination of biological processes that could remove iron, bioavailable organic carbon and other microbial nutrients/energy sources including phosphorus and arsenic, and also oxidize 4 mg ammonium-N/L. This well water, when allowed to oxidize chemically, produced 0.5 million particles/mL, but biological oxidation (with the bacteria gaining energy from the oxidation process) prevented this particle formation and particle levels after biological filtration were less than 100 particles/mL. The objectives of the biological treatment were therefore to remove microbial nutrients and energy sources as well as preventing the formation of particles. Because of high salt levels and refractory dissolved organic material, the biologically treated water needed to be polished with reverse osmosis (RO) membranes. Particles can directly foul RO membranes and bacterial energy or nutrient sources promote the growth of bacteria with high risk of membrane biofouling. An integrated biological treatment process was developed to deal with all of these issues. Three biofilters operated in series provided different conditions suitable for the sequential purification of this extremely poor quality water source. Air was introduced ahead of the three filter units. This was to increase the redox potential to a

 Recent Progress in Slow Sand and Alternative Biofiltration Processes edited by Rolf Gimbel, Nigel J.D. Graham and M. Robin Collins. ISBN: 9781843391203. Published by IWA Publishing, London, UK.

range where bacteria can start the biological oxidation of iron. The full-scale water treatment process is a copy of one of the pilot systems and continued development in the full-scale plant has further increased its efficiency. Despite $6^{\circ}C$ water, all of the ammonium is oxidized to nitrate, most of the arsenic and phosphorus is removed, and other bioavailable material (e.g., bioavailable DOC) is removed. After the biological filtration the water is biologically stable resulting in low biological fouling of the RO membranes with up to 18 months between membrane cleanings. The Yellow Quill Water Treatment Plant has been providing high quality drinking water to its 1,000 people for more than 2 years and the developed process has now been successfully piloted in three other water treatment plants. Two of these plants, Gordon and Pasqua First Nations, have now replaced their manganese greensand filtration material with biological support material. The developed process offers the following advantages compared with the most commonly used conventional groundwater treatment process, manganese greensand pre-treatment ahead of RO membranes: 1) The biological filtration uses no chemicals, 2) The process is robust with few operator interventions, 3) It is highly cost-effective due to overall low chemical use, low antiscalant ahead of membranes and low chlorine doses required, 4) Inexpensive and long-lasting filtration material (Filtralite® Expanded Clay), 5) Continuous operation (can be >30 days), 6) Long time interval between membrane cleanings (>1 year), 6) Increased longevity of RO membranes, 7) Time interval between filter backwashes >10 times greater and <5% backwash water is required, 8) Improved work quality for operators as backwashing and chemical handling is much reduced.

Keywords Aboriginal communities, ammonium, arsenic, bioavailable dissolved organic carbon, biological filtration, ground water treatment, iron, nitrification, Reverse Osmosis

Introduction

The regulation of drinking water quality has received increasing attention during the past decade with major waterborne disease outbreaks in North America and elsewhere. However, production of high quality drinking water at the treatment plant meeting all health and aesthetic guidelines may still not provide acceptable quality water at the kitchen tap. The reason for this is deterioration of quality in the distribution system (Geldreich, 1996). There are no guidelines anywhere that deal effectively with the cause of this problem, the presence of microbial nutrients in the treated water. Any compound capable of encouraging microbial growth is a concern including phosphorus, nitrogen, dissolved organic material, iron, manganese, and arsenic.

While some microbial nutrients, such as iron has an aesthetic guideline associated with it, even when this guideline is met low iron levels can provide an essential nutrient requirement for microbes growing in the distribution system. Others, such as nitrate and arsenic, have health guidelines, but again the microbes can make use of these compounds at levels that are even lower than these guidelines. Several, however, such as ammonium, phosphorus and dissolved organic material have no direct guidelines associated with them. Yet, these compounds are key requirements for the microbial communities that establish themselves as biofilms in drinking water distribution systems (Geldreich, 1996).

With no effective guidelines or regulations to curb the release of microbial nutrients in the distribution system it can be anticipated that even with better processes to treat the water, the quality of the delivered water can still be a concern. Indeed, virtually no presently used treatment processes are capable of effectively removing all microbial nutrients. It was long assumed that tight membrane technologies, such as nanofiltration and reverse osmosis, were capable of eliminating microbial nutrients in an effective fashion. It has, however, now been clearly shown that nanofiltration membranes provide a poor barrier for bioavailable organic material with proliferation of microbes in the treated water (Liikanen et al. 2003). Even a tight Reverse Osmosis (RO) membrane was only able to remove 42% of the compounds causing Biological Oxygen Demand (Al-Wazzan et al. 2002). This is especially interesting as the RO membrane rejected much smaller ions, such as sodium, by 99%.

Indeed, some conventional and advanced treatment technologies, such as ozonation generate microbial nutrients. Even chlorination before distribution can cause oxidation of DOC making it more bio-available (Prevost et al. 2005). Therefore most conventional and advanced technologies are not able to produce finished water that is devoid of microbial nutrients with oxidative disinfection practices exacerbating the problems. The ensuing microbial growth can cause loss of chlorine residuals, development of taste and odour, and it can facilitate the proliferation of disease-causing microorganisms, such as *Legionella*, and *Mycobacterium* as well as repair and re-growth of coliforms etc., in the distribution system (Geldreich, 1996).

A major limiting factor when membrane technologies are implemented is microbial fouling of the membranes, which will decrease the quality of produced water and cause increased requirements for membrane cleanings and premature failure of the membranes (Baker and Dudley 1998). As tight membranes (Reverse Osmosis and nanofiltration membranes) are important parts of most advanced water treatment processes it is essential to remove compounds that can support microbial growth ahead of the membranes. Clearly, no conventional or advanced technologies can effectively deal with this. Emerging water treatment processes that incorporate microbial growth can, however, remove most of these compounds. It comes down to either grow the microbes in a controlled way in the water treatment plant or suffer the consequences of uncontrolled growth in membrane processes and distribution systems.

We have worked on including microbial water treatment processes for poor quality groundwater sources in aboriginal communities in Central Canada. A concern in implementing such treatments is the low temperature of the raw waters, which range from 5°C–10°C. With well depths from 100–200 m the water temperature is stable with no more than 1°C seasonal variation for an individual well. Theoretically these low temperatures should result in much decreased microbial activities compared with higher temperatures impeding the development of biological water treatment systems. However, we have during the past 4 years developed treatment processes that are not only effective in removing microbial nutrients, they are also completed without chemical additions. These processes have also been integrated with reverse osmosis (RO) membrane treatment. A 20 month pilot and development study resulted in the design of an Integrated Biological and RO Membrane Treatment Plant, which was completed in December 2003 at Yellow Quill. On-site research has enabled optimization of the processes and two more plants, Pasqua and Gordon First Nations, were commissioned at the end of 2005.

Methods

For the pilot study two refrigerated truck trailers (2.4 by 14.4 m) were outfitted with a laboratory, pilot-scale treatment units, office, living and eating quarters. These fully insulated trailers can withstand winter conditions down to −40°C and they were parked next to Yellow Quill's well heads. An insulated connection to one well was made and from July 2002, to March 2004 well water was continuously pumped at 200 L/min into the pilot trailer. This water was distributed by a header pipe on the wall close to the ceiling allowing simultaneous piloting of different processes. Pressure and gravity filters, chemical injectors systems, ozone and UV units were all easily connected in series to find which combination of treatments were most suitable. We had most success with a biological treatment combination consisting of one contactor followed by three filters (containing different types of aggregates for microbial adhesion) and the discussion in this paper centers around the trials with these units, which were followed by RO membranes. Due to the aggressive nature of this water we used plastic material wherever possible.

Samples for routine parameters, such as pH, redox potential, temperature, dissolved oxygen, iron, manganese, turbidity, colour, ammonium, and nitrate were carried out within minutes after samples had been collected in the pilot system using mainly Hach equipment and methodologies. Our Hach equipment included oxygen, turbidity, redox and pH meters as well as several spectrophotometers (Hach DR 890, DR 2000, DR 3000, and DR 4000). For extensive metal analyses, ion coupled plasma-atomic emission spectrometry (ICP-AES) was used by an outside analytical laboratory. Dissolved organic carbon was analyzed using persulfate oxidation with a phenolphthalein colour reagent. Sulphate analysis was carried out using a turbidometric/spectophotometric analysis with barium sulphate. Analyses were carried out according to Standard Operating Procedures almost all adhering to the APHA.

Results and discussion

Background

This work originated at Yellow Quill First Nations, a native community of around 1000 people in Central Canada. Yellow Quill has had the longest boil-water-advisory in Canada for this size of community. The boil-water advisory was put in place in 1995 and it could not be lifted with the existing surface water source and water treatment equipment. A search for better quality source waters was not successful as good quality water that could possibly be used was one hundred kilometers away. Because communities like Yellow Quill are frequently remote with low populations, pipeline costs are high and while extensively used, heavy government subsidies have been the rule rather than the exception for such developments. But, as many of these water sources contain ample microbial nutrients, which are sometimes increased rather than decreased by conventional water treatment processes, the growth of disease-causing microbes and loss of chlorine residuals in long pipelines makes remote treatment and distribution less attractive.

The potential for local groundwater sources to supply these rural communities is positive in terms of quantity, but quality is frequently extremely poor. Despite concerns at Yellow Quill with high levels of iron, manganese, arsenic, dissolved organic carbon, ammonium, hardness and high TDS levels, it was decided that a pilot-facility be established to determine which treatment processes could possibly render this water drinkable. Conventional technologies, such as manganese greensand, were ineffective and while some other technologies were promising, such as ozonation, the floc generated resulted in short filter runs. While biological filtration was attractive, the temperature of the well water was only 6°C and its usefulness was not known.

Yellow Quill's ground water and other similar sources rapidly generate both scaling and fouling problems, which must be taken into consideration when designing the full-scale treatment systems. This also translates into problems with analytical equipment, such as various probes requiring frequent cleanings. Daily calibration of the redox probes was essential, and other probes were also frequently calibrated. Indeed, more than 20 redox probes were poisoned during the Yellow Quill pilot study. Even industrial in-line probes, such as Endress and Hauser redox probes, rapidly failed when exposed to marginally treated Yellow Quill groundwater. Before on-line probes are used it is essential to moderate the fouling properties of the water.

Yellow Quill's ground water source

Yellow Quill's wells are around 100 m into the ground resulting in rather cold temperatures year-round, around 5.5°C. Deeper wells have typically water temperatures 8–10°C. The water is high in TDS, and specific ions including iron, manganese, arsenic, ammonium, sulphate, as well as hardness and dissolved organics. The chemical composition of the Yellow Quill raw water is shown in Table 1.1 with the Canadian Water Quality Guideline for the different compounds. This was the original water analysis from which we were basing our treatment needs. The ammonium-N level was above 10 mg/L, but after the wells were operational this decreased to around 4.0 mg/L. In contrast, manganese levels stabilized at higher rather than lower levels after the wells were drawn upon. The compounds that are above the Canadian Water Quality Guidelines in the table have been highlighted (DOC has also been highlighted as this level is too high to safely chlorinate without generating other problems).

Biological water treatment

In biological treatment of water microbes attached to filter media carry out the desired removal. The goal with biological removal of different compounds is to move these compounds into a particulate state either as a part of the microbial biomass or as organic/inorganic flocs that can be easily separated out. We have worked with GAC as well as expanded clay material (Filtralite®).

The processes used here is aerated biological treatment where oxidizing processes (bio-oxidation) are used and the microbes gain energy when they transform the ions from a reduced to an oxidized state (i.e., from Fe^{2+} to Fe^{3+}, Mn^{2+} to Mn^{3+}, As^{3+} to As^{5+}, and NH_4^+ to NO_3^-, Mouchet 1992). Biological treatment as discussed here is centred on establishing microbes that will need to obtain energy (bio-oxidation) from the conversion and sequestration of compounds that we would like to remove from the water.

Therefore a key factor in biological treatment is the provision of oxygen which was introduced ahead of the first filter in a series of three biological filters (both air and oxygen

Table 1.1 The composition of the Yellow Quill Raw water in the spring of 2001 in comparison to the Canadian Water Quality Guidelines (Health Canada, 1996).

Constituent	Unit	Reserve Aquifer	GCDWQ
Inorganic Parameters			
Alkalinity	mg/L	427	500
Aluminum	mg/L	<0.005	0.1
Arsenic	mg/L	0.011	0.025
Barium	mg/L	0.008	1
Boron	mg/L	0.74	5
Cadmium	mg/L	<0.001	0.005
Chromium	mg/L	<0.001	0.05
Copper	mg/L	<0.001	1
Hardness	mg/L	972	800
Iron	mg/L	7.6	0.3
Lead	mg/L	<0.002	0.01
Magnesium	mg/L	92	200
Manganese	mg/L	0.21	0.05
Nitrate	mg/L	<0.01	45
pH		7.68	6.5–8.5
Sodium	mg/L	238	00
Sulfate	mg/L	1010	500
TDS	mg/L	2098	500
Zinc	mg/L	<0.005	5
Organic Parameters			
DOC	mg/L	10	Not in guideline

was used). The majority of the oxygen consumed occurred in the last filter, which is the nitrification unit. Oxygen consumption rates >10 mg/L in 30 min are similar to oxygen consumption rates over a 3.5 month period in ice-covered lakes in Central Canada making the biological activity in Filter 3 some 5000 times greater than a highly biologically active natural aquatic system in a cold climate.

Biological iron and manganese removal

The key factors to be controlled during biological iron removal include establishing the correct redox potential and oxygen concentrations. To remove iron a redox potential of around 0 mV is desired, but the groundwater at Yellow Quill is around −100 mV. In addition, the water smells metallic and it was considered essential to remove the gasses giving rise to this smell.

Table 1.2 The iron content, iron removal, % removal, and redox potential in the raw water and after the different treatment units (average of determinations for three different dates with standard deviation within brackets, water temperature 6°C)

Filter Unit	Iron level (mg/L)	Iron removed (mg/L)	% Removed	Redox
Raw Water	8.15	0	0	−82
One	2.30 (0.52)	5.85	72	−13 (12)
Two	0.20 (0.09)	2.10	26	60 (12)
Three	0.01 (0.01)	0.19	2	239 (8)

Various means of changing the redox potential and removing the gasses were tested and the issue became how to introduce oxygen and what contact time was required. Several contactors were designed before one that allowed for an optimum increase of the redox potential was found. In the full-scale treatment plant several modifications to the contactor have been made. Iron removal rates can be variable and three determinations over a two-week period have been averaged in Table 1.2. All iron is removed by the biological process and even when challenged the residual iron levels after filter 3 are less than 0.05 mg Fe/L or >99.4% removal. Increasing redox potentials indicate complete iron removal and oxidation of ammonium (Table 1.2).

Biological removal of arsenic, ammonium, and bioavailable organic carbon

The As compound of most concern is inorganic As^{3+}, which is more difficult to remove from solution using physical and chemical removal techniques than oxidized As^{5+}. Conventional processes use oxidizing compounds, such as chlorine, ozone and potassium permanganate to obtain this transformation. Microorganisms can, however, also carry out the transformation from As^{3+} to As^{5+} gaining energy in the process. While As^{3+} is poorly removed by RO membranes As^{5+} is effectively removed (Ning 2002). The biological process will convert As^{3+} to As^{5+} and when followed by RO membranes any remaining As will be removed. Arsenic can also co-precipitate with iron. The arsenic level in the Yellow Quill raw water was 10–12 μg/L and removal to levels close to our detection limit of 2 μg/L was achieved by the biological filters. Phosphorus is a compound that is similar to arsenic and arsenic's anion, arsenate, can be mistaken by microbes for the phosphorus' anion, phosphate, which is an essential nutrient. The phosphate level is close to our limit of detection at Yellow Quill even in the raw water.

Ammonium ions play a key part in water treatment because they need to be removed before breakpoint chlorination can be achieved. Oxidizing reagents, such as ozone, ClO_2, chloramines, and potassium permanganate cannot remove ammonium ions while chlorine reacts with the ammonium ions forming chloramines that are 10–100 times less potent than chlorine in terms of disinfection capacity (Degremont 1991). However, for every mg ammonium removed, the chlorine demand is around 10–15 mg making it a totally unsuitable process for Yellow Quill. Biological removal can be achieved by oxidation of ammonium to nitrate (nitrification). We were unable to generate a nitrifying population using the existing well water at Yellow Quill. Instead, we started searching for suitable nitrifying populations from uncontaminated natural water sources. This was successful and ammonium oxidation rates around 80% were achieved in the pilot units. During full-scale operation nitrification was discouraged (and no additions of any microbes were carried out), but after 8 months of operation nitrifying activity was evident. Modifications to the full-scale plant set out to encourage nitrification has now resulted in >98% removal of ammonium. Yellow Quill's ground water is "old" and only a small part of the DOC is bioavailable. Around ten percent of the DOC and 20% of the colour and UV-254 absorbing compounds were removed by the biological filters.

The biologically treated water was further polished in various Reverse Osmosis membrane combinations ranging from single stage to three stage units. Low fouling resulted in few opportunities to clean membranes and our practical knowledge in this area is still poor.

The Integrated biological and RO treatment process

From the different pilot trials the groundwater treatment process that we developed for Yellow Quill First Nations include three biological filters run in series followed by an RO membrane treatment unit.

A summary of the chemistry of the groundwater coming in, the biologically treated water, and the membrane treated water is shown for some of the compounds that we were testing (many of them on a daily basis) in Table 1.3.

In terms of removals the biological filtration removed all iron, arsenic and most particles from the water. In addition ammonium was converted to nitrate and bioavailable DOC was

Table 1.3 The chemistry of the raw water, biotreated water and membrane treated water. Compounds affected strongly by biological filtration are shown in grey, while compounds that are affected strongly by membrane filtration are shown in white.

Chemical	***Raw***	***BioTreated***	***Membrane Treated***
Iron (mg/L)	8.15	0.01	0.00
Arsenic (μg/L)	12	<2	<2
Ammonium-N (mg/L)	4.25	0.87	0.03
Nitrate-N (mg/L)	0.02	2.94	1.08
Particles (#/mL)	428484	62	0
Turbidity (NTU)	99.6	0.2	0.1
TDS (mg/L)	1853	1766	123
Alkalinity (mg/L)	424	384	20
Sulphate (mg/L)	1022	1015	1.6
DOC (mg/L)	9.9	8.9	0.5
Colour (TCU)	10.7	8.5	2.4
UV-254	0.093	0.075	0.002
Manganese (mg/L)	0.350	0.354	0.019

also removed (although it is a small part of total DOC). Because of these biological removals we were able to run the membranes with extremely low fouling. The membranes could then effectively remove TDS, alkalinity, sulphate, DOC, colour and manganese. Under the conditions we were using we did not want manganese removal by the biological filtration units, but instead we used the membranes to remove the manganese. We have worked with several scenarios where ammonium removal rates have been optimized or indeed minimized as this ion can also be removed by some membranes.

A water treatment plant was constructed incorporating the developed biological and RO membrane processes. After 9 years of continuous boil water advisories Health Canada was able to lift the boil water advisory in March 2004 and the community has enjoyed high quality drinking water since that time. The plant has operated well with few challenges and two more plants were commissioned in the winter of 2005.

Acknowledgements

The authors thank Yellow Quill First Nations and the Department of Indian and Northern Affairs Canada for the opportunity to carry out this work. The authors also acknowledge the contributions of the Yellow Quill Water Project Team including Associated Engineering.

References

Al-Wazzan Y., Safar, M., Ebrahim, S., Burney, N. and Mesri, A. (2002) Desalting of subsurface water using spiral-wound reverse osmosis (RO) system: technical and economic assessment. *Desalination* **143**:21–28.

Baker, J.S. and Dudley, L.Y. (1998) Biofouling in membrane systems – A review. *Desalination* **118**: 81–90.

Degremont 1991. Water Treatment Handbook. Sixth Edition, Vol. 1, F Berne and Y Richard (Chief Editors), Lavoisier Publishing, Paris, France.

Geldreich, E. 1996. Microbial quality of water supply in distribution systems. CRC Press Inc., Boca Raton, Florida 33431, 504 pp.

Health Canada. (1996) Guidelines for Canadian Drinking Water Quality (Sixth Edition). Canada Communication Group Publishing, Ottawa, 90 pp.

Liikanen, R., Miettinen, I. Mi. and Laukkanen, R. (2003) Selection of NF membrane to improve quality of chemically treated surface water. *Water Research* **37**:864–872.

Mouchet, P. (1992) From conventional to biological removal of iron and manganese in France. *Journal of the American Water Works Association*, **84**:158–167.

Ning, R.Y. (2002) Arsenic removal by reverse osmosis. *Desalination* **143**:237–241.

Prevost, M., Laurent, P., and Servais, P. (Editors) 2005. Biodegradable organic matter in drinking water treatment and distribution, American Water Works Association, Denver, USA.

52 Heterotrophic denitrification in drinking water treatment - results from pilot plant experiments in Mashhad/Iran

Oliver Dördelmann[1], Patrick Buchta[1], Stefan Panglisch[1], Ferdinand Klegraf[2], Ahmad Moshiri[3], Ali Emami[3]

[1]IWW Rhenish-Westfalian Institute for Water Research, Moritzstr.26, 45476 Mülheim an der Ruhr, Germany (o.doerdelmann@iww-online.de)
[2]VA Tech Wabag, Vienna, Austria [3]Mashhad Water and Wastewater Co. Iran

Abstract This paper summarizes the research results of a biological denitrification pilot plant employing two fixed bed reactors operated parallel in downstream with acetic acid as substrate. The two reactors were filled with two different types of expanded clay. The nitrate concentration of the well which supplied the pilot plant was 115 mg/l. In the reactors a reduction to less than 40 mg/l nitrate could be established for a long time. Due to effective post-treatment (aeration, dual media and activated carbon filtration) the water quality of the pilot plant's effluent was excellent, as there were no nitrite, sulphide or ammonia detectable. Furthermore TOC, turbidity and bacterial count in the pilot plant's effluent were low as well. The nitrate degradation rates of the reactors were about 7 kg NO_3/m^3*d. The overall water demand for flushing the reactors and the dual media filter amounted to 7 % of the feed water.

Keywords Drinking water, ground water treatment, nitrate removal, biological denitrification

Introduction

Mashhad, the 2nd largest city in Iran, has a population of approx. 2.5 million people and its growth of population is high. The city is located in a dry area (~250 mm rainfall per year) in the north-east of Iran and at present Mashhad's water supply depends mainly on groundwater sources. In some of the groundwater wells used for drinking water supply the nitrate concentrations are higher than 100 mg/l – sometimes even higher than 250 mg/l. Due to the risk of methaemoglobinaemia for infants ("blue baby syndrome") the WHO recommends a guideline value for nitrate in drinking water of 50 mg/l.

Besides a chlorination, at present there exists no groundwater treatment in Mashhad. Therefore an international research project was initiated, funded by the Iranian Ministry of Energy & Water and the German Ministry of Education and Research (BMBF). Project partners are: Mashhad Water & Wastewater Co. Iran, VA Tech Wabag Co. and IWW Institute

 Recent Progress in Slow Sand and Alternative Biofiltration Processes edited by Rolf Gimbel, Nigel J.D. Graham and M. Robin Collins. ISBN: 9781843391203. Published by IWA Publishing, London, UK.

Figure 1 Biological denitrification pilot plant in Mashhad

for Water Research, Mülheim, Germany. Since October 2004 three pilot plants, designed and constructed in Germany, are in parallel operation in Mashhad (**Figure 1**). The three processes tested for nitrate removal in pilot scale are:

- Ion Exchange,
- Reverse Osmosis and
- Biological Denitrification.

Currently a team of Iranian and German Engineers is operating these plants, probably until the end of 2006. The objectives of this research project are to gain experiences about the operation of the nitrate removal processes under the specific conditions of Iran (e.g. dry climate, supply of energy and chemicals, education of staff). Furthermore it is intended to adapt, to optimise and to evaluate the processes for planning of a full-scale drinking water treatment plant (Panglisch et al. 2005).

Biological denitrification processes in drinking water treatment

In the biological denitrification processes denitrifying bacteria reduce nitrate to nitrogen by the following equation:

$$NO_3 + [H] \rightarrow N_2 \ (+\text{Biomass})$$

For this reduction process a substrate or "electron donator" [H] is required. In autotrophic processes the bacteria use inorganic compounds (e.g. CO_2 and hydrogen) as an electron donator, whereas in heterotrophic denitrification the bacteria use organic compounds. Due to very low concentrations of these compounds in drinking water treatment the denitrification process usually can only be operated when one of the above mentioned substrates is added to the water (e.g. by dosing acetic acid or ethanol). If acetic acid is used as electron donator, the biological denitrification proceeds as follows (Rohmann and Sontheimer, 1985):

$$NO_3^- + 0{,}82\ CH_3COOH \rightarrow 0{,}47\ N_2 + 0{,}3\ CO_2 + 0{,}9\ H_2O + HCO_3^- + 0{,}07\ C_5H_7O_2N$$

In drinking water treatment different processes are applied for biological denitrification and several plants are operating in full scale, mainly in Germany (Overath, 2000). In most of these processes after substrate dosage the feed water flows through a fixed bed reactor. In this reactor an inert material acts as surface for bacterial growth. Typical materials for fixed bed reactors are anthracite, activated carbon, sand, expanded clay and organic polymers (e.g. polystyrene). The grain sizes of these materials vary in a wide range and depend on the operating conditions of the reactor. Some processes are operated in upstream, others in downstream (Uhl and Overath, 2004). During the denitrification process the amount of biomass on the surface material rises and the differential pressure of the reactor increases. Thus, in most processes the reactor has to be backwashed or flushed from time to time. There are some other processes tested in full scale, in which a certain amount of surface material is discharged continuously and then backwashed outside of the reactor, before it is transferred back to the process. Nitrate degradation rates between 0.1 and 5.5 kg NO_3/m^3d are reported in the literature for full scale plants (Sonnenburg and Urban, 2005).

For some years there is a new and promising process applying biodegradable polymers such as Polycaprolactone (Müller et al., 2001). This polymer acts both as surface for bacteria and as a substrate. The process development is still under way and mainly intended as a "simple process" for developing countries. Currently a simultaneous removal of nitrate (by denitrification) and pesticides (by adsorption) is investigated for this process (Boley et al. 2005). Furthermore, in recent years several "new" denitrification processes (e.g. membrane bio reactors) were developed and tested both in laboratory and in pilot scale (McCleaf and Schroeder, 1995; Fuchs et al. 1997; Ergas and Reuss, 2001; Mohseni-Banpi et al. 1999; Nuhoglu, 2002; Wright et al. 2005). It is unclear if one of these "new" processes in the meantime is applied in a full-scale drinking water treatment plant.

Finally it is important to mention that any biological denitrification process applied in drinking water treatment needs a very effective post-treatment following the biological stage for the removal of substrate, gas exchange, particle removal (filtration) and disinfection.

Methods

Description of the biological denitrification pilot plant

The essential part of biological denitrification plant, the two fixed bed bio-reactors were operated downstream and in parallel. The operation pressure was 1.5 bar. The flow rates could be varied from 0.5 to 3.0 m^3/h for each reactor. Expanded clay (Liapor®) with two different grain sizes in the two reactors was employed as growth material. **Table 1** shows the main design parameters of the two biological reactors.

A 12 % aqueous solution of acetic acid was dosed as electron donator into the feed of the pilot plant. Additionally ferrous sulphate and disodiumhydrogenphosphate were dosed as nutrients (see **Figure 2**). The bio-reactors were followed by a gas exchange, a dual media filtration (filter stage 1) a second gas exchange and an activated carbon filtration (stage

Table 1 Design parameters of the biological reactors

Growth Material	—	expanded clay
Grain size Reactor 1	mm	3–6
Grain size Reactor 2	mm	2–5
Reactor diameter	m	0,5
Reactor height	m	5,0
Bed depth	m	3,6
Reactor Volume (Fixed bed)	m^3	0,7
Flow rate	m^3/h	0,5–3,0
Filter velocity	m/h	2,5–15,3
Flow direction	—	downstream
$c(NO_3)_{Feed}$	mg/l	110–115
$c(NO_3)_{Effluent}(Goal)$	mg/l	40

2). From time to time the bio-reactors and the dual media filter had to be backwashed for removing biomass and reducing the differential pressure. The pilot plant did not have a disinfection stage because the effluent was not used for drinking water purpose.

The biological pilot plant worked fully automated with automatic flushing and flow control. Several process data were measured and logged continuously:

- nitrate concentration (feed and effluents of two reactors, dual media & carbon filter)
- turbidity (effluents of dual media filter & carbon filter)
- pH and temperature (after 1st and 2nd gas exchange)
- differential pressure (two reactors, dual media filter & carbon filter)

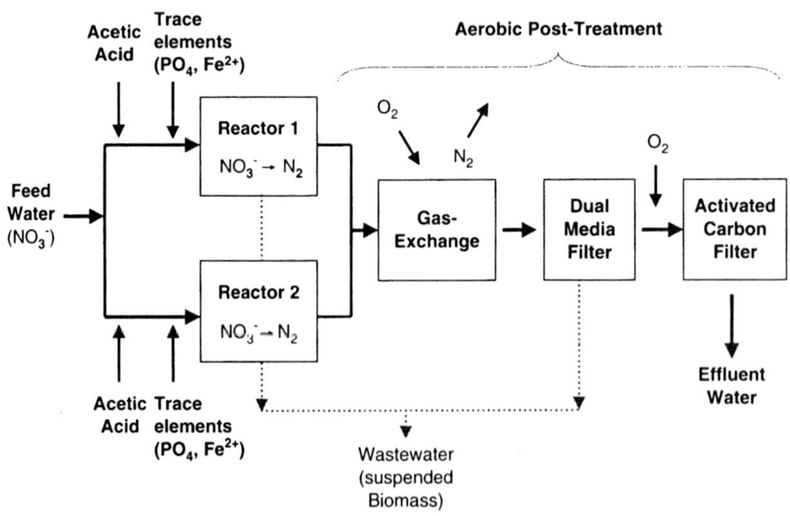

Figure 2 Process Scheme of the biological denitrification pilot plant

- absolute pressure (two reactors, dual media filter & carbon filter)
- water level (two reactors, dual media filter & carbon filter)
- flow rate (two reactors, dual media filter & carbon filter, flushing flow)

Analysis

Besides the above mentioned online-measurement the following parameters were analysed five days per week on-site by hand measurement with instruments (WTW® or Hach®) or by rapid tests (Merck Reflectoquant® and Aquaquant®):

- pH and temperature (hand measurement)
- Electrical conductivity (hand measurement)
- Dissolved Oxygen (hand measurement)
- Turbidity (hand measurement)
- Nitrate (rapid test strips)
- Nitrite (rapid test strips and cuvette)
- Ammonia (rapid test cuvette)
- Phosphate (rapid test cuvette)
- Sulphide (rapid test cuvette)

The parameter Total Organic Carbon (TOC) was measured daily on-site with a Shimadzu TOC 5050 from the feed and the effluents of the two reactors and from dual media & carbon filter effluents.

Additionally the following parameters were analysed weekly by the central laboratory of Mashhad Water and Wastewater Co:

Calcium, Magnesium, Potassium, Sodium, Chloride, Sulfate, Nitrate, Nitrite, Bicarbonate, Phosphate, Iron, Bacterial count (by HPC-Method)

Results and discussion

Corresponding to the project's goals, a nitrate reduction from 115 mg/l (feed) down to 40 mg/l (effluent) was desired. Due to a stoichiometry factor of 0.82 (mole acetic acid / mole nitrate) this amount of nitrate reduction required a dosage of approximately 60 mg/l acetic acid (24 mg/l DOC).

Figure 3 shows a vertical profile measurement of the two bio-reactors.

The nitrate feed concentration was 115 mg/l. It can be seen that in reactor 2 the predominant amount of the nitrate was reduced in the upper part of the reactor because in the sample taken from a bed depth of 115 cm the nitrate concentration was already down to 55 mg/l nitrate. Simultaneously the TOC concentration was reduced from 29 to 7 mg/l. However, in the middle and lower part of reactor 2 the decline of nitrate and TOC concentrations was less intensive. In contrast to this, in reactor 1 the nitrate and TOC curves declined more uniformly. This means that in reactor 1 the nitrate degradation rates in the upper and lower part of the reactor were more or less the same, whereas in the upper part of reactor 2 the nitrate degradation rate was significantly higher than at the bottom, probably because the substrate was limited at the bottom of reactor 2.

During this profile measurement nitrite was detected in two samples taken at bed depths of 60 and 115 cm from reactor 2, obtaining concentrations of 0.7 respectively 0.8 mg/l.

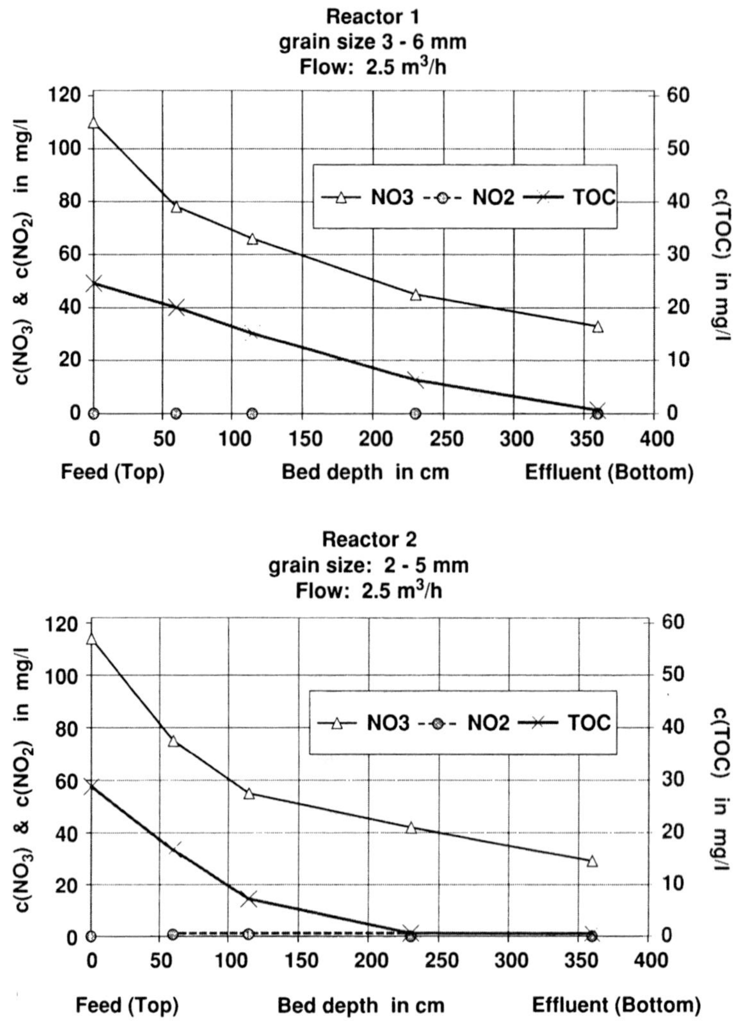

Figure 3 Vertical profile measurement of nitrate-, nitrite- and TOC-Concentration in the two reactors

In the other profile samples and particularly in the reactor's effluent samples there was no nitrite at this time. However, during the pilot plant investigations sometimes nitrite was detected in the reactor effluents, for example 2 mg/l. But in those cases nitrite was always removed surely by aeration and oxidation during the following two aerobic filter stages. Likewise sulphide was never detected in the effluent of the pilot plant. **Table 2** shows the composition of the feed water (before dosage) and the effluent water (after activated carbon filter).

Table 2 shows a significant decrease in nitrate concentration between the feed and the effluent of the pilot plant. Simultaneously there was an increase of pH, mainly caused by

Table 2 Chemical Analysis of feed and effluent of the biological pilot plant

		Feed of Biological Pilot Plant	Effluent of Biological Pilot Plant
Date of Sampling		19.11.05	19.11.05
pH*	—	7,24	7,65
Conductivity (25 °C)*	μS/cm	1516	1472
Temperature*	°C	17	17
Dissolved O_2*	mg/l	2,97	4,55
Turbidity*	FNU	0,38	0,33
Turbidity (online)	FNU	—	0,08
TOC	mg/l	0,37	0,38
Ca	mg/l	139	136
Mg	mg/l	43,7	43,1
Na	mg/l	109	106
K	mg/l	1,8	1,7
Cl	mg/l	184	185
NO_3	mg/l	115	40
SO_4	mg/l	150	147
HCO_3	mmol/l	5,36	6,32
NH_4**	mg/l	<0,2	<0,2
S^{2-}**	mg/l	—	<0,02
NO_2**	mg/l	—	<0,02
PO_4**	mg/l	<0,1	<0,1
Fe	mg/l	<0,01	<0,01

* on site analysis
** analysis by rapid test

bicarbonate production during the denitrification process. The concentrations of the other major cations (Ca, Mg, Na, K) and anions (Cl, SO_4) were not affected by the denitrification process. Due to the dosage of acetic acid the TOC was increased significantly during the treatment process. However, at the end of the process the TOC concentrations of the effluent sample (after activated carbon filter) and the feed sample (before dosage) were 0.4 mg/l both. This means that the acetic acid substrate (~25 mg/l TOC) was degraded completely by the denitrifying bacteria. Nitrite, sulphide or ammonia were not detectable in the effluent sample and also the turbidity was very low - 0.08 FNU (online) respectively 0,33 (hand measurement). The water temperature during the whole process was 17 °C.

Between September and December 2005 six samples were taken from the pilot plant effluent and analysed by HPC-Method. The bacterial count of these samples varied between

Table 3 Nitrate degradation rates and flushing characteristics of the two reactors

		Reactor 1	Reactor 2
grain size	mm	3–6	2–5
Flow rate (feasible)	m^3/h	2,5	2,5*
Filter velocity	m/h	12,7	12,7*
NO_3-degradation rate	kg N/m^3 d	1,4	1,6*
NO_3-degradation rate	kg NO_3/m^3 d	6,4	7,0*
Flushing Time	h	20	16
diff. Press. before flushing	mbar	300	300
diff. Press. after flushing	mbar	150	180
Water demand for Flushing	%	7 %**	7 %**
Water Recovery Rate	%	93 %	93 %

* Flow & NO_3-degr. rate can be increased (result: higher pressure drop)
** approx. 3 % for reactor + 4 % for post treatment (dual media filter)

35 and 300 per ml (average: 180 ml^{-1}), representing a good drinking water quality with regard to this parameter.

One of the most difficult parts of the operation was the optimization of the flushing procedure (flow rate and duration of air and water flushing) of the two reactors. On the one hand an intensive cleaning of the filter material was desired for reducing the differential pressure. But on the other hand a too intensive flushing caused a big loss of biomass resulting in a low nitrate degradation rate in the initial start-up phase after flushing. At optimum conditions flushing of the bio-reactors was required after 20 hours of operation (reactor 1) respectively after 16 hours (reactor 2). **Table 3** gives an overview about the performance and the flushing characteristics of the two reactors.

In reactor 1 filled with expanded clay (grain size 3–6 mm) the long-term nitrate degradation rate was 6.4 kg NO_3/m^3d whereas in Reactor 2 (expanded clay, grain size 2–5 mm) 7.0 kg NO_3/m^3d was found. With the finer material of reactor 2 even higher nitrate degradation rates were feasible for short periods, e.g. by increasing the acetic acid dosage or the flow rate, but this resulted in a much higher pressure drop.

For both reactors the water consumption for flushing was about 3 % of the feed water. In addition to the 4 % water consumption for flushing the dual media filter the total water demand for backwash water was about 7 %.

The spent backwash water contains biomass but no harmful chemicals. Recently IWW has submitted an application for an additional research project focussing on the treatment (and reuse) of this backwash water by membrane filtration.

Conclusions

The biological denitrification pilot plant could be operated very effectively in Mashhad/Iran. There were two fixed bed reactors working in parallel, both filled with expanded clay and operated in downstream. The main difference between the two reactors was the grain size of the expanded clay which was employed as growth surface for the denitrifying bacteria. The material in reactor 1 had a grain size of 3–6 mm, the material in reactor 2 had a grain size of 2–5 mm and therefore had a bigger surface. The feasible flow rate was 2.5 m^3/h for each of the two reactors. The nitrate concentration could be decreased from 115 mg/l (feed) to 40 mg/l in the effluent of reactor 1 and to less than 35 mg/l in the effluent of reactor 2. This means the nitrate degradation rate of this process was 6.4 respectively 7.0 kg NO_3/m^3d. In the literature the highest value reported for nitrate degradation in (full scale) drinking water treatment plants is 5.5 kg NO_3/m^3d. One reason for the remarkable performance of the biological pilot plant in Mashhad could be the good properties of the expanded clay material, which was used in drinking water denitrification for the first time. The second reason is probably the higher temperature of Mashhad's groundwater (~17 °C). In contrast to the typical groundwater temperature in Germany (~11 °C) the biological activity in the bio-reactors in Mashhad could be significantly higher.

Due to effective post-treatment (aeration, dual media and activated carbon filtration) the water quality of the pilot plant effluent was very good, as there were no nitrite, sulphide or ammonia detectable. Furthermore TOC, turbidity and bacterial count in the pilot plant's effluent were low as well.

It has to be mentioned that every biological denitrification process needs very reliable and well trained staff members for operating and maintaining of the plant and for conducting the chemical and microbiological analyses. Fortunately the staff members of Mashhad Water and Wastewater Co. were just excellent.

On account of the very good results obtained at the pilot plant in Mashhad (e.g. high performance rate, good effluent water quality and low water demand for flushing) the biological denitrification process can be regarded as a very promising technology for application in full scale in water treatment plants in Iran and in other countries affected by high nitrate concentrations in drinking water sources.

Acknowledgements

We would like to thank the Iranian Ministry of Energy and the German Ministry of Education and Research (BMBF) for supporting this research project. Our special thanks go to Mr. Naseri and Mr. Ansari who operated the pilot plants.

References

Boley, A., Unger, B., Müller, W.-R., Kuch, B. (2005). Simultane Elimination von Nitrat und Alpha-Endosulfan aus Trinkwasser mit Hilfe von Polycaprolacton (PCL); Poster 02, Proceedings of 71. Jahrestagung der Wasserchemischen Gesellschaft - Fachgruppe in der Gesellschaft Deutscher Chemiker., 161–165. Technical University Berlin

Ergas, S.J., Reuss, A.F. (2001). Hydrogenotrophic denitrification of drinking water using a hollow fibre membrane bioreactor. Journal of Water Supply: Research and Technology-Aqua. **50**.3. 161–171

Fuchs, W., Schatzmayer, G. Braun. R. (1997). Nitrate removal from drinking water using a membrane-fixed biofilm reactor. Appl. Microbiol. Biotechnolgy, **48**, 267–274

McCleaf, P., Schroeder, E. (1995). Denitrification using a membrane-immobilized biofilm. Journal AWWA. March 1995. 77–86

Mohseni-Bandpi, A., Elliott, D.J., Momeny-Mazdeh, A. (1999). Denitrification of Groundwater using Acetic Acid as Carbon Source. Wat. Sci. Technol. **40** (2), 53–59

Müller W.-R., Muller C., Boley A. (2001). Trinkwasseraufbereitung mit biologisch abbaubaren Polymeren - ein Einfachverfahren zur Denitrifikation und Elimination organischer Schadstoffe. In: Der Stickstoff im Wasser/Abwasser. Stuttgarter Berichte zur Siedlungswasserwirtschaft., Institut für Siedlungswasserbau (ed.), vol 166, Oldenbourg Industrieverlag, München, 197–216

Nuhoglu, A., Pekdemir, T., Yildiz, E. (2002). Drinking water denitrification by a membrane bio-reactor. Water Research, **36**, 1155–1166

Panglisch, S., Dördelmann, O., Buchta, P., Klegraf, F., Höll, W.H. (2005). Nitrate Removal from Groundwater in Mashhad/Iran. Poster & Presentation. IFAT München. BMBF-Workshop: Forschung für innovative und nachhaltige Wassertechnologien. Projektträger Forschungszentrum Karlsruhe, Bereich Wassertechnologie und Entsorgung (PTKA-WTE) 25–26, 73

Overath, H. (2000). Stand der Technologie zur Nitratentfernung im Trinkwasserbereich. Technology of Nitrate Removal from Drinking Water: Actual State and Recent Developments. Vom Wasser, **94**, 267–298

Rohmann, U., Sontheimer, H. (1985). Nitrat im Grundwasser – Ursachen, Bedeutung, Lösungswege. DVGW-Forschungsstelle. University Karlsruhe

Sonnenburg A., Urban W. (2005). Untersuchungen zur Denitrifikation in der Wasseraufbereitung mit Hilfe des biologisch abbaubaren Kunststoffs PCL. GWF Wasser Abwasser, **146** (1), 33–39

Uhl W., Overath H. (2004). Biologische Verfahren der Trinkwasseraufbereitung. In: Wasseraufbereitung - Grundlagen und Verfahren, DVGW Lehr- und Handbuch Wasserversorgung (ed.), vol 6, Oldenbourg, München, 339–402

Wright, W.F., Bowman, R.H., Chung, J., Rittmann, B.E. (2005). Economical nitrate reduction for contaminated drinking water sources using hydrogen-based membrane biofilm reactor. AWWA. Annual Conference. Proceedings CD ROM. WED 15–4. Inorganic Potpourri Session 1–11

53 A biological filtration process for denitrification with polycaprolactone as solid substrate in a rotating reactor

A. Boley*, I. Frommert* and W.-R. Müller*

*University of Stuttgart, Institute for Sanitary Engineering, Water Quality and Solid Waste Management, Chair of Hydrochemistry and Hydrobiology, Taskgroup Biology, Bandtäle 2, D-70569 Stuttgart, Germany
(E-mail: angela.boley@iswa.uni-stuttgart.de, w-r.mueller@iswa.uni-stuttgart.de)

Abstract A denitrification system based on the biodegradable polymer Poly-ε-Caprolactone (PCL) was tested in a semi-technical system with a rotating reactor ("Roto-Bio-Reactor"). The objective was the examination of the feasibility and performance of this process for drinking water treatment. A surprising oscillation in the denitrification performance was observed.

Keywords Denitrification, Polycaprolactone (PCL), rotating biological reactor, oscillating performance

Introduction

The intense use of fertilizers in agriculture is a main source of deterioration of ground water in industrialized and developing countries. Nitrate removal in drinking water treatment is therefore still a problem which in practice has not always found simple and satisfactory solutions.

Denitrification with PCL (Poly-ε-Caprolactone) is a process where micro-organisms use the polymer pellets as biofilm carrier and simultaneously as water insoluble carbon source for denitrification. This polymer and others are accessible mostly by enzymatic attack and have already been investigated in the past, Boley et al. (2000), Müller et al. (2001), Hiraishi and Khan (2003). An advantage of PCL in contrast to other biodegradable material is its intrinsic operational security, since it is only biodegradable under anoxic (denitrifying) but not under anaerobic conditions (NO_3^--N < 0.1 mg/L). A low technology realization with respect to maintenance and monitoring is possible. In contrast to conventional denitrification processes with dosage of soluble organic substrates, high performance control mechanisms are not necessary. In water treatment, the use of biodegradable polymers is not confined to denitrification purposes but in addition, a simultaneous removal of organic pollutants as e.g. pesticides is possible, Boley et al. (2003), Deger (2005), Boley et al. (2005).

 Recent Progress in Slow Sand and Alternative Biofiltration Processes edited by Rolf Gimbel, Nigel J.D. Graham and M. Robin Collins. ISBN: 9781843391203. Published by IWA Publishing, London, UK.

One of the problems in former experiments with fixed bed reactors was the occurrence of short-circuiting or blocking due to biomass production. This resulted in frequent backwashing intervals, and led to diminution of the denitrification rates when oxygenated water was used. The intention of this work was the examination and evaluation of an alternative reactor configuration. One of them is the "Roto-Bio-Reactor" (Ambs GmbH, Emmendingen, DE), a slowly rotating drum reactor with three separated compartments in series, Overath et al. (1988). We expected a continuous operation with the biomass-production to be controlled by the shearing forces acting on the continuously moved bed of the granules.

The experiments described here were performed to investigate performance and boundary conditions of the process with regard of the effluent water quality, the susceptibility of the process, and practical aspects of the operation and maintenance.

Methods

The "Roto-Bio-Reactor" (RBR) a semi-technical lab pilot-plant (60 L void volume) was filled with 33 kg PCL (TONETM P787, DOW Chemical Comp.), s. Tab. 2.1 and Fig. 2.1.

The influent was ground water, s. Tab. 2.2, supplemented with NO_3^-, PO_4^{3-} and MoO_4^{2-}, stemming from a stock solution introduced in the mixing tank, in a mass relation of N / P / Mo as 1000/100/1. Iron was added to maintain a concentration range of 0,1 . . . 0,2 mg/L Fe^{2+} all over the experiments. Nitrite was not detectable in the influent from the mixing tank.

For monitoring the water quality in the influent ("in") and the effluent ("out"), computer controlled dosing pumps were installed. The water was then pumped alternately from the mixing tank (influent) and the effluent vessel into an armature for bypass operation, where a UV-Spectrophotometer was installed (spectro::lyser, S::CAN Messtechnik GmbH, Vienna, AT) UV: 200–380 nm, path: 35 mm). Corresponding NO_3^--N and NO_2^--N concentrations were collected, stored via PC equipped with special software (ana::pro V4.0, S::CAN,

Table 2.1 Dimensions of RBR

Reactor	
Void volume	60 L
Length	0.80 m
Diameter	0.31 m
Compartment 1 & 3 (C1 & C3)	
Length, respectively	0.30 m
Volume, void, respectively	22.5 L
Compartment 2	
Length	0.20 m
Volume, void	15.0 L
Separating walls equipped with centered wedge-wire screen	
Diameter	0.11 m
Rotational speed of drum	9.6 h^{-1}

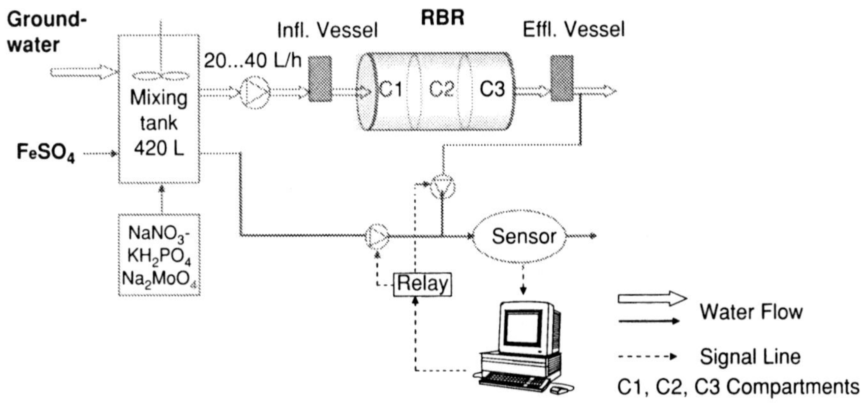

Figure 2.1 Scheme of the denitrification system with a RBR = "Roto-Bio-Reactor" as central unit.

Vienna, AT). Global calibration (transformation of the UV spectra into concentration values) was provided by the manufacturer, local calibration procedures (improved adaptation to local water composition) as described in the software manual were carried out with the ground water spiked with nitrate and nitrite stock solutions, Boley and Müller (2004). Additional parameters, as flow rates (Q), temperature (T), pH, oxygen (O_2) were also measured continuously. O_2, pH, and T were obtained using Multiline P4 devices (WTW, Weilheim, DE). Flow rates were measured with a flowmeter (MAGFLO 5000, MAG 1100, Danfoss, Nordborg, DK).

Table 2.2 Selected ground water parameters (ranges)

Parameter	Value
pH	7.2... 7.4
Temperature	14... 16.5 °C
Conductivity (25°C)	950... 987 μS/cm
NO_3^-N	4.5... 4.6 mg/L
NO_2^-N	<0.3 mg/L
NH_4^+-N	<0.019 mg/L
P_{tot}	1,7 mg/L P
Iron	<0.045 mg/L
Sulfate	40.6... 44.4 mg/L
Chloride	79.8... 98.3 mg/L
Calcium	119... 135 mg/L
Magnesium	36.1... 36.5 mg/L
Total Hardness	4.45... 4.87 mmol/L
Carbonate hardness	3.27... 3.37 mmol/L
Color, Odor	color and odorless

Manual determination of NO_3^--N, NO_2^--N, PO_4^{3-}-P and Fe was performed by photometry, using cuvette tests as described in the specific instructions (Dr. Lange, Düsseldorf, DE). DOC was measured by Multi-N/C (thermocatalytic combustion principle, Analytik Jena, Jena, DE).

A presumable inhibition of the bacteria by water constituents was tested with the luminescent bacteria test (DIN EN ISO 11348-3, 1998) with test kits and device LUMIStox (Dr. Lange, Düsseldorf, DE).

Results and discussion

After a short initial phase of 6 d denitrification started. In the following months denitrification rates were not stable, they varied from 5 to 15 mg/(m^2*h) NO_3^--N (related to the surface area of the PCL-granulate). Technical problems occurred such as migration of the PCL-granules out of compartment 1 back into the inlet and from compartment 3 to the outlet vessel. After solving these "teething problems", the system was running technically stable without any clogging.

After six months, the system showed a strange kind of stabilization with periodical oscillations of the effluent quality at a period of 2 to 4 d (Fig. 3.1). The reasons for these variations are not yet known.

We supposed quality changes of the influent water but another denitrification system with PCL and the same influent water did not show any inhibition or oscillation. Sometimes the groundwater supply (technical problems) had to be replaced for short periods to tap water and the system was faced with a sudden raise of O_2 conc. from 6 to 12 mg/L. Consequently, the nitrate reduction decreased but the variations continued.

As possible explanation for this oscillation, we actually assume a "short" breakdown of the biocenosis performance for yet unknown reasons.

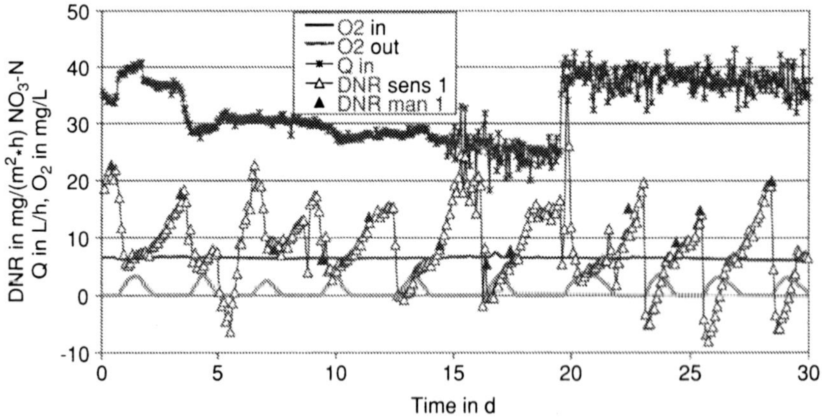

Figure 3.1 Oxygen concentrations (O_2), Denitrification rates (DNR . . . 1) and flow rates (Q in) after 6 month of operation. In: influent, out: effluent of the reactor, sens: sensor, man: manual. Temp. in: 14 °C, out: 16 °C. (Negative values of "DNR sens" due to interference with NO_2^-).

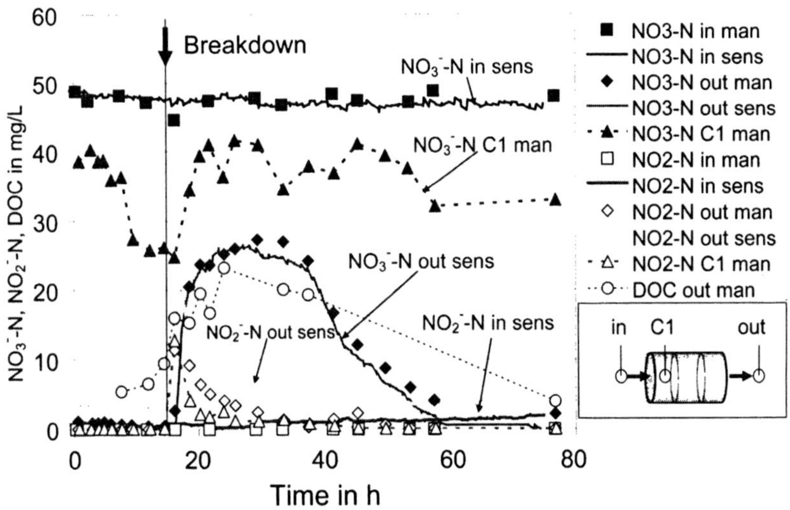

Figure 3.2 Breakdown incidents in the "Roto-Bio-Reactor". Sampling points see sketch: in = influent, out = effluent, C1 = compartment 1, man = manual, sens = sensor. Temp. in: 12.5°C, out: 14.5°C, Q=39 L/h, average retention time referred to void water volume: 1.5 h.

To obtain an insight into this phenomenon, samples were drawn in narrow intervals from the influent, the effluent and out of the first compartment (C1) of the RBR and analyzed (s. Fig. 3.2 + 3.3). In parallel nitrate and nitrite were measured by the UV-Spectrophotometer (sensor). O_2, pH and temperature measurements were also performed automatically.

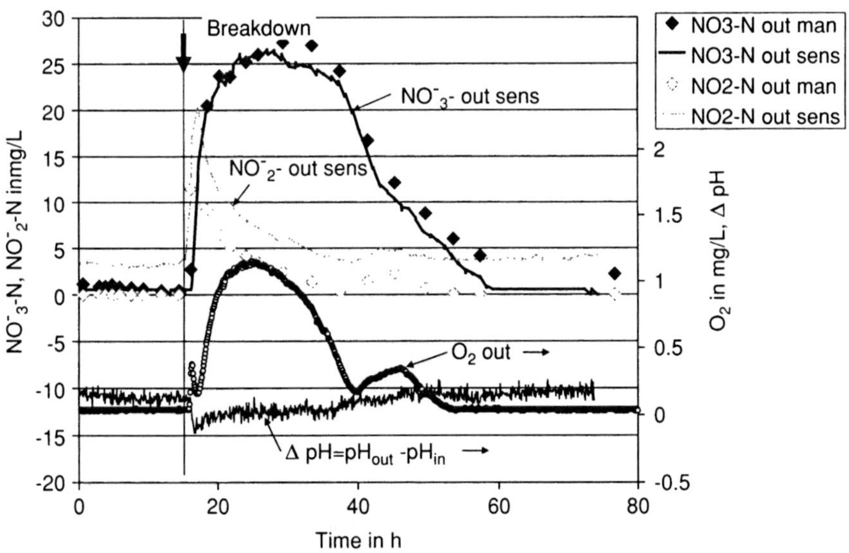

Figure 3.3 Same breakdown incident as in Fig. 3.2 with Δ pH = pH out - pH in and O_2 out, measured by sensors. O_2 in: 6.0 . . . 7.2 mg/L.

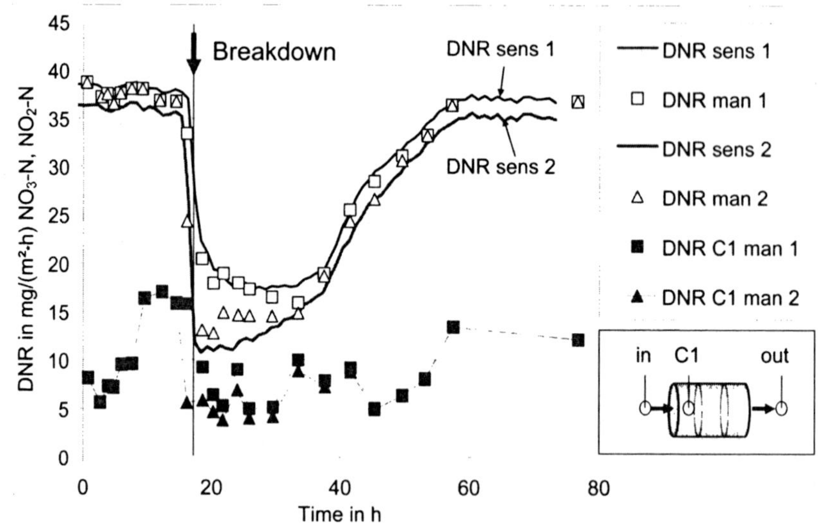

Figure 3.4 Breakdown incidents as in Fig. 3.2 & 3.3. DNR: denitrification rates; sens: sensor; man: manual; C1: compartment 1; DNR...1: $NO_3^- \rightarrow NO_2^-$; DNR...2: $NO_2^- \rightarrow N_2$.

The surprising result was that nearly all parameters measured had a sudden change at the same time ("breakdown"). We can identify here an acute inhibition of the biocenosis in the aerobic part of the reactor, which resulted in higher oxygen concentrations right up to the effluent. The failure in the denitrifying part was indicated by higher nitrite and nitrate concentrations in the effluent, and, due to the lower denitrification activity, lower pH-differences between influent and effluent. The parameter changes which preceded the breakdown were surprisingly enough the depletion of the nitrate concentration in C1 and the increase of the DOC concentration in the effluent (Fig. 3.2) and C1 (not shown).

The effect of the breakdown on the reaction velocity is visualized in Fig. 3.4. Five hours before the overall denitrification rate steeply decreased a strong nitrate removal in compartment C1 can be determined. When these data are transformed into velocity rates without considering biomass production a relevant increase in velocity can be observed which then - with the overall breakdown event - again deteriorates back to the rates 5 h before.

Samples from influent, effluent and C1 around the breakdown incident were subjected to luminescent bacteria tests. In none of these samples any inhibition effect could be found. For the sake of comparison monomer caprolactone was tested, which is the parent compound for the production of PCL and a possible metabolite of its biodegradation. This compound slightly affected the luminescent bacteria > 2.5 mg/L.

Conclusions

The "Roto-Bio-Reactor" proved to be a very comfortable system for denitrification with biodegradable polymer granules of PCL (Poly-ε-Caprolactone) in that sense that no back-washing process as with a fixed bed reactor is necessary. The granules are continuously and gently moved in the different chambers of the reactor. By this way, no clogging with biomass in the bed continuously moved occurred and we never remarked a "mud ball" formation of the granules. The flow-scheme corresponds rather well to a series of complete mixed tanks. Compared to a fluidized bed reactor the energy input is considerably less. After solving the "teething problems" mentioned it seemed to be a reliable system for denitrification.

However, during the operation we observed an oscillation in the performance of the reactor. Until now, with the analytical tools employed, we could not detect its origin. It was not possible to identify any inhibitory effect on luminescent bacteria; therefore a toxic effect of the influent water is unlikely.

We suppose some relevant and rapid change in the biocenosis of the biofilm as the turbidity increased considerably when the breakdown occurred (not shown). However, this is in opposite to all our previous experience where the biocenosis for denitrification proved to be very stable, Mergaert et al. (2001).

Acknowledgement

This work was carried out with the financial support of the State of Baden-Württemberg, Germany, Az. 23-7532.24-30-14/1.

List of abbreviation and symbols

C1, C2, C3	Compartment 1, 2 and 3 of RBR	
conc.	Concentration	
Deni	Denitrification	
DNR C1 man 1	Deni-rate in C1, analyt. data (seq. $NO_3^- \rightarrow NO_2^-$, assumed)	in mg/($m^2 \cdot h$) NO_3^--N
DNR C1 man 2	Deni- rate in C1, analyt. data (seq. $NO_2^- \rightarrow N_2$, assumed)	in mg/($m^2 \cdot h$) NO_2^--N
DNR man 1	Overall Deni-rate RBR, analyt. data (seq. $NO_3^- \rightarrow NO_2^-$, assumed)	in mg/($m^2 \cdot h$) NO_3^--N
DNR man 2	Overall Deni-rate RBR, analyt. data (seq. $NO_2^- \rightarrow N_2$, assumed)	in mg/($m^2 \cdot h$) NO_2^--N
DNR sens 1	Overall Deni-rate RBR, sensor data (seq. $NO_3^- \rightarrow NO_2^-$, assumed)	in mg/($m^2 \cdot h$) NO_3^--N
DNR sens 2	Overall Deni-rate RBR, sensor data (seq. $NO_2^- \rightarrow N_2$, assumed)	in mg/($m^2 \cdot h$) NO_2^--N
DOC in man	Dissolved Organic Carbon, in, analyt. data	in mg/L C
DOC out man	Dissolved Organic Carbon, out, analyt. data	in mg/L C
in	Influent	
man	Manual sampling and determination of respective value analytically	
NO2-N C1 man	Nitrite, compartment 1, analyt. data	in mg/L NO_2^--N
NO2-N in man	Nitrite, influent, analyt. data	in mg/L NO_2^--N
NO2-N in sens	Nitrite, influent, sensor data	in mg/L NO_2^--N
NO2-N out man	Nitrite, effluent, analyt. data	in mg/L NO_2^--N

NO2-N out sens	Nitrite, effluent, sensor data	in mg/L NO_2^--N
NO3-N C1 man	Nitrate, compartment 1, analyt. data	in mg/L NO_3^--N
NO3-N in man	Nitrate, influent, analyt. data	in mg/L NO_3^--N
NO3-N in sens	Nitrate, influent, sensor data	in mg/L NO_3^--N
NO3-N out man	Nitrate, effluent, analyt. data	in mg/L NO_3^--N
NO3-N out sens	Nitrate, effluent, sensor data	in mg/L NO_3^--N
out	Effluent	
PCL	Poly-ε-Caprolactone	
RBR	"Roto-Bio-Reactor"	
sens	Sensor, value based on sensor measurement	
seq.	Reaction sequence, no biomass production considered	

References

Boley, A., Müller, W.-R. and Haider, G. (2000) Biodegradable polymers as solid substrate and biofilm carrier for denitrification in recirculated aquaculture systems. *Aquacult. Eng.* **22**, 75–85.

Boley, A. and Müller, W.-R. (2004) Experiences with a multi-parameter submersible UV/VIS spectrometer for monitoring a denitrification reactor system, *Proceed. 2nd Internat. IWA Conf. Automation in Water Quality Monitoring, Vienna Austria, 19-20.04.2004*, pp. 321–325.

Boley, A., Mergaert, J., Muller, C., Lebrenz, H., Cnockart, M.C., Müller, W.-R. and Swings, J. (2003) Denitrification and Pesticide Elimination in Drinking Water Treatment with the Biodegradable Polymer Poly(ε-caprolactone) (PCL), *Acta hydrochim. hydrobiol.* **31**, 1–9.

Boley, A., Unger, B., Müller, W.-R. and Kuch, B. (2005) Simultane Elimination von Nitrat und Alpha-Endosulfan (ein Pflanzenschutzmittel) aus Trinkwasser mit Hilfe von Polycaprolacton, *Proceed. 71. Jahrestag. Fachgr. Wasserchemie GDCh, Bad Mergentheim, 2. -4.05.2005*, Poster P 02, pp. 161–165.

Deger, A. (2005) Entfernung von organischen Schadstoffen aus Wasser mit Hilfe von Poly-ε-caprolacton. *Schriftenr. Ber. Wasserchem., Engler-Bunte-Inst. Uni Karlsruhe*, vol. 43, Karlsruhe.

DIN EN ISO 11348-3 (1998) Water quality - Determination of the inhibitory effect of water samples on the light emission of Vibrio fischeri (Luminescent bacteria test) - Part 3: Method using freeze-dried bacteria (ISO 11348-3:1998); German version EN ISO 11348-3:1998

Hiraishi, A. and Khan, S.T. (2003) Application of Polyhydroxyalkanoates for denitrification in water and wastewater treatment, *Appl. Microbiol Biotechnol.* **61**, 103–109.

Mergaert J., Boley A., Cnockaert M.C., Müller W-R. and Swings J. (2001) Identity and Potential Functions of Heterotrophic Bacterial Isolates from a Continuous-Upflow Fixed-Bed Reactor for Denitrification of Drinking Water with Bacterial Polyester as Source of Carbon and Electron Donor. *Syst. & Appl. Microbiol.* **24**, 303–310.

Müller, W.-R., Muller, C. and Boley, A. (2001) Trinkwasseraufbereitung mit biologisch abbaubaren Polymeren, ein Einfachverfahren zur Denitrifikation und Elimination organischer Schadstoffe. 1. Hydrochem. Hydrobiol. Kolloq., Universität Stuttgart, 29.11.2001 in *Stuttg. Ber. Siedl.Wass.Wrtschft*, vol. 166, pp. 197–216, Oldenbourg Industrieverl., München.

Overath, H., Raphael, T., Salhani, N. and Soeder, C. J. (1988) Denitrifikation von Trinkwasser in Festbett-Drehrohr-Reaktoren, *Wass. u. Boden*, **40**(4), 199–202.

54 Biological denitrification of ground water – 8 years full scale experiences with the BIODEN-process

F. Hell

VA TECH WABAG Siemensstrasse 89 A-1210 Wien Austria

Abstract In order to offer optimum solutions for each individual case VA TECH WABAG developed a biological as well as a physical process for the nitrate removal from drinking water sources. The BIODEN-process applies heterotrophic denitrification and the ENR-process uses electrodialysis for the selective nitrate separation. Both processes are meanwhile in full-scale application for nearly one decade. The main advantages of the BIODEN-process are the degradation of the pollutant and the high water recovery. The strong points of the ENR-process are the additional hardness removal and the quick availability, so that the plant can be started and stopped on demand. Furthermore the ENR-process shows high nitrate selectivity compared to other physical processes.

Keywords Nitrate removal, drinking water, biological denitrification, biological filtration, dual media filtration, electrodialysis

Introduction

Nitrate as a drinking water pollutant became a problem in the last quarter of the 20th century for many water works especially in western and central Europe (Packham 1991; Packer et al. 1995). Beside sanitation programs and blending strategies, suitable nitrate removal technologies were necessary. In principle there are desalination processes, such as reverse osmosis, ion exchange and electrodialysis as well as biological denitrification processes applicable (Rohmann and Sontheimer 1985). Furthermore the catalytically nitrate degradation should be also mentioned in this respect (Tacke and Vorlop 1991).

Degrading processes, such as autotrophic respectively heterotrophic biological denitrification and catalytic nitrate reduction solve the problem by nitrate destruction (Gimbel et al. 2004). The physical processes separate the nitrate from drinking water. Hence a suitable discharge possibility for the brine stream is required. In the case of ion exchange, excepting the CARIX-process (Hagen 1991), an additional salt load results from the regenerating chemicals.

The BIODEN-process is based on the natural biological denitrification, which takes place in soil and ground water. However, in the BIODEN-process the denitrification is enhanced under controlled conditions in a fixed bed biofilter. In order to meet drinking

 Recent Progress in Slow Sand and Alternative Biofiltration Processes edited by Rolf Gimbel, Nigel J.D. Graham and M. Robin Collins. ISBN: 9781843391203. Published by IWA Publishing, London, UK.

water requirements, the denitrification process needs an aerobic post-treatment. So the BIODEN-process consists of the following main process steps (see Figure 1):

The BIODEN-process

- Substrate dosing
- Denitrifying biofilter
- Aeration
- Flock-filtration
- Polishing filter
- Safety disinfection
- Backwash system and backwash water recycling

As substrate a carbon and phosphate source are dosed into the raw water. Normally all other nutrients are normally sufficiently present in the raw water.

The nitrate degradation is based on the heterotrophic biological denitrification. This process takes place in a pressurised fixed bed biofilter. The choice of granulation of the biofilter material observes the optimum between biomass retention and built-up of filter headloss.

In the aeration, the denitrified water is saturated with oxygen. Thus aerobic activities in the flock-filters and polishing filters are supported and the final drinking water has sufficient oxygen content.

In the flock-filtration the low turbidity, caused by a minor biomass washout is removed by flocculation and dual media filtration.

Any accidentally occurring residual substrate is aerobically degraded in the dual media filters and / or subsequent polishing filters.

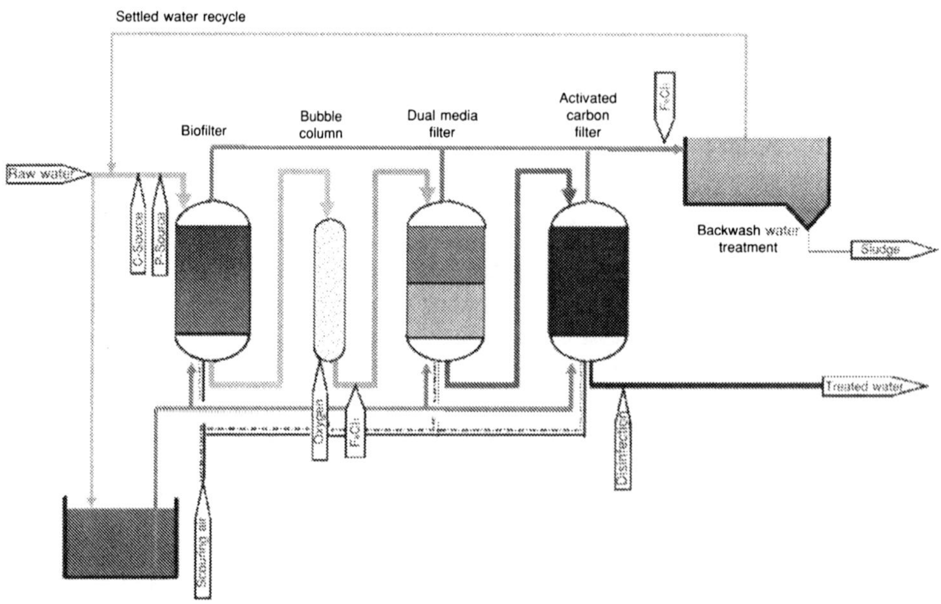

Figure 1 Principal scheme of the BIODEN-process

The last stage of the treatment train is a safety disinfection. However, long term experiences of the full scale plant showed, that the treated water already fulfilled drinking water standards only after the polishing filters.

Biofilters, dual media filters and polishing filters are regenerated by means of backwashing. The used backwash water is collected and recycled after clarification. Excessive biomass and particles from flock-filtration are removed as a concentrated liquid or even dewatered sludge.

The BIODEN-process has several strong points. As a main point the degradation of the pollutant shall be mentioned. The fixed bed biofilter has an excellent retention for the denitrifying biomass. Hence the loading of the subsequent aerobic post-treatment is low. The backwash water of the filters is clarified by simple sedimentation. As the clarified water contains only the denitrifying biocenosis it is directly recycled to the plant inlet. Thus a water recovery of practically 100% is accomplished. The technology is based on standard filtration processes and does not need any special equipment. Compared to other nitrate removal technologies, the BIODEN-process is very cost effective.

The only probable disadvantage is the necessity of continuous operation in order to maintain the biological activity. However, performance increase needs only short time, as long as the biological activity is maintained at a minimum level.

The BIODEN-process was developed and tested in a two year pilot test trail (Lahnsteiner and Hell 1993).

Results

The BIODEN-plant obersiebenbrunn

The first full scale plant with a capacity of 180 m^3/h was started up in Austria in 1997. This plant is now running for 8 years with excellent results. The plant consists of two parallel lines. Each line is equipped with substrate dosing, biofilter, bubble column for aeration, flock-filtration and polishing filters. Ethanol is dosed as carbon source and diluted phosphoric acid is added as phosphate source. The plant is operated continuously at nominal capacity. The treated water is blended with untreated water from other wells of the same well field. When the drinking water demand is low, the treated water is infiltrated into the catchment area of the un-treated wells. The substrate dosing is controlled in order to reach an average nitrate value of 5 mg/l in the denitrified water. In the case, that slightly elevated concentrations of nitrite and / or organic matter could be detected in the effluent of the biofilter, these substances were always safely degraded in the dual media filter. The colony count in the effluent of the biofilter already falls below 1000 cfu/ml. After dual media filters and polishing filters, the colony count is fare below 100 cfu/ml, in the most cases below 10 cfu/ml.

The backwash water from all filters is collected and clarified in two parallel settling basins. The clear water is recycled to the plant inlet. The concentrated sludge is discharged into the local sewage treatment plant. The sludge discharge lies below the guaranteed value of 25 m^3/d. Hence the water recovery is greater than 99 %.

During start-up of the plant special attention was drawn to the optimisation of the flock filtration. As turbidity is rather low, the iron concentration was utilised for assessment of the filter ripening. Figure 2 shows an iron peak after backwash of a dual media filter.

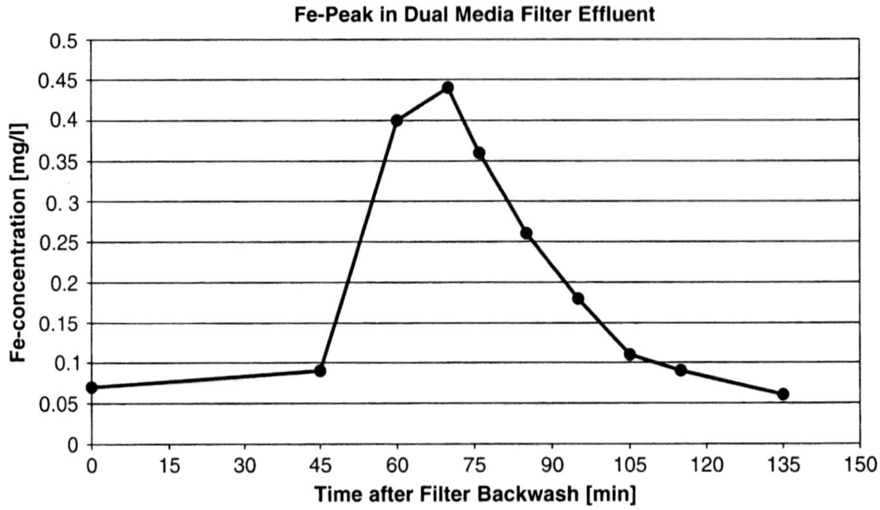

Figure 2 Iron peak after backwash of dual media filter

The maximum iron concentration appears more than one hour after filter restart. For that reason product discharge after filter restart was not applied. However, that iron peak was markedly buffered by the subsequent polishing filter and the iron peak maxima decreased in the following month of operation.

The average annual production rate during the last four years was 1 381 000 m^3/a. Based on annual costs the following specific expenditures were calculated.

- Chemicals 0.026 €/m^3
- Electric power 0.015 €/m^3
- Wastewater disposal 0.013 €/m^3
- Spare parts 0.005 €/m^3
- Service costs 0.005 €/m^3

Hence the operating costs except manpower amounts to 0.064 €/m^3. The expenditure of time is about 22 to 25 hours per week.

Other BIODEN-references

Two treatment plants in Italy are also based on the BIODEN-process. One demonstration plant with a capacity of 50 m^3/h was also started up in 1997 and one 540 m^3/h plant was started up in 2004. These two plants are operated with acetic acid as carbon source. Deviating from the BIODEN-process these plants are not equipped with activated carbon filters and backwash water is not recycled. The 540 m^3/h plant consists of three lines. The lines were started up one by one. For the first line full nitrate degradation was accomplished approximately 1 week after start-up. The raw water nitrate concentration was 80 mg/l. The design concentration was 100 mg/l. The treated water nitrate concentration is adjusted to 5 mg/l. After optimisation of substrate dosing and backwash procedure, the plant is running satisfyingly at full capacity.

A 500 m^3/h BIODEN-plant is started up in January 2006 in Czestochowa (Poland). This plant consists of 3 lines of biofilters, bubble columns, flock-filtration, polishing filters and backwash- and sludge-treatment. The capacity of each line is 50% of the nominal plant capacity. Hence the effective capacity would be 750 m^3/h. As there is no sewer connection available, the special requirement of the Czestochowa plant was 100% water recovery. For that reason the plant is also equipped with a sludge dewatering unit.

The ENR-process

Description of the ENR-process

The principle of electrodialysis involves the removal of ionic components from aqueous solutions through ion exchange membranes using the driving force of an electric field. The water to be treated is pumped through a membrane stack, which consists of alternately placed anionic and cationic selective membranes. Separated by gasket frames and spacers, the membranes are fixed between two end plates, which contain the electrodes producing the electric field. In order to transfer the electric current and to remove gases produced by the electrode reactions, the electrode chambers are rinsed with an electrolyte solution. In the compartments of the membrane stack, the ion content is diluted or concentrated according to the ion penetration through the membranes (see **Figure 3**). Identical compartments are connected by a distributing and collecting system. Thus the raw water is separated into a diluate and a brine stream. Both streams flow at the same velocity through the membrane stack. As a rule the water to be diluted passes the membrane stack once whereas the brine stream is recycled. The brine concentration is adjusted by means of dilution with raw water. Thus the brine stream is operated in a feed and bleed mode.

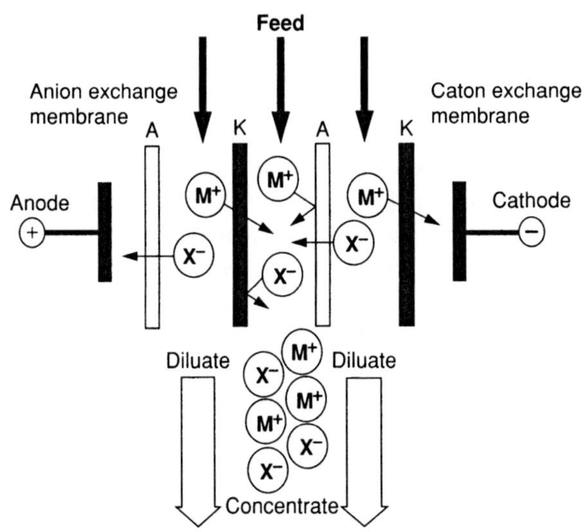

Figure 3 The electrodialysis principle

The degree of desalination is determined by the ion content of the raw water, the applied stack voltage, the type of membranes, the geometry of the membrane stack and the cross-flow velocity.

The application of monovalent selective anion exchange membranes results in an excellent nitrate selectivity (Mizutani 1990). Thus the brine quality and water recovery is much better compared to reverse osmosis (Rautenbach et al. 1986).

The ENR-plant Kleylehof

The ENR-plant Kleylehof is designed for a hydraulic capacity of 40 l/s (144 m^3/h) with three hydraulic stages and a maximum of 160 mg NO_3/l in the raw water. The guaranteed value for the treated water, in the case of maximum nitrate concentration in the raw water, is less than 50 mg NO_3/l. The special requirement for this plant was seasonal operation. The plant is remote controlled and works fully automatically. The plant was started up 1997 (Hell and Lahnsteiner 2002).

The main components of the plant are:

- Raw water intake
- Electrodialysis unit
- Disinfection unit
- Brine loop
- Electrode rinsing loop
- Concentrate discharge
- Dosing station

The raw water is abstracted from a well and pre-filtered with bag filters. According to the water demand, one, two or three hydraulic stages are selected by the remote control. The PLC controls the corresponding flow rate and distributes the raw water to one, two or three electrodialysis stacks.

The stacks are equipped with standard cation exchange membranes and monovalent selective anion exchange membranes. Each stack is allocated to one rectifier. The voltage of the rectifiers is manually adjusted according to the raw water conductivity and the requested nitrate content in the treated water. The polarity of the stacks is changed on a regular time basis by the PLC.

The effluent on the diluate side of the electrodialysis stacks passes an UV-disinfection unit. The quality of the product is checked by means of conductivity measurement. The treated water is recycled to the well, if a maximum value is exceeded. The product water flows into a buffer tank. From there the water is pumped into the distribution system.

The quality of the brine is controlled by means of conductivity measurement. According to the requested conductivity, the brine loop is replenished with raw water. The circulation tank overflow is connected to the circulation tank of the electrode rinsing loop. Thus the concentrate serves as the electrolyte solution for rinsing of the electrodes. The brine effluent of the electrode rinsing loop flows into a buffer tank. From this buffer tank the brine is pumped into a storage pond.

Normally the electrodialysis stacks are cleaned automatically with a hydrochloric acid solution once a month.

Table 1 Results of the Kleylehof ENR plant

Parameter	Unit	Raw water	Product	Brine
pH-value	–	7.4	7.5	7.6
Conductivity	μS/cm	850	640	2820
TDS	mg/l	704	550	2387
Total hardness	meq/l	9.04	6.97	31.86
Alkalinity	meq/l	4.21	3.97	8.51
Langelier index	–	0.0	0.0	1.0
Calcium	mg Ca/l	115	85	433
Magnesium	mg Mg/l	40	33	124
Sodium	mg Na/l	11	11	15
Potassium	mg K/l	2.6	2.5	4.6
Chloride	mg Cl/l	45	25	270
Bicarbonate	mg HCO_3/l	257	242	519
Sulphate	mg SO_4/l	113	110	132
Nitrate	**mg NO_3/l**	**120**	**41**	**889**

In the first three years, the plant was only operated from May to October. During winter, the plant was mothballed. Later on this concept was changed. During seasons, where the water demand is low, the treatment plant is only shortly operated from time to time. Thus stagnation of the water is avoided and the plant is available at all times.

Results of the Kleylehof plant

When the plant was started the nitrate content in the raw water was 120 mg NO_3/l. A maximum Nitrate removal of approximately 100 mg NO_3/l was achievable. However, the stack voltage was adjusted to a nitrate removal down to 40 mg NO_3/l. In proportion to the nitrate removal, the hardness of the drinking water was reduced by approximately 23%. Table 1 shows an analysis for the raw water, product water and brine.

Water quality is also sufficient from hygienic view point. The colony count of the raw water and the treated water was zero.

The regular plant service is done after the winter season. The necessary membrane change was always below the guaranteed value of 10% per annum. Also the lifetime of the electrodes is much higher than expected.

The desalination performance is constant and no decrease in plant performance could be detected.

There are two possibilities for brine discharge. Both, blending with irrigation water and discharge into the local sewage treatment plant, proved to be suitable.

Comparison between BIODEN and ENR-process

Both processes have their strong points, so that the choice between the two processes depends on the particular circumstances of each individual case. The main advantages of the BIODEN-process are the degradation of the pollutant, the almost full water recovery and the biological stabilisation of the water – due to the bioactive filters all biodegradable substances are degraded. Due to the nearly complete nitrate degradation, it is only necessary to treat a part stream of the raw water with blending afterwards. On the other hand, biological processes require constant operation and some ripening time in case of marked changes in the operational point.

The strong points of the ENR-process are the possibility to change the operational points rapidly, the additional removal of hardness, the selective nitrate removal and the high brine quality. Unfortunately electrodialysis is not as wide spread as reverse osmosis, therefore membrane prices did not decrease as for reverse osmosis.

Investment costs are nearly the same for BIODEN- and ENR-process. Costs for electromechanical equipment are lower for the biological treatment. However the space requirement for electrodialysis is much lower. The effort for supervision is approximately the same for both processes. Whereas, due to costs for membrane replacement and brine discharge, total operating costs are normally higher for the ENR-process.

Conclusions

After more than eight years of operational experiences it can be concluded, that both the BIODEN- and the ENR-process are very stable processes and highly suitable for waterworks operation. The operational effort is in a reasonable range. The decision for biological or physical treatment requires proper assessment of the circumstances of each individual case. As a rule of thumb ENR is mainly suitable for small scale waterworks and applications, which require quick changes in operation such as stop and go operation. BIODEN is mainly suitable for medium and large sized waterworks and relatively constant operation.

References

Packham, R.F. (1991) Public health and regulatory aspects of inorganic nitrogen compounds in drinking water. Int. Workshop on Inorganic Nitrogen Compounds and Water Supply, Hamburg, Germany 27–29 Nov.

Packer, P.J. Caygill, C.P.J. Hill, M.J. and Leach, S.A. (1995) Regional variation in potable water nitrate concentration and its effect on total dietary nitrate intake. J. Water SRT - Aqua 44: 5, 224–229.

Rohmann, U. and Sontheimer H. (1985) Nitrat im Grundwasser Ursachen Bedeutung Lösungswege. Engler-Bunte-Institut der Universität Karlsruhe.

Tacke, T. and Vorlop, K.D. Die katalytische Nitratreduction (KNR) als Methode zur Nitratentfernung in der Wasseraufbereitung. DECHEMA-Monographien Band 122 – VCH Verlagsgesellschaft.

Hagen, K. (1991) The CARIX-process. Int. Workshop on Inorganic Nitrogen Compounds and Water Supply, Hamburg, Germany 27–29 Nov.

Gimbel, R. Jekel, M. Ließfeld, R. Uhl, W. Overath, H. (2004) Wasseraufbereitung – Grundlagen und Verfahren. Lehr- und Handbuch Wasserversorgung Bd.6.

Lahnsteiner, J. and Hell, F. (1993) Pilot experiences with the biological denitrification at the Leibnitz Waterworks. in Proc. of European Water Filtration Congress, Oostende 15–17 march.

Mizutani, Y. (1990) Ion exchange membranes with preferential permselectivity for monovalent ions. J Membr Sci 54: 233–257.
Rautenbach, R. Kopp, W. Opbergen, G. Peters, Th. Helekes, R. (1986) Prozeßvergleich von Umkehrosmose und Elektrodialyse am Beispiel der Nitratentfernung aus Grundwässern. Chem.-Ing.-Tech. 58 Nr.12: 938–945.
Hell, F. and Lahnsteiner, J. (2002) The Application of Electrodialysis for Drinking Water Treatment. in Springer monograph "Water Resources Quality" pp.: 315–327.

55 Alternative low density media for use in biological roughing filtration prior to slow sand filtration

R. G. W. Bayley*, M. J. Chipps*, M. Steele*, R. White*, A. Mikol, C. S. B. Fitzpatrick****

*Thames Water/RWE, R&D, Kempton Park AWTC, Feltham Hill Road, Hanworth, Middx. TW13 6XH, UK.
**Dept.Civil & Env.Eng., UCL, Gower St., London WC1E 6BT, UK.

Abstract Existing rapid filtration back-wash rates for air and water may be sub-optimal for sand and this may require expensive and regular external cleaning. It was postulated that low-density materials might be cleaned by existing low wash rates. Experiments tested alternatives to the sand-only medium commonly used for roughing filtration prior to slow sand filtration. Two back-wash regimes were used, both of which were sub-optimal. Early trials have been successful but demonstrated that wash rates remain critical.

Keywords Biological, filtration, alternative media, pumice, anthracite, expanded clay.

Introduction

Most of London's water supply is derived from the R.Thames and R.Lea and impounded in pumped storage reservoirs. Rapid gravity [biological] filters (RGF) were installed at various times throughout the 20^{th}C. to remove a proportion of the resultant lentic phytoplankton biomass (without any chemical addition), and thus allow the downstream slow sand filters (SSF) to be up-rated.

The RGFs were constructed using a variety of designs; the most recent incorporating: sand and anthracite 'dual media'; 'collapse-pulsing' wash rates (Amirthirajah, 1993, Fitzpatrick, 1993); and a final high rate water phase to remove deposits/reclassify the layers. The older designs use sand as the filtration medium; this is not collapse-pulse washed and as a result it accumulates deposits. This increases back-wash frequency and requires expensive external cleaning. When clean, the sand medium is vulnerable to surface blinding by filamentous phytoplankton; e.g. *Melosira granulata*, but as deposits accumulate, other taxa may cause very short run durations; either within the depth of the medium or via surface blinding. Whilst correctly washed dual media filters operate well when fed 'algal-rich' water, it has proved difficult to retrofit this design into older, usually shallower, filter shells. High reclassification rates may also cause waste-water handling issues.

It was postulated that lower density granular materials might perform better than sand (Mörgeli and Ives, 1979) even though available wash rates were 'sub-optimal'. A preliminary

Table 1 Back-wash rates applied to mimic pilot plants. Filtration rates were 6 m/h for works 'A' filters and 10 m/h for works 'B' filters. Back-wash rates stated as rise velocities (m/h).

	Works mimic 'A'		**Works mimic 'B'**	
Wash Phase	Rate m/h	Duration mins	Rate m/h	Duration mins
Air	22	1	27	1
Air + Water	22A/12W	15	27A/8W	10
Water	17	3	30	2

trial was set up using an automatic three-column pilot plant. Columns were filled with either sand ('control'), pumice or anthracite. This pilot plant was then set to mimic one water treatment works (works 'A'). A later automatic pilot plant was constructed to mimic another works (works 'B'). This latter trial used four columns; one each of sand, pumice, anthracite and 'expanded clay'. The backwash conditions are given in Table 1.

The objectives of the trials were to: increase filter productivity; reduce the volume of water required for back-washing; reduce the volume of waste water; maintain the filtration media in a clean state and to improve filtrate quality (if possible). To achieve this it would be necessary to allow biofilms to develop but to control them such as to prevent excessive growth (e.g. formation of 'mud-balls'). Whilst it was desirable to improve filtrate quality, this was not deemed essential i.e. some 'feed' material passed to the SSFs.

Methods

The media tested, with size range and Specific gravities (SG), were as follows:

Sand BSS 14/25 0.6–1.18 mm; S.G. 2.65 (source Leighton Buzzard; UK).

"Expanded clay" 0.8–1.6 mm; S.G. 1.65 (source Filtralite^{TM}).

Pumice 0.6–1.3 mm; S.G. 1.1–1.2 (source Pumex^{TM}).

Anthracite 0.6–1.2 mm, S.G. 1.4 (source Betwys, Wales).

All columns were designed with plenums and nozzles (0.25 mm slits), the latter being covered with pea shingle. Works 'A' columns were filled with 600 mm of media whilst works 'B' filters were filled with 950 mm. Works 'A' columns were 450 mm internal diameter and run at 6 m/h filtration rate; whilst Works 'B' columns were 500 mm internal diameter and run at 10 m/h.

Raw and filtrate water quality were continuously monitored using turbidimeters and particle counters. Batch samples were taken for particulate organic carbon, chlorophyll-*a* and identification/relative frequency of phytoplanktonic taxa. Hydraulic data recorded were filtration rates and pressure differentials (head-losses) at one-minute intervals. Temperature was checked twice per week and, together with recorded flow rate, used to normalise the head-loss values. Media depth was checked periodically as was the calibration and cleaning of the on-line instruments.

Media samples were obtained from mimic 'B', after eight months; fixed, dehydrated and stored, and then 'viewed' using scanning electron microscopy (SEM).

Table 2 Total number of back-washes required during test periods.

Media	Works 'A' Total Washes	Works 'B' Total Washes
Sand	131	75
Pumice	112	44
Anthracite	93	45
Expanded clay		46

Results

The raw water turbidity and algal 'loads' were consistently low, and at times resulted in many columns back-washing once every 7 days (time triggered).

Differences between the media were more apparent on the 'B' mimic with the greatest variability being in run duration, as assessed by backwash frequency (table 2).

Clearly, lower density media out-perform the sand although the behaviour was different for the different mimics and thus back-wash regime.

Differences in starting normalised head-losses (SNHL) were also apparent. The 'A' mimic exhibited a slightly higher SNHL for sand *cf.* pumice and anthracite. After seventeen months of operation the sand SNHL had increased to nearly 0.5 m (from 0.25 m) whilst the pumice and anthracite showed much lower increases.

The 'B' mimic initially showed low SNHLs that then increased markedly due to a fault in the system. This was rectified and the SNHLs reduced once again by October 2004 (figure 1). The back-wash air was lost during November and the SNHLs increased again; markedly so with the sand column. With the air supply returned in December, the SNHLs reduced and remained fairly stable until the end of March 2005. At this time there was little difference between the SNHLs of the three low-density media at *circa.* 0.3 m *cf.* the sand showing consistently higher SNHLs of 0.4–0.5 m. The behaviour of the sand was also consistent with a steadily rising trend frequently observed on full-scale operational filters.

There were no clear differences in filtrate quality between the media types for the 'A' mimic. The 'B' mimic, however, showed large differences in filtrate quality, particularly with turbidity and particle counts. These differences were apparent after the first few days

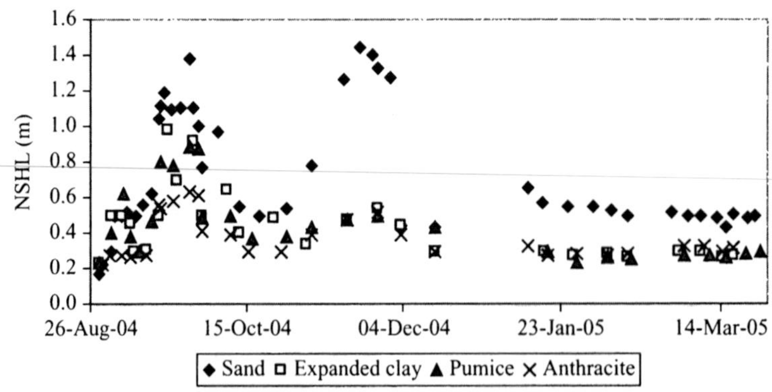

Figure 1 Normalised starting head-loss for works 'B' trials.

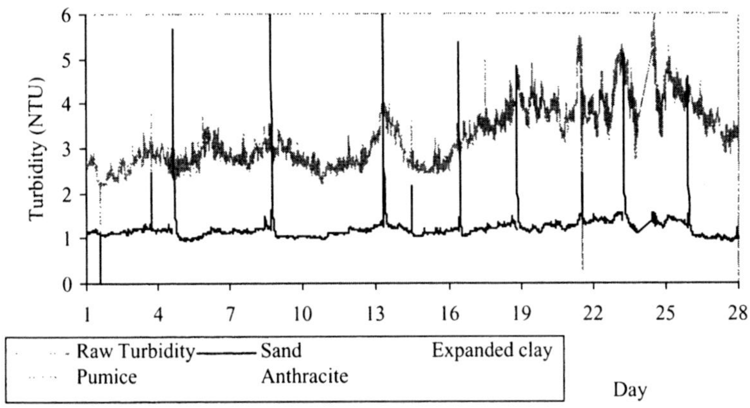

Figure 2 Turbidity for works 'B' mimic. March 2002.

of operation with the sand turbidity gradually increasing *cf.* the low-density media. This trend continued and by March 2005 the difference between the low-density media and sand was marked (figure 2). The particle counter trends were similar but with differences of greater magnitude. The other quality parameters showed little variation between the media, probably due to the low values in the feed water during the trial period.

Discussion and conclusions

Performance variations are postulated to be related to the biofilm on the grain (surface morphology & charge). The anthracite is more angular *cf.* sand, and the pumice and expanded clay show a rougher surface. It is likely that the alternative media grains are supporting a large biomass of microorganisms but that the back-wash conditions prevent the linking of grains and thus formation of mud-balls and other adverse phenomena associated with "dirty" filters.

Figure 3, shows the development of a considerable biomass of material in the alternative media that, unlike sand, does not appear to cause progressive and harmful accumulation of material bridging grains together.

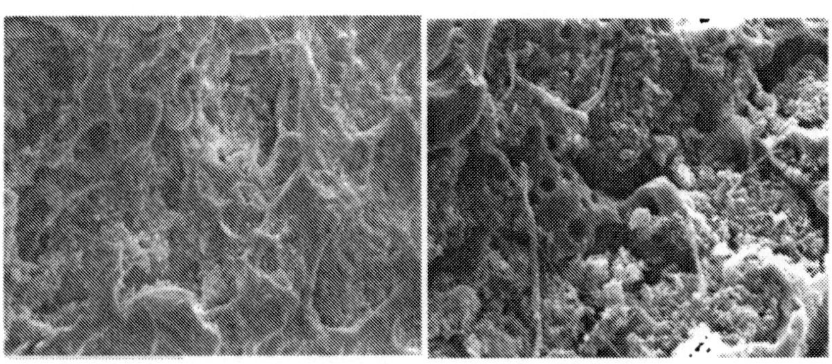

Figure 3 SEM images of pumice (left) and expanded clay showing surface and deposits.

Acknowledgements

The authors would like to thank RWE-Thames Water for permission to publish this paper. The views expressed by the authors are not necessarily those of RWE-Thames Water. Andrew Mikol is supported by a UK EPSRC Eng.D. award and Thames Water. We are grateful to Nicky Morden of the Eastman Dental Institute, UCL for the SEM work.

References

Amirtharajah, A., (1993). Optimum Backwash of Filters with Air Scour. Water Science and Technology, **17**(10), 195–211

Fitzpatrick, C. S. B. (1993). Observations of particle detachment during filter backwashing. Water Science and Technology, **27** (10), pp 213–221.

Mörgeli, B; Ives, K.J. (1979) New media for effluent Filtration. Water Research. **13**, pp 1001–1007.

56 Assessment of roughing filtration for pre-treatment of urban wetland waters

E. Lin*, J. Hutson*, C. Le Gal La Salle*, P. Dillon, D. Page**, and P. Pavelic****

*Flinders University, School of Chemistry, Physics, and Earth Sciences, GPO Box 2100, Adelaide SA 5001, Australia
**CSIRO Land and Water, PMB2, Glen Osmond, SA 5064, Australia.

Abstract Demand for water in many cities of the world is approaching the limits of sustainable supply, and often a significant amount of high quality imported water is used for non-potable purposes. Engineered urban wetlands established to reduce the threat of flooding, improve water quality and enhance local amenity, represent a potentially valuable water resource. This research is aimed at assessing passive treatment methods, including roughing filtration followed by slow sand filtration, to improve wetland water quality to levels considered acceptable for sustainable aquifer storage and recovery (ASR) in siliceous aquifers in Australia. If the program proves successful, recharged wetland water will be recovered during the dry season for landscape irrigation and industrial uses. This paper presents results from the first phase of this investigation, the assessment of roughing filtration for pre-treatment of urban wetland waters, and shows that colloid filtration theory can be used to determine site specific design parameters for roughing filters to be effective in producing a water quality suitable for slow sand filtration.

Keywords Urban wetlands, roughing filters, managed aquifer recharge.

Introduction

One method that has been used successfully in Adelaide to capture and store stormwater for re-use is aquifer storage and recovery (ASR). ASR represents a specialised form of managed aquifer recharge, where surplus surface water is injected into an aquifer through a well during the wet season and recovered from the same well, when water is needed during the dry season (Pyne 1995).

One of the most common operational challenges in ASR is the management of well clogging, which over time may lead to declining well efficiency (Dillon and Pavelic 1998). Physical and biological clogging may be caused by elevated levels of suspended solids, nutrients, and organic constituents in the stormwater (Rinck-Pfeiffer *et al.* 2000). This is managed easily in carbonate aquifers by regular well purging (Herczeg *et al.* 2004) but is more problematic in siliceous aquifers. Well clogging highlights the need to improve stormwater quality intended for ASR, the most effective strategy for which is the pre-treatment of stormwater prior to well injection.

Multistage granular filtration, including roughing filtration and slow sand filtration, has been identified as a promising method for pre-treating stormwater intended for ASR. Slow sand filtration is recognised as a well established water treatment technology capable of reducing levels of labile organic carbon found in stormwater (Logsdon *et al.* 2002). However, a significant limitation of slow sand filtration is the high source water quality required to prevent rapid filter clogging (Collins 1991). Maximum turbidity allowances for source waters range from 5 to 50 NTU (Cleasby 1991) with 10 to 20 NTU considered a conservative range to allow for reasonably long filter runs (Collins 1994). Because stormwater retained in urban wetlands frequently has high turbidity pre-treatment to remove suspended solids is required prior to slow sand filtration.

The emphasis of this research is on the application of roughing filtration for the removal of suspended solids in stormwater prior to slow sand filtration. Previous studies have shown that particle removal efficiency in roughing filtration is dependent on filter design, particulate, and water quality parameters (Boller 1993; Wegelin 1996; Collins 1994). However, site-specific trials are needed to determine if roughing filtration is a suitable pre-treatment technology for stormwater.

The specific objectives of this research were to:

(1) determine the influence of key filter design parameters on steady-state particle removal efficiency by conducting bench-scale upflow column experiments using a model inorganic (kaolinite clay) suspension;
(2) evaluate results using colloid filtration theory (Yao *et al.* 1971); and
(3) compare filter performance for the kaolinite clay suspension and natural stormwater and recommend a final roughing filter design.

Methods

Experimental waters were injected through four acrylic columns (6.3 cm diameter, 60 cm length) packed with sand and gravel media and connected in series to simulate a 2.4-m upflow roughing filter in series (URFS) (Figure 1).

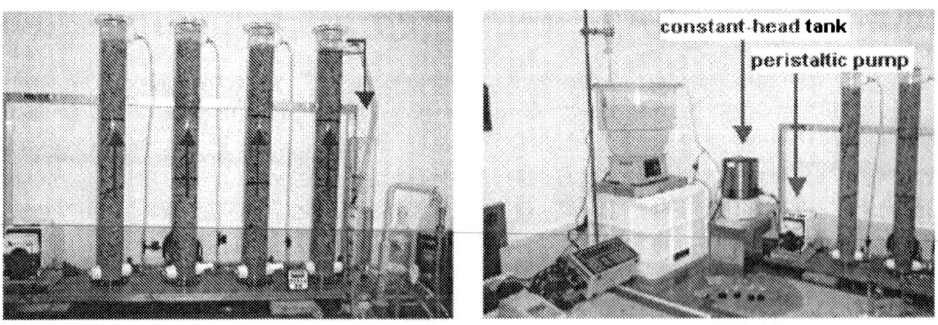

Figure 1 Experimental setup of bench-scale upflow roughing filter in series (URFS). The left photograph shows a close-up of the four up-flow columns. The right photograph shows the reservoir containing the kaolinitic clay solution with stirrer, constant head tank and peristaltic pump.

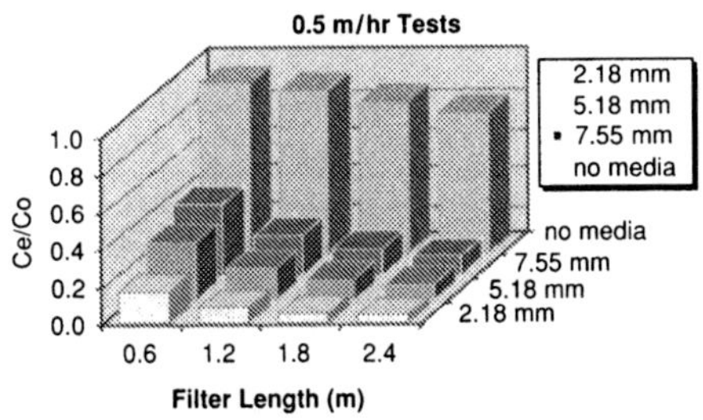

2a

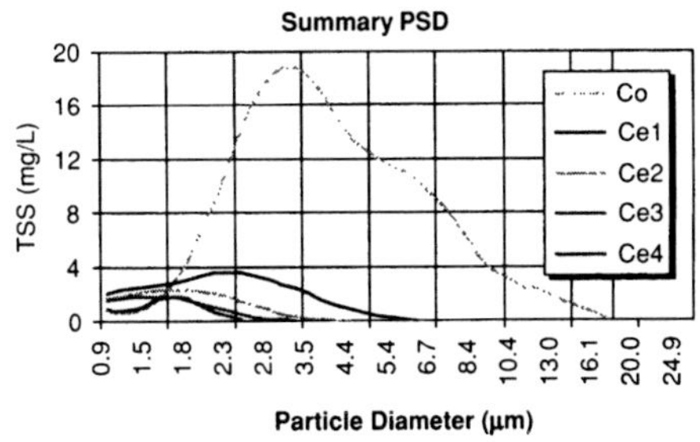

2b

Figure 2a Cumulative removal efficiencies (Ce/Co) over increasing filter length for 0.5-m/hr URFS experiments using model kaolinite clay suspension. **2b.** PSD analysis of one URFS experiment (2.18 mm, 0.5 m/hr) using model kaolinite clay suspension. Co represents the initial influent PSD; Ce represents the effluent from each successive 60-cm column.

The effects of design parameters and influent particle concentration on steady-state removal efficiencies for a kaolinite clay suspension were evaluated using three grades of media (2.18, 5.18, and 7.55 mm diameter), three hydraulic loads (0.5, 1.0 and 1.5 m/hr), four filter lengths (0.6, 1.2, 1.8, and 2.4 m), and three initial influent particle concentrations (100, 400, and 700 NTU). Two experiments were conducted using natural stormwater samples for comparison. After completion of experiments, selected water samples were characterised by particle size distribution (PSD) and suspension stability tests to improve understanding of particle removal mechanisms and water quality changes in roughing filters.

Results and discussion

Removal efficiencies for URFS experiments with kaolinite clay revealed a remarkably consistent pattern (Figure 2a), whereby incremental removal efficiencies declined with increasing filter length due to the preferential removal of larger particles as supported by PSD analyses (Figure 2b).

Improved removal efficiencies were correlated to smaller media sizes, lower hydraulic loading rates, and longer filter lengths. Increases in hydraulic loading rate and filter media size resulted in smaller PSD changes between initial and filtered samples. Influent particle concentrations had no effect on removal efficiencies for total suspended solids.

Removal efficiency and PSD results were used to model particle removal processes using colloid filtration theory. Attachment factors for the URFS experiments were calculated over increasing filter length using theoretical single collector efficiency values as described by Yao, *et al.* (1971). Calculated attachment factors were relatively consistent for all experiments ($\alpha = 0.128 \pm 0.027$). Similar to Collins (1994), decreasing collector efficiencies were observed with increasing filter length, although differences in the magnitude of decrease were dependant on the media size.

URFS experiments with natural stormwaters revealed poorer removal efficiencies compared to kaolinite clay, which was attributed to the slow settling velocity of the organic-inorganic particle assemblages in the stormwater (Figure 3).

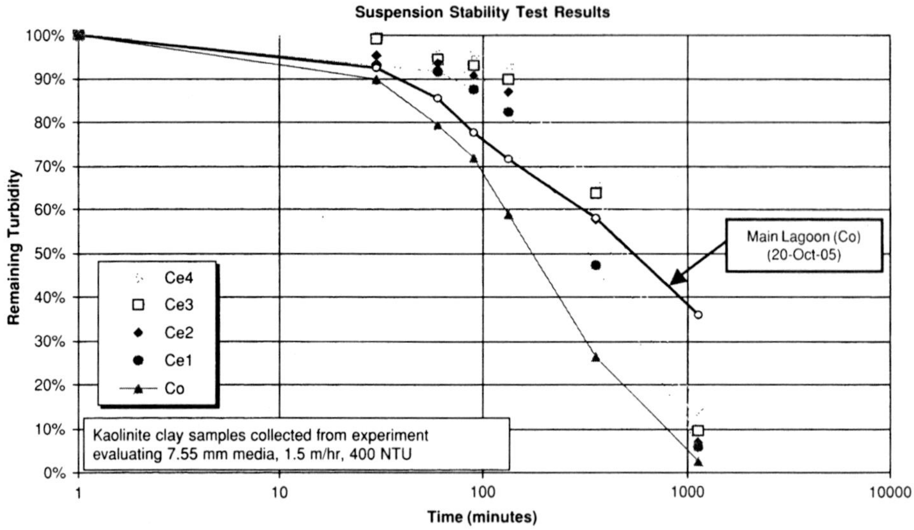

Figure 3 Relationship of remaining turbidity over time for kaolinite clay samples collected from one URFS experiment (7.55 mm, 1.5 m/hr) and untreated stormwater (main lagoon water). Turbidity measured 3.5 cm below water surface.

An optimal roughing filter design to meet source water turbidity requirements for slow sand filtration and ASR was determined from results of URFS experiments with natural stormwater. The URFS design to be adopted consists of a 2.4-m long filter packed with 7.55 × 5.18 × 2.18 × 2.18 mm diameter media in 60-cm filter segments. This configuration provides a filtrate water quality consistently below the 10 NTU and hence is suitable for use in conjunction with the requirements for operation of slow sand filters.

Conclusions

The conclusions of this study included:

(1) That the relationship between roughing filtration design variables and steady-state particle removal efficiencies using bench-scale column experiments can be determined.

(2) Modelling using colloid filtration theory resulted in an average attachment factor of 0.128 across the 2.4 m column.

(3) Pilot-scale roughing filters can be designed using colloid filtration theory and experimentally determined attachment factors to ensure that treated waters meet quality targets (turbidity < 10 NTU) suitable for slow sand filtration and use for ASR applications.

References

Boller, M. (1993), Filter mechanisms in roughing filters, *Aqua*, **42**(3), 174–85.

Cleasby, J.L. 1991, Source water quality and pre-treatment options for slow sand filters, in Slow Sand Filtration, Logsdon, G. (eds), American Society of Civil Engineers, Virginia, USA.

Collins, M.R., Eighmy, T.T., and Malley, J.P. (1991), Evaluating modifications to slow sand filters, *J. AWWA*, vol. 83, no. 9, pp. 62–70.

Collins, M.R., Westersund, C.M., Cole, J.O., and Roccaro, J.V. (1994), Evaluation of roughing filtration design variables, AWWARF, USA.

Dillon, P. and Pavelic, P. (1996), Environmental guidelines for aquifer storage and recovery: Australian experience, In: Artificial recharge of groundwaters, Peters *et al.* (eds), Balkema, Rotterdam.

Herczeg, A.L., K.J. Rattray, P.J. Dillon, P. Pavelic, K.E. Barry (2004) Geochemical processes during five years of Aquifer Storage Recovery, *Ground Water* **42**(3), 438–445.

Logsdon, G., Kohne, R., Abel S., and La Bonde, S. (2002), Slow sand filtration for small water systems, *J. Env. Eng. Sci.*, **1**, 339–48.

Pyne, R.D. (1995), Groundwater recharge and wells: A guide to aquifer storage recovery, Lewis Publishers, USA.

Rinck-Pfeiffer, S.M., Ragusa, S.R., Sztajnbok, P. and Vandevelde, T. (2000) Interrelationships between biological, chemical and physical processes as an analog to clogging in aquifer storage and recovery (ASR) wells. *Water Research* **34**(7), 2110–2118.

Wegelin, M. (1996), Surface water treatment by roughing filters - a design, construction and operation manual, Swiss Federal Institute for Environmental Science and Technology (EAWAG).

Wegelin, M., Bolller, M., and Schertenleib, R. (1987), Particle removal by horizontal-flow roughing filtration, *Aqua*, **2**, 80–90.

Yao, K.Y., Habibian, M. T., and O'Melia, C.R. (1971), Water and Waste Water Filtration: Concepts and Applications, *Env. Sci. Tech.*, **5**(11), 1105–12.

57 Treatment of highly turbid water by direct horizontal roughing filtration (DHRF)

A. H. Mahvi, M. Ahmadi Moghaddam

School of Public Health, Center for Environmental Research, Tehran University of Medical Sciences, Tehran, Iran

Abstract Vertical or horizontal flow gravel beds can be used in water treatment as roughing filters. In order to improve the performance of horizontal flow roughing filtration (HRF) this process has been modified earlier by applying a constant coagulant dose prior to filtration named direct horizontal roughing filtration (DHRF). In this research a lab scale DHRF pilot plant was used for investigate DHRF performance. The study results indicated that DHRF (6.5 m long consisting of 2.5 m first compartment with 12–18 mm diameter grain, 2 m/s compartment with 8–12 mm diameter and 1.5m third compartment with 4–8 mm diameter) was systematically yielded good effluent quality (<2NTU, C.I = 0.95) with raw water quality of 200–400 NTU. DHRF is a versatile pretreatment process capable of handling wide fluctuation in raw water turbidity (200–400 NTU) while operating condition such as coagulant dose [2 mg/L of Fe (III)], mixing intensity, time and filtration rate remained constant.

Keywords Direct horizontal roughing filter, treatment, highly turbid water

Introduction

The demand for safe water is increasing drastically in the urbanizing areas of the developing countries. Many tropical rivers show a wide fluctuation in suspended solids content and other water quality characteristics; water treatment plant drawing raw water from such rivers are facing growing problems in providing desired treated water quality as well as quantity. This problem is compounded for the towns and small cities in developing countries due to their limited financial resources (Wegelin and Schertenleib, 1993).

The horizontal – flow roughing filter (HRF) is claimed to have the advantage of being able to tackle relatively high raw water turbidity, whilst at the same time offering long filter run time and a simple technology (Galvis et al, 1993). Its applicability is, however, limited, e.g. due to its low filtration rate and limited capability to remove stable colloids and color. In order to improve the HRF performance a modified configuration (DHRF), which combines it with other water treatment processes can be considered. DHRF can overcome the limitation of HRF and on the other hand retain its advantages, notably long run time and ease in operation and maintenance, would be advantages (Pardon, 1992).

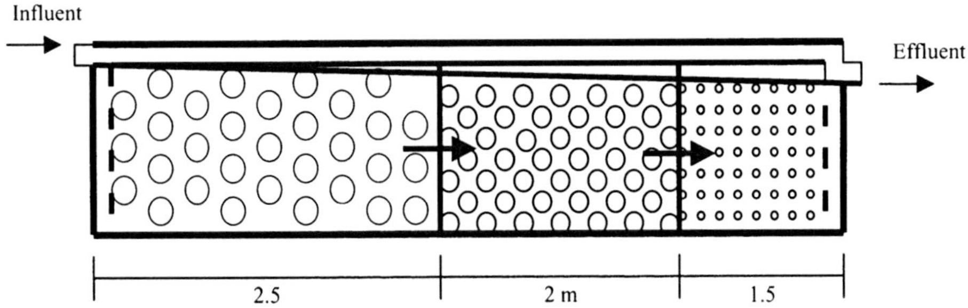

Figure 1 The schematic of the pilot

Methods

Laboratory scale pilot plant was operated, which consisted of a raw water preparation system, a coagulant dosing system and a roughing filter unit. Tap water was fed continuously to the constant head water tank, from there water was passed though a flow meter into the clay mixing tank where it was mixed with clay slurry continuously added by a peristaltic pump. The required clay slurry was prepared every two days. The stack slurry was kept under continuous and vigorous mixing with two constant speed mechanical stirrers was used in the clay-mixing tank. The turbid suspension was passed to a plain sedimentation tank (l h detention time) to eliminate coarse particles. The required amount of Fe (III) was pumped to the rapid mixing unit and mixed by the same type with the clay suspension. The coagulant suspension was then passed on to the roughing filter unit. The container of the filter bed was made of Plexiglas with 6 m length, 10 cm width and 10 cm height and had 3 compartments (Figure 1). The length and grain size of 1st, 2nd and 3rd compartment were 2.5, 2, 1.5 m, 12–18, 8–12 and 4–8 respectively. Turbidity as the main indicator of the removal performance was expressed in NTU (APHA, 1998). After reaching to head loss around 20 cm, filter cleaning through hydraulic flushing was started.

Results

The effluent turbidity from DHRF was <2 NTU (C.I = 0.95) with raw water quality of 200–400 NTU and surface over flow rate of 1.5–5 m/h. (Fig.2).

The effluent turbidity of the filter run with the influent turbidity varying over time was compared with that of 2 filter runs, the first with 200–250 NTU and second with 350–400 NTU. All the parameters remained identical. Effluent turbidity fluctuation in the filter run with the variable influent turbidity remained limited ($< \pm 0.3$ NTU) despite the drastic influent turbidity change. In filtration rate 1.5–5 m/h at high influent turbidity (200–250 NTU) and under identical process conditions the average effluent turbidity was from 1.5 ± 0.2 to 1.7 ± 0.2. Also in turbidity 300–350 NTU the average effluent turbidity was from 1.6 ± 0.3 to 1.7 ± 0.3 NTU. The results of the experiments with different Fe (III) dose are summarized in Table 1. At coagulant dose of less than 1 mg/L of Fe (III) the effluent turbidity was good to excellent. With dose of ≥ 4 mg (III)/L the turbidity was very low (0.2–0.3 NTU), however, filter run time decreased ($<$ 10.4 h). At dose 2 mg/L of Fe

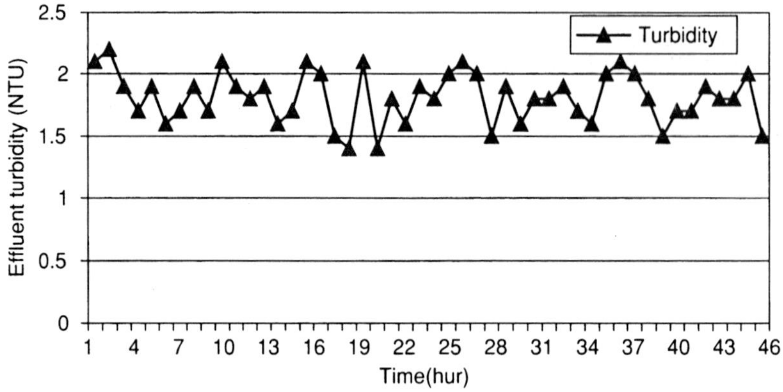

Figure 2 DHRF performance in treatment of highly turbid water

Table 1 DHRF performance as a function of coagulant dose

Fe (III)dose	Ripening period	Filtration run time	Total head loss	Average effluent turbidity ± sd*
Fe mg	(h)	(h)	(Cm)	(NTU)
1	9	80	13.9	5.7 ± 1.1
2	4	96	16.7	1.9 ± 0.2
4	2	75	16.1	0.3 ± 0.1
8	1.5	45	10.4	0.2 ± 0.1

*Standard deviation of turbidity between ripening period and filter break-through.

(III) the effluent turbidity was low (1.9±0.2 NTU) and the filter run time was increased (96 h).

The filter run time generally decreased with increasing coagulant dose. Head loss in all cases developed linearly and was within a reasonable limit of 20 cm. The filter ripening period decreased with increasing Fe (III) dose (Table 1). Total filter length was 6.5 m, initial turbidity was between 250 to 300 NTU, pH was 7.2 ± 0.2 and filtration rate was 5 m/h.

Conclusions

DHRF is systematically yielded good effluent quality (<2 NTU, C.I = 0.95) with raw water quality of 200–400 NTU it showed to be a versatile pretreatment process in a handling wide fluctuation in raw water turbidity while the process condition remained unchanged (Guibai and Gregrgy, 1991; Aph, 1998). In the case where the operational parameters such as coagulant dose and filtration rate require to remain unchanged for reasons of simple and reliable operation, DHRF will be an effective pretreatment process to produce stable effluent quality even if the raw water quality fluctuates widely or shock loads are received. Therefore less operational control, such as adjust coagulant dose, pH and rapid mixing intensity and time, is necessary in DHRF over a wide range of the raw water quality, turbidity and pH (Ahsan, 1995, Fazeli, 2003, Guibai et al., 1988).

Simple DHRF operation suggests that it could be an appropriate pretreatment technology for towns and small cities in developing countries.

References

Ahsan, T. (1995) Process analysis and optimization of direct horizontal – flow roughing filtration. Ph.D. thesis, Department of Water Engineering, Univ. Of Delft, Holland.

APHA., et al. (1998) Standard methods for the examination of water and wastewater. 20 rd edn., APHA, N.W, Washington D.C.

Fazeli, M. (2003) Determination of design basic and development of DHRF. Ph.D. thesis Department of Environmental Engineering, Univ. of Tehran, Iran.

Galvis, R., et al. (1993) Pretreatment alternatives for drinking water supply. Occasional paper. No. 22. IRC, the Hagne.

Guibai, L., et al. (1988) Optimum GT value in mixing flocculation for highly turbid water. Proc. Eng. Foundation conf, palm coast, Florida.

Guibai, L., Gregrgy, M. (1991) Flocculation and sedimentation of high—turbidity waters. *Wat. Res.* **25** (9):1137–1143.

Pardon, M. (1992) Research, development and implementation of roughing filtration technology in Peru. Workshop on roughing filters for water treatment, Zurich, Switzerland.

Wegelin, M., Schertenleib, R. (1993) Roughing filters for water treatment. *IRCWD News*, No. 27, PP. 1–8.

58 Retention of hygienically relevant microorganisms from storm water effluents by sand filters

Grobe, S., Petry-Hansen H., Uhl M.*, Flemming H.-C.

IWW, Institute for Water, Moritzstr. 26, D-45476 Mülheim a. d. Ruhr, Germany
*University of Applied Sciences Münster, Laboratory of Water Resources Management, Corrensstraße 25, 48149 Münster, Germany

Abstract Experimental studies of vertical flow sand filters for storm water treatment were performed in semi scale to investigate the dependency of processes and treatment efficiencies on filter material characteristics and operational boundary conditions. Sands with grain sizes 0/2 mm and carbonate additives offer the best material conditions for high elimination rates. Artificial throttling of the outflow to constant 0.01 l/(s × m^2) proved to be essential for high treatment efficiency for faecal bacteria. Long impoundage did not effect the efficiencies of the filter. Too long dry periods impair the treatment abilities due to reduced bacteria population in the filter. Hygienically relevant microorganisms can be reduced about 1–2 log steps.

Keywords Biological filtration, sand filters, enhanced treatment, storm water, faecal coliforms, faecal streptococci

Introduction

Dominant effects of urban runoff to receiving water ecosystems are hydraulic stress to benthic organisms, toxicity due to ammonia and O_2-depletion and loads of heavy metals. Microbiological emissions from storm water can seriously affect human water related leisure activities and chemical or hygienical quality standards for drinking water reservoirs. To meet the requirements of the EU WFD individual objectives of enhanced treatment can be developed for the receiving waters. Sand filters are meanwhile increasingly used in the FRG for enhanced treatment of combined sewer overflows and heavily polluted runoff of separate systems and highways (Hagendorf and Hahn (1994), Grotehusmann et al. (1999), Hagendorf and Diehl (2001)). Important processes are retention, filtration, adsorption, biodegradation of bacteria, organics and nitrification of ammonium. The experimental developments based on lysimeters in semi scale. The treatment experiences of selected full scale plants lead to new technical guidelines (Uhl et al. (2003), DWA M 178 (2005), Uhl et al. (2005)). The paper focuses on the results of a experimental study contributing to fundamental understanding of treatment processes in vertical flow sand filters (Grobe et al. (2005)). The semi scale study

Table 1 Grain size distribution of the investigated filter material (sand+15 vol-% carbonate).

	grain size				
	silt, clay <0.06 mm	**fine sand 0.06–0.2 mm**	**medium sand 0.2–0.6 mm**	**coarse sand 0.6–2.0 mm**	**fine gravel >2.0 mm**
	weight-%	**weight-%**	**weight-%**	**weight-%**	**weight-%**
LFU1	0	12	48	37	3
NRW1	0	12	60	25	3
NRW2	0	8	55	34	3
NRW3	0	6	41	47	6

was carried out by the authors with the following objectives for practical aspects:

(1) quantification of the elimination rates of hygienically relevant microorganisms in the filters depending on configuration and operation methods

(2) evaluation of suitable sands and carbonate supplements

Methods

The study was carried out with 8 lysimeters with 1.065 m diameter and a 0.75 m sand layer above a drainage layer of 0.1 m fine gravel (2/8 mm) and 0.1 m coarse gravel (8/16 mm). Pumps controlled each outflow with constant outflow rates of 0.01 $l/(s \times m^2)$ or 0.03 $l/(s \times m^2)$. The investigated sands meet recommendations of existing guidelines (LFU 1) or cover the spectrum of available material (NRW 1-3) in North Rhine Westfalia (NRW), FRG. To ensure sufficient carbonate supply for the nitrification process 15 vol-% carbonate material (0/2 mm) was mixed homogeneously with the sand (see Table 1). Five lysimeters (see Table 2) were operated in two phases over 2.5 years. In phase 1 the lysimeters were fed by artificial combined sewage (0.5 h settled wastewater diluted by drinking water) following

Table 2 Characteristics of the lysimeters.

		lysimeter no.				
		1	**2**	**3**	**4**	**5**
filter material		LFU1	NRW2	NRW2	NRW3	NRW1
outflow rate	$l/(s \times m^2)$	0.01	0.03	0.01	0.01	0.01
phase 1						
number of applications		29	29	29	29	29
sum of application	m^3/m^2	37.7	47.3	37.7	37.7	37.7
phase 2						
number of applications		30	5	28	30	28
sum of application	m^3/m^2	35.2	20.6	36.4	38.8	37.3

a feeding plan representing different boundary conditions. In phase 2 the lysimeters were fed by combined sewage flow from a storm water tank following the natural occurance of overflow events. Lysimeter 2 was taken out of operation earlier in phase 2 because the tendency of clogging and lower treatment efficiency. Flow proportional mixed samples were taken from the outflow over the whole duration of emptying the lysimeters. The in- and outflow of the lysimeters were analysized for microbial (faecal coliforms, faecal streptococci, salmonella, clostridia, enteropathogenic virus, protozoa) and typical chemical parameters.

Results and discussion

The in- and outflow of sand filters were investigated with regard to several hygienically relevant and general microbiological parameters (culturable heterotrophic bacteria, faecal and total coliforms, faecal streptococci, clostridia, salmonella, enteropathogenic virus, Cryptosporidia, Giardia). This paper focuses on the parameter "faecal coliforms" and "faecal streptococci". The faecal coliform and faecal streptococci inflow concentration of phase 1 and 2 showed a median of 10^6 MPN/100 ml and 10^5 MPN/100 ml, respectively (Figure 1). In phase 1 and also in phase 2 the concentration of faecal coliforms in the outflow of the sand filters showed a median between 10^4–10^5 MPN/100 ml. For most cases the guideline (10^2 MPN/100 ml) and the mandatory value (2.0×10^3 MPN/100 ml) for the parameter faecal coliforms of the directive 76/160/EEC "Bathing water quality" could be achieved in single cases. The faecal streptococci outflow concentrations were different in phase 1 and 2. In phase 1 the concentration showed a median between 10^3 MPN/100 ml and 10^4 MPN/100 ml. The concentration of faecal streptococci in phase 2 was in maximum one

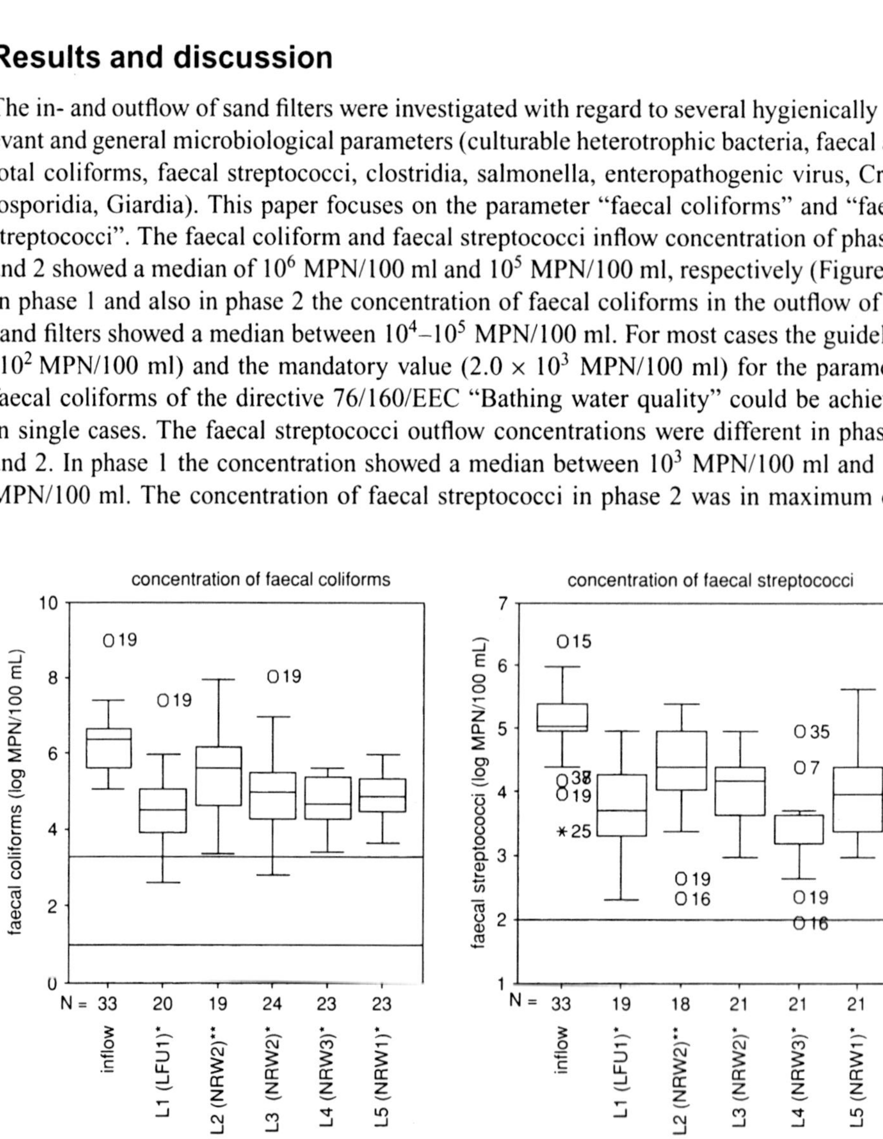

Figure 1 Concentration of faecal coliforms and faecal streptococci in the inflow and outflow of different sand filters (*filter velocity 0.01 l/(s × m²); **filter velocity 0.03 l/(s × m²)).

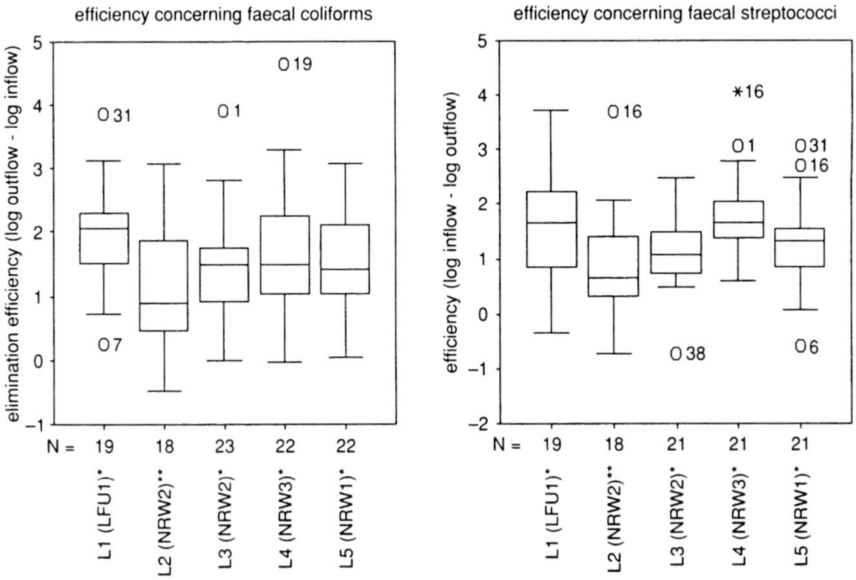

Figure 2 Efficiency of faecal coliform and faecal streptococci removal with sand filters.

log step higher. The existing guideline for the parameter faecal streptococci of the directive 76/160/EEC (100 MPN/100 ml) could only be reached in one case.

The efficiency of the sand filters depends in phase 1 on the type of filter material. Filters with sand LFU1 and NRW3 shows the best elimination rates (1.5–2.1 log for faecal coliforms and 0.6–1.8 log for faecal streptococci; figure 2). Furthermore, higher infiltration rates (filter L2, NRW2, $q_d = 0.03$ l/(s × m^2) and filter 4, NRW2, $q_d = 0.01$ l/(s × m^2)) result in a 36–44 % lower efficiency. In phase 2 (data not shown) the difference between the eliminations rates are not significant. Recently analyzed data of a reed covered filter with material NRW3 and a filter velocity 0.01 l/(s × m^2) indicate best elimination rates for faecal coliforms (2.0 log) and faecal streptococci (1.82 log). Long times of impoundage has no effect. Long charging brakes influence the efficiency negatively. This might be an effect of decreased activity of the biofilm population. Higher bacterial inflow concentrations lead to a more effective elimination of bacteria. This could be an effect of adsorption capacity and bioflocculation.

Conclusions

The experiments confirmed the vertical flow sand filters to be efficient tools for enhanced treatment of storm water ingredients as hygienically relevant microorganisms, ammonium and COD. High treatment efficiencies could be observed with respect to fine organic and anorganic particles and associated bacteria as well as soluble substances and hygienically relevant microorganisms in the free water phase. Bacterial concentration of hygienically relevant microorganisms can be reduced about 1–2 log steps but the requirements of the directive 76/160/EEC "Bathing water quality" can not be reached in a reliable way. Other additional biological, physical and chemical disinfection methods must be used to ensure low bacteria concentrations.

Relevant processes during the impoundage are filtration and adsorption. Biodegradation of organics can be observed during and after impoundage. Nitrification of adsorbed ammonium essentially happens after emptying of the filter under oxic conditions.

The experiments show the important role of the filter velocity and retention time on the treatment efficiency. An artificial reduced outflow down to constant 0.01 $l/(s \times m^2)$ causes low filter velocities and therefore longer retention times in the filters compared to free outflow. This supports the adsorption of bacteria and soluble substances (ammonium, soluble COD) and the filtration of bacteria adsorbed to fine particles (particulate COD). Reducing the outflow seems to contribute to process stabilization and slightly smoothes material born differences in adsorption capacity. Long times of impoundage do not negatively influence retention of bacteria but can negatively influence the retention of chemical substances. Significant influences were observed from about 40 h impoundage for ammonium. Ongoing analyses of the mass balances show limitations of the adsorption capacity depending on filter material characteristics as specific surface, carbon content and pore volume.

After each feeding the sand filters have to be emptied to aerate the filter. After 1–2 d the regenerative processes as nitrification of adsorbed ammonium is finished so that full treatment capabilities can be offered again. Long charging brakes however influenced the efficiency for all parameters negatively. This might be an effect of decreased mass and activity of the biofilm population.

Acknowledgements

The lysimeter study was funded by the Ministry for the Environment, Nature Conservation, Agriculture and Consumers Protection, North Rhine Westfalia, FRG.

References

DWA M 178 (2005). Empfehlungen für Planung, Konstruktion und Betrieb von Retentionsbodenfilteranlagen zur weitergehenden Regenwasserbehandlung im Misch- und Trennsystem. Gesellschaft zur Förderung der Abwassertechnik (GFA) Hennef, 2005, in work.

Grobe, S., Uhl M., Henrichs M., Merkel W. (2005). Weitergehende Mischwasserbehandlung mit Bodenfiltern – Untersuchungen zum Filteraufbau und Filterbetrieb zur Verbesserung der Reinigungsleistung insbesondere hinsichtlich hygienisch relevanter Mikroorganismen. Abschlussbericht zum Untersuchungsauftrag des Ministeriums für Umwelt, Naturschutz, Landwirtschaft und Verbraucherschutz des Landes Nordrhein-Westfalen.

Grotehusmann, D., Schütte, M., Schiedt, L. and Kasting, U. (1999). Wissenschaftliche Begleitung Bodenfilteranlage Alsdorf. Abschlussbericht über den 2. Untersuchungszeitraum.

Hagendorf, U. and Hahn, J. (1994). Untersuchungen zur umwelt- und seuchenhygienischen Bewertung von naturnahen Abwasserbehandlungsanlagen. UBA-Texte 60/94 Berlin.

Hagendorf, U. and Diehl, K. (2001). Keimelimination in bewachsenen Bodenfilter. Wasser und Boden 53 (2001), pp. 16–18.

Merkel, W., Hiekel, S. and Overath, H. (2000). Einleitung von Kläranlagenabläufen in kleine Fließgewässer: Bewertung der Ablaufqualität nach der EG-Badegewässerrichtlinie.

Abschlussbericht zum Untersuchungsauftrag des Ministeriums für Umwelt, Naturschutz, Landwirtschaft und Verbraucherschutz des Landes Nordrhein-Westfalen.

Uhl, M., Fuchs, S., Grothehusmann, D., Kasting, U., Lambert, B., Liebeskind, M. and Zech, H. (2003). Retentionsbodenfilter – Handbuch für Planung, Bau und Betrieb. Ministerium für Umwelt und Naturschutz, Landwirtschaft und Verbraucherschutz (MUNLV) des Landes Nordrhein-Westfalen, Düsseldorf, ISBN 3-9808617-1-6, 2003.

Uhl, M., Dittmer, U. and Fuchs, S. (2005). Advanced CSO Treatment with Soil Filters. Paper submitted to the 10th ICUD.

59 Submerged Biological Aerated Filter for Pretreatment of Potable Water in China

J. J. Chen, Ph.D., DEE, MBA

Severn Trent Services, 3177 Ross Road, Palo Alto, CA 94303 USA

Abstract Despite the fact that water is one of the most important natural resources to China. The country ranks at approximately the 13th poorest country in water resources, with a per capita water availability of 6 m3/day. Man-made pollution has contaminated the already limited potable water source. Some cities such as Jiaxing City in Jhejiang province are forced to use the polluted water near its City for drinking water uses. In order to upgrade the Class 4 water source to Class 1 potable water standard a fixed-growth biological treatment technology was selected for the pretreatment. The submerged biological aerated filter (SAF) by Severn Trent Services was pilot tested for 60 days, including low water temperature environment to verify its technical viability and cost effectiveness. The pilot study confirmed that with a hydraulic retention time as low as 49 minutes the average influent NH4-N of 10.1 mg/L was reduced to 0.19 mg/L with a 98% removal, while the influent BOD was decreased from 12 mg/L to 2 mg/L with an 83% removal. The pilot study has led to the design of a full-scale water treatment plant of 25,000 m3/day capacity using three SAF reactors, each with a dimension of 3.56 m W × 25.6 m L × 6.1 m gravel media. The water treatment plant is currently under construction and plans to deliver potable water during the first quarter of 2006.

Keywords Fixed-growth, Biofiltration, Nitrification, Micro-pollutants, Potable Water Pretreatment

Introduction

Water is one of the most important natural resources to any country. In the past, water resources were considered to be abundant and inexhaustible. However, with explosive population growth as well as the progressive agricultural and industrial development, water resources are becoming more precious and scarce on a global level.

China is experiencing a severe water shortage problem. The country is faced with limited quantity in water resources, and is confronting a simultaneous deterioration of their limited water resources from man-made pollution. According to a report by BBC News on June 30, 2005, more than 300 million people are left without clean drinking water as a result of the clean water crisis existing in rural China; due in part to the rapid economic growth and corresponding pollution. With scarce potable water sources, some municipalities are forced to use polluted river water near their cities and towns for drinking water uses.

Jiaxing City in Zhejiang Province is one of the fastest growing cities in China. In order to meet the drinking water demand for this rapidly expanding city, Jiaxing Harbor Area Water Company made the hard decision to build a potable water treatment plant at Zhapu using the polluted river water from the nearby Yen Ping Tung River. The water treatment plant with a capacity of 25,000 m3/day will be constructed at an accelerated pace. According to the river water monitoring record for the last three years, the river water has been moderately contaminated by some "micro-pollutants" such as ammonium and oxidized nitrogens, organics (BOD/COD), iron, manganese, and turbidity. The river water is categorized to fall between the Class 3 and 4 in the potable water standard, meaning it is not ideally suited for potable water use. However, the depletion of underground water sources and lack of other surface water sources have forced the City to find ways to make use of the contaminated river water.

Methods

The City and its consulting engineers, Northwest Municipal Design Institute, started an investigation for selecting the most efficient and cost effective treatment approach for the pretreatment of the contaminated river water to meet potable water supply needs. While conventional physical/chemical treatment methods are effective at removing iron, manganese and turbidity, they are not cost effective in reducing ammonium nitrogen and organics. Therefore, a biological treatment approach was considered for the pretreatment of the river water to address this need.

Conventional biological treatment methods such as activated sludge process would not be suitable for the pretreatment of the river water since it only contained moderate concentration of ammonium and organics. Maintaining a steady level of the required biomass would be difficult, if not impossible, at such minor concentration of contaminants for any suspended-growth treatment systems. Therefore, after a thorough investigation of the best available treatment alternatives to the activated sludge process, Northwest Municipal Design Institute and the City selected the submerged biological aerated filter (TETRA® SAF™), a fixed-growth biological treatment process by Severn Trent Services to be the best treatment option.

In order to verify that the SAF™ process would be a technically viable and cost effective biological treatment process for the task, a 60-day pilot study using the system was initiated in the winter of 2004. Cold-water temperature is very critical to the nitrification reaction; therefore, undertaking the pilot study in the winter was the best way to ascertain the success of the treatment process throughout the year.

The pilot plant utilized for this study at Zhapu had a bioreactor vessel with a dimension of 0.4 m in diameter and an overall height of 5.3 m. The reactor underdrain was a proprietary nozzleless air and water distribution device. The selected silica gravel, hard and round, with sizes between 20 and 40 mm and 3.4 m depth was used as the bio-growth media. As the river water and compressed air entered at the bottom of the bioreactor, the biomass on the surfaces of the gravel would closely contact with the water and air such that the microorganisms using the oxygen in the air would oxidize the contaminants in the water. The biochemical reaction and bio-flocculation by the biomass were fast and complete. The purified effluent would exit from the top of the bioreactor.

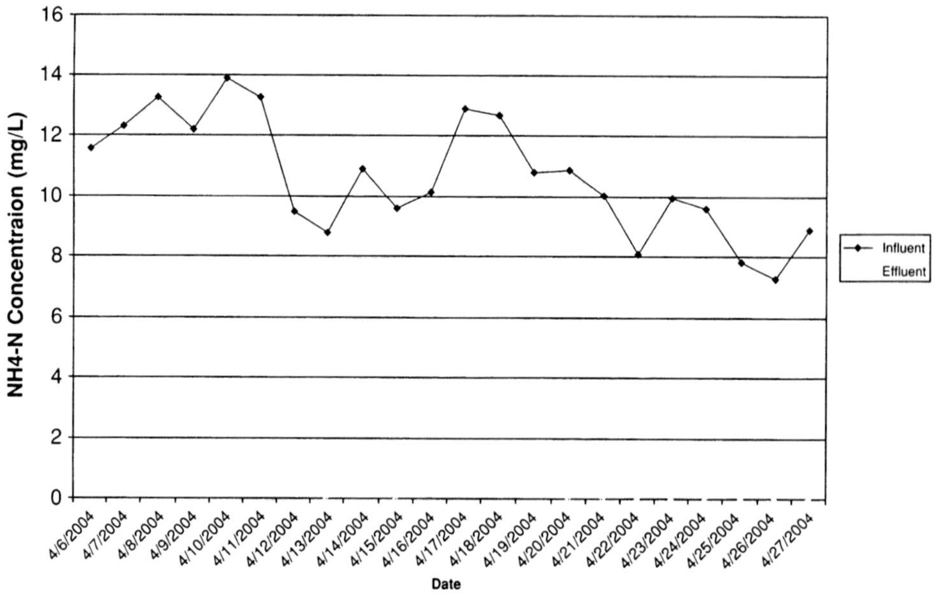

Figure 1 Ammonium nitrogen profile graph

Results and discussion

During the SAF pilot study period the water temperature varied in a range between 11 and 18 degree C. The steady state hydraulic retention time (HRT) was 98 minutes while in the hydraulic shock loading tests the HRT was reduced to 49 minutes. The results of the pilot study were very positive in that the steady state test period the average influent NH4-N of 10.1 mg/L was reduced to 0.19 mg/L in the SAF effluent, which was below the required effluent standard of 0.5 mg/L, with a nitrification efficiency of over 98%. And at the same time the average influent BOD of 12 mg/L was decreased to 2 mg/L that was less than the required 3 mg/L with a removal efficiency of more than 83%. Figure 1 depicts the influent and effluent NH4-N profiles during this test period. Even during the hydraulic shock loading tests with the doubling of the flow rate the average influent NH4-N of 8.60 mg/L was reduced to 0.06 mg/L or more than 99% removal rate. The tolerance capacity of the SAF bioreactor to the hydraulic shock loading was clearly demonstrated.

It is important to make interpretation of the pilot test data in order to derive the useful design parameters for the process design. As shown in Figure 2 with the plot of the specific NH4-N mass loading to the SAF bioreactor versus the effluent NH4-N concentration one may determine a design NH4-N loading for a selected effluent standard.

For the Zhapu plant the required effluent NH4-N is 0.5 mg/L, therefore, an NH4-N mass loading of 22.5 lb/kft/day (0.36 kg/m3/day) may be selected for the plant design.

The pilot study in effect had demonstrated that the SAF treatment process was able to transform the contaminated river water with the Class 3–4 potable water standard to a good drinking water source with the Class 1 standard. The successful pilot study led to the design of the 25,000 m3/day potable water treatment facilities at Zhapu. The final proposed potable

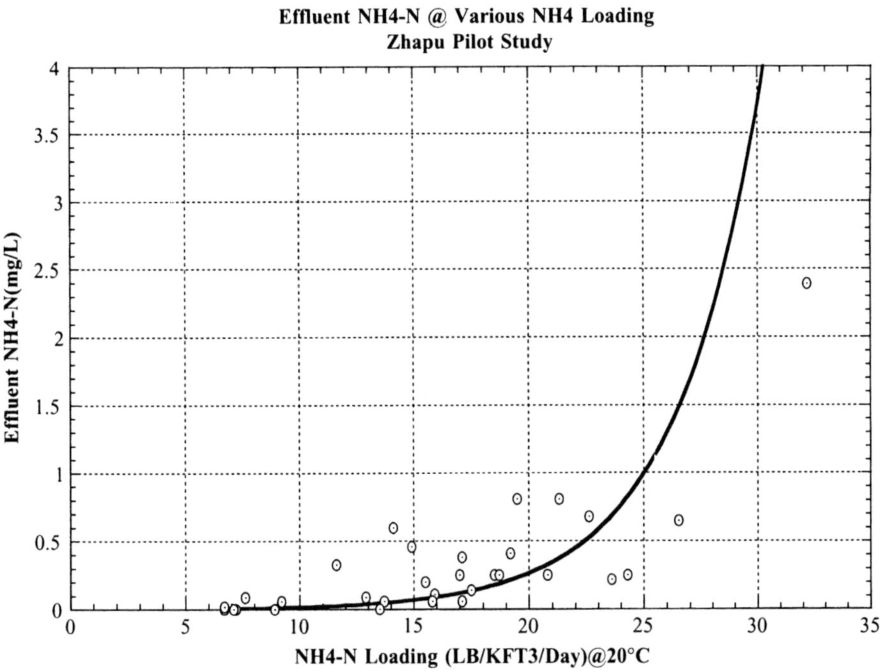

Figure 2 Specific NH4-N mass loading to the SAF bioreactor versus the effluent NH4-N concentration

water treatment plant will consist of TETRA SAF for the pretreatment followed by pre-ozonation, chemical coagulation, sedimentation, sand filtration, post-ozonation, granular activated carbon adsorption and final chlorination for disinfection. For the 25,000 m3/day capacity plant three SAF bioreactors will be utilized for the pretreatment, each with a dimension of 3.56 m W × 25.6 m L × 6.1 m gravel media. The water treatment plant is currently under an accelerated construction schedule at the time. It will be completed and plans to start delivering potable water in the first quarter of 2006.

Conclusions

(1) The SAF pilot plant study at Zhapu demonstrated that the SAF bioreactor was very effective in removing NH4-N and organics from the polluted river water and transforming it to meet the required potable water standard.
(2) The SAF bioreactor was very tolerant to the hydraulic shock load when the flow rate to the bioreactor was increased to 200% of the designed hydraulic load.
(3) The operation of the SAF bioreactor was very simple and without requiring much input from the operators. The operation may be automated with a minimum instrumentation.

60 Effects of inlet type and reactor style on biological aerated filter backwashing

H. Han, W. Ma, J. Huang, D. Zhong, M. Xue, Z. Wang

Harbin Institute of Technology, Harbin, 150090, China, damayingxiong@163.com

Abstract The effects of inlet type and reactor style on biological aerated filter (BAF) backwashing were investigated. The results showed that impulse backwashing effect was better than that of air-water continuous backwashing at the same backwashing intensity. Contrast to uniform flow, expanded flow had better backwashing effect and higher velocity of biological filtering layer recovery at the impulse backwashing intensity. The optimal impulse backwashing operation parameters of expanded flow BAF were determined as follows: the backwashing airflow intensity was 8–10 L/(s·m^2), the backwashing water-flow intensity was 2–4 L/(s·m^2), and the backwashing duration was 5 min.

Keywords Air-water continuous backwashing, biological aerated filter (BAF), expanded flow, Impulse backwashing, uniform flow

Introduction

BAF is a wastewater treatment process with high disposal efficiency and low energy cost. Backwashing of BAF is to release intercepted suspended solids and renovate biofilm, and it is the key step to keep BAF in high disposal efficiency. Several issues faced in the process of BAF backwashing are discussed in this paper, which are included as below: the inlet way of backwash; the comparison between different BAF styles; the operation parameters.

Methods

In this paper, the impulse backwashing (impulse air and continuous water flow) and continuous backwashing (continuous air and water flow) was taken. In the process of impulse backwashing, the air was pumped into filter layer instantly that made filter layer stay in varied speed expansion phase. The expanded flow and uniform flow BAF were taken to carry out continuous backwashing and impulse backwashing comparison test. The equipment and materials of the two types of BAF were shown in Table. 1.

Table 1 Equipment and materials in test

Reactor style	Diameter in inlet end (mm)	Diameter in outlet end (mm)	Height (mm)	Filter material	Accumulation specific gravity (g/cm^3)	Average particle size (mm)
Uniform flow	300	300	1800	Light porous ceramsite	0.83	4.57
Expanded flow	200	500				

Results and discussion

Comparison test of different BAF with two backwashing style

The different BAF reactors were both carried out the continuous and impulse backwashing test. The intensity of airflow was 10 L/(s • m^2) and that of water flow was 2.3 L/(s • m^2). Total solids content of backwash liquor was taken to indicate the state of backwashing.

Data showed that in the first 3mins, the difference of backwashing effect between the two models was inconspicuous and then became obvious gradually. It was because that the aging biofilm and the carried impurity attached to the filter bed loosely, both models can wash off these substances quickly. Contrast to continuous backwashing, impulse backwashing presented higher backwash effect and less backwash time. In addition, the continuous backwashing only made the lower filter layer disturb intensely, but the upper filter layer even kept moveless. The impulse backwashing made the whole filter layer disturb like snake moving and impulse frequently. This phenomenon was more obvious in expended flow BAF.

Backwashing efficiency and recovery of COD removal rate in different BAF

The backwashing efficiency and recovery of COD removal rate of impulse backwashing were studied: airflow intensity 10 L/(s • m^2), water flow intensity 2.21 L/(s • m^2).

As shown in Fig.1 and Fig.2, the expanded flow had less backwashing time and higher recovery rate than that of uniform flow. It can be explained that in different sections, the upper part velocity was higher than that of lower part; in the same section, middle part

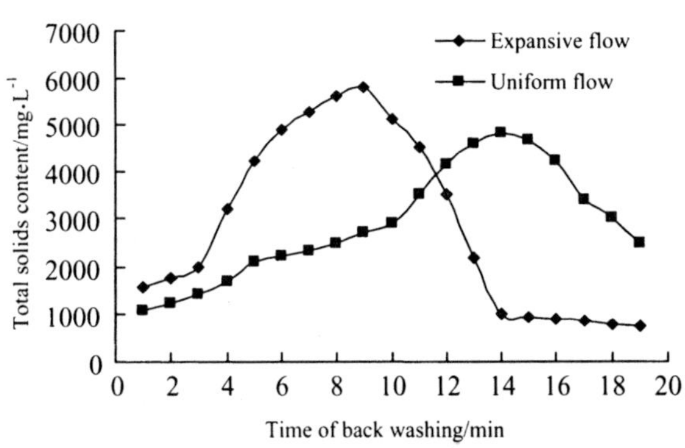

Figure 1 Relation between total solids of backwash liquor and backwash duration

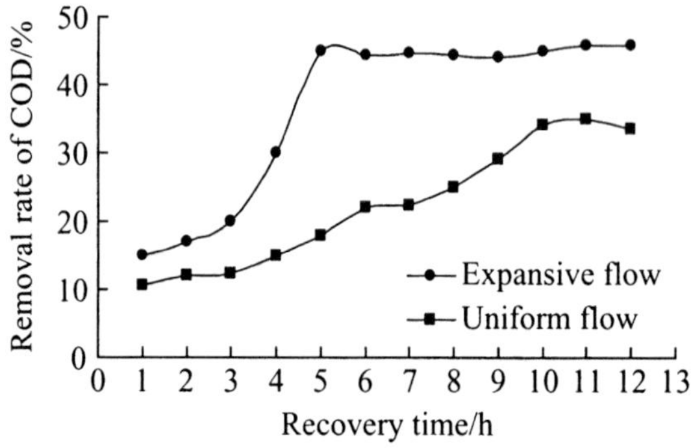

Figure 2 Relation between COD, removal and filter bed recovery time

velocity was higher than that of surrounding part. So it leads to the bulk transposition and circulation movement from middle part to surrounding. On the one hand, this movement promoted fluidization and impulse action and enhanced impact-friction action among particles. On the other hand, it caused transposition between bottom filter material with abundant activated strains and top filter material with few inactive strains. The washed activated biofilm accumulated in top part of reactor properly. So the top filter layer showed the removal ability after a brief stabilization period and assured whole removal effect.

Test of impulse backwashing intensity and time in expanded flow BAF

Test of air and water intensity of backwashing

The air intensity (6–8, 8–10, 10–12 L/(s • m^2)) was taken while the water flow intensity was 2.3 L/(s • m^2). Then the water intensity (1–2, 2–4, 4–6 L/(s • m^2)) was taken while the airflow intensity was 10 L/(s • m^2)

As shown in Fig.3 and Fig.4, airflow intensity variety had larger effect on backwashing efficiency than water flow. When airflow intensity was 6-8L/(s • m^2), the filter bed had no

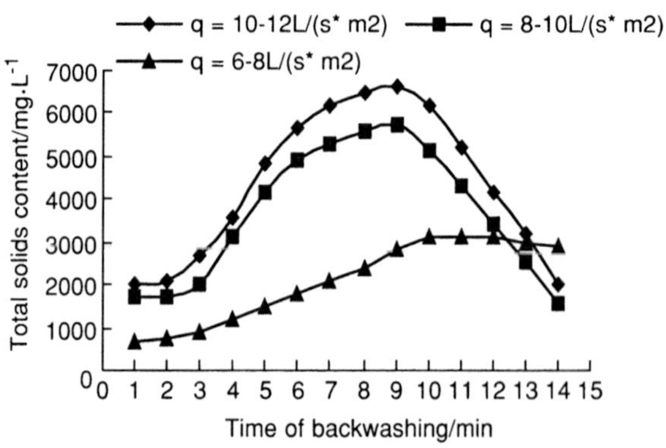

Figure 3 Relation between total solids of backwash liquor and time for different backwash airflow

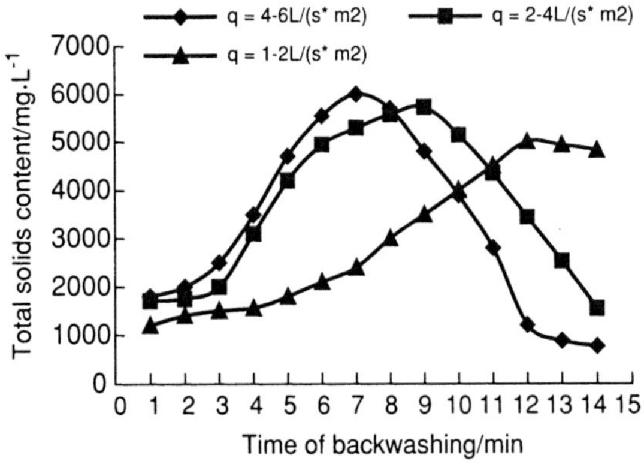

Figure 4 Relation between total solids of backwash liquor and time for different backwashwaterflow

agitating phenomenon and the biofilm desquamation only depended on low intensity water shearing and small range impact-friction action of filter material. When airflow intensity was 8–10 L/(s • m^2), big air bubble was formed immediately at the bottom and moved up discontinuous that caused impact-friction of the whole filter layer. The filter layer with fluidized impulse and bulk transposition had high backwashing efficiency. When the intensity of airflow was increased to 10–12 L/(s • m^2), filter bed fluidized and pulsed more intensely. A mass of filter materials was taken out of reactor by airflow.

The fixed airflow intensity was 10 L/(s • m^2). When the water flow intensity was 1–2 L/(s • m^2), the water shearing and impurity transport ability declined. When the water flow intensity was 2–4 L/(s • m^2), both impurity transport ability and airflow effect were promoted. The backwashing efficiency was improved obviously. When the water flow intensity was increased to 4–6 L/(s • m^2), filter materials was taken out of reactor by water flow with high speed.

Test of backwashing time

The recovery time and relative content of organic/inorganic matter was taken to determine the backwashing time. The results were shown as Fig.5 and Fig.6.

As shown in Fig.5, at the beginning of backwashing, the ratio of organic/inorganic matter declined rapidly and reached the minimum value at 5 min. In the first 3 mins, the organic, inorganic and total solids content kept steadily, and then increased rapidly. After 5min, the organic matter content kept on increasing with constant speed, but the speed of inorganic content increasing reduced. It was considered that at the beginning of backwashing, the light and aging biofilm was washed off from filter layer firstly. Compared with the inorganic solids adsorbed and intercepted by filter material, it had more organic matter and steady ratio of organic/inorganic matter. After that, tightly attached biofilm was washed off. The external biofilm was desquamated firstly which was mainly composed of mass of extracellular polymeric substances and inorganic impurity. As the high content of inorganic matter in extracellular polymeric substances, the ratio of organic/inorganic matter declined gradually. When the inner biofilm was desquamated, the inorganic content decreased because of lower

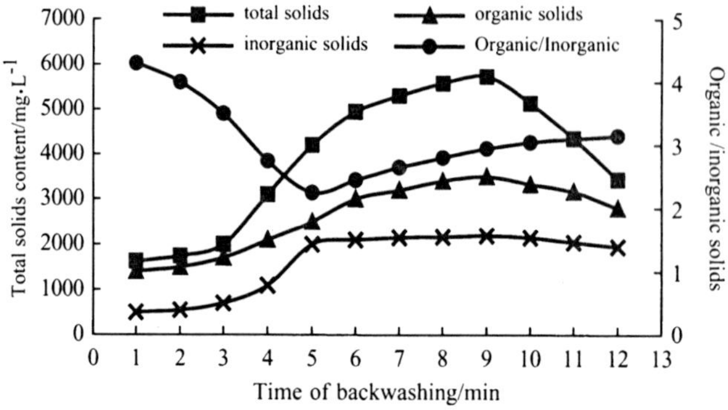

Figure 5 Relation between COD removal rate and filter bed recovery time

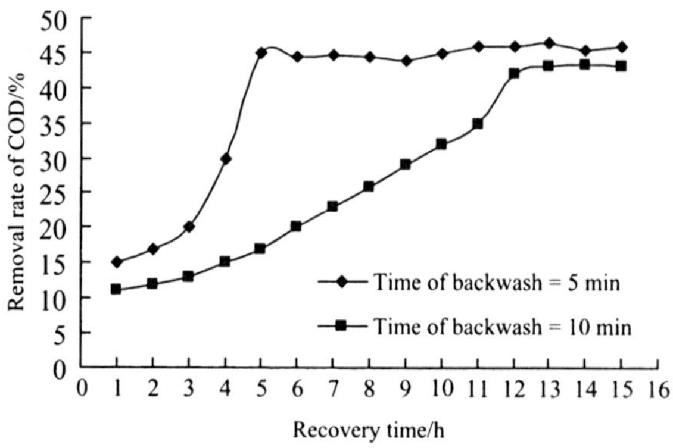

Figure 6 Relation between the backwash time and the properties of backwash liquor

interspaced ability of the inner compact bioflim and weak osmotic ability of inorganic particulate. The content of organic matter kept constant that made the ratio rise continuously. It can be concluded that 5 min of backwashing was the start point of inner organic biofilm desquamation. As shown in Fig.6, the recovery time of 5 min backwashing was 9 h faster than 10 min backwashing. So it can be concluded that organic activated biofilm of filter bed surface layers was still preserved properly and the backwashing time was 5 min.

Conclusions

1. The impulse backwashing had higher efficiency than air-water continuous backwashing.
2. The expanded flow BAF presented higher backwashing efficiency and recovery rate than that of uniform flow BAF.
3. The backwash parameter of in expanded flow BAF was determined as: airflow intensity 8 ~ 10 L/(s • m^2); water flow intensity 2 ~ 4 L/(s • m^2); backwashing time 5 min.

References

David Hall, Caroline S, B Fitzpatrick. Research note -spectral analysis of pressure variations during combined air and water backwash of rapid gravity filters. War, Sci. Tech. 1999, 33(17):3666–3672

Lu Rongguang, Luo Huirong. Research on Mechanism of Biological Aerated Filter Air-water Backwashing. Journal of Chongqing Jianzhu University. 1998, 20(6):7–11

Spath R, H C Flemwing, S Wuertz. Sorption Properties ofBiofil ms.War, Sci. Tech., 1998, 37(4–5):207–2

Wu Guangchun, Tang Chuangxiang. Research and Design on Biological Aerated Filter Air-water Backwashing. Chemical Water & Wastewater Design. 1995, 4:1–6

Yuan Zhiyu, Chen Xiaoru. Study on the Mechanism of Suspended Moving Simultaneous Air water Backwashing. Water & Wastewater Engineering. 1999, 25(5):21–2.

Zhang Junzhen, Deng Cailing. Mathematical Model of Air-water Backwashing of filter. China Water & Wastewater. 1997, 13(3):10–13

Zou Weiguo, Sun Qun. Research on Development of New BIOS MEDI Biological Filter. China Water & Wastewater. 2001, 17(1):1–4

61 Intermittent filtration of bacteria and colloids in porous media

A. A. Keller and M. Auset

Bren School of Environmental Science and Management, University of California, Santa Barbara.

Abstract Intermittent filtration through porous media used for water treatment can achieve high pathogen and colloid removal efficiencies. To predict the removal of bacteria in these systems, the effects of cyclic infiltration and draining events were investigated. We visualized the intermittent transport of *Escherichia coli* in unsaturated micromodels. Column experiments provided quantitative measurements of the phenomena observed at the pore scale. Breakthrough behavior between colloids and a conservative dissolved tracer differed significantly, reflecting the differences in transport processes. Advancement of the wetting front remobilized bacteria which were held in thin water films, attached to the air-water interface (AWI), or entrapped in stagnant pore water between gas bubbles. In contrast, the tracer was only remobilized by diffusion from immobile to mobile water. Remobilization led to successive concentration peaks of bacteria and tracer in the effluent, but with significant temporal differences. Colloids were essentially irreversibly attached to the solid-water interface, which explained the high removal efficiency of microbes in the porous media. Straining, cluster filtration, cell lysis, protozoa grazing and bacteriophage parasitism could also contribute to the removal efficiency of bacteria.

Keywords Wastewater treatment, intermittent filters, unsaturated flow, pathogen, micromodel

Introduction

Intermittent infiltration through permeable soils has been increasingly used for the treatment of wastewater effluents (Brissaud and Lesavre, 1993; Stevik et al., 2004) because of its efficiency, innocuousness and low cost. The surface of the filter is sequentially dosed with effluent that percolates in a single pass through the unsaturated porous medium (US EPA, 1999). If properly designed, infiltration systems offer a high capacity to remove pathogenic microorganisms from wastewater. The mechanism of microbial removal can be attributed to the combination of physical filtration, adsorption and inactivation. The purpose of this study was to assess the effect of cyclic infiltration events on the fate of bacteria during intermittent filtration of wastewater. The research was conducted at two scales: pore scale (micromodels) and macroscale (columns). The pore scale permits microscopic visualization and column experiments provide quantitative measurements of the phenomena taking place.

Material and methods

Micromodels

PDMS (polydimethylsiloxane) micromodels containing a realistic pattern of pore networks were used (Auset and Keller, 2004). *Escherichia coli* ATTC 25922 and 5 μm-diameter microspheres, suspended in partially treated wastewater, were injected into the micromodels. The injection cycles were as follows: 2 minutes wastewater infiltration/8 hours drainage. A pulse of bacteria was injected in a 2 min pulse and the micromodel was allowed to drain for 8 hr. Subsequent pulses of wastewater without bacteria were applied, allowing drainage after every pulse. Continuous real-time images were recorded from selected regions in the micromodel.

Column

A cylindrical column (length: 150 cm; diameter: 19 cm) was filled with sand ($d_{10} = 0.36$ mm, $d_{50} = 0.91$ mm, $d_{60}/d_{10} = 2.72$, bulk density = 1.57 g/cm^{-3} and porosity = 0.34). The pumping rate (0.4 m/d) and the number of flushes per day (6), were set to match a typical intermittent sand filter. The application rate of each flush was 2.2×10^{-4} m/s and each flush lasted 5 minutes. DAPI-stained bacteria and a conservative tracer (NaI ($2H_2O$)) were injected within the initial pulse, followed by wastewater flushes with no tracers. Effluent samples were continuously collected for 4 days.

Results and discussion

This study applied two complementary approaches to understand the transport of colloids in general, and bacteria in specific, in transient unsaturated flow. First, we employed micromodels at the pore scale. Second, we used columns to simulate realistic 3-D filtration. Micromodels allowed direct mechanism visualization while columns provided quantitative measurements.

The micromodel observations showed that bacteria traveled with the mobile water by advection and dispersion. During their transport, the colloids interacted with the air and solid phases, resulting in temporary or permanent immobilization. Between flushes, colloids were retained in the AWI and solid-water interface (SWI), as well as in thin water films.

The mechanisms observed in the micromodels for bacteria and microspheres were very similar; therefore we present in Figure 1 examples of the 5 μm colloids because they are clearer in printed format than the images of 1 μm Escherichia coli. Figure 1a shows the micromodel after almost 8 hours of drainage following the injection of microspheres. A high proportion of the pore space was filled with air, which was seen as the phase with the thick interface (AWI). All the colloids observed in this image were suspended in stagnant water, trapped at the AWI or attached to the SWI. A number of colloids were trapped in very thin water films along the pore walls (film straining). Figure 1b presents the micromodel 20 seconds after flushing. As the infiltration front advanceed from left to right, air was displaced, carrying colloids attached to the AWI. Pores became filled with water, reconnecting immobile water and contributing to the advance of the hydraulic pulse.

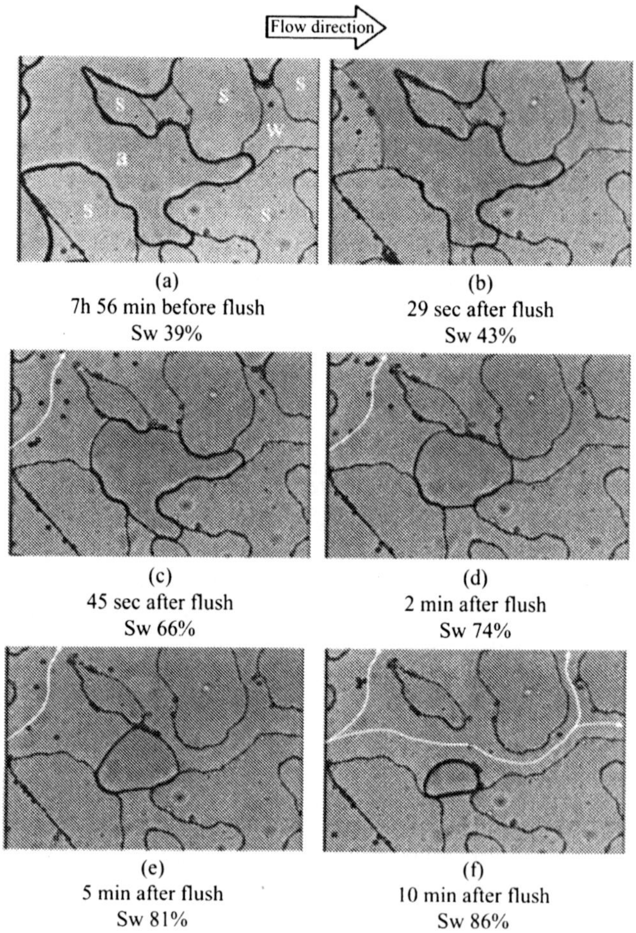

Figure 1 Transport of 5-μm colloids in the micromodel during infiltration/drainage events. S_w = water saturation, s = solid; a = air; w = water. (a) After drainage; (b–f) immediately after flush.

Colloids trapped in stagnant water regions were remobilized (Figures 1c–f). A significant number of colloids remained attached to the SWI, which resulted in essentially irreversible sorption at these sites under the experimental conditions. It is also possible that some of these colloids were temporarily attached to the triple interface, air-water-solid.

Detailed concentration histories of iodide and DAPI-stained bacteria from column effluent are shown in Figure 2. The breakthrough curves of iodide and bacteria highlighted the differences in transport processes between solutes and bacteria. The iodide displayed an asymmetrical breakthrough curve with significant tailing, typical of non-equilibrium phenomena with mobile and immobile flow regions.

The breakthrough of bacteria occurred in successive pulses associated with each hydraulic flush. During the sequential pulses, the concentrations of eluted microorganisms

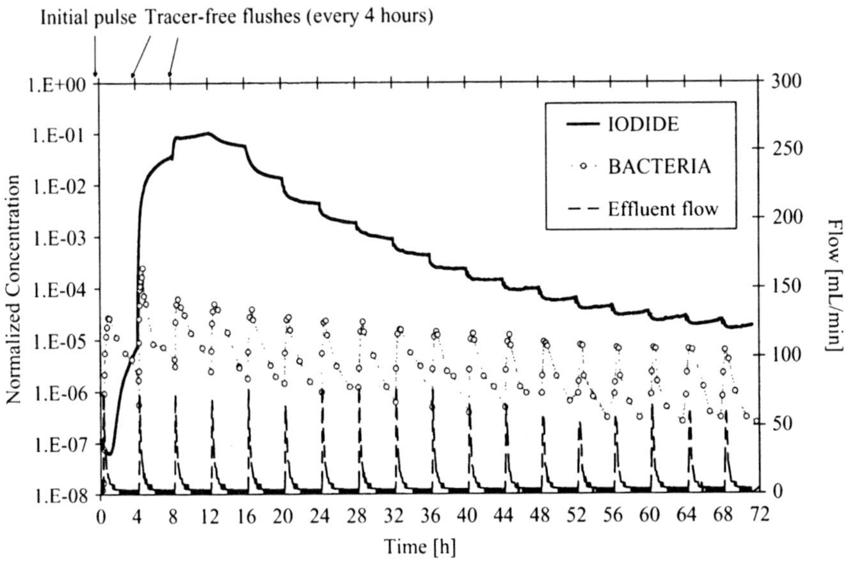

Figure 2 Breakthrough curves for iodide and bacteria in log ordinate axis.

fluctuated up and down, with an oscillation of almost two $\log_{10}$ units between peak and valley and a gradual reduction in concentration. These pulses were due to the mobilization of bacteria by the advancing wetting front from thin water films and the disappearing AWI, as well as from stagnant water, as observed in the micromodel studies. Only a very small fraction of the injected bacteria was recovered at the column effluent during the 4 days of the experiment (*E. coli* recovery rate = 0.028%, standard deviation = 0.008), compared to that of the dissolved tracer (Iodide recovery rate = 96%).

Both dissolved tracer and tagged bacteria exhibited persistent tailing and could still be detected in the column effluent after 72 h. However, the elution of bacteria was much slower than the elution of iodide.

Conclusions

The results from both the pore and column scales suggested that bacterial and colloidal processes during intermittent infiltration were quite different from processes for dissolved compounds. Both the conservative tracer and the colloids advanced during each flushing event as connectivity between mobile and immobile water increaseed, allowing solutes and colloids to mobilize. However, the behavior of the solute was dominated by slow diffusion from stagnant water regions, while transport of bacteria was controlled by sorption on AWI and SWI, as well as storage in thin water films. Remobilization of colloids depended on the movement of the air phase, the dissolution of existing air bubbles that released colloids from the AWI, and/or the rewetting of thin water films which also released colloids. The correlation between the pore scale observations and the column scale studies reinforced our understanding of these transport mechanisms.

Bacteria were removed to a high extent in the column. Irreversible sorption at the SWI and straining in narrow pore throats were dominant removal mechanisms. In addition, other loss mechanisms such as cell lysis, protozoa grazing and bacteriophage parasitism might reduce the load of bacteria that exited the system. This confirmed the usefulness of intermittent filtration through porous media as a means to achieve high removal of colloids, and in particular bacteria, from wastewater.

References

Auset, M. and Keller, A.A. (2004) Pore-scale processes that control dispersion of colloids in saturated porous media. *Water Resour. Res.* 40 (3): Art. No. W03503.

Brissaud, F. and Lesavre, J. (1993) Infiltration percolation in France: 10 years experience. *Water Sci. Technol.* 28, 73–81.

Stevik, T.K., A. K., Ausland, G. and. Hanssen, J.K (2004) Retention and removal of pathogenic bacteria in wastewater percolating through porous media: a review. *Wat. Res.* 38 (), 1355–1367.

United States Environmental Protection Agency, Wastewater Technology Fact Sheet (1999) Intermittent Sand Filters, Office of Water Washington, D.C. EPA 832-F-99-067.

62 Technical evaluation of rainwater harvesting filtration systems in India

D. Khare*, Ramakant, C.S.P. Ojha*****

* Associate Professor and **Research Scholar, Department of WRD&M, Indian Institute of Technology Roorkee, Roorkee 247 667, India; E-mail: kharefwt@iitr.ernet.in
**Professor, Department of Civil Engineering, Indian Institute of Technology Roorkee, Roorkee 247 667, India; E-mail: cspojha@rediffmail.com

Abstract There are many filters installed in the field for rainwater harvesting. Technical evaluation of the installed filters in different conditions reveals that the design of the sand filters is on adhoc basis and no design guidelines are available. Hsiung and Cleasby have suggested models for prediction of effluent quality and critical head loss for water filtration, based on the experiments conducted by continuously adding Ferrous Sulfate in hard well water. The filter influent and effluent quality were evaluated by their iron contents. The present study deals with different sand filter modules in the laboratory operated with rainwater. The compatibility of Hsiung and Cleasby models for rainwater followed by formulation of new models has been investigated in the present work.

Keywords Rainwater, turbidity, sand filter, effluent quality and head loss

Introduction

Rainwater harvesting is a technology used for collecting and storing rainwater from rooftops, the land surfaces, steep slopes, road surfaces or rock catchments using simple techniques such as pots, tanks and cisterns as well as more complex techniques such as underground check dam (Makoto, 1999). A number of studies have investigated the prevalence of microbiological and chemical contaminants in roof-collected rainwater (Fujioka *et al.*, 1991). The presence of turbidity can have significant effects on both the microbiological quality of drinking water and the detection of bacteria and viruses in the water (Brock, 1966). Studies reported indicate bacteriological count reductions with decreasing filtrate turbidity and practically complete removal of algae and coliform bacteria with a 0.1 NTU effluent (Logsdon and Lippy, 1982). Only by treating the water samples by filtration prior to chlorination could the water be rendered safe to drink.

At present different types of filters are in use for rainwater harvesting in India and hardly any design information is available for these filters (Figs. 1-3). Hence, the design description and technical parameters evaluation of sand filters are required. Hsiung and Cleasby (1968) have given models which are based on the influent and effluent iron concentration for performance prediction of sand filters and have been recommended for all types of

 Recent Progress in Slow Sand and Alternative Biofiltration Processes edited by Rolf Gimbel, Nigel J.D. Graham and M. Robin Collins. ISBN: 9781843391203. Published by IWA Publishing, London, UK.

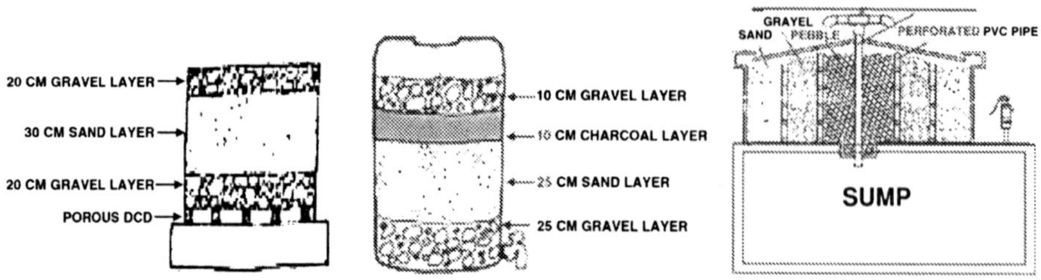

Figure 1–3 Critical sand filters being used in the field for rainwater harvesting (RWH manual, 2001)

water filtration system. In the present study, turbidity, an indicative parameter has been considered to check the compatibility of Hsiung and Cleasby (1968) sand filter models for rainwater harvesting. Hence, the study was carried out at different sand filter modules installed in the laboratory to restrict the installation of filters in field, particularly for domestic purposes.

Methodology

Lab-scale sand filter modules of different depth were installed in the laboratory. The filters were made of HDPE pipe circular in shape to know the variability of the performance of the filters. The experiments were run in two different cases, one with natural roof-top rainwater and another with Fullers' earth (China clay) suspension water solution. The concentrations of suspension of Fuller's earth suspension water were 25 and 30 mg/l which were higher than the concentration of turbidity of roof-top rainwater (17 mg/l). Filter flow rates were selected based on rainfall intensity and duration. These filters were run for 36 hours in both the cases. Mean size of sand filter media were selected as 3.647, 2.366, 1.673, 1.091, 0.714 and 0.505 mm and flow rates were maintained at 1.65, 3.30, 4.95, 6.60 and 8.25 $m^3/hr/m^2$. Sand filters were operated for 76 runs using Fuller's earth suspension water and 34 runs using rainwater in different combinations of flow rates and filter grain sizes and the details are given in doctoral thesis (Ramakant, 2006).

Model verification and development

Lumped data obtained from the laboratory for various grain sizes, flow rates and influent turbidity concentrations were used to predict the performance of sand filter using Hsiung and Cleasby (1968) models, as given in Eqs. (1) and (2).

$$\log\left(\frac{U}{L}\right) = -0.208 + 1.950\log\left(\frac{G}{L^{1.2}}\right) - 0.645\left[\log\left(\frac{G}{L^{1.2}}\right)\right]^2 \tag{1}$$

$$\log\left(\frac{R}{L^{1.6}}\right) = -3.250 + 1.013\log\left(\frac{G}{L^{1.2}}\right) - 0.036\left[\log\left(\frac{G}{L^{1.2}}\right)\right]^2 \tag{2}$$

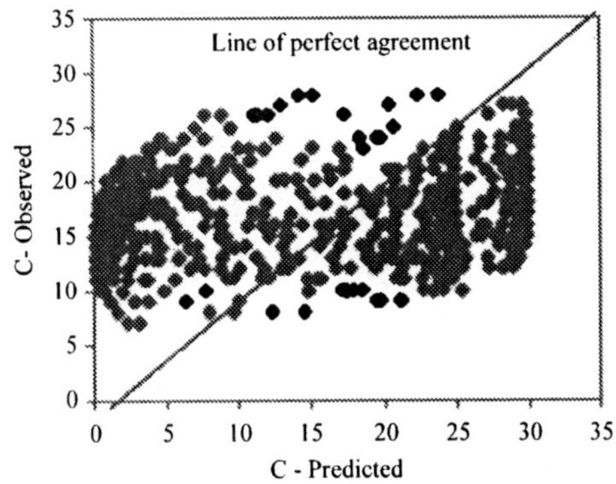

Figure 4 C observed and C predicted using Hsiung and Cleasby (1968) model

where, U, variant, is a function of (C/C_o), C_o is influent concentration of turbidity in mg/l, C is effluent concentration of turbidity in mg/l, L is depth of filter media in inch, $G = Q^{0.29}d^{0.62}t$, Q is flow rate in gpm/ft^2, d is diameter of sand grains in mm and t is the time of filter run in hr., $R = d^{2.5}\,(H_t - H_o)\,/Q^{1.2}C_o^{1.4}$, (Ht- Ho) is critical head loss in feet.

Initially, Hsiung and Cleasby (1968) model was used to predict the effluent turbidity concentration but the scattering of the data for Fuller's earth suspension water reveals that this model is not compatible to predict effluent quality at different conditions (Fig. 4). Hence, a new model (Eq. 3) was developed for lumped data for various grain sizes, flow rates and influent turbidity concentrations using Fuller's earth suspension water solution and the model was tested against rainwater.

In the same fashion, Hsiung and Cleasby (1968) model was also used to predict the critical head loss for sand filter. Scattering of head loss data reveals that the model is not compatible for the performance prediction (Fig. 5). Hence, a new Model (Eq. 4) was developed for lumped data for various grain sizes, flow rates and influent turbidity concentrations using Fuller's earth suspension water solution and tested against rainwater for head

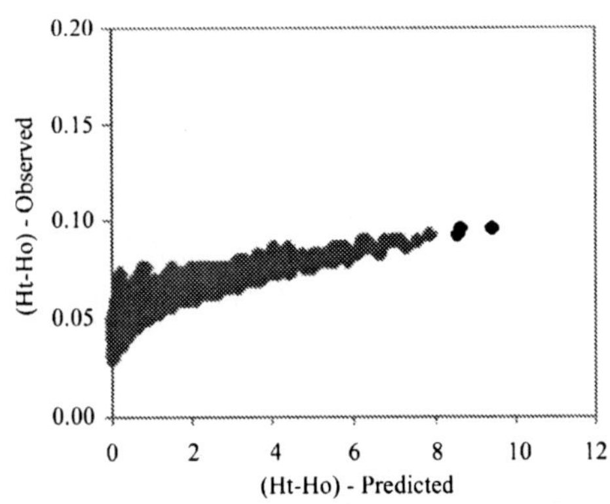

Figure 5 Head loss observed and predicted using Hsiung and Cleasby (1968) model

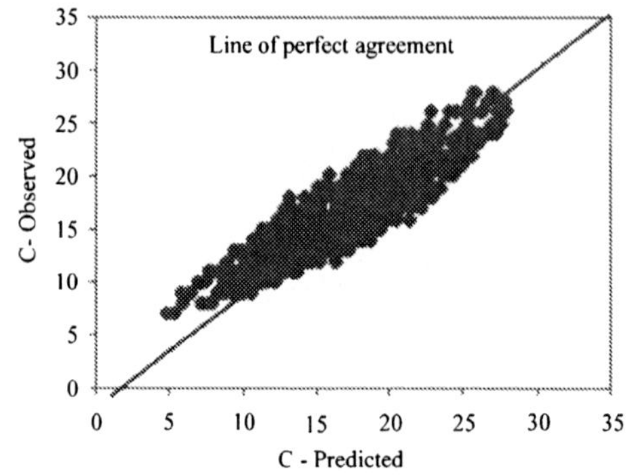

Figure 6 C observed and C predicted using new developed model

loss prediction.

$$\log\left(\frac{U}{L}\right) = -0.840 + 1.497\log\left(\frac{G'}{L^{1.5}}\right) - 0.137\left[\log\left(\frac{G'}{L^{1.5}}\right)\right]^2 \quad (3)$$

$$\log\left(\frac{R'}{L^{1.3}}\right) = -2.463 - 0.477\log\left(\frac{G'}{L^{1.5}}\right) + 0.211\left[\log\left(\frac{G'}{L^{1.5}}\right)\right]^2 \quad (4)$$

where, U, variant, is a function of (C/C_o), C_o is influent concentration of turbidity in mg/l, C is effluent concentration of turbidity in mg/l, L is depth of filtering media in m, $G' = Q^{0.12}d^{0.35}t$, Q is flow rate in $m^3/hr/m^2$, d is diameter of sand grains in mm and t is the time of filter run in hr., $R' = d^{1.2}(H_t - H_o)/Q^{0.19}C_o^{1.3}$, $(H_t - H_o)$ is critical head loss in m.

The use of Eq. (3) for effluent quality prediction gives an average percent error of 5.63 and 6.21% and removal efficiency of 60 and 60% at 2.366 mm sand grain and 1.65 $m^3/hr/m^2$ flow velocity for 25 and 30 mg/l turbidity concentration respectively while the use of Hsiung and Cleasby model gives an average percent error of 52.41 and 51.69% and removal efficiency of 26 and 26% respectively. The scattering of data from the line of perfect agreement for 2.366 mm sand grain shows that the new model is more compatible to predict the performance of sand filters using Fuller's earth suspension solution (Fig. 6).

In the same fashion, the use of Eq. (4) for head loss prediction gives an average percent error of 68.36 and 67.32% at 2.366 mm sand grain and 1.65 $m^3/hr/m^2$ flow velocity for 25 and 30 mg/l turbidity concentration respectively while the use of Hsiung and Cleasby model gives an average percent error of 379.84 and 413.21% respectively. The scattering of the data from the line of perfect agreement for 2.366 mm sand grain shows that the new model is much more compatible to predict the sand filter performance than Hsiung and Cleasby (1968) model for critical head loss (Fig. 7).

The turbidity removal efficiency was high at high average percentage error for Hsiung and Cleasby (1968) model and removal efficiency was high at low average percentage error for the new developed model. Higher value of average percentage error shows high degree of scattering of data from line of perfect agreement. The details of testing of the new models for rainwater samples and various filtration conditions are given elsewhere (Ramakant, 2006).

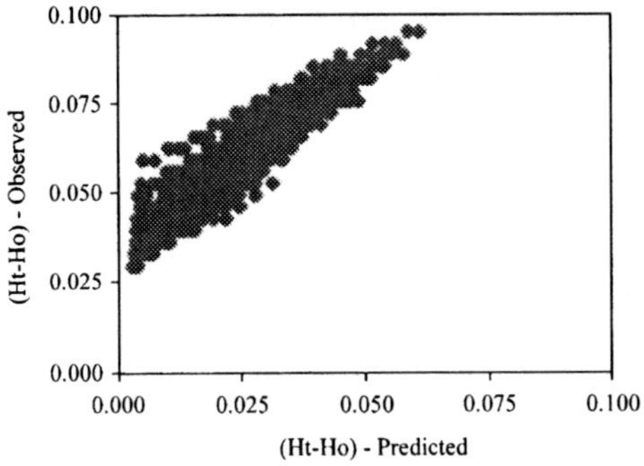

Figure 7 Head loss observed and predicted using new developed model

Dimensionless model analysis is also studied for Fuller's earth suspension water solution and rainwater and is given elsewhere (Ramakant, 2006).

Conclusions

Available models for the prediction of sand filter performance in different conditions are not compatible for effluent turbidity concentration and critical head loss. Hence, new models have been developed for effluent turbidity concentration and critical head loss considering different filtration conditions in the laboratory. New models seem much more compatible than previously developed model. The behaviour of filters depends on filter grain size, depth of filter media, flow rate and its orientation, particularly for rainwater harvesting.

References

Brock, T.D. (1966) Principles of microbial ecology. Prentice-Hall, NJ. 72.

Fujioka, R., Inserra, S. and Chinn R. (1991) The bacterial content of cistern waters in Hawaii. *Proc. 5th Inter. conf. on rainwater cistern systems,* Keelung, Taiwan. 33–45.

Hsiung, K. and Cleasby, J.L. (1968) Prediction of filter performance. *Sanit. Engg. Div., Proc. ASCE*, SA6, 1043–1069.

Makoto, M. (1999) Creating rainwater utilization based society for sustainable development. *Proc. Inter. symp. efficient water use in urban areas*, UNEP International Environment Technology Centre, Osaka, Japan, 107.

Rainwater Harvesting (RWH) Manual for urban areas: case studies from Delhi (2001). Centre for Science and Environment, New Delhi, India.

Ramakant (2006) Design and technical evaluation of rainwater harvesting filtration systems. Doctoral thesis, I.I.T. Roorkee, Roorkee, India.

63 Biological filtration of organic solid materials from municipal wastewater with the aerated constructed wetland

D.-S. Kim, J.-H. Yoon, S.-C. Kim and D.-K. Lee

Dept. of Chemical and Biological Engineering, Environmental Biotechnology National Core Research Center (EBNCRC), Environmental and Regional Development Institute, Gyeongsang National University, Kajwa-dong 900, Jinju, Gyeongnam 660-701, Korea

Abstract Aerated constructed wetland was employed for the removal of organic solid materials in raw municipal wastewater. The wetland was equipped with a specially designed natural air draft system. More than 99% of the organic solid materials could be successfully removed through the filtration by the sand and gravel beds and the following biochemical oxidation. This combined filtration and biochemical decomposition made the wetland accomplish the continuous successful removal of organic solid materials in the 100m^3/day municipal wastewater for 3 years without blockage of the sands and gravels.

Keywords Aerated constructed wetland, municipal wastewaters, organic solid materials, dissolved oxygen

Introduction

Wastewater contains a variety of organics solid materials, and the separation and subsequent decomposition of organic solid materials have received a great attention to meet strict discharge regulations. Constructed wetland is an eco-friendly alternative municipal and industrial wastewater treatment (Crites and Tchbanoglous, 1998; DWLC, 1998; Leonard and Swanson, 2001; Reed et al, 1988; U.S. EPA, 1993). The environment within a constructed wetland is mostly either anoxic or anaerobic, because there is no direct contact between the water column and the atmosphere. Therefore typical constructed wetland systems are not good for the treatment of raw wastewaters, because the pollutant loadings are too high to be treated successfully by the biological elements of the wetland.

Experience shows that removal of solid materials and ammonia from a wastewater in a typical constructed wetland is likely to be limiting design factors, because solid material decomposition and nitrification are always limited by oxygen availability and natural aeration is a really slow process in wastewater wetland. Poor decomposition of organic solid materials will result in the blockage of filtration bed within constructed wetland, and thereby constructed wetland might be taken out of service for several days or weeks. By making the

 Recent Progress in Slow Sand and Alternative Biofiltration Processes edited by Rolf Gimbel, Nigel J.D. Graham and M. Robin Collins. ISBN: 9781843391203. Published by IWA Publishing, London, UK.

Figure 1 The natural air draft system installed inside the constructed wetland.

wetland aerobic, faster and more efficient organic solid biological decomposition together with nitrification is expected to occur within the wetland.

In this study, a constructed wetland was designed to remove organic solid materials effectively from 100 m^3/day municipal wastewater, and the continuous performances of the constructed wetland were investigated for three years.

Methods

The constructed wetland could remain aerobic owing to the continuous supply of air through the natural air draft system (Figure 1) whose driving force for air flow was the temperature difference between the ambient air and the air inside the wetland (Schroeder and Tchobanoglous, 1976). The wetland had 145 m^2 area and 1.0 m depth, and accordingly the theoretical hydraulic residence time (HRT) in the wetland was 1.45 days. From the bottom of the tank spherical gravels (25 mm diameter) were packed upto the depth of 0.7 m above which tiny sands (2.5 mm diameter) were packed again with 0.3 m depth. Reeds were planted on the sand level.

100 m^3/day municipal wastewater was distributed onto the surface of the wetland by using the wastewater distribution lines as shown in Figure 2. The distributed wastewater flows vertically under gravity force. During the flows through the wetland, the wastewater contacts directly air supplied by the natural air draft system. The treated wastewater was collected at the bottom of the wetland.

Results and discussion

The changes in DO concentration can be clearly seen from the data in Figure 3. The DO concentrations in the influent wastewater were about 0.3 mg/L. DO concentrations in the effluent were, however, more than ten times higher than the corresponding values in the

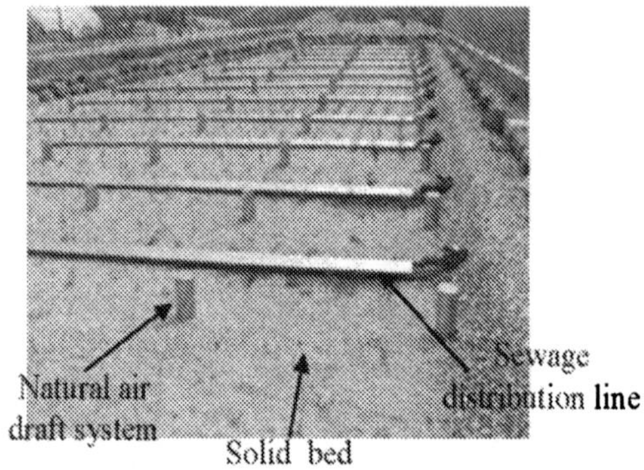

Figure 2 The detailed photograph of the wastewater distribution lines onto the surface of the wetland.

influent. This result indicates that the natural air draft system supplies the constructed wetland with sufficient oxygen.

In addition DO concentrations at the effluent of the wetland were highly proportional to the absolute temperature difference between the ambient air and the air inside the natural air draft system (Figure 4). This provides an evidence that the driving force for air flows is temperature difference between ambient air and the air inside the wetland.

Figure 5 shows the efficiencies of SS removal in the aerated constructed wetland. More than 99% of the initial SS was removed continuously and stably upto 3years operation.

Organic solid materials concentrations attached onto the surface of the sands and gravels in the aerated constructed wetland were measured at 120 different locations and at different times (3, 6, 12, 18, 24 and 36 months). In Table 1 are summarized average concentrations of the attached organic solid materials onto the sands (0.2 m depth) and gravels (0.5 and 0.8 m depths). Concentrations onto the sands are somewhat higher than those onto the gravels.

This might be due to the more efficient filtration of the organic solid materials in the sands rather than gravels. An interesting feature, when looking the results in Table 1 over

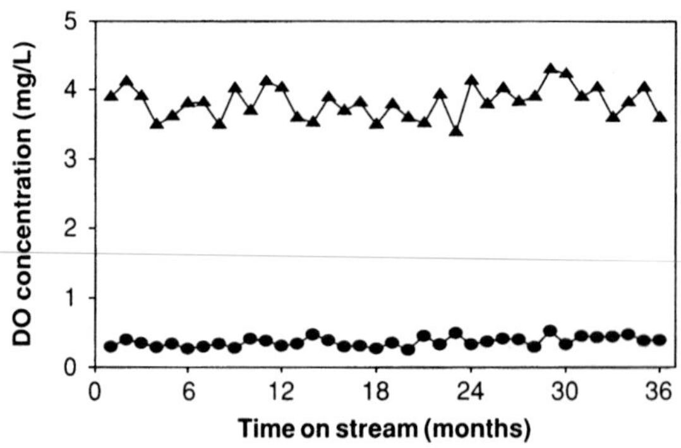

Figure 3 DO concentration in the influent (•) and the effluent (▲) from the aerated tank of the constructed wetland.

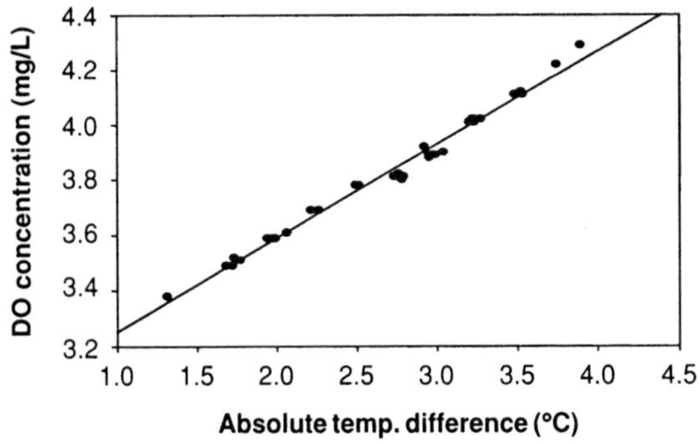

Figure 4 Dependence of DO concentration on the temperature difference between the ambient air and the air inside the wetland

more carefully, is that the concentrations of the organic solid materials increased gradually upto 12 month operation and then remained almost constant upto 36 month operation. These changes in concentrations with time become evident from Figure 6 which shows the time dependence of the total amounts of the attached organic solids on sands and gravels. No more accumulation of organic solid materials within the wetland after 12 month operation means that the input loading rate of organic solid materials nearby equals to the disappearance rate of organic solid materials. In other words, the aerobic biochemical decomposition rate of the solids equilibrates the input loading rate.

The total equilibrated amounts of the attached solids onto sands and gravels were calculated to be about 120 kg and 215 kg, respectively. When taking the density of the organic solids to be 0.5 g/cm^3, the total volumes of the solids on sands and gravels became to be 0.24 m^3 and 0.43 m^3, respectively, which corresponded to just 1.6% and 1.1% of the initial void volumes of sands (15.23 m^3) and gravels (37.56 m^3). These values seem to be low enough not to block the pathways of water inside the wetland.

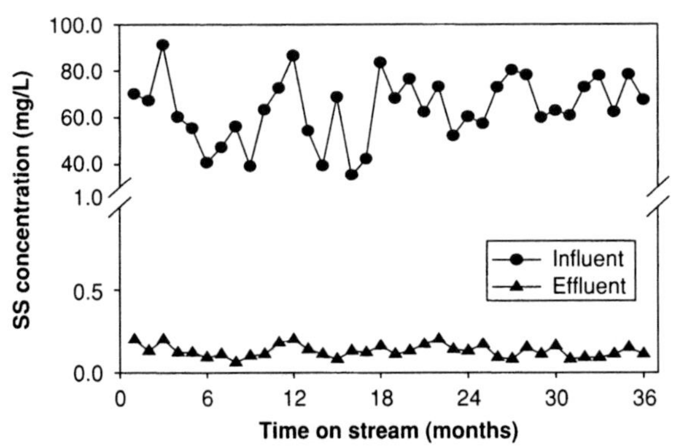

Figure 5 The influent and effluent concentrations of SS in the constructed wetland.

Table 1 Average concentrations of organic solid materials attached onto the surface of sands and gravels at different operation times.

Time on stream (month)	Depth (m) from the top	Concentrations (mg/kg)
3	0.2 0.5 0.8	0.122×10^4 0.110×10^4 0.103×10^4
6	0.2 0.5 0.8	0.182×10^4 0.172×10^4 0.169×10^4
12	0.2 0.5 0.8	0.204×10^4 0.177×10^4 0.170×10^4
18	0.2 0.5 0.8	0.204×10^4 0.175×10^4 0.172×10^4
24	0.2 0.5 0.8	0.206×10^4 0.174×10^4 0.170×10^4
36	0.2 0.5 0.8	0.202×10^4 0.173×10^4 0.169×10^4

In addition to the successful removal of organic solid materials, other encouraging results could be obtained. About 92.5% of BOD_5 was removed through biochemical oxidation. Even in the case of total nitrogen (T-N) about 31.2% of the input T-N was removed. When considering the fact that more than 94% of the organic nitrogen and ammonia nitrogen were nitrified successfully into nitrate nitrogen, which is indicative of the successful biological nitrification, about 30% denitrification in the aerated wetland is somewhat surprising, because

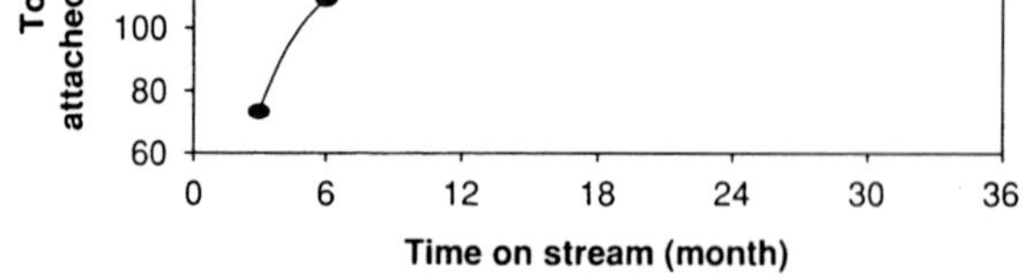

Figure 6 Time dependence of the amounts of the attached organic solids

the biological denitrification proceeds under anoxic condition. This partial denitrification is believed to occur at the local anoxic area within the aerated wetland, which seems to be placed somewhat far away from the natural air draft system.

Conclusion

A constructed wetland was designed to remove organic solid materials effectively from 100 m^3/day raw municipal wastewater, and the continuous performances of the wetland were investigated for three years.

Sufficient air was supplied to the constructed wetland through a specially designed natural air draft system whose driving force for air flow was temperature difference between ambient air and the air inside the wetland.

More than 99% of the organic solid materials could be successfully and stably upto three year operation. After 12 month operation no more accumulation of organic solid materials within the wetland was observed, which was proved to be due to the equilibrium between the input loading rate of the organic solid materials and their biochemical decomposition. This equilibrium made the wetland be operated successfully without blockage of the sands and gravels.

Acknowledgement

This work was supported by a grant from the KOSEF/MOST to the Environmental Biotechnology National Core Research Center (grant #: R15-2003-02002-0).

References

Crites R. and Tchbanoglous G. (1998). Small and Decentralized Management Systems, McGraw-Hill, New York, NY.

The Constructed Wetlands Manual (1998). 1st edn, Department of Land and Water Conservation (DLWC), New South Wales, Australia.

Leonard K. M., and Swanson G. W. (2001). Comparison of Operational Design Criteria for Subsurface Flow Constructed Wetlands for Wastewater Treatment. Water Science and Technology, **43**(11), 301–308.

Reed S.C., Crites R.W. and Middlebrooks E.J. (1988). Natural Systems for Waste Management and Treatment, McGraw-Hill, Inc., New York, NY.

U.S. Environmental Protection Agency (1994). Subsurface Flow Constructed Wetlands for Wastewater Treatment; A Technology Assessment, Report EPA/ 832-R/93/001.Washington, D.C.

Schroeder E. D. and Tchbanoglous G.. 1976. "Mass Transfer Limitations in Trickling Filter Designs" Journal of Water Pollution Control Fedration, **48**, 771–775.

64 Method for calculation of filtration in layered filters (linear mass exchange kinetics)

V.L.Polyakov, V.B.Sidor

National University of Building and Architecture, Kiev, Ukraine; e-mail: yura_pma@i.com.ua

Abstract An effective theoretical method is proposed for calculation of concentration and filtration characteristics in multilayered filter media. Area of its application is restricted by constant filtrate discharge and linear mass-exchange kinetics. The relationships presented admit presence of initial deposit. Characteristic changes in concentrations of the suspended and attached particles are shown in examples.

Keywords Filtration modelling, linear mass exchange kinetics, multi-layer filter, theoretical performance

Introduction

Efficiency of ordinary filters with uniform medium proves frequently to be insufficiently high. The point is that almost all solid suspended particles are retarded in the upper part of a filter. As a consequence, resistance on the part of porous (grain) structure increases, head loss grows. If there is a need to keep the filter performance at the initial level then one has to enhance the inlet head. But such operation is often difficult to be carried out or isn't profitable from the economical viewpoint. Along with it, other way of extending effective filter plant work exists. It consists in employment of non-uniform media. Ideally, their physico-chemical properties should vary smoothly in depth and corresponding changes should agree with pollution composition. Unfortunately, construction of filters with similar media is too complicated and expensive business. So just layered media have been of wide use in practice of mechanical purification by filtration. And the question arises immediately here on correct determination of purification characteristics, correct choice of filter parameters. To reply this question properly, reliable calculated methods are necessary. However, such methods, at least, for multilayered media have been absent until now. Our report includes some statements and relationships dealing with precisely multilayered media, mainly, two-layer and three-layer ones. Reliability of the relationships proposed is ensured by strictness of the analytical solution obtained of the appropriate mathematical task. The model forming the basis of the task and taking into account accurately mass transfer and mass exchange, two-phase flow through variable porous space comes from a general model of filtration thanks to accepting a number of reasonable assumptions. These assumptions allowed to simplify basic formalisms essentially but errors due to them appeared to be minimal in calculation of the desired characteristics. Serious limitation of the above relationships is related to a

 Recent Progress in Slow Sand and Alternative Biofiltration Processes edited by Rolf Gimbel, Nigel J.D. Graham and M. Robin Collins. ISBN: 9781843391203. Published by IWA Publishing, London, UK.

single form of mass exchange (between solid and liquid phases of filter medium) kinetics – only linear kinetics is considered.

Statement

The model in question consists of N constituent parts and each part includes, in one's turn, hydrodynamic and deformation submodels. The latter takes for $i-th$ layer of N-layer medium the following form

$$n_{ei}\frac{\partial C_i}{\partial t}+V\frac{\partial C_i}{\partial z}+\frac{\partial S_i}{\partial t}=0,\qquad \frac{\partial S_i}{\partial t}=\alpha_i C_i-\beta_i S_i,\quad N\geq i\geq 1, \tag{1}$$

where n_{ei} is the equivalent porosity; C_i, S_i are the volumetric concentrations of the suspended and attached fine particles; V is the constant filtrate discharge, α_i, β_i are the coefficients of the attachment and detachment rates; z is the depth from the filter upper boundary; t is the time. Due to neglecting diffusive (dispersive) mechanism of suspended particles transport, boundary conditions will be

$$z=0,\quad C_1=C_0;\quad z=z_i,\quad C_{i-1}=C_i;\quad N\geq i\geq 2. \tag{2}$$

New dimensionless variables were introduced:

$$\tau_i=\bar{t}-\sum_{j=1}^{i-1}\bar{n}_j\bar{l}_j-\bar{n}_i\left(\bar{z}-\sum_{j=1}^{i-1}\bar{l}_j\right),\quad \nu_i=\bar{z}-\sum_{j=1}^{i-1}\bar{l}_j,\quad N\geq i\geq 1, \tag{3}$$

where $\bar{t}=Vt/(n_0L), \bar{z}=z/L, \bar{n}_i=n_{ei}/n_0, n_0$ is the weighted mean initial porosity (clean medium), $n_0=\frac{1}{L}\sum_{j=1}^{N}n_{0j}l_j$, n_{0j} is the initial porosity of $j-th$ layer, $\bar{l}_i=l_i/L$; L, l_i are the thicknesses of total filter medium and $i-th$ layer. By the way, if $\bar{t}\gg 1$, as is seen from (3), then $\tau_1\approx\tau_2\approx\cdots\approx\tau_N\approx\bar{t}$.

As a result, the model (1), (2) is reduced to

$$\frac{\partial\bar{C}_i}{\partial\nu_i}+\frac{\partial\bar{S}_i}{\partial\tau_i}=0,\qquad \frac{\partial\bar{S}_i}{\partial\tau_i}=\bar{\alpha}_i\bar{C}_i-\bar{\beta}_i\bar{S}_i, \tag{4}$$

$$\begin{gathered}\nu_i=0,\quad \bar{C}_i=\bar{C}_{0i}(\tau_i)=\bar{C}_{i-1}(\bar{l}_{i-1},\tau_{i-1})=\bar{C}_{i-1}(\bar{l}_{i-1},\tau_i);\\ \tau_1=0,\quad \bar{S}_1=0;\quad \tau_i=0,\quad \bar{S}_i=0,\quad N\geq i\geq 2,\end{gathered} \tag{5}$$

where $\bar{C}_i=C_i/C_0,\quad \bar{S}_i=S_i/n_0C_0,\quad \bar{\alpha}_i=\alpha_iL/V,\quad \bar{\beta}_i=\beta_in_0L/V.$

Results and discussion

Solution of the problem (4), (5) was found by the Laplace transform and expressed by the relations

$$\bar{C}_i(\nu_i,\tau_i)=e^{-\bar{\alpha}_i\nu_i}\left[\bar{C}_{0i}(\tau_i)+\sqrt{\bar{\alpha}_i\bar{\beta}_i\nu_i}\int_0^{\tau_i}\frac{e^{-\bar{\beta}_i(\tau_i-\xi)}\cdot I_1\left(2\sqrt{\bar{\alpha}_i\bar{\beta}_i\nu_i(\tau_i-\xi)}\right)}{\sqrt{\tau_i-\xi}}\bar{C}_{0i}(\xi)\,d\xi\right],$$

$$\overline{S}_i(\nu_i,\tau_i)=\overline{\alpha}_i e^{-\overline{\alpha}_i\nu_i}\int\limits_0^{\tau_i} e^{-\overline{\beta}_i(\tau_i-\xi)}\cdot I_0\left(2\sqrt{\overline{\alpha}_i\overline{\beta}_i\nu_i(\tau_i-\xi)}\right)\cdot\overline{C}_{0i}(\xi)\,d\xi. \qquad (6)$$

As a consequence, the outlet concentration $\overline{C}_e=\overline{C}_N\left(\overline{l}_N,\tau_{N*}\right)$ is equal to

$$\bar{C}_e(\bar{t})=e^{-\bar{\alpha}_N\bar{l}_n}\left[\bar{C}_{0N}(\tau_{N*})\right.$$
$$\left.+\sqrt{\bar{\alpha}_N\bar{\beta}_N\bar{l}_N}\int\limits_0^{\tau_{N*}}\frac{e^{-\bar{\beta}_N(\tau_{N*}-\xi)}\cdot I_1\left(2\sqrt{\bar{\alpha}_N\bar{\beta}_N\bar{l}_N(\tau_{N*}-\xi)}\right)}{\sqrt{\tau_{N*}-\xi}}\bar{C}_{0N}(\xi)\,d\xi\right], \qquad (7)$$

where $\tau_{N*}=\bar{t}-\sum_{j=1}^{N}\bar{n}_j\bar{l}_j.$

Because of characteristic properties of the solution presented above, calculation of the sought concentrations in lower layers should be performed successively starting from the first layer and passing down as far as N-layer. In particular, the concentrations within the upper layer change according to the relationships

$$\overline{C}_1(\nu_1,\tau_1)=e^{-\overline{\alpha}_1\nu_1}\left[e^{-\overline{\beta}_1\tau_1}I_0\left(2\sqrt{\overline{\alpha}_1\overline{\beta}_1\nu_1\tau_1}\right)+\overline{\beta}_1\int\limits_0^{\tau_1}e^{-\overline{\beta}_1\lambda}I_0\left(2\sqrt{\overline{\alpha}_1\overline{\beta}_1\nu\lambda}\right)d\lambda\right]$$
$$+\overline{\beta}_1\int\limits_0^{\nu_1}e^{-\overline{\alpha}_1(\nu_1-\xi)}\cdot\overline{S}_0(\xi)\,e^{-\overline{\beta}_1(\tau+\overline{n}_1\xi)}\cdot I_0\left(2\sqrt{\overline{\alpha}_1\overline{\beta}_1(\nu_1-\xi)(\tau_1+\overline{n}_1\xi)}\right)d\xi, \qquad (8)$$

$$\overline{S}_1(\nu_1,\tau_1)=\overline{S}_0(\nu_1)\,e^{-\overline{\beta}_1(\tau_1+\overline{n}_1\nu_1)}+\overline{\alpha}_1e^{-\overline{\alpha}_1\nu_1}\int\limits_0^{\tau_1}e^{-\overline{\beta}_1\eta}\cdot I_0\left(2\sqrt{\overline{\alpha}_1\overline{\beta}_1\overline{\nu}_1\eta}\right)d\eta$$
$$+\overline{\alpha}_1\overline{\beta}_1e^{-\overline{\beta}_1(\tau_1+\overline{n}_1\nu_1)}\int\limits_0^{\nu_1}e^{-(\overline{\alpha}_1-\overline{\beta}_1\overline{n}_1)(\nu_1-\xi)}\cdot\overline{S}_0(\xi)\cdot\int\limits_{\overline{n}_1(\nu_1-\xi)}^{\tau_1+n_1\nu_1}I_0\left(2\sqrt{\overline{\alpha}_1\overline{\beta}_1(\nu_1-\xi)}\right)d\zeta d\xi. \qquad (9)$$

The expressions (8), (9) generalize the well-known formulae for the concentrations C, S to case of initial deposition in a filter. The deposit is distributed in depth in the following way

$$\bar{t}=0,\quad \overline{S}_1=\overline{S}_0(\overline{z}). \qquad (10)$$

As follows from (9) at $\overline{S}_0=0$,

$$\overline{C}_{01}(\tau_2)=e^{-\overline{\alpha}_1\overline{l}_1}\left[e^{-\overline{\beta}_1\tau_2}\cdot I_0\left(2\sqrt{\overline{\alpha}_1\overline{\beta}_1\overline{l}_1\cdot\tau_2}\right)+\overline{\beta}_1\int\limits_0^{\tau_2}e^{-\overline{\beta}_1\lambda}I_0\left(2\sqrt{\overline{\alpha}_1\overline{\beta}_1\overline{l}_1\lambda}\right)d\lambda\right]. \qquad (11)$$

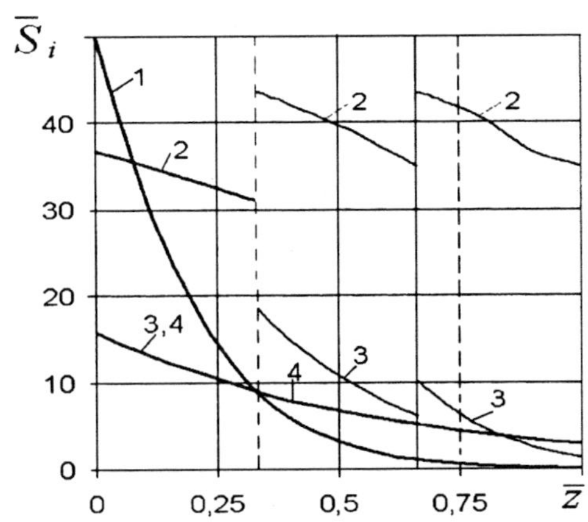

Figure 1 Profiles of deposit concentration: 1, 4 – unif.; 2,3 – 3-layer; *2* – $\bar{t} = 50$, *1, 3, 4* – $\bar{t} = 10$

Substitution of the expression (11) into (6) enables to determine, first, content if suspension particles in both physical states within the second layer, secondly, boundary condition for C_3, S_3. Continuing this procedure, $N-th$ layer is reached in the end and after that determination of concentrations dynamics becomes possible along the whole filter depth. Volume of particles delayed in $i-th$ layer W_i by time $\bar{t}$is calculated as follows

$$\overline{W}_i = \int_{L_{i-1}}^{L_i} \overline{S}_i\left(\bar{z}, \bar{t}\right) d\bar{z}, \qquad (12)$$

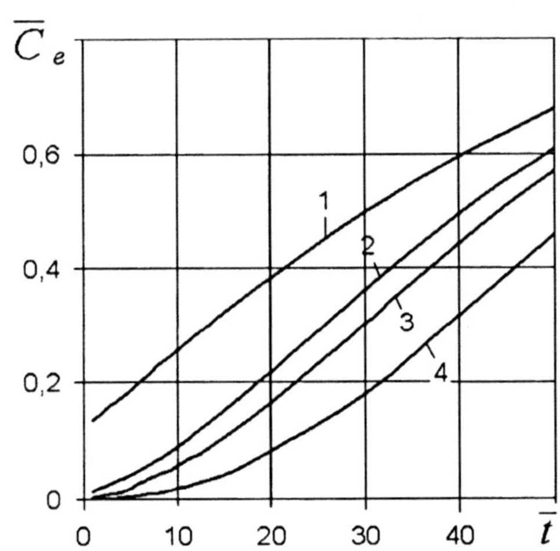

Figure 2 Outlet curves of the suspended particles concentration: *1* – $\bar{l}_1 = 1$, *2* – $\bar{l}_1 = 0,5$, *3* - $\bar{l}_1 = 0,333$, *4* – $\bar{l}_1 = 0$

where $\overline{W}_i = \dfrac{W_i}{n_0 C_0 L}$, $\overline{L}_i = \sum_{j=1}^{i} l_j$. It should be also known to correct $\overline{n}_i$. To keep constant discharge V, the relative inlet head $\overline{H}_u$ should increase in such way

$$\overline{H}_u\left(\overline{t}\right) = \frac{H_u - H_l}{H_{u0} - H_l} = \sum_{j=1}^{N} \frac{\overline{n}_{0j}{}^{m}}{\overline{k}_{j0}} \int_0^{\overline{l}_j} \frac{d\xi}{\left[\overline{n}_{0j} - C_0 \overline{S}_j\left(\xi, \tau_{j*}\left(\overline{t}\right) - \xi\right)\right]^m}. \tag{13}$$

Here H_u, H_{u0} are the actual and initial heads at the upper filter boundary, H_l is the head at the lower boundary; $\overline{n}_{0j} = n_{0j}/n_0$; $\overline{k}_{j0}$ is the relative hydraulic conductivity of $j-th$ clean layer. At last, a power law is applied to represent the hydraulic conductivity k_j as a function of $\overline{S}_j$.

Three-layer filter design was taken to illustrate the developed method. Layer parameters were chosen to provide filter operation by its total volume: $\overline{\alpha}_1 = 2, \alpha_2 = 5, \alpha_3 = 10$; $\overline{\beta}_1 = 0.05, \overline{\beta}_2 = 0.1, \overline{\beta}_3 = 0.16$. Accumulation of deposit by the moments $\overline{t} = 10$ and 50 is shown in fig.1 and has almost uniform character during the whole filtration period. Advantage of a layered medium (see curves 2, 3) over one-layer medium (curve 1) is visually obvious. Quality of the filter work is best of all judged by examination of the outlet particles concentration $\overline{C}_e$. Proper curves $\overline{C}_e\left(\overline{t}\right)$ are given in fig.2 for different values of the relative upper layer thickness $\overline{l}_1$ (thicknesses of other layers were taken equal). The smallest values $\overline{C}_e$ take place, certainly, at $\overline{l}_1 = 0$ (4). But the fastest growth of the inlet head will occur in this case so that thorough analysis is necessary to choose optimal operation parameters.

Conclusions

The calculated method proposed is effective and reliable thanks to completeness of the initial mathematical model and exactness of its solution. It allows to perform calculations correctly at any number of medium layers. Linear mass exchange kinetics and constant filtrate discharge are responsible for restriction of the formulae presented.

65 Multi-stage filtration (MSF) to prevent biofilm growth in a distribution network

L.D. Sánchez, L.M. Burbano and A. Sánchez*

*Universidad del Valle, Instituto Cinara, A.A 25157, Cali, Colombia

Abstract In a study carried out by the Instituto Cinara, at the Universidad del Valle (2003–2004), in a community located in the rural zone of the municipality of Cali, when physiochemical and microbiological water quality was evaluated at the source, a TOC concentration of 14mg/l was found, with average values of 25,000 CFU/100 ml for faecal coliforms, 121,000 CFU/100 ml for total coliforms and 190,000 CFU/100 ml for mesophile. Due to design and operation problems, these were not removed in the treatment plant which used chemical coagulation. The contaminants were allowed to penetrate into the distribution network and favour the formation of biofilm. This water quality deterioration increased the microbiological risk to water users. Considering these conditions, research was carried out with the aim of taking corrective measures and avoiding health risks to users. These measures included technological changes in the treatment plant, projecting a multi-stage filtration system in combination with chemical coagulation, upflow roughing filter and slow sand filters with activated carbon. The MSF plant in combination with chemical coagulation has a removal efficiency of 97% for turbidity, 81% for true colour, and for mesophile, faecal coliforms, and total coliforms were 99%. This led to lower risks for users as well as lower system operation and maintenance costs.

Keywords Multi-stage filtration, biofilm, removal efficiency, risk, water quality

Introduction

Protecting the distribution system is essential to ensuring a supply of safe water to consumers. Water treatment systems therefore need to have a guaranteed supply of good quality water to avoid the formation of biofilm, which could alter the microbiological water quality for users. Supply sources are the first barrier to guaranteeing large quantities of high quality water. Domestic waste can have high levels of organic matter and faecal micro-organisms that cause water borne diseases, which in turn generate high percentages of morbidity and mortality in the population. They also cause problems of biofilm formation in the distribution network, decreasing its capacity for transport and creating ideal conditions for many kinds of micro-organisms to remain in the pipes, thus increasing the risk of disease in the recipient communities (Galvis *et al.,* 1999).

This study analyses the solution to a microbiological re-contamination problem in the water in a distribution network that supplies water to a community of 70 homes in the rural

edited by Rolf Gimbel, Nigel J.D. Graham and M. Robin Collins. ISBN: 9781843391203. Published by IWA Publishing, London, UK.

Table 1 Water quality in different points of the system

Sampling point	Faecal coliforms CFU/100 ml	Total coliforms CFU/100 ml	Mesophiles CFU/100 ml	Turbidity NTU	True Color CPU	pH
Raw water	1,145	45,000	22,100	0.8	10	7.4
Conventional plant effluent	10	28	27,500	0.97	3	7.3
Network: water from tank – Point 1	1	2	3,600	0.42	2	7.4
Pipes – Point 2	100	500	252,400			
Pipes – Point 3	300	455	131,200			
Pipes – Point 4	1	0	114,800			

zone of Cali in the Valle del Cauca. The community had a treatment plant functioning with the processes of coagulation, flocculation, sedimentation, rapid filtration and disinfection with chlorine and ultraviolet light. However, in the studies completed by the Instituto Cinara at the Universidad del Valle on water quality in the distribution network, it was found that the tubes had biofilm, which generated problems of re-growth and contamination of micro-organisms in the water supplied to users (see Table 1). It was also identified that the treatment system contributed to biofilm formation due to its failure in its functioning, allowing solids and micro-organisms to pass through. The treatment system thus had to be redesigned and rebuilt.

Multi-stage filtration (MSF) technology, in combination with chemical coagulation, was the technology built. Each one of its treatment barriers progressively removes pollutants and organic material present in the surface source, producing good quality water and minimising risk of disease and biofilm formation in the distribution network (Lloyd *et al.*, 1991). Due to the problems associated with the presence of organic material and bacteria in the surface source, a study was carried out to develop a new treatment system.

Methods

During the study, follow-up was made to water quality at the source, in the treatment system and at different points on the distribution network. The increase of coliforms was verified on different points on the network and systematic faults in the conventional treatment system process were identified. Once the new treatment system was built, follow-up was made to the affluent water quality at the treatment plant. For this, treatment barrier behaviour was analysed for removal of turbidity, true colour, faecal coliforms, total coliforms and mesophile bacteria. Follow-up was also carried out in the distribution network for the same parameters as well as for residual chlorine, under normal operation conditions. Figure 1 shows a general diagram of the plant of the treatment plant.

The characteristics of the treatment system implemented are the following: dynamic roughing filters (DRF) with a filtration rate of 2 m/h, a bed length of 0.60 m, a coagulation unit, upflow roughing filters (URF) with a filtering rate of 0.60 m/h, a bed length of 1.2 m

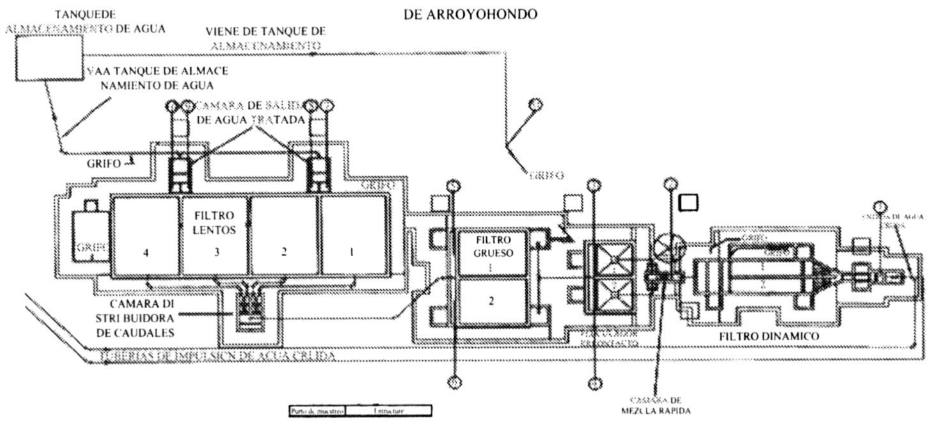

Figure 1 Treatment Plant General Layout

and slow sand filters (SSF) with a filtering rate of 0.15 m/h, sand thickness of 0.80 m of sand and a layer of 0.05 m of granular activated carbon located at 0.20 m depth.

Results and discussion

Raw water quality

The source has a high level of microbiological contamination as shown in Figure 2. The average value of mesophile, total coliforms and faecal coliforms are 190,000 CFU/100 ml, 121,000 CFU/100 ml and 25,000 CFU/100 ml, respectively. These results indicate that the water source has a microbiological risk catalogued by the World Health Organisation as very high.

The average concentration of total organic carbon in the source was 14 mg/L and as shown in Figure 3, with respect to the physical and chemical risk, the source has turbidity between 1.5 and 23 NTU, and true colour between 8 and 34 colour units. Given the quality of the untreated water, it was necessary to dose it with ferric chloride as a coagulator at 2%, with a variation of the dose used between 6 and 10 mg/L.

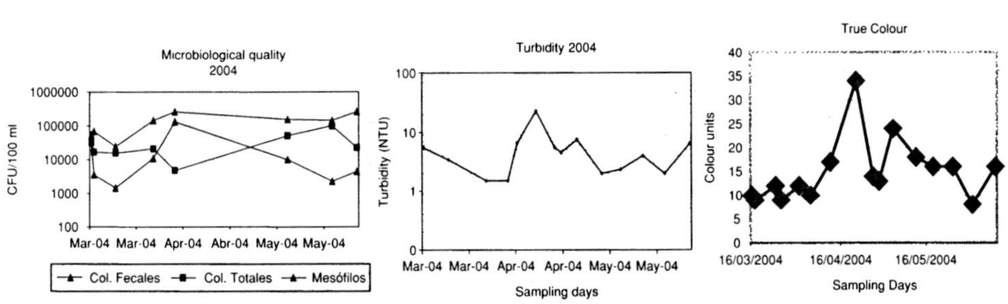

Figure 2 Variations Microbiological and physiochemical Quality in the source

Table 2 Water quality monitoring (2003–2004)

	URF		SSF		Network	
Parameter	**Average**	**STD**	**Average**	**STD**	**Average**	**STD**
Turbidity (NTU)	1.3	0.72	0.33	0.12	0.29	0.11
Colour (CPU)	9	3	4	2	3	1
Faecal coliforms (CFU/100 ml)	10,000	31,000	7	5	0	0
Total coliforms (CFU /100 ml)	31,800	62,000	28	11	0	0
Mesophile (CFU/100 ml)	87,000	196,000	1,700	4,800	0	0

Water quality effluent

The evaluation results demonstrate the efficiency of the treatment barriers. The parameters evaluated are presented in Table 2.

Values for the URF effluent showed that faecal coliforms, total coliforms and mesophile were removed at 60%, 74% and 54%, respectively, and were removed at 76% of turbidity and 40% of colour. These values are within the typical reductions for this type of upflow roughing filter, according to Galvis *et al.*, 1999. The SSF have removal efficiencies of 99.9% for faecal coliforms, and for total coliforms, 98% for mesophile, 75% turbidity and, 56% for colour. Although the removal efficiencies are good for this kind of system, disinfectant must be used as the final treatment and protection barrier in the distribution networks. As observed in Table 2, the distribution network has no microbiological contamination, and therefore disinfection is efficient in this treatment system. In contrast, given the good functioning of the treatment plant, there were no changes in the water quality in the distribution system.

As regards the physiochemical behaviour of the parameters evaluated, it has been observed that the system responds efficiently. At the end of the system, there was a microbiological removal for faecal coliforms, total coliforms, and mesophile of 99.9%, 99.9% and 99.91%, respectively. With respect to turbidity, the removal was of 97% and for colour, 81%. In the distribution network, there was no recontamination of the water produced by the treatment units, and therefore, after disinfection, the water consumed by the population supplied had measurements of turbidity between 0.15 and 0.50 NTU, colour between 2 and 7 CPU and faecal coliforms, total coliforms and mesophile 0 UFC/100 ml. The concentration of total organic carbon varied between 2 to 4 mg/L. In addition, during the study, there were no problems identified with the filter beds in the slow filter units due to obstruction. On average, the filtering runs lasted 50 to 70 days, with a maximum headloss of 70 centimetres.

Conclusions

The treatment system is efficient as it has total removal percentages for faecal coliforms, total coliforms and mesophile of 99.9%. For turbidity, the removal was 94% and for colour it was 73%. The good quality of the treatment systems effluent water decreased the probability of biofilm formation in the distribution network. In addition, the solid accumulation in pipes was decreased and 100% of micro-organisms were removed with disinfection. The selection of this treatment plant and its operation improved water quality. This led to lower risks for users as well as lower system operation and maintenance costs.

References

Galvis, G. and Latorre, J. (1999) Filtración en Múltiples Etapas, Tecnología Innovativa para el Tratamiento de Agua. Cinara, Universidad del Valle, IRC. Colombia, 70–160.

Lloyd, B (1991) Functional Microbial Ecology of Slow Sand Filters. University of Surrey, England.

Part VIII

RIVER BANK FILTRATION AND GROUND WATER RECHARGE

66 Bank filtration and groundwater recharge for treatment of polluted surface waters

Prof. Dr.-Ing. Martin Jekel and Steffen Gruenheid

Technical University Berlin, Institute for Environmental Engineering, Department of Water Quality Control, KF 4, Strasse des 17. Juni 135, 10623 Berlin, Germany (e-mail contact: martin.jekel@tu-berlin.de)

Abstract Investigations on the behavior of different bulk organics and trace organic compounds at a bank filtration site at Lake Tegel in Berlin, Germany, and in a long retention soil column system are reported. Objective of the research was to assess important factors of influence for the degradation of bulk and trace organics. More than two years of monitoring for the bulk parameter DOC proved that the redox conditions significantly influence the DOC-degradation kinetic but not necessarily the residual concentration. LC-OCD measurements confirmed that the change in character is comparable for aerobic and anoxic/anaerobic infiltration. Only the fraction of polysaccharides shows a better removal under aerobic conditions. Furthermore, adsorbable organic iodine (AOI) measurements revealed a more efficient degradation of AOI and AOBr under anoxic/anaerobic conditions. The monitoring of the single organic pollutants Iopromide, Sulfamethoxazole and naphthalenedisulfonic acids showed that the redox conditions have an influence on the degradation behavior of some of the monitored compounds. Iopromide was efficiently removed at all times, but no evidence for a dehalogenation under oxic conditions was found. Sulfamethoxazole showed a better removal under anoxic/anaerobic conditions. The very stable 1.5-naphthalenesulfonic acid was not removed under either redox conditions.

Keywords AOI; bank filtration; DOC; LC-OCD; trace pollutants

Introduction and objectives

Bank filtration and groundwater recharge are traditional methods for drinking water treatment and extraction in central Europe, but the processes for removal of dissolved organic substances are not yet well understood, especially for the emerging compounds of different origin, like industrial chemicals and pharmaceutical substances. This research study is part of the NASRI-project of the Center of Competence for Water Berlin (NASRI, 2003). As part of this project the Department of Water Quality Control at the Technical University of Berlin (DWQC) studies the removal of bulk organics (dissolved organic carbon (DOC), effluent organic matter (EfOM)) and selected trace organics at various field sites with different characteristics in the Berlin, Germany area and in soil column systems. In the city

 Recent Progress in Slow Sand and Alternative Biofiltration Processes edited by Rolf Gimbel, Nigel J.D. Graham and M. Robin Collins. ISBN: 9781843391203. Published by IWA Publishing, London, UK.

of Berlin, 56% of the drinking water is derived from bank filtration and 14% from artificial recharge (BWB, 2003). The water utility is using well operated bank filtration and artificial recharge facilities to provide high quality water. With the growth of the city a semi-closed urban water cycle has been managed. At some bank filtration sites the surface water is strongly influenced by well-treated domestic waste water effluent (e.g. 15–30% in Lake Tegel) (Ziegler, 2001). Despite this indirect potable reuse, the bank filtration system is providing high quality water which needs little further treatment and is distributed without chlorination. Since the processes during infiltration are very complex, it is difficult to predict bulk organic composition in the bank filtrate or to estimate important factors of influence for the degradation. There are indications that bank filtration under anoxic/anaerobic conditions provides different removal of organics than aerobic bank filtration (Gruenheid et al., 2004). In addition to the redox state, factors such as retention time, temperature, initial degradable carbon concentration, soil properties, and hydrogeological conditions may affect the final concentration. The factors of influence and the fate of these bulk organics are studied in the field and in long term experiments.

Methods

The paper focuses on the fate of bulk and trace organics at a bank filtration site in Berlin-Tegel. Additionally, results of experiments on a long retention column system which was operated with the same water under oxic and anoxic conditions will be presented. The monthly analytical program of field and column samples is comprised of DOC, UVA, LC-OCD, differentiated AOX (AOI, AOBr)-analysis and solid phase extraction (SPE)/ HPLC-analysis of three groups of trace compounds. The groups are X-ray contrast media (Iopromide); bacteriostatica (Sulfamethoxazole) and the industrial chemicals naphthalenedisulfonates (1,5-NSA; 1,7-NSA; 2,7-NSA). Additionally, all samples were analyzed for general hydrochemistry parameters to characterize the infiltration process more detailed.

Field site. The examined field site is operated with water from Lake Tegel which is influenced by the discharge of the wastewater treatment plant Schönerlinde. Based on LC-OCD analysis, the DOC-composition of Lake Tegel is characterized by high values (~7.5 mg/l), of a mostly natural origin: fulvic acids associated with the background groundwater (4–5 mg/l DOC), effluent organic matter with a contribution of 1–2 mg/l DOC and algal cellular products. The dominant soil type at the Lake Tegel bank filtration site is sand. In the direct infiltration zone near the lake the hydraulic permeability of the clogged sand ($k = 5*10^{-6}$ m/s) (Fritz et al., 2002) is lower than in the deeper aquifer ($k = 28*10^{-4}$ m/s) (Pekdeger et al., 2004). The sampling points for the bulk organic analysis at the field site were selected to reproduce the flow path of the infiltrating water to the production well. The individual retention times for each monitoring well were determined by the shift of the seasonal variation of the ratio of oxygen isotopes (δ^{18} O) in the surface water (Pekdeger et al., 2004). At the bank filtration site some dilution with deeper groundwater was observed and quantified with the help of boron data (Pekdeger et al., 2004). Data for retention times and dilution with deeper groundwater are included in Figure 1. A detailed description of the field site and a distribution of the used monitoring wells are available at Pekdeger et al., (2004).

At the bank filtration site it was found that most of the surface water is infiltrating under oxic conditions. Because of mineralization of DOC and sedimentary bound POC

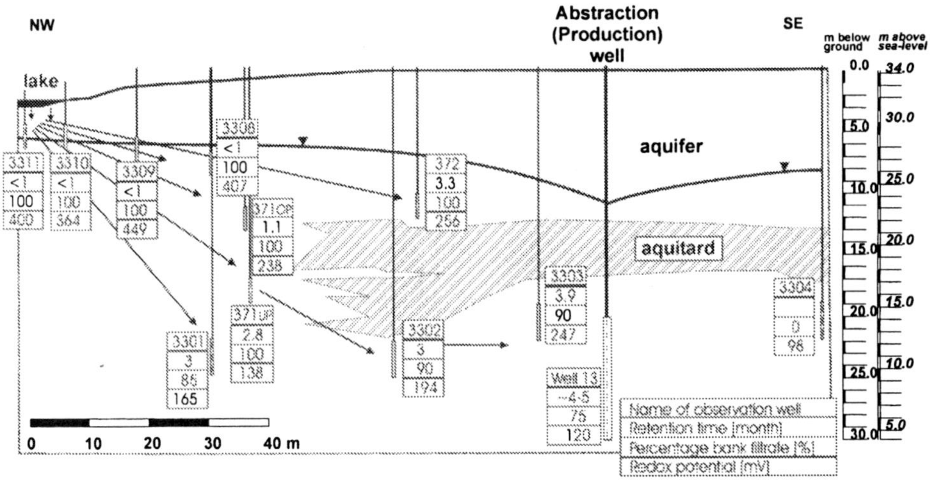

Figure 1 Cross section of bank filtration site at Lake Tegel

(~0.5 % w/w) during infiltration oxygen is used up quickly and most of the 4–5 month long infiltration (100 m) to the production well is usually taking place under anoxic and anaerobic conditions (iron and manganese reduction). The extension and position of the redox zones varies seasonally and is horizontally stratified (Pekdeger et al., 2004). The average nitrate concentration in Lake Tegel is 1.83 mg/l NO_3-N. During the first sixteen months (05/02–09/03) of the observation period the nitrate level was significantly reduced during infiltration (Average NO_3-N: 3302: 0.07 mg/l and 3303: <LOQ).

The summer of 2003 was very dry in Berlin, and the combination of the low water level in Lake Tegel and the heavy pumping of the production well pumps led to an expanded zone of unsaturated infiltration under the lake. After passage of the biologically very active water-sediment interface the bank filtrate was aerated again during a short unsaturated infiltration. Beginning in October 2003 until June 2004 elevated oxygen and nitrate concentrations were observed in the deeper monitoring wells and the production well (Ø-NO_3-N: 3302: 1.16 mg/l and 3303: 0.47 mg/l). For nine month of the observation period the dominant redox conditions at the field site changed from anoxic/anaerobic with Fe/Mn-reduction to oxic. The differences in the fate of bulk and trace organics between the two redox conditions are investigated.

Column system. A long retention soil column system was used to simulate a one dimensional aquifer. The study was conducted on a 30 m column system, which was operated with spiked water from Lake Tegel and had a retention time of 30 days (Figure 2). The columns were filled with natural fine sand from the Berlin area. Flow and hydraulic loading rate were selected to fit the conditions at the field site. The columns were sampled at different depths to investigate the degradation kinetics.

An online dosage system for five trace compounds was installed to assure a constant conc-entration of these five compounds in the influent. The spiking concentrations were selected to be approximately ten times higher than the concentration in Lake Tegel. Table 1 gives an overview of the dosages of trace compounds to the lake water.

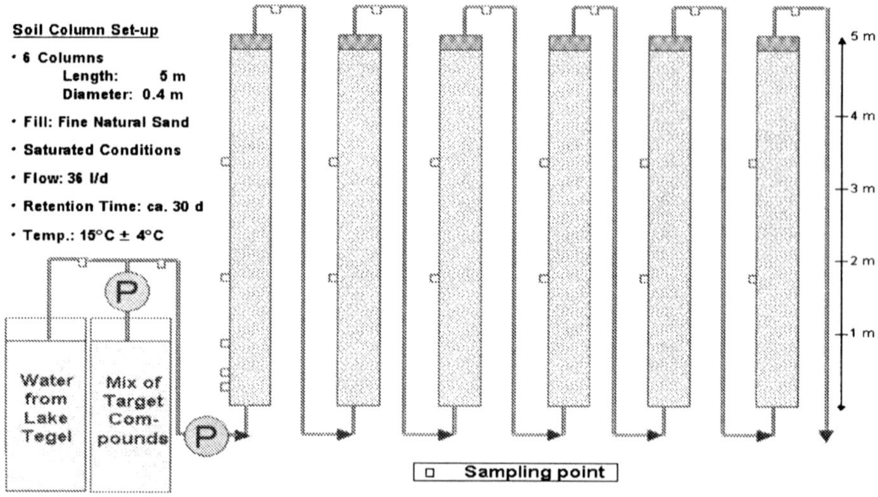

Figure 2 Experimental set-up of the long retention columns

Table 1 Spiking concentrations for the long retention columns in Marienfelde

Trace Compound	Iopromide	Sulfamethoxazole	1,5-NSA	1,7-NSA	2,7-NSA
Spiking Concentration	10 μg/l	2.5 μg/l	2.5 μg/l	2.5 μg/l	2.5 μg/l

During the first part of the experiment oxic conditions were established in the soil columns. The oxic period lasted twelve months until May 2004. During this period oxygen was present in the whole column and no denitrification was observed. From May 2004 the redox status of the column system was switched to anoxic (NO_3-reducing) by sparging the influent with nitrogen (O_2 in influent: ~1mg/l). Denitrification was observed but nitrate was not completely depleted from the system. No oxygen was detectable in the sampling ports. Monitoring of the DOC, UVA and redox-conditions ensured stable conditions over the time. Results from both phases will be presented.

Analytics. At the field site the surface water, monitoring wells, and abstraction (production) well were sampled monthly over a period of 27 months for the bulk organic parameters and for 16 months for trace pollutants. The wells were purged for 30 minutes at pumping rates of 1–2 m^3/h before sampling to ensure stable conditions (on line measurement of temperature, redox potential, dissolved O_2, pH, and conductivity). Afterwards the samples were stored in glass bottles at 4°C in the dark and if necessary filtered through 0.45 μm cellulose nitrate membrane filters. DOC was determined after 0.45 μm filtration with an elementar HighTOC analyzer (Hanau, Germany). UV absorbance at 254 nm (UVA_{254}) was measured using a Perkin-Elmer Photometer Lambda 12 (Berlin, Germany). For the LC-OCD, a liquid chromatography method that quantitatively distinguishes between different fractions of DOC, a system of the DOC-Labor Dr. Huber (Karlsruhe, Germany), was used (Huber and Frimmel, 1996). Furthermore, the analytical program is comprised of differentiated AOX-analysis and trace organic analysis. The protocol for the differentiated

Table 2 Details on trace organic compound analysis

	Iopromide	Sulfamethoxazole	Naphthalenesulfonic acids
SPE-Cartridge	Lichrolute EN	Lichrolute EN	Lichrolute RP-18
Liquid chromatography	H_2O/MeOH (Gradient)	H_2O/MeOH (Gradient)	Ion pair chromatography
Detector	MS/MS (Micromass)	MS/MS (Micromass)	Fluorescence (Gynkotek)
Limit of detection	5 ng l^{-1}	5 ng l^{-1}	10 ng l^{-1}
Limit of quantification	20 ng l^{-1}	20 ng l^{-1}	30 ng l^{-1}
Method described in	Putschew et al. 2001	Hartig 2000	Storm et al. 1999

analysis of the organohalogens consists of coupled combustion and ion chromatography and is described in Oleksy-Frenzel et al. (2000).

The trace organic compounds Iopromide, Sulfamethoxazole and the isomers of naphthalenesulfonic acids were all extracted by different solid phase extractions (SPE) and measured with standard addition in high performance liquid chromatography with MS/MS- and FLD-detectors. Table 2 shows more details on the trace compound analysis.

Results and discussion

Results of $2^1/_4$ years of field monitoring at Lake Tegel and more than two years of column studies are summarized in this paper. The results are presented with the focus on the influence of the redox conditions on the removal of bulk and trace organics.

Dissolved organic carbon (DOC). The monthly DOC-monitoring (05/2002–08/2004) for the bank filtration site confirmed stable surface water concentrations in Lake Tegel of 7–7.5 mg/l. The results of the monitoring well sampling are plotted in Figure 3a as box plots with the median value (horizontal centre line), the 25%- and 75%-quartiles (box) and the minimum and maximum values (vertical line). The removal rates are calculated from

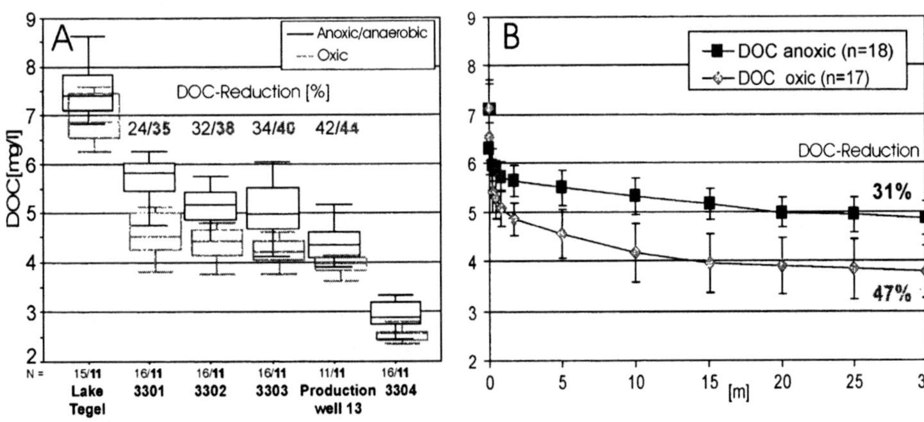

Figure 3a/b DOC at bank filtration site for two observed redox conditions (a) / DOC of column experiment (b)

the arithmetic means of all measurements. Under oxic and anoxic/anaerobic conditions the DOC is reduced by 2.8 and 2.6 mg/l from the surface water to the monitoring well 3303, respectively. The degradation kinetic indicates a more rapid removal of DOC under oxic conditions and a slower but continuing removal under anoxic/anaerobic conditions. Under oxic conditions 35% of the DOC is degraded from the surface water to the first deep monitoring well 3301. During further infiltration the fraction of degraded DOC is increasing by only 5 % (3303) or 9% (Well 13, includes some dilution). Under anoxic/anaerobic conditions 24% of the initial DOC is degraded in monitoring well 3301. The continuing removal of DOC leads to reduction rates of 34% (3303) and 42% in the production well (Figure 3). Comparing the concentrations of DOC during oxic and anoxic/anaerobic infiltration reveals that under both conditions similarly 3.1 mg/l DOC were removed between the surface water and the production well. The dilution in the production well is assumed to be constant (25% background groundwater) and the background monitoring well 3304 shows low DOC-concentration during the observation period (oxic: 2.5 mg/l; anoxic/anaerobic: 2.9 mg/l). Under anoxic/anaerobic conditions the slower process of DOC-mineralization demands the entire retention time, whereas under oxic conditions the efficient removal during the initial infiltration is followed by a plateauing of the DOC levels and a slower removal. Under both conditions a similar residual DOC concentration can be achieved, if sufficient retention time is allowed under anoxic/anaerobic conditions.

The field results are consistent with finding of the soil columns experiments (Figure 3b). The simulation experiments allowed a more precise investigation of the fate of DOC during the first month of infiltration depending on the dominant redox conditions. In the oxic soil columns 47% of the DOC was removed. Under anoxic conditions the initial oxic phase which occurs invariably in the field because of oxygen saturated surface water was reduced to the lowest practicable dimension (<0.21m). Only 31% of the source DOC was removed during 30 days of infiltration. The results support the field results that the DOC removal under oxic conditions is faster than under anoxic or anaerobic conditions. The removal rates are within the range of the rates observed in the field. A slightly higher removal can be explained by a higher average water temperature in the soil columns (15°C, field: 12°C). Because of their limited retention time the column experiments could not demonstrate that the residual not-degradable DOC-concentration is similar under oxic and anoxic/anaerobic conditions.

LC-OCD. Figure 4a/b shows LC-OCD chromatograms of samples from the soil column experiment, and indicates that the character of the DOC partly changed after infiltration. The first peak corresponds to the polysaccharide (PS; elution time at 35–45 min) peak, the second corresponds to the humic substances (HS; elution time 52 min) and HS building blocks (secondary peak; 57 min), and the third to low molecular weight acids (LMA; 62 min).

LC-OCD measurements of samples from bank filtration site (not shown) clearly indicated that the fraction of polysaccharides is completely removed very fast under both redox conditions, whereas other fractions (humics, humic hydrolysates) exhibited only partial removal. At no time the polysaccharides were detected in the first observation well (3311). It is assumed that this fraction is mineralized instantly in the oxic infiltration zone. The results of the soil column experiments support the assumption of a rapid biodegradation for this DOC-fraction, because in an abiotic column only minor filtration effects were observed. Generally, LC-OCD chromatograms of the field site were comparable to

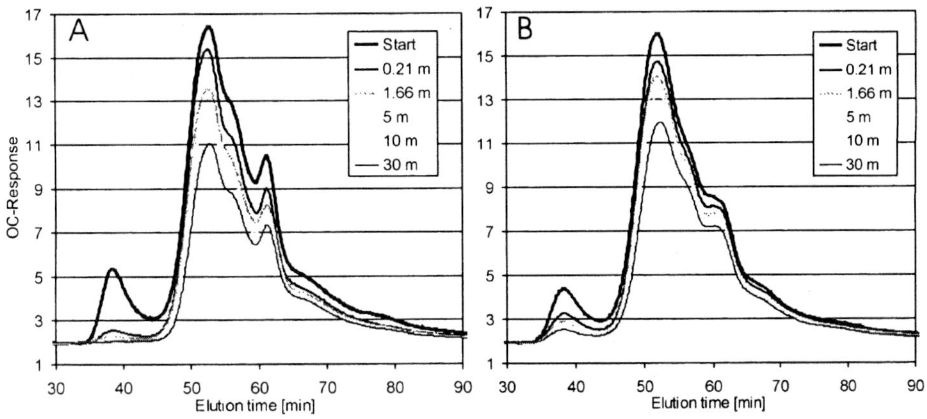

Figure 4a/b LC-OCD chromatogram of soil column experiment - oxic (a) / anoxic (b) Furthermore, the fractions of HS, HS building blocks and LMA were only partially degraded. It is noteworthy that even a part of the HS peak was removed, generally assumed to be mostly nondegradable. Under anoxic conditions in the column system a slower degradation of the polysaccharide fraction was observed. PS were still detectable after 30 days of anoxic infiltration (Figure 4b). The fractions of HS, HS building blocks and LMA showed a similar partial removal as under oxic conditions. Because of the slower DOC degradation kinetics the chromatograms are more pooled together in the anoxic period. However, despite of the PS-fraction a difference in the degradability of the single fractions under oxic or anoxic/anaerobic conditions was not observed.

the chromatograms of the oxic period of the long retention column system and showed the same trends (Figure 4a).

Both, in the field and in the columns the most fundamental change during infiltration was observed for the fraction of the polysaccharides. The largest molecular weight fraction, interpreted as polysaccharides, was efficiently removed in the first 0.21 m of oxic infiltration in the soil columns (Figure 4a).

Adsorbable organic iodine (AOI). Adsorbable organic halogens (AOX) are an issue for the aquatic environment for several years. Triiodinated X-ray agents, which are very stable and hydrophilic, contribute to large amounts (>50%) of the total AOI in Berlins municipal wastewater and are introduced to the surface waters. The long term AOI-monitoring confirmed the seasonal changes of AOI-concentration in Lake Tegel due to variation in the dilution of the sewage treatment plant effluents. Additionally, a general increase in AOI concentration in spring 2003 was observed, which was due to an extension of the treatment plant Schönerlinde that discharges into Lake Tegel.

Therefore, the AOI surface water concentrations were considerably higher during oxic conditions at the bank filtration site (oxic: 13.3 μ/l; anoxic: 9.6 μg/l) (Figure 5a). The comparison of the removal rates of the anoxic/anaerobic infiltration and the oxic infiltration at the same field site points towards a more efficient removal of AOI under anoxic/anaerobic conditions. During the anoxic/anaerobic part of the observation period 60–70% of the source water AOI were removed during infiltration. The AOI of the background groundwater at

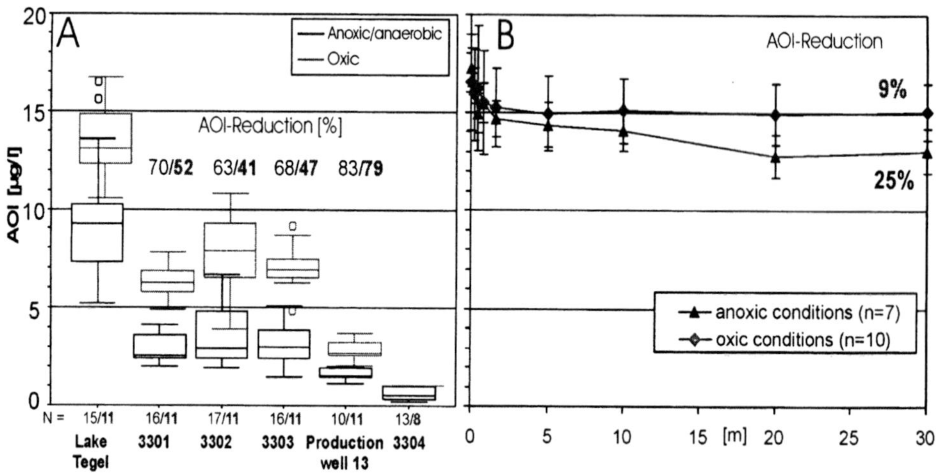

Figure 5a/b AOI at bank filtration site for two observed redox conditions (a) / AOI of column experiment (b)

the bank filtration site is very low (1 µg/l) and the mixture in the production well leads to an AOI-concentration in the extracted raw water of 1–2 µg/l. The time series of the AOI measurements (not included) shows the reaction of the monitoring wells to the change in redox conditions in fall 2003. The deep monitoring wells (3301, 3302, and 3303) react with an increase in AOI-concentration and significantly lower reduction rates. During the oxic period only 40–50% of the surface water AOI is removed during infiltration. It was found that under oxic conditions the AOI level of the extracted raw water was slightly increased by 1.5 µg/l. The results give evidence that there is a more efficient reduction of AOI during infiltration under low redox potentials. These results are consistent with findings at other field sites in Berlin. A similar dependency of the AOI-concentration from the dominant redox conditions was observed at a bank filtration site at Lake Wannsee. AOI reduction rates of app. 65% were observed during one month of anaerobic infiltration, whereas under oxic conditions only 30–40% of the AOI was removed.

At the oxic artificial recharge facility in Berlin-Tegel the AOI-reduction rate was 30% (retention time: 50 days). This inverse correlation between AOI-removal and redox potential most probably originates from the initial step of AOI-mineralization - reductive dehalogenation. It is known that the degradation of halogen substituted organics is more effective in soil passages with low redox potential (Mohn &Tiedje 1992). The results from the field site Berlin-Tegel show evidently that low redox potentials are essential for an efficient degradation of AOI. In case of a permanent transition of the redox conditions at the bank filtration Tegel site to oxic, considerably higher AOI-concentrations can occur in the extracted raw water.

Figure 5b displays the fate of AOI in the long retention column system under two different enforced redox conditions. The high standard deviation of the measurements originates from the seasonal variations in the used Lake Tegel water. Under oxic conditions only 9% of the source water AOI is dehalogenated, but under anoxic conditions 25% are degraded.

These results support the field studies indicating a better removal under anoxic/anaerobic conditions. A more efficient reduction of AOI in 30 days might be possible under strictly anaerobic conditions which were not tested. Under anoxic conditions some iron and manganese reduction occurred towards the end of the column, but nitrate was still present and the redox potential remained high (180–240 mV).

Trace pollutants. The single compound analysis for the trace pollutants Iopromide, Sulfamethoxazole and different naphthalenesulfonic acids was started in spring 2003, providing data for the time period May 2003 to August 2004. The monitoring showed that these trace organic compounds which stand for different groups of persistent pollutants behave differently during infiltration. For some compounds an influence of the changing redox conditions on the degradability in the field and in the columns could be revealed.

Iopromide, the X-ray contrast agent, occurred in the highest concentrations in the surface water (970 ng/l, $n = 16$, $\sigma = 270$ ng/l). Due to the fact that Iopromide as a triiodinated benzene derivate is part of the bulk parameter AOI (share in surface water ~5%), it was expected that the removal mechanisms for both parameters are similar. This could not be confirmed. Contrary to the fate of AOI the degradation of Iopromide showed no dependency on the dominant redox conditions. At the bank filtration site in Tegel Iopromide was very efficiently degraded under both redox conditions (oxic and anoxic/anaerobic: 99% removal). This was consistent with findings at other field sites. In the soil column experiments 99% and 98% of the initial Iopromide concentration were removed under oxic and anoxic conditions, respectively. The measurement of AOI parallel to the single compound analysis revealed that Iopromide is not mineralized but metabolized during infiltration. In the soil column experiment the spiking of 10 μg/l Iopromide increased the AOI-concentration in the feed by ~4.81 μg/l (~28%). But only a removal of 1.5 μg/l (9%, Figure 4b) was observed under oxic conditions. This proves the assumption that the Iopromide molecule is rapidly metabolized, but not mineralized during infiltration. It still remains an iodinated organic molecule and is detectable as AOI. The chemical structure of the metabolite/s is unclear. Under anoxic conditions 4.3 μg/l AOI (25%, Figure 4b) were removed during 30 days of infiltration in the column system. The share of Iopromide-originated AOI on this removal remains unclear.

The effluent concentration of the bactriostatic Sulfamethoxazole in Berlin treatment plants varies between 370–1200 ng/l. Because of its high stability it is also found in the surface waters at bank filtration sites. The degradation of Sulfamethoxazole seems to be redox dependent as well. In Berlin-Tegel Sulfamethoxazole shows a better removal under anoxic/anaerobic conditions were removal rates of 80–90% were observed. Under oxic conditions the removal decreased to rates of 50–70% and considerably higher concentrations were found in the production well. The better removal of Sulfamethoxazole under anoxic/anaerobic conditions is consistent with findings of Schmidt et al. (2004) at other bank filtration sites in Germany. Naphthalenedisulfonates are well known polar contaminants of treated wastewater and surface water. The different isomers behave nearly similar in the environment but show different degradation properties. Consistently with findings of Stüber et al. (2002) who reported the 1.5-NSA as very stable in wastewater treatment plants, the 1,5-NSA-concentration remained nearly constant during oxic and anoxic/anaerobic infiltration in the field and the column system. For the 1,7- and the 2,7-NSA isomers, the degradation was more efficient under oxic conditions. These results show that some naphthalenedisulfonic acids (1,5-NSA) are very poorly biodegradable and it is not to be expected to eliminate these isomers during bank filtration. Other biodegradable isomers (1,7- and 2,7-NSA) display a more efficient reduction during oxic infiltration.

Conclusions

The results of the field monitoring program give new insights into the mechanisms of bulk and trace organic removal. Aerobic and anoxic subsurface conditions during infiltration can lead to approximately the same residual DOC but the study indicates significant differences in the kinetics of DOC-removal depending on the redox conditions. Aerobic conditions result in a fast degradation of BDOC within a month, while anoxic conditions appear to require 3–6 months. LC-OCD analysis reveals that the change in character is comparable. Under oxic conditions the fraction of polysaccharides is removed more efficiently than under anoxic conditions. The fractions of humic substances, building blocks and low molecular weight acids are degraded partially. Significant differences were observed in AOI removal during aerobic and anoxic/anaerobic bank filtration. Dehalogenation of AOI occurred under reducing conditions, whereas during oxic infiltration no efficient dehalogenation was observed. The trace compound monitoring revealed a different behavior of the examined trace compounds during infiltration. For some compounds (Sulfamethoxazole) an influence of redox conditions was observed. In addition to rapidly removed compounds (e.g. Iopromide) and very persistent compounds (1,5-NSA), the study determined a group of medium persistent compounds (Sulfamethoxazole, 1,7-NSA, 2,7-NSA) that require special conditions for optimal degradation.

References

Berliner Wasser Betriebe, (2003): http://www.bwb.de.

Fritz B., Sievers J., Eichhorn S. and Pekdeger A. . Geochemical and hydraulic investigations of river sediments in a bank filtration system. In: Dillon, P. (ed.): Management of Aquifer Recharge for Sustainability, Balkema, Adelaide, Australia (2002) 95–100.

Grünheid, S., Schittko, S. and Jekel, M. Behaviour of bulk organics and trace pollutants during bank filtration and groundwater recharge of wastewater-impacted surface waters. Proceedings of the annual meeting of the Water Chemical Society, Bad Saarow, pp. 75–80 (2004).

Hartig C. (2000) Analytik, Vorkommen und Verhalten aromatischer Sulfonamide in der aquatischen Umwelt. (Analysis and behaviour of aromatic sulfonamides in the aquatic environment). Doctoral Dissertation accepted by: Technical University of Berlin, School of Process Sciences and Engineering, 2000-08-29, http://edocs.tu-berlin.de/diss/2000/hartig_claudia.htm.

Huber, S. and Frimmel, F. (1996): Gelchromatographie mit Kohlenstoffdetektion (LC-OCD): Ein rasches und aussagekräftiges Verfahren zur Charakterisierung hydrophiler organischer Wasserinhaltsstoffe, Vom Wasser, Vol. 86, pp. 277–290.

Mohn W. W. and J. M. Tiedje (1992) Microbial reductive dehalogenation. Microbiol Rev. 1992; 56(3): 482–507.

NASRI, (2003): http://www.kompetenz-wasser.de.

Oleksy-Frenzel J., Wischnack S. and Jekel M. (2000) Application of ion-chromatography for the determination of the organic-group parameters AOCl, AOBr and AOI in water. Fresen. J. Anal. Chem., 366, 89–94.

Pekdeger A., Massmann G., Ohm B. (2004) 2nd Periodic Report: Hydrogeological-hydrogeochemical processes during bank filtration and groundwater recharge using a multitracer tracer approach. NASRI-Projekt, 80 pages.

Putschew A., S. Schittko, M. Jekel (2001) Quantification of triiodinated benzene derivatives and X-ray contrast media in water samples by liquid chromatography-electrospray tandem mass spectrometry. J. Chromatogr. A, 930, 127–134.

Schmidt C.K., F.T. Lange; H.-J. Brauch. (2004) Assessing the impact of different redox conditions and residence times on the fate of organic micropollutants during riverbank filtration. Proceedings “4th International Conference on Pharmaceuticals and Endocrine Disrupting Chemicals in Water”, 13–15.10.2004, Minneapolis, Minnesota.

Storm T., T. Reemtsma, M. Jekel (1999) Use of volatile amines as ion-pairing agents for the high-performance liquid chromatographic-tandem mass spectrometric determination of aromatic sulfonates in industrial wastewater. J. Chromatogr. A, 854, 175–185.

Stüber M., T. Reemtsma, M. Jekel (2002) Determination of naphthalene sulfonates in tannery wastewater and their behaviour in a membrane bioreactor. Vom Wasser, 98, 133–144.

Ziegler D. (2001) Untersuchungen zur Nachhaltigkeit der Uferfiltration und künstlichen Grundwasseranreicherung in Berlin. (Investigations on the sustainability of bank filtration in Berlin's water cycle), Doctoral Dissertation accepted by: Technical University of Berlin, 2001-04-24. http://edocs.tu-berlin.de/diss/2001/ziegler_doerte.htm.

67 Classification of riverbank filtration sites and removal capacity

C. Skark*, F. Remmler*, N. Zullei-Seibert**

*Institut für Wasserforschung GmbH, Zum Kellerbach 46, D-58239 Schwerte, Germany
skark@ifw-dortmund.de, remmler@ifw-dortmund.de
**Westfälische Wasser- und Umweltanalytik GmbH, Willy-Brandt-Allee 26, D-45891 Gelsenkirchen, Germany, ninette.zullei-seibert@wwu-labor.de

Abstract In an empirical investigation middle European sites with riverbank filtration (RBF) were classified and the site intrinsic removal capacity for dissolved organic carbon (DOC) was considered. By a literature study and an inquiry among waterworks data for 33 RBF sites in several countries were collected. The classification by cluster analysis was based upon data about local hydrogeology and surface water chemistry, distinguished 4 site groups and was confirmed by discriminant and factor analysis. By multivariate regression calculation model functions regarding the DOC removal efficiency during the underground passage were derived. The hydro-chemical changes in terms of DOC removal at a new site can be estimated with these regression functions. The classification of a new RBF site is done with discriminant functions and confirms the applicability of the regression model function at the new RBF site.

Keywords Riverbank filtration, water treatment, drinking water, DOC, elimination, classification statistics, surface water, hydrogeology

Introduction

The riverbank filtration (RBF) with a subsequent underground passage has been used in Europe as a close to nature, low cost and reliable purification method for drinking water supply since long. Investigations on various RBF systems revealed that the behaviour of substances during the underground passage depends on many hydrogeological, hydraulic and physical-chemical factors.

Up to now there is a lack of a purification orientated guideline for the planning, dimensioning and operating of RBF sites for drinking water production. Particularly the reaction of RBF to surface water with rather different concentrations of chemical substances than in central Europe are not well understood as well as the impact of differing extreme climate conditions. In the project "Technical concepts and adjusted operational modes for an optimal adaptation of riverbank filtration to local boundary conditions", as part of a joint research project of the German Ministry of Education and Research (BMBF), fundamental variables for planning, dimensioning and running water production plants were compiled (Lenk et al., 2005).

 Recent Progress in Slow Sand and Alternative Biofiltration Processes edited by Rolf Gimbel, Nigel J.D. Graham and M. Robin Collins. ISBN: 9781843391203. Published by IWA Publishing, London, UK.

In other parts of the joint research project with respect to RBF the change of DOC composition and the fate of anorganic and organic micropollutants during infiltration and the subsequent underground passage were studied in laboratory and field scale (Marschke and Wichmann, 2005; Schmidt and Lange, 2005). Particular attention was paid to the influence of different temperature, flow velocity and DOC concentration during the RBF process (Schönheinz and Worch, 2005). In addition the use of manganese oxides as specific electron acceptors during RBF redox processes and the elimination of microorganisms assessed by molecular biological methods were investigated (Bloem and Jekel, 2005, Geis and Obst, 2005).

Materials and methods

Knowledge from literature was supplemented with a representative data inquiry among water supply companies in Germany, France, Austria and Switzerland. Thus 33 case studies of RBF sites in middle-Europe, comprising 19 different surface water sources were collected. All the data from the literature and the survey were interpreted using various multi-variate statistical methods (Backhaus et al., 1994; Otto, 1997, Hartung and Elpelt, 1999, Brosius, 2002).

The aim of the study was to classify central European type situations of RBF sites on the basis of selected local boundary conditions. The classification of sites was computed by a hierarchic cluster analysis (euclidic distance; WARD algorithm). The results were confirmed by discriminant and factor analysis. To cope with the different dimensions of the considered parameters transformed values of the variables were used so that every variable has a mean = 0 and a standard deviation = 1 over the whole sample (z-transformation).

The degradation capacities of the subsurface passages were characterized using correlation and regression analysis between local boundary parameters and the regarded systems purification capacity to relevant water solutes (e.g. dissolved organic carbon, DOC).

Results and discussion

For the classification of RBF type situations the following local hydrogeological site parameters were considered: hydraulic conductivity, thickness and transmissivity of the aquifer. Local surface water was characterized by following parameters: maximum temperature, pH-value and mean concentrations of O_2, NO_3^-, NH_4^+, SO_4^{2-}, and DOC. These parameters determine the redox induced transformation process of substances during the infiltration and the subsequent underground passage.

By the cluster analysis four different middle European site groups, regarding hydrochemical and hydrogeological parameters of RBF, could be classified, which are shown in figure 3.1 in the form of a dendrogram providing the relative distance of the case studies. The classification in 4 site groups was chosen as this solution showed the most significant reduction of the cluster immanent variance. Each cluster group has characteristic combinations of the considered local boundary conditions. There are significant differences for each group between the chemical situation of the surface water and the characteristics of the adjacent aquifer. Figure 3.2 provides the average values for selected variables of the different site groups. Site group 4, for instance is characterized by a low permeability and

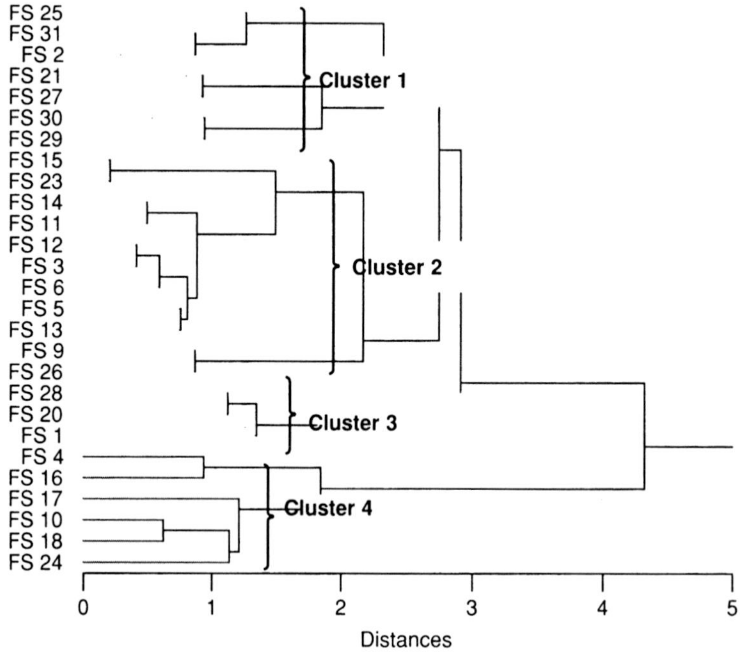

Figure 3.1 Classification by cluster analysis (FS = case study)

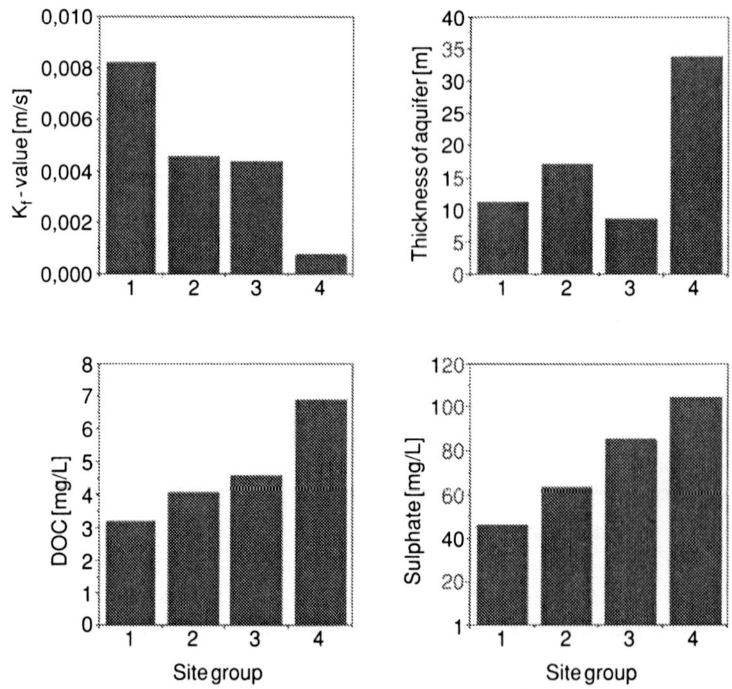

Figure 3.2 Selected properties characterizing the 4 site groups (group mean values)

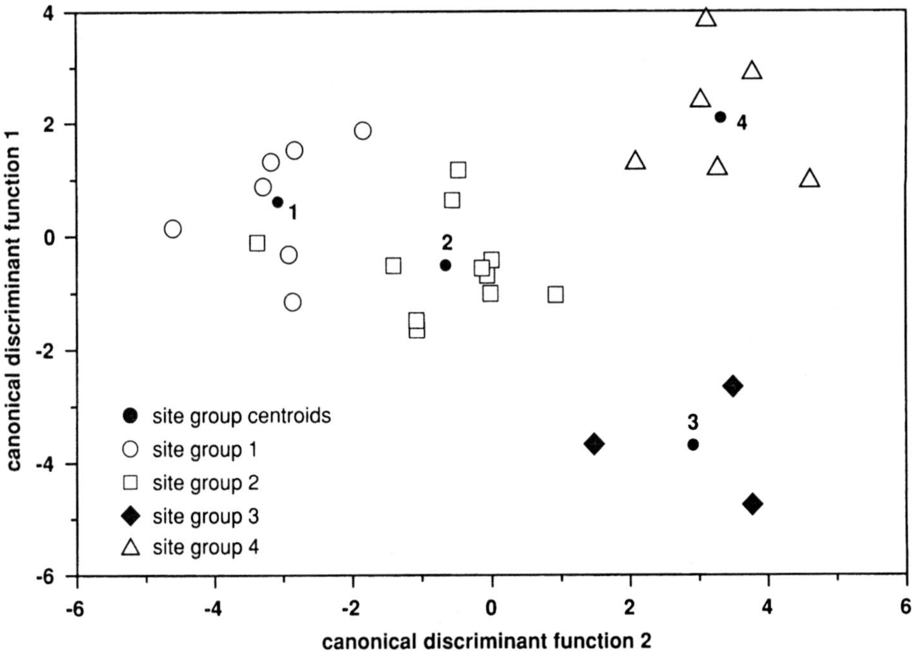

Figure 3.3 Score plot as a result of a discriminant analysis

a high thickness of the aquifer as well as a high DOC and sulphate concentration in the surface water. The high DOC concentration is often the result of a huge proportion of treated waste water in the surface water. As a consequence of the high initial DOC concentration anaerobic redox conditions are established in the aquifer during the RBF process in water works of site group 4.

In a further step the quality of the classification by cluster analysis was corroborated by a discriminant analysis (figure 3.3). The discrimination process yields a set of coefficients for the linear combination of the considered variables. The discriminant coefficients are determined in a way that the ratio of the variance between the classification groups to the variance within the classification groups gains a maximum. Four site groups are represented by 3 discriminant functions. For 2 of these 3 functions the scores are shown in figure 3.3.

The different site groups represented by their group means (centroids) as well as the single case studies are fairly separated in this plot. These 2 functions explain 85% of the whole variance. The quality of the separation is significantly high as Wilk's Lambda is obtained with 0.011 ($p = 0.000$).

Furthermore a factor analysis of the data was performed in order to answer 2 questions:

(1) Can the variety of variables used for the classification process be reduced to a small number of comprehensive, hidden factors ?
(2) Does every variable have the same impact on the differentiation between the site groups?

In the process of a factor analysis every variable should be expressed as a linear combination of several factors that has been drawn during the analysis. In preparation of this process the linear correlation between the considered variables is assessed by the mean of the measure of sampling adequacy (MSA, ratio of the sum of squared simple correlation coefficients to the sum of simple and partial correlation coefficients). Several variables, as pH-value, O_2−, NO_3^-−, NH_4^+-concentration in surface, do not comply the MSA criteria and have to be excluded from further analysis. The remaining 5 variables for the local hydrogeology and the surface water chemistry (hydraulic conductivity, thickness of the aquifer, maximum temperature of the surface water, DOC- and sulphate-concentration in the surface water) can be reduced to 2 factors. Factor 2 represents dominantly the physical property temperature of the surface water whereas factor 1 comprises the information about local hydrogeology and hydrochemistry of surface water (figure 3.4). The linear combination of both factors explains between 29% and 73% of the variance of the considered variables. Thus the factor analysis model can only reproduce the real variance of the variables in a restricted manner as the residuals of unexplained variances are relatively great.

By the transposed factor load matrix the factors can be expressed as functions of the different variables (Q-Modus) which is depicted in figure 3.4 for the considered variables as well as for the different case studies. The separation of the site groups is not as clear

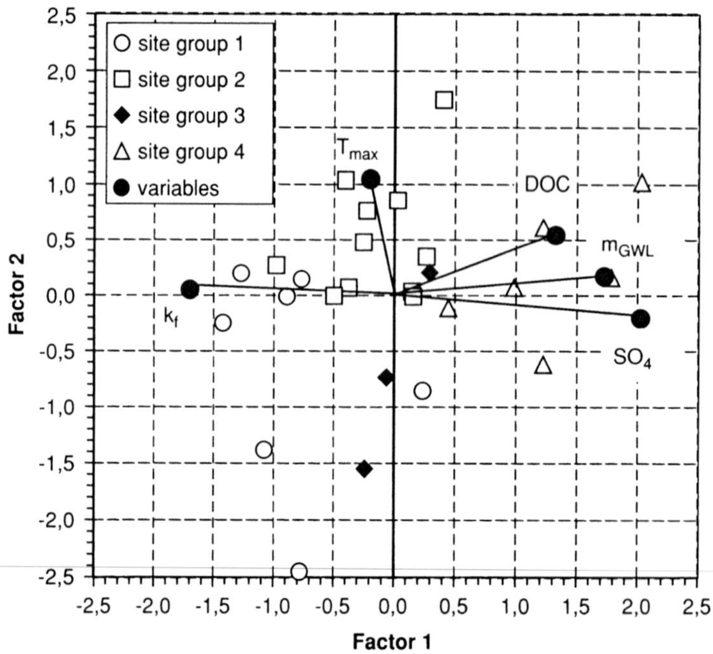

Figure 3.4 Factors as a function of variables and case studies – result of factor analysis (kf hydraulic conductivity, m_{GWL} – aquifer thickness, T_{max} – maximum surface water temperature, DOC dissolved organic carbon in surface water, SO_4 – sulphate in surface water)

as by cluster and discriminant analysis which is due to the smaller number of considered variables and the reduction to 2 background factors. Nevertheless a certain distinction between the site groups can be seen. Thus the classification process by the cluster analysis is confirmed. As the main separating variables hydraulic conductivity, aquifer thickness, temperature, DOC- and sulphate concentration of the surface water can be seen. So the local hydrogeological condition is described by the transmissivity, the product of hydraulic concuctivity and aquifer thickness. The separation of site group 1 is linked to the hydraulic conductivity whereas the temperature of the surface water has the greatest influence for site group 2. Characteristic values of the aquifer thickness, DOC and sulphate concentration in surface water lead to the distinction of site group 4.

The reduction of organic compounds is one of the central goals of the water purification for drinking water purposes. The RBF as a pre-treatment step in the purification process for drinking water and as a part of a multi-barrier concept has essential advantages compared to the direct treatment of surface waters. Organic substances will be reduced and degraded in the biological high active infiltration zone and in the subsequent underground passage. By this it is possible to reduce the amounts of needed energy and chemical substances at subsequent treatment steps.

The data from the case studies provide the possibility to investigate the DOC elimination by means of bi- and multivariate correlation- and regression analysis. As the essential determining variables for the measured DOC-elimination at the examined locations, (i) the initial DOC concentration in surface waters, (ii) the residence time of the riverbank filtrate in subsurface passage and (iii) the transmissivity of the adjacent aquifer could be found.

The best fitting bivariate models can be combined to a multiple, non-linear regression function:

$$Y = -0{,}503 + 0{,}811 * \ln(X_1) + 7{,}428 * (X_2) + 0{,}236 * (X_3)^{0{,}437} \qquad \text{(Eq. 1)}$$

with: Y = DOC-elimination [mg/L];
X_1 = DOC-concentration of the surface water [mg/L],
X_2 = transmissivity of the aquifer [m^2/s];
X_3 = residence time in the underground [d]

The model function in equation 1 can explain 74% of the monitored variance in the whole sample regarding the DOC-elimination ($n = 43$). The mean deviation between observed field data and model results is below 0.5 mg/L DOC-elimination. If the sample is reduced to sites with aerobic conditions in the subsurface filtration process ($n = 23$) the proportion of the explained variance will increase to 87% using the same model function type. This clear differentiation between sites with varying abundance of oxygen has to be expected. This can also be seen in the accompanying field investigation which reveals a certain difference in the elimination of organic micropollutants with the variation of the redox condition and confirms the site classification in this study (Schmidt and Lange, 2005).

Apart from describing the compiled data set the regression equation serves also as a predictive model which can estimate the degree of DOC-elimination at unknown sites. This can be done either by calculation or by the means of the nomograms shown in figure 3.5. A valid estimate can be obtained under the assumption that the characteristic values of the

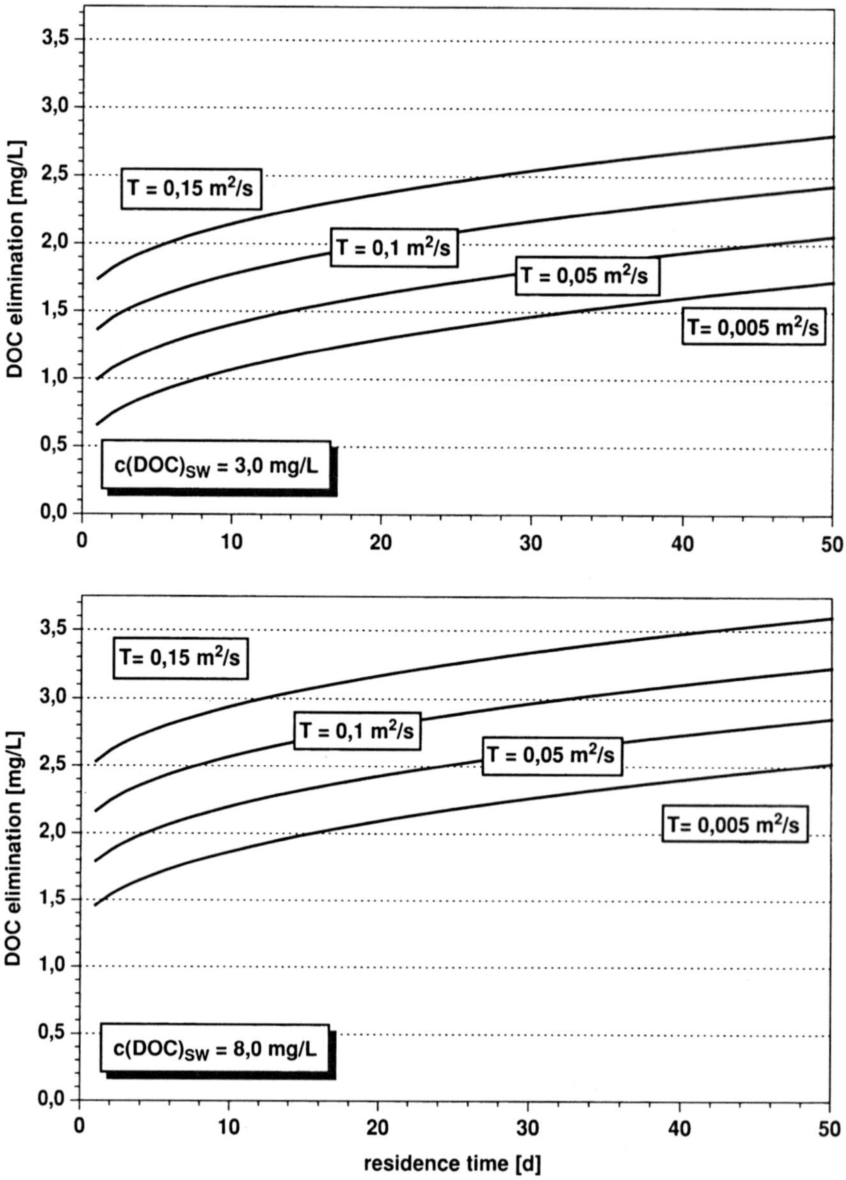

Figure 3.5 Nomogram for a prognosis of the DOC elimination at the riverbank filtration and underground passage on the base of equation (1) (SW = surface water, T = transmissivity)

new site will meet the value range of the variables used for the derivation of the regression model.

As a refinement for this validation the classification of a new site can be studied by a further application of discriminant analysis. For that purpose the known site groups are supplemented by additional site groups which are characterized by a high DOC concentration in surface water as well as a combination of low and high surface water temperature and transmissivity. So this additional site groups are determined by a variation of the main variables found in factor analysis. For this supplemented data set the FISHER coefficients are calculated by a discriminant analysis using the original data. By this FISHER coefficients a new site can be classified. In the case of a classification in a known site group the regression model for DOC elimination can be used as a valid estimator whereas in the case of a classification in an additional site group the regression model estimate may not meet the real DOC decrease during subsurface passage at the new site. The specific DOC elimination may be changed particularly by the seasonal variation of temperature and flow velocity as shown in accompanying column studies (Schoenheinz and Worch, 2005).

Conclusions

By cluster analysis the considered case studies can be classified to 4 typical site groups which show a significant distinction of their hydrogeological site properties and the hydrochemical characteristics of the infiltrated surface water. This classification is confirmed by the means of discriminant and factor analysis. The discriminant analysis provides also a set of FISHER coefficients by which unknown sites can be classified.

Essential purification processes like DOC-elimination are determined to a high extent by three parameters: the initial DOC concentration in the surface water, the residence time of the riverbank filtrate in the subsurface passage and the transmissivity of the aquifer. In a non-linear model DOC-elimination can be described by a multivariate regression equation. The empirical regression function can be used in a prognostic manner for the estimation of the DOC-elimination at new planned sites.

The findings of the joint project were compiled in a guideline which helps the planning engineer in the design of new RBF sites and the optimization of existing water works (DVGW-TZW, 2005).

Acknowledgements

This study was conducted with subsidy from the German Ministry of Education and Research (BMBF) and is part of the joint research project "Adjusted water treatment technologies and water distribution under regional conditions". We appreciate the support by the participating waterworks.

References

Backhaus, K., Erichson, B., Plinke, W. and Weiber, R. (1994) *Multivariate Analysemethoden*, Springer-Verlag., Berlin, Heidelberg.

Bloem, D. and Jekel, M. (2005) *Rolle des Redoxmilieus für die Aufbereitungswirkung der Uferfiltration (B3)*. In *Exportorientierte F&E auf dem Gebiet der Wasserver- und -entsorgung Teil I: Trinkwasser – Leitfaden* (ed. Deutsche Vereinigung des Gas- und Wasserfaches – Technologiezentrum Wasser (DVGW-TZW)) Vol. 2, Karlsruhe.

Brosius, F. (2002) *SPSS 11*, mitp-Verlag, Bonn.

Deutsche Vereinigung des Gas- und Wasserfaches – Technologiezentrum Wasser (DVGW-TZW, ed., 2005) *Exportorientierte F&E auf dem Gebiet der Wasserver- und –entsorgung Teil I: Trinkwasser – Leitfaden,* Vol. 2, Karlsruhe.

Geis, M. and Obst, U. (2005) *Typisierung der Uferfiltration anhand mikrobieller Einflussgrößen sowie optimierte Kontrolle pathogener Bakterien und Parasiten (B2).* In *Exportorientierte F&E auf dem Gebiet der Wasserver- und –entsorgung Teil I: Trinkwasser – Leitfaden* (ed. Deutsche Vereinigung des Gas- und Wasserfaches – Technologiezentrum Wasser (DVGW-TZW)) Vol. 2, Karlsruhe.

Hartung, J. and Elpelt, B. (1999) *Multivariate Statistik*, R. Oldenbourg Verlag, München, Wien.

Lenk, S., Remmler, F., Skark, C. and Schulte-Ebbert, U. (2005) *Technische Konzepte und abgestimmte Betriebsweisen zur optimalen Anpassung der Uferfiltration an lokale Randbedingungen (B1).* In *Exportorientierte F&E auf dem Gebiet der Wasserver- und –entsorgung Teil I: Trinkwasser – Leitfaden* (ed. Deutsche Vereinigung des Gas- und Wasserfaches – Technologiezentrum Wasser (DVGW-TZW)) Vol. 2, Karlsruhe.

Marschke, M. and Wichmann, K. (2005) *Einfluss von Randbedingungen auf die nutzbare Reinigungsleistung einer optimierten Uferfiltration (B5).* In *Exportorientierte F&E auf dem Gebiet der Wasserver- und –entsorgung Teil I: Trinkwasser – Leitfaden* (ed. Deutsche Vereinigung des Gas- und Wasserfaches – Technologiezentrum Wasser (DVGW-TZW)) Vol. 2, Karlsruhe.

Schmidt, C.K. and Lange, F.T. (2005) *Reinigungsleistung der Uferfiltration hinsichtlich der Eliminierung organischer Schadstoffe unter standortspezifischen Randbedingungen (B6).* In *Exportorientierte F&E auf dem Gebiet der Wasserver- und –entsorgung Teil I: Trinkwasser – Leitfaden* (ed. Deutsche Vereinigung des Gas- und Wasserfaches – Technologiezentrum Wasser (DVGW-TZW)) Vol. 2, Karlsruhe.

Schoenheinz, D. and Worch, E. (2005) *DOC-Entfernung bei der Uferfiltration unter Berücksichtigung extremer Temperaturbedingungen und Belastungsschwankungen (B4).* In *Exportorientierte F&E auf dem Gebiet der Wasserver- und –entsorgung Teil I: Trinkwasser – Leitfaden* (ed. Deutsche Vereinigung des Gas- und Wasserfaches – Technologiezentrum Wasser (DVGW-TZW)) Vol. 2, Karlsruhe.

Otto M. (1997) *Chemometrie*, VCH, Weinheim.

68 Efficiency of riverbank filtration considering the removal of pathogenic microorganisms of the River Rhine

H.-P. Rohns, C. Forner, P. Eckert and R. Irmscher

Stadtwerke Düsseldorf AG, Höherweg 100, 40233 Düsseldorf

Abstract Riverbank filtration (RBF) has been applied successfully in the Düsseldorf waterworks since 1870. RBF is the first step in a multi-barrier concept combined with ozone treatment and activated carbon filtration to obtain safe drinking water. According to the hydrogeological conditions the production wells discharge raw water consisting of up to 90 % of bank filtrate. In order to achieve detailed information of the purification processes a research program was conducted in the year 2003/2004 including the extreme low water event in the summer of 2003 and the following flood event in 2004. With rising temperatures the increasing biological activity was leading to higher microbial oxygen consumption in the bank filtrate. The degradation capacity of the microorganisms was limited by decreasing oxygen concentrations in the infiltrating water. In this period anoxic conditions occurred in the bank filtrate. Independent of the varying content of total organic carbon (TOC) in the river water the TOC in the bank filtrated water remained constant at a level of 1 mg/l. In periods with normal water level the total cell number of bacteria was reduced by 2 to 3 log-steps during river bank filtration. The hydraulic residence time of the water was about 35 days. During a flood the residence time of the water decreased to 5 to 7 days. TOC was metabolised incompletely and could be detected in the bank filtrated raw water in concentrations up to 2 mg/l. Together with the detected TOC peak the oxygen consumption decreased. The elimination capacity regarding coliforms was reduced to about 2 log-units. But there was no evidence for any viruses or protozoa in the raw water during low and high water conditions. The multi protective barrier concept including the natural purification by bank filtration combined with technical oxidation and adsorption facilities had been proven to be a reliable and effective system for the production of safe drinking water. The enhanced understanding of the bank filtration processes is important to assess the possibilities and the limitations for their application.

Keywords Riverbank filtration, temperature, oxygen, TOC, microorganisms, high water event

Introduction

Riverbank filtration (RBF) is a process widely established in Germany for drinking water production (Kühn and Müller, 2000). Since 1870 bank filtration is applied in the waterworks

 Recent Progress in Slow Sand and Alternative Biofiltration Processes edited by Rolf Gimbel, Nigel J.D. Graham and M. Robin Collins. ISBN: 9781843391203. Published by IWA Publishing, London, UK.

of Düsseldorf and responsible for a raw water quality with a low amount of particulate material. Organic compounds are reduced by microorganisms in the active filtration zones below the river bed. Fluctuating concentrations of compounds in the river Rhine are equalized by various processes in bank filtration. As a result of retention times between one and six weeks according to the water levels in the River microorganisms are eliminated by natural processes (Schubert, 2000). As a consequence the bank filtrate is already of very good quality. Up to the 1950s RBF was used only in combination with disinfection. After World War II further treatment steps had become necessary as a consequence of the growing industrialisation and the lack of adequate sewage purification systems. Today the bank filtration process is a combination of ozone treatment, biological and activated carbon filtration, and a low dosage of disinfectants for a safe transport of the drinking water. RBF is still the first and very important step in this multi-barrier system ensuring drinking water of high quality.

The raw water originates from bank filtrate of a quaternary aquifer. Water from the river Rhine seeps through the riverbed and is pumped out at the river bank. According to the hydro-geological condition and the varying Rhine water level the raw water consists of 50 % up to 90 % of bank filtrated water (Eckert and Irmscher, 2005).

Hiscock and Grischeck (2002) give an overview about attenuation of groundwater pollution by bank filtration. They pointed out that although RBF is a widely used process in drinking water production it's processes to reduce pollutants and to eliminate microorganisms are not completely understood by now. Since twenty years the Stadtwerke Düsseldorf AG are collecting data to gain detailed information about the hydraulic conditions and the capacity of the RBF purification processes at the Rhine in Düsseldorf (Sontheimer, 1991; Schubert, 2000; Schubert, 2002; Eckert *et al.*, 2003, Eckert and Irmscher, 2005). In order to achieve a profound knowledge of the purification processes during bank filtration a research program was performed in the year 2003/2004. The period of investigation included the extreme low water event in summer 2003 and the following flood event in January 2004. For the acquisition of physical, chemical and biological data a great number of water samples were taken from a multi-level well at the bank side (B1-3) and one at the land side (C1-3) of the vertical filter well gallery used for the extraction of raw water from the river bank. To obtain data from the former unsaturated zone during a high water event additional levels (4-6) have been installed in the multi-level well B between the river Rhine and the production well (**Figure 1**).

Methods

Due to the hydraulic conditions during operation of the extraction wells the raw water consists of an mixture of bank filtrate and groundwater according to the water level of the Rhine. The water in the multi-level well B situated between the river and the abstraction wells of the vertical filter well gallery consists of 100% bank. The water in the multi-level well C consists normally only of groundwater.

During the monitoring program a large set of physical and chemical as well as biological data was collected from the river Rhine, the bank filtrate (multi-level B), the groundwater (multi-level C) and the raw water. The analytical methods were carried out according the standards of the Deutsche Verfahren zur Wasser-, Abwasser- und Schlammuntersuchung

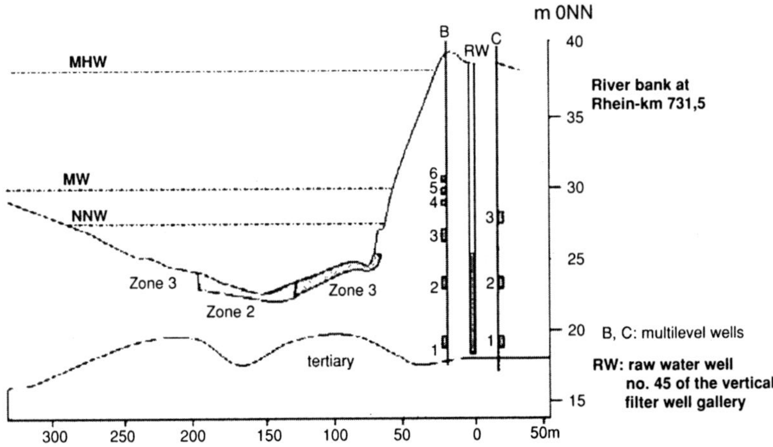

Figure 1 Overview of the observation site in the water works "Flehe" of the Stadtwerke Düsseldorf AG.

(1993). The total cell numbers were obtained in the plate method on nutrient-poor R2A agar-medium by cultivating for 7 days at 20°C.

Results and discussion

Oxygen

The oxygen content in water is directly correlated to the water temperature. In winter the oxygen concentration in the river water varied between 11 and 13 mg/L and decreased during summer time to 7 mg/L (**Figure 2**). Additionally the microbial activity also reduces the oxygen content during the bank filtration. Due to the conversion of total organic carbon (TOC) into CO_2 and energy oxygen is consumed. The microbial and enzymatic activity also increases with the temperature. The rising water temperature led to a higher oxygen consumption by the microorganisms and resulted in a higher consumption rate of TOC and inorganic carbon. But the degradation capacity of the microorganisms is in summer limited by the oxygen available in the infiltrating water (**Figure 2**). As a consequence of the limited oxygen concentration anoxic conditions prevailed in the bank filtrate in this period where anaerobic degradation of micro-pollutants might occur. Denitrification was observed, but a mobilisation of iron and manganese could not be detected.

Total organic carbon (TOC)

In the bank filtration the varying TOC-concentrations of 2–4 mg/l in the river water had been reduced to a stable level of 1 mg/l under normal hydraulic conditions. The average residence time of the water during bank filtration was about 35 days in summer 2003. Investigations about the fractions of TOC in the river water and the bank filtrate by Ms. Firch of the TU Hamburg Harburg revealed the absence of polysaccharides in the bank filtrate. The concentration of humic substances is about half of concentration in Rhine. There is

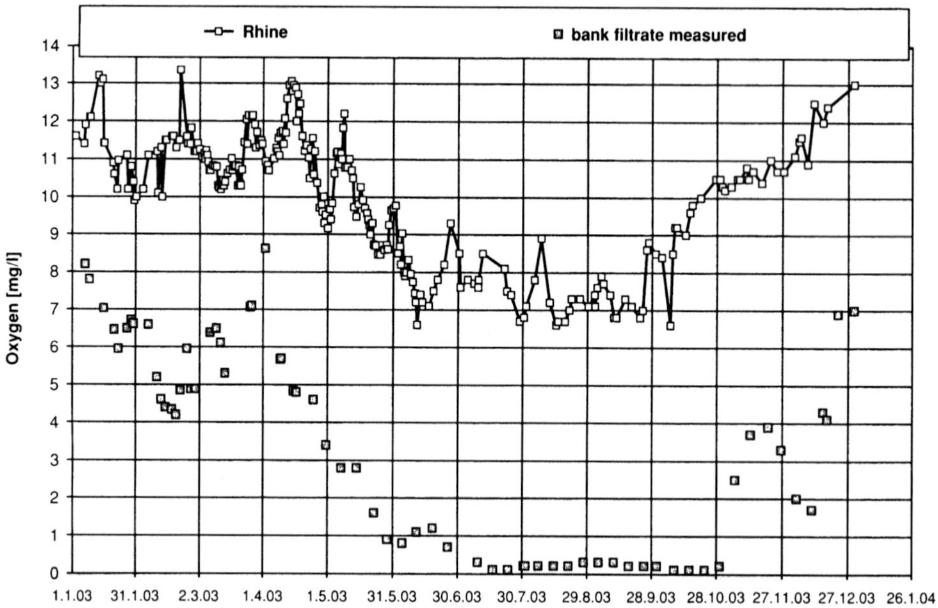

Figure 2 Comparison of the oxygen concentration of the river water and the bank filtrate.

significant reduction of metabolic building blocks and neutral and amphiphilic substances in bank filtrate in contrast to river water.

Under high water level conditions the hydraulic pressure on the riverbank increased and the residence time was reduced to 5 and 7 days. In this short time the microbial community was not able to consume all of the biodegradable TOC. This resulted in a breakthrough of TOC in the bank filtrate (**Figure 3**).

Microorganisms

Already the bank filtrated water mostly met the European Water Standards concerning pathogenic microorganisms . The analysis of water samples from the river Rhine and the production well regarding total cell counts showed a significant reduction of 2 to 3 log-units under normal water level conditions (**Figure 4**). Assuming that *E. coli* has a mortality rate of 0.2 log/d at a water temperature of 15°C the detected concentration of 10^4 *E. coli*/100ml in the Rhine water was completely eliminated within 20 days. This means that a retention time of 35 days under normal water level conditions was sufficient to eliminate all considered pathogenic bacteria. In the high water event in January 2004 a reduction of the removal efficiency of the river bank was observed. Nevertheless about 2 log-units of coliform bacteria were retained (**Figure 5**). The breakthrough of bacteria showed that the application of ozone remains essential at least under high water conditions.

In general the fecal contamination of the river Rhine is drastically reduced by bank filtration. During the high tide of the river the water in the production was investigated in order to detect pathogenic protozoa. Mr. Gornik from the University of Bonn in Germany,

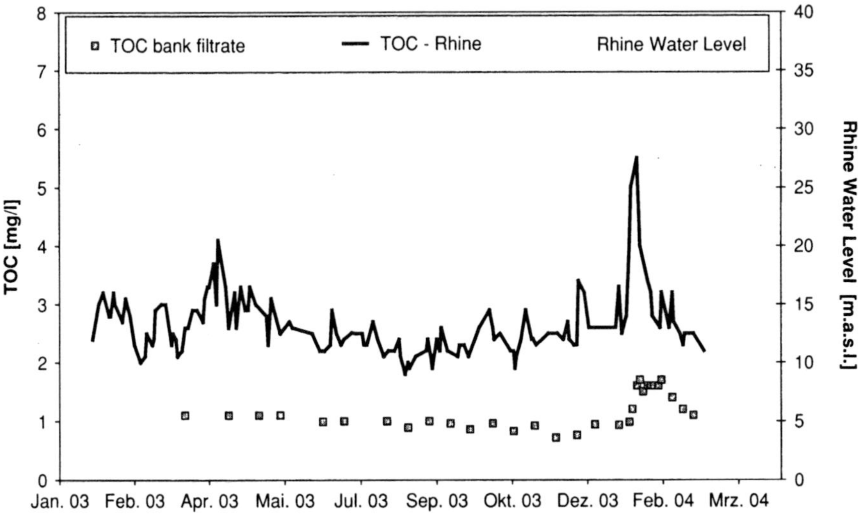

Figure 3 Rhine water level and the total organic carbon (TOC) concentration of the river water and the bank filtrate.

a specialist for the detection of cysts deriving from *Giardia* and *Cryptosporidia* recovered samples by a special filtration technique. The data of the concentration in the river Rhine is shown in **Figure 6**.

The contamination of cysts in the river was totally removed by bank filtration. The filtration capacity of the bank filtration was the major process concerning the sizes of the pathogenic protozoa. But instead of the cysts from *Giardia* and *Cryptosporidia* several

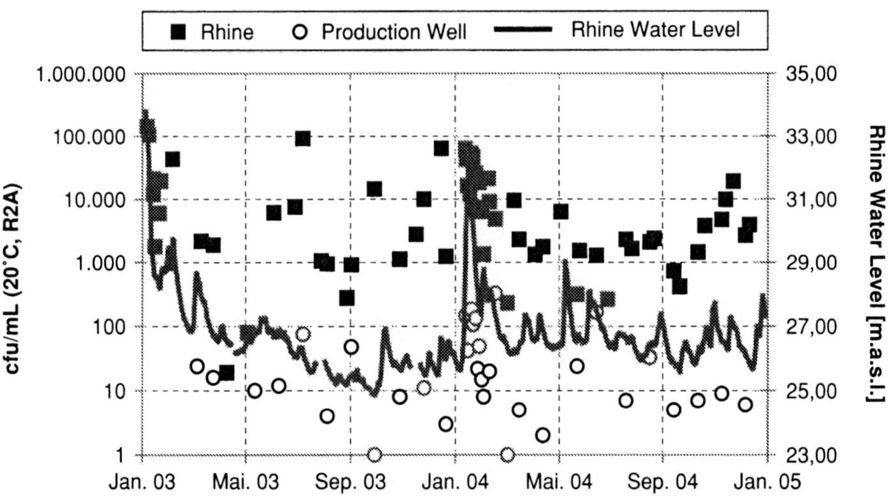

Figure 4 Total cell counts in the river Rhine and the production well

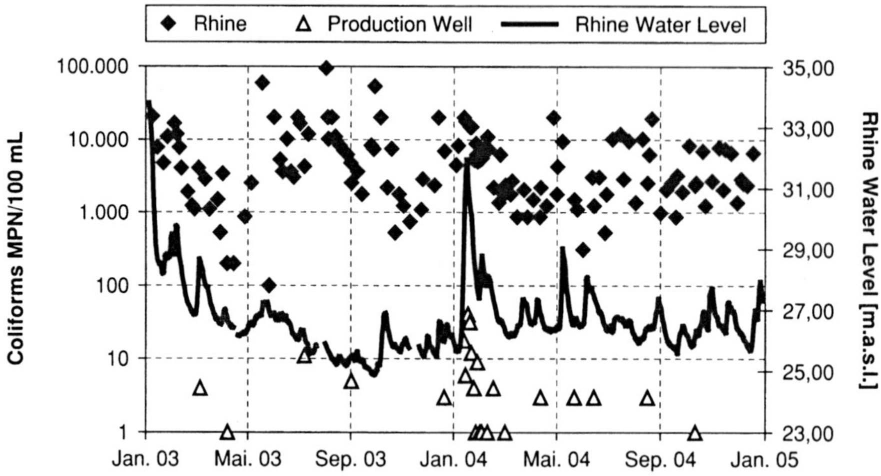

Figure 5 Coliforms in the river Rhine and the production well

micro algae could be detected. The size between 3 and 10 µm of these micro algae made obvious that the protozoic cysts generally have a slight chance to cross the passage, but couldn't be detected in this investigation.

In order to assess the reduction capacity of bank filtration regarding pathogenic microorganisms a number of water samples had been investigated by the National Institute for Public Health and Environment in the Netherlands to elucidate the number of somatic phages and viruses (**Table 1**).

Somatic phages are bacteria specific viruses that inject their genetic information into bacterial cells. The F-specific phages therefore use the F-pili connecting two bacterial cells for the transfer of information. Enteroviruses are representing a group of pathogenic viruses

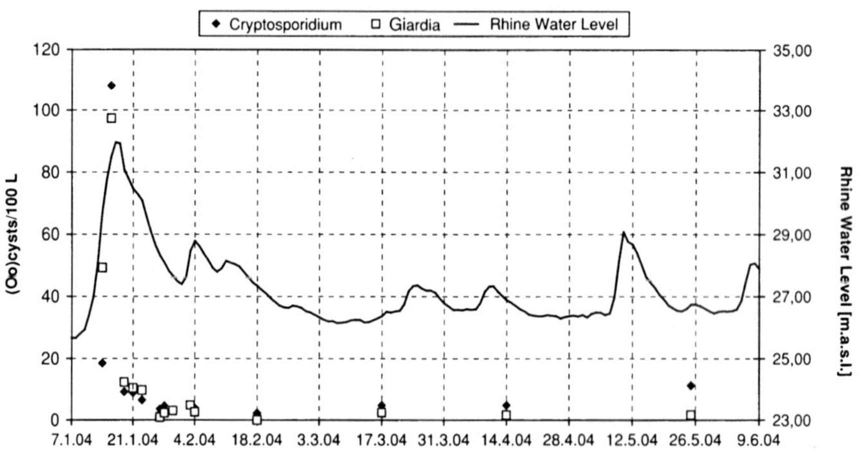

Figure 6 Pathogenic protozoa in the river Rhine

Table 1 Viruses in Rhine, raw water of the production well and after ozone tretment

	26.01.2004			02.02.2004		
	Rhine [pfu/l]	prod. well [pfu/l]	after ozone [pfu/l]	Rhine [pfu/l]	prod. well [pfu/l]	after ozone [pfu/l]
somat. Phages	6753,5	2	0	6831,5	0,52	0
F-spec. Phages	1060,2	0,16	0	924,7	0,018	0
Enteroviruses	0,26	n.d.	n.d.	0,17	n.d.	n.d.
Reoviruses	0,85	n.d.	n.d.	10,9	n.d.	n.d.
Noroviruses	0	n.d.	n.d.	248,1	n.d.	n.d.
Rotaviruses	0	n.d.	n.d.	15,9	n.d.	n.d.

which can be transmitted by water. The group of enteroviruses consists of Polio-, coxsackie-, echoviruses and hepatitis-A-viruses. The reoviruses can infect the respiration tract and the stomach of children and cause severe illnesses. Noroviruses, the former norwalkviruses cause a stomach flu. Rotavirus can cause severe diarrhea by babies and young children. These viruses are not protected by a protein-shelter. These kind of viruses are very resistant under environmental conditions without loosing their ability to infect human beings. But the results represented in **Table 1** made evident that the concentration of viruses could be decreased by 3.5 to 4 log-units in the processes of the bank filtration.

Conclusions

The multi protective barrier concept including the natural purification step of bank filtration combined with technical oxidation and adsorption facilities has proven to be a reliable and effective procedure for the production of drinking water. The enhanced understanding of the bank filtration processes is important to assess the possibilities and the limitations for their application.

The goals of the IAWR-Memorandum 2003 demanding a water quality of the Rhine suitable for the drinking water production only with the help of natural purification processes cannot be achieved by now. Additional purification remains necessary for meeting standards of drinking water regulations at least under high water conditions regarding micropollutants and microorganisms.

References

Anonymus (1993) Deutsche Verfahren zur Wasser-, Abwasser- und Schlamm-Untersuchung (DEV). Beuth-Verlag, Berlin.

Eckert, P., Rohns, H.-P. and Irmscher R. (2003) Periodische Schwankungen hydrogeochemischer Prozesse bei der Uferfiltration. In *ARW-Jahresbericht 2003,* 87–103, Arbeitsgemeinschaft der Rhein-Wasserwerke e.V., Köln, Germany.

Eckert, P., and Irmscher R. (2005) Over 130 years of experience with riverbank filtration in Düsseldorf, Germany. AQUA, in press.

Hiscock, K.M., and Grischek, T. (2002) Attenuation of groundwater pollution by bank filtration. J. Hydrol. **266**, 139–144.
Kühn, W., and Müller, U. (2000) Riverbank filtration an overview. J. AWWA Dec. 2002, 60–69.
Schubert, J. (2000) Entfernung von Schwebstoffen und Mikroorganismen sowie Verminderung der Mutagenität bei der Uferfiltration. Wasser – Abwasser **141**(4), 218–225.
Schubert, J. (2002) Hydraulic aspects of riverbank filtration – field studies. J. Hydrol. **266**, 145–161.
Sontheimer, H. (1991) Trinkwasser aus dem Rhein? Bericht über ein Verbundforschungsvorhaben zur Sicherheit der Trinkwassergewinnung aus Rheinuferfiltrat. Academia-Verlag, Sankt Augustin.

69 Changes in DOC fractions in the flow regime of a riverbank filtration system

K. Wichmann, C. Schlinke, M. Marschke

DVGW-Forschungsstelle TUHH, Institute for Water Resources and Water Supply, Hamburg University of Technology, Germany, (email: Wichmann@tuhh.de)

Abstract The changes of DOC fractions (analyzed by the LC-OCD method) in the flow regime of a riverbank filtration system from the anoxic/aerob type at the river Rhine in Germany are described. In the flow path of the bank infiltrate the overall DOC concentration is reduced and the partial concentrations of DOC fractions are changed. The chromatogrammes of the natural groundwater inflowing from the land side is significantly different from the bank infiltrate, so that a mixing rate in the production wells of 75% bank infiltrate and 25% natural ground water can be estimated.

Keywords Riverbank filtration, DOC, DOC fractions, LC-OCD method

Riverbank filtration system

The described riverbank filtration system in Figure 1 is from the *anoxic/aerob type*, operated as water catchment site for a municipal water work at the river Rhine in Germany. From the production wells groundwater consisting of approx. *75% bank infiltrate* and approx. *25% natural groundwater* is pumped.

Groundwater monitoring wells with filter positions at 3 different levels situated between the river and the production well (Br. 45) and also to the side of inflowing natural groundwater allows water sampling with measuring and analysis to investigate the changes in water quality, especially concerning DOC.

DOC fractioning: LC-OCD method

DOC values give information on the summary contents of dissolved NOM and anthropogenic DOC (e.g. in the case of river water: effluent from waste water treatment plants WWTP) in water samples. To gain more information on the behaviour of different DOC fractions in the flow regime of a riverbank filtration system, the LC-OCD method (*Liquid Chromatography – Organic Carbon Detection*) (s. Figure 2) after HUBER and FRIMMEL (1996) can be used. With the use of a mathematical evaluation programme (FIFFIKUS

 Recent Progress in Slow Sand and Alternative Biofiltration Processes edited by Rolf Gimbel, Nigel J.D. Graham and M. Robin Collins. ISBN: 9781843391203. Published by IWA Publishing, London, UK.

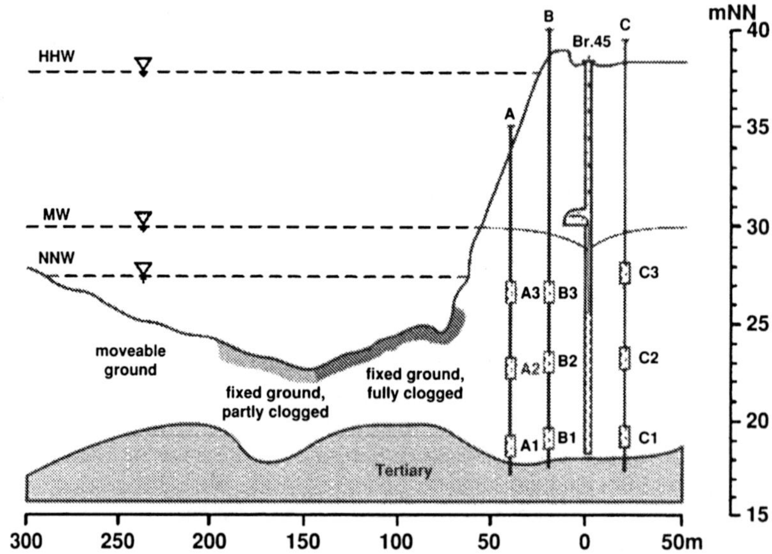

Figure 1 Cross section of the investigated riverbank filtration system with zones of riverbed classifications (changed, after SONTHEIMER 1991)

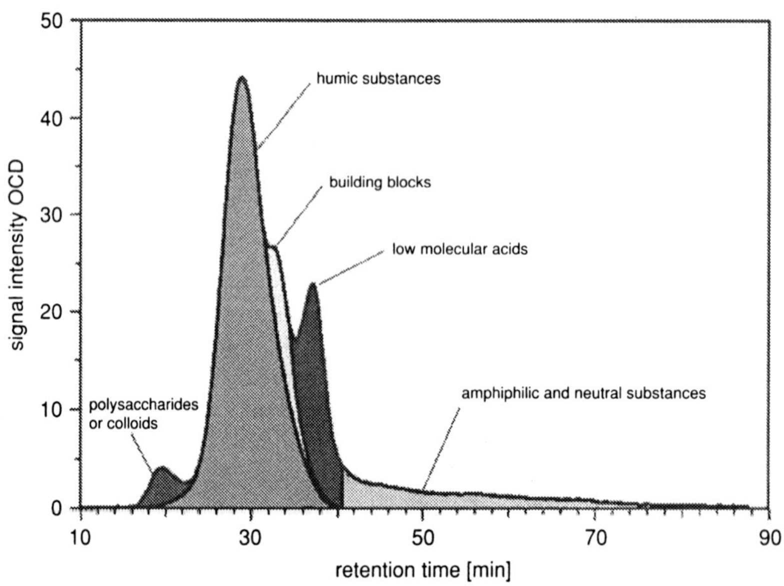

Figure 2 DOC fractions of a natural surface water sample detectable with LC-OCD

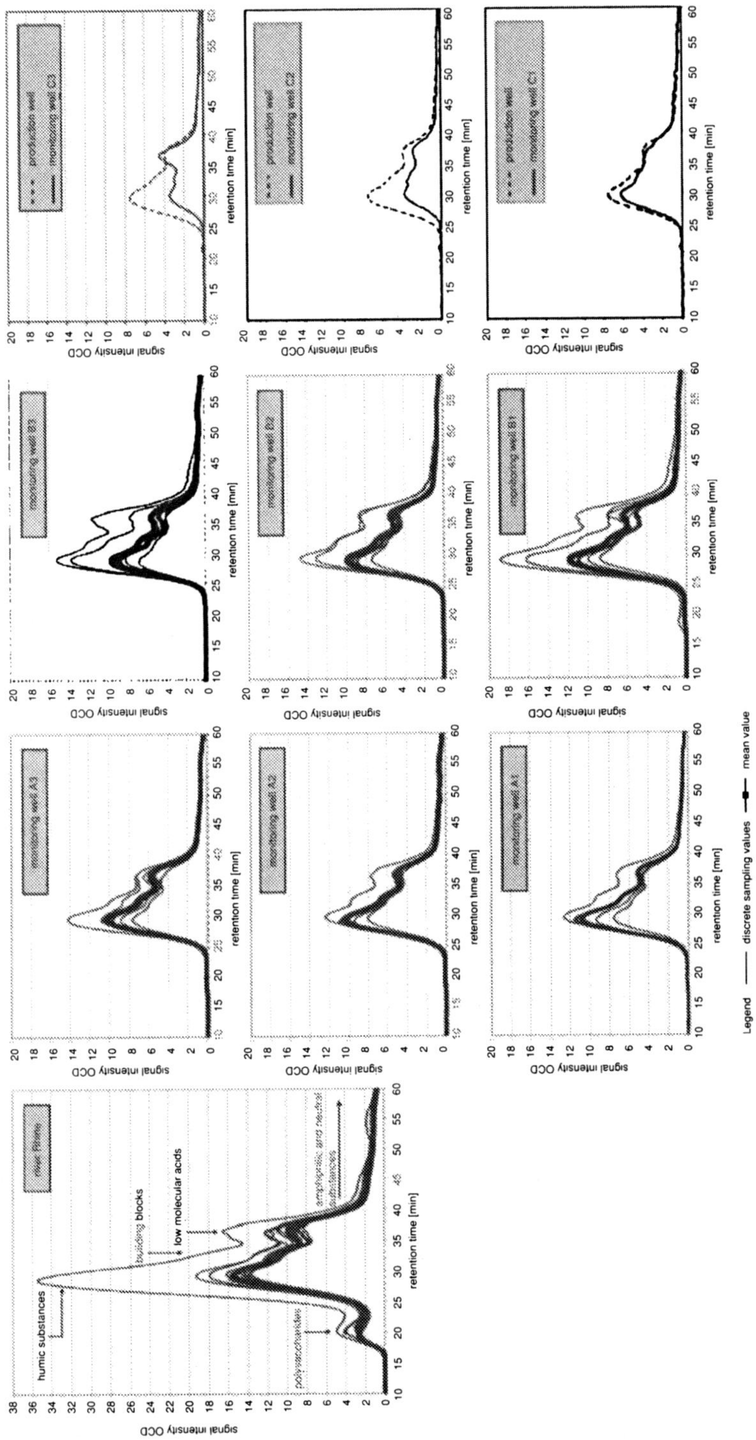

Figure 3 LC-OCD detectable DOC fractions of river Rhine water, monitoring wells A, B, C water (dependant of sampling depths) and the production well water samples→ Please notice that all chromatogrammes are given in the same scale. So the peak height of the different fractions gives a rough impression of concentration changes.

by HUBER) mass values of DOC fractions are calculated. Fractions to be found in river water are:

- polysaccharides
- humic substances
- building blocks
- low molecular acids
- amphiphilic and neutral substances.

Characterization of DOC riverbank infiltrate and natural ground water

The reduction of DOC in riverbank infiltration systems is mostly depending on microbiological degradation (SONTHEIMER 1991).

In the flow path of river water through bank infiltration and underground passage there are *significant changes in the concentrations of DOC fractions* to be found from analysis by the

LC-OCD method. The polysaccharide fraction typically present in river water (Figure 3) is eliminated with bank infiltration. The fraction of humic substances is reduced with bank infiltration and with further underground passage. The fractions of building blocks and low molecular acids are also reduced with bank infiltration and are decreasing in the following underground passage nearly at the same rate as the humic substances.

The LC-OCD signature of the natural groundwater is significantly different compared with the bank infiltrate (Figure 3). From the chromatogrammes of the different ground water and the pumped raw water from the production wells a ration of 3:1 (bank infiltrate to natural ground water) can be deduced.

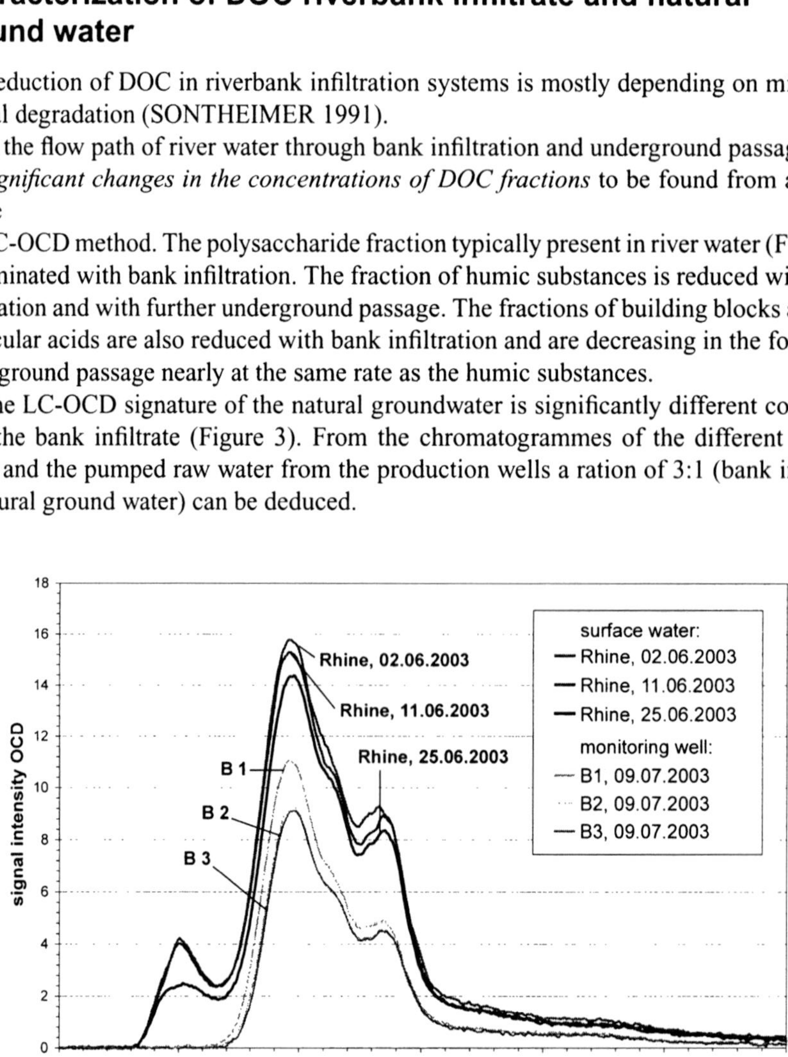

Figure 4 Time corresponding samples from river Rhine and matching monitoring wells

Seasonal influences in river water and bank infiltrate quality are to be noticed (Figure 3 and 4).

References

Huber, S.A. and Frimmel, F.H. (1996) Gelchromatographie mit Kohlenstoffdetektion (LC-OCD): Ein rasches und aussage-kräftiges Verfahren zur Charakterisierung hydrophiler organischer Wasserinhaltsstoffe. *Vom Wasser*, 86, S. 277–290

Marschke, M.; Schlinke, C. and Wichmann, K. (2005) *Einfluss von Randbedingungen auf die nutzbare Reinigungsleistung einer optimierten Uferfiltration.* Abschlußbericht zum Teilprojekt B5 im BMBF-Forschungsvorhaben Exportorientierte Forschung und Entwicklung auf dem Gebiet der Wasserver- und –entsorgung Teil I: Trinkwasser. TU Hamburg-Harburg, DVGW Forschungsstelle TUHH

Sontheimer, H. (1991) *Trinkwasser aus dem Rhein?* – Academica Verlag, St. Augustin

These investigations were supported by the German Federal Ministry of Education and Research and are part of the cooperative project Exportorientierte Forschung und Entwicklung auf dem Gebiet der Wasserver- und Entsorgung Teil I: Trinkwasser".

70 Behaviour of *Cryptosporidium* oocysts and *Giardia* cysts during artificial groundwater recharge

U. Hütter*, G. Preuß*, N. Zullei-Seibert**

*Institut für Wasserforschung GmbH, Zum Kellerbach 46, 58239 Schwerte, Germany
**Westfälische Wasser- und Umweltanalytik GmbH, Willy-Brandt-Alle 26, 45891 Gelsenkirchen, Germany

Abstract To guarantee the quality of drinking water produced during the treatment of surface water via artificial groundwater recharge, adequate removal of human pathogenic parasites has to be ensured. Information about the effectivity of this drinking water treatment process was obtained by monitoring a number of water treatment plants on the banks of the upper and middle Ruhr over a period of several years. The elimination of the parasites from surface water by different filtration stages during artificial groundwater recharge was investigated. Additional information about the parasite separation efficiency of different filter sands was obtained by investigations at semi-technical test plants. The combination of sedimentation, respectively gravel pre-filtration, slow sand filtration and underground passage showed parasite elimination efficiencies comparable to other treatment processes (e.g. membrane filtration) under real operating conditions.

Keywords *Clostridium perfringens*, *Cryptosporidium*, *Giardia*, groundwater recharge, parasites, slow sand filtration, surface water

Introduction

Cryptosporidium parvum (diameter 4 to 6 μm) and *Giardia lamblia* (diameter 11 to 16 μm) are recognized as waterborne human pathogens which can cause severe diarrheal illness. These waterborne parasitic protozoa exist in either the free-swimming trophozoite form or the dormant oocysts form (*Cryptosporidium*) respectively cysts form (*Giardia*). The dormant forms - being an environmentally stable lifeform - can survive for several months in the aquatic environment and are resistant to the drinking water disinfection with chlorine.

Cryptosporidium and *Giardia* are characterized by wide-ranging host organisms (human, farm and wild animals), and are released with the faeces of infected animals and manure (slurry). Via sewage and surface run-off from pastures, stables and manure-treated agricultural areas the parasites are spread into the environment and can contaminate surface water and groundwater (Gornik and Exner, 1997; Moulton-Hancock *et al.*, 2000). The occurrence of *Giardia* cysts is mainly connencted with the input of sewage plants whereas cattle

 Recent Progress in Slow Sand and Alternative Biofiltration Processes edited by Rolf Gimbel, Nigel J.D. Graham and M. Robin Collins. ISBN: 9781843391203. Published by IWA Publishing, London, UK.

ranching and agricultural influence are considered as the main source for *Cryptosporidium* oocysts (Bischoff and Feuerpfeil, 2001).

To guarantee the quality of drinking water produced during the treatment of surface water via artificial groundwater recharge, adequate removal of human pathogenic parasites has to be ensured. In several studies the elimination of the parasites from surface water by different filtration stages during artificial groundwater recharge was investigated. Information about the effectivity of this drinking water treatment process was obtained by monitoring a number of water treatment plants on the banks of the upper and middle Ruhr over a period of several years. Supplementary investigations were carried out at a semi-technical test plant simulating gravel pre-filtration and slow sand filtration. This study examined the influence of different filter sands and filter operating stages on the parasites elimination (Hütter and Preuß, 2001).

Materials and methods

Water treatment plants

Investigations at the water treatment plants were carried out from 1998 to 2004. Spot samples were taken during characteristic seasonal situations of the river Ruhr (minimum flow, mean flow and peak flow) to assess their impact on the water catchment areas. Sampling points were the river Ruhr at different water works, the effluent from sedimentation basins and gravel pre-filters, the recharged groundwater after one-stage slow sand filtration or a second infiltration and subsequent underground passage, and at last the purified water and the drinking water.

Pilot plant

The method of artificial groundwater recharge via gravel pre-filtration and slow sand filtration without subsequent underground passage was reproduced at a semi-technical pilot plant (figure 2.1). The gravel pre-filter received the surface water (river Ruhr) with its natural inconstant oocysts concentrations; the produced pre-filtrate was used as influent to the sand filters. The double stage filtration at the pilot plant simulated realistic field conditions. Only the depth of the sand filter was low in order to investigate optimization possibilities. Three filter sands with different effective particle sizes (d_{10}) and particle-size distributions were investigated concerning their elimination ability for *Cryptosporidium* oocysts (table 2.1). During three operating stages with defined conditions close to practice (start-up stage, supernatant stage, restart stage) samples of surface water, pre-filtrate and filtrate (5 consecutive samples) were taken. After the start-up of the sand filters they were intermittently operated (2-day-rhythm). The supernatant phase was investigated after a filter running time of 3 to 4 weeks, and the restart phase a further 3 weeks later.

Analyses

Microbiological parameters (e.g. *Cryptosporidium parvum*, *Clostridium perfringens*, *Giardia lamblia*, total coliforms) and relevant hydrochemical parameters (e.g. pH factor, turbidity) were measured during the different experiments. Sampling procedures and

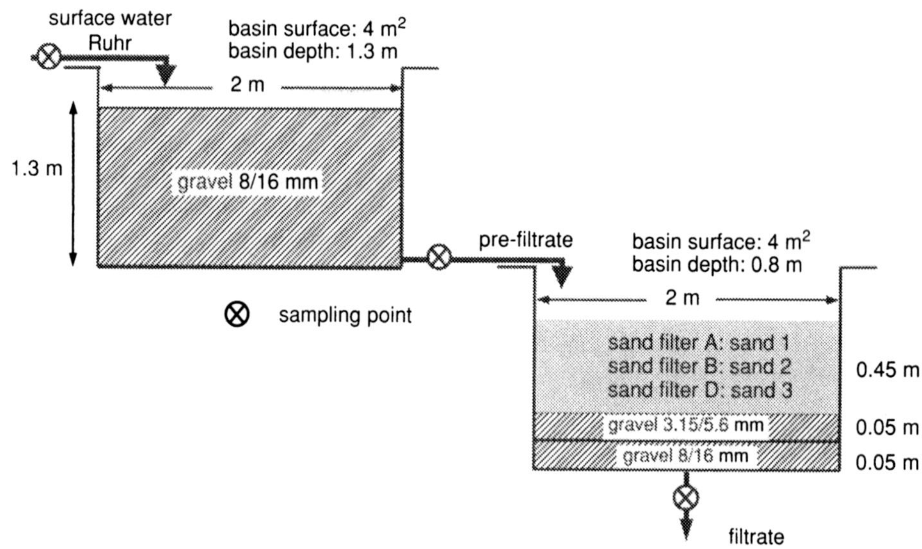

Figure 2.1 Scheme of the double stage filtration (cross section)

determination of microbiological and chemical parameters were accomplished according to German standard methods (Drinking Water Ordinance, 2001). In addition *E. coli* and total coliforms were detected on ENDO-Agar (DIFCO) after membrane filtration and incubation at 36°C (± 1°C) for 20 hours. The protozoan parasites *Cryptosporidium* and *Giardia* were determined according to the USEPA method 1623 (US-EPA, 1999) by using filter capsules (Envirochek®), immunomagnatic separation technique (IMS), fluorescent monoclonal antibodies (FITC) and DAPI staining. Caused by cross reactions this method detects the human pathogenic species as well as the not human pathogenic *Giardia* and *Cryptosporidium* species. In addition it is not possible to assess the vitality or infectiveness of the detected dormant forms. The recovery of this method of detection was from 70 to 95% depending on the type of water and the species. The presented results do not include the recovery rate.

Results and discussion

Microbial characterization of the river Ruhr and results of the treatment plants

The chemical and microbiological conditions in the river Ruhr depend on regional and seasonal effects like e.g. rainfall and effluents from sewage plants or agricultural areas.

Table 2.1 Characteristics of the applied filter sands

	Sand A	Sand B	Sand D
particle size at 10% sieve pass d_{10} [mm]	0,28	0,12	0,15
particle size at 60% sieve pass d_{60} [mm]	0,68	0,42	0,40
unconformity coefficient (d_{60}/d_{10}) [-]	2,43	3,50	2,67

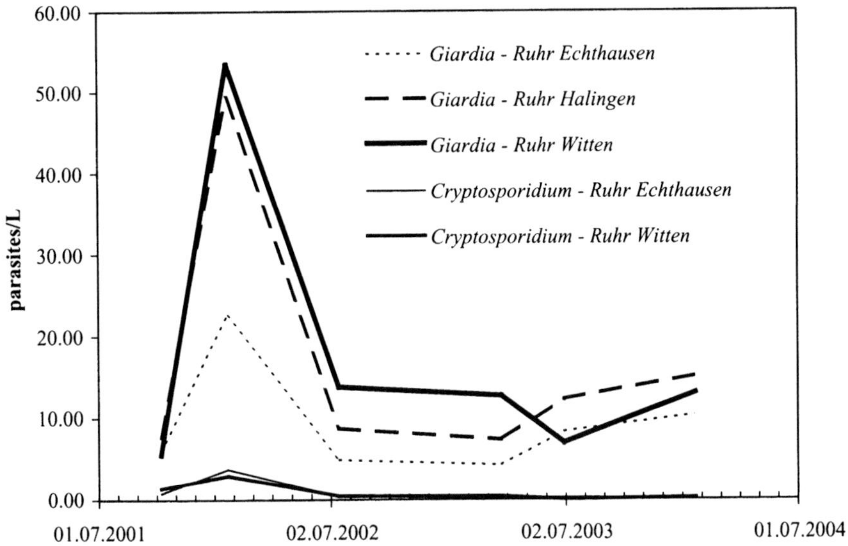

Figure 3.1 Concentrations of parasites in the river Ruhr

Investigations in the past years achieved during different flows of water in the river Ruhr, showed a permanent but varying presence of protozoan parasites (*Cryptosporidium* and *Giardia*) (figure 3.1) (Hütter *et al.*, 2005). Positive results for *Cryptosporidium* were detected between 0.04 and 3.79 oocysts per litre. There was no great difference in the *Cryptosporidium* concentration between the sampling point Echthausen upstream and the sampling point Witten downstream. The four sampling points between Echthausen and Witten also showed comparable concentration levels. The maximum oocysts concentrations were measured during a high water period in January 2002. The detected low levels of Cryptosporidium oocysts coincide with the comparatively low agricultural cattle ranching in the catchment area of that part of the river Ruhr.

Giardia cysts were measured in higher concentrations compared to *Cryptosporidium* oocysts (4.30 to a maximum of 53.47 cysts per litre, figure 3.1). Between Echthausen and the downstream sampling point Halingen (distance only 14.3 km) a significant concentration difference was detected. This indicates in accordance with other publications on the influence of the existing sewage plant. Further downstream the impact of the additional sewage plant effluent is suppressed by the influx of running water bodies and the resulting higher flow of water in the river Ruhr. Comparable to the *Cryptosporidium* results the maximum concentration was found during the high water period in January 2002.

Most investigations were carried out during a normal flow of water in the river Ruhr. With an increased flow of water (40 to 170 m^3/s) the positive results showed a greater variety. The high water in January 2002 corresponded with the highest concentration in each case (figure 3.1). With the available data no significant correlation between flow of water and parasite concentrations can be drawn.

The chemical and biological composition of the river Ruhr depends strongly on the local and seasonal conditions. Sporadically measured numbers of parasites and bacterial

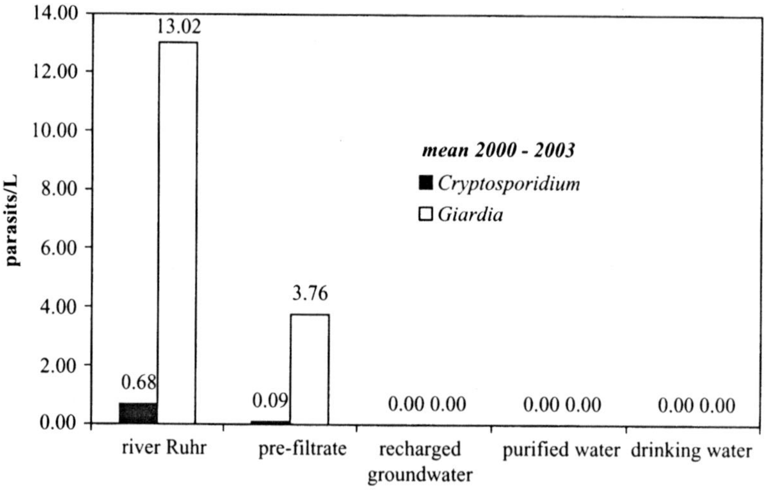

Figure 3.2 Mean reduction of parasites during artificial groundwater recharge on the banks of the middle Ruhr

faecal indicators are not representative for the complete running water body. The input of substances and hygienic relevant microorganisms into the river Ruhr depends on the temperature and rainfall situation which has an impact on the supply of sewage plants and agricultural influenced tributaries.

A number of water treatment plants at the upper and middle Ruhr including several treatment steps (e.g. sedimentation or gravel pre-filtration, slow sand filtration, further infiltration steps and underground passage) were monitored over a period of several years. Samples were taken at different filtration stages of the treatment process. In all about 160 samples were analysed for *Cryptosporidium* and about 180 for *Giardia* in this time period. The measured mean concentration reduction of parasites during the treatment stages of the artificial groundwater recharge is presented in figure 3.2. The evaluation shows that parasites were found in the pre-filtrate. But the following purification processes during slow sand filtration and subsequent underground passage led under normal operating conditions always to an elimination of *Cryptosporidium* and *Giardia* below the lower limit of detection (figure 3.2). These results which were achieved under real operating conditions prove that the investigated water treatment processes (primary treatment, slow sand filtration and underground passage) lead to an elimination of the parasites below the detection limit. Because of the relatively low input concentrations an elimination of 2 log-stages is sufficient to reach this aim. This high performance of the treatment processes is also confirmed by the long-term bacteriological quality monitoring of the treatment stages in the water works.

Pilot plant

During the investigations at the semi-technical test plant oocysts concentrations from 0.29 to 1.28 per L (mean value 0.70 per L) were detected in the surface water (river Ruhr). The measured concentrations in the **pre-filtrate** (= sand filter influent) laid between 0.03 and 0.28 oocysts per L (mean value 0.14 oocysts per L). These results confirm in accordance with the results achieved at the water works an obvious removal of oocysts in the gravel pre-filter (particle size 8–16 mm). The concentrations of *Cryptosporidium* oocysts, *C. perfringens* and total coliforms in the sampling points of sand filter A, B and D during the different operating stages are shown in the figures 3.3, 3.4 and 3.5.

In the **start-up stage** (filtration rate 3.2 m/d) only in one filtrate of sand filter A a small amount of oocysts could be detected. The high amount of *C. perfringens* in the filtrates of sand filter A exceeds in all samples the concentration in the influent. This effect is probably caused by the leaching of organisms which were brought in during the sand transport and storage. The samplings in the **supernatant stage** (filtration rate 2.4 m/d) were carried out after a filter running time of 12 to 15 days. In 10 of 15 filtrates oocysts concentrations from 0.01 to 0.03 per L were detected. Sand filter D ($d_{10} = 0.15$ mm) showed the fewest positive results. After a filter running time of 22 to 25 days samplings were carried out in a **restart stage** (filtration rate 3.3 m/d). Positive results were observed in 7 of 15 filtrates. Whereas in sand filter B ($d_{10} = 0.12$ mm) no oocysts were detected in the filtrate samples, sand filter A ($d_{10} = 0.28$ mm) showed the highest concentrations. In filtrate 1 0.19 oocysts per L were determined. This concentration decreased rapidly in the following filtrate samples. Sand filter D ($d_{10} = 0.15$ mm) showed the same effect. These results prove that in restart stages with higher filtration rates oocysts concentrations especially in the first filtrates can increase. *C. perfringens* and coliforms also showed higher levels in the first filtrates. The relationship between the numbers of oocysts and *C. perfringens* was not statistically significant.

The interpretation of these results has to consider the small filter depth chosen fort these pilot experiments. Furthermore pilot scale investigations can not be directly transferred to full scale water treatment plants. Table 3.1 shows the *Cryptosporidium* oocysts reduction

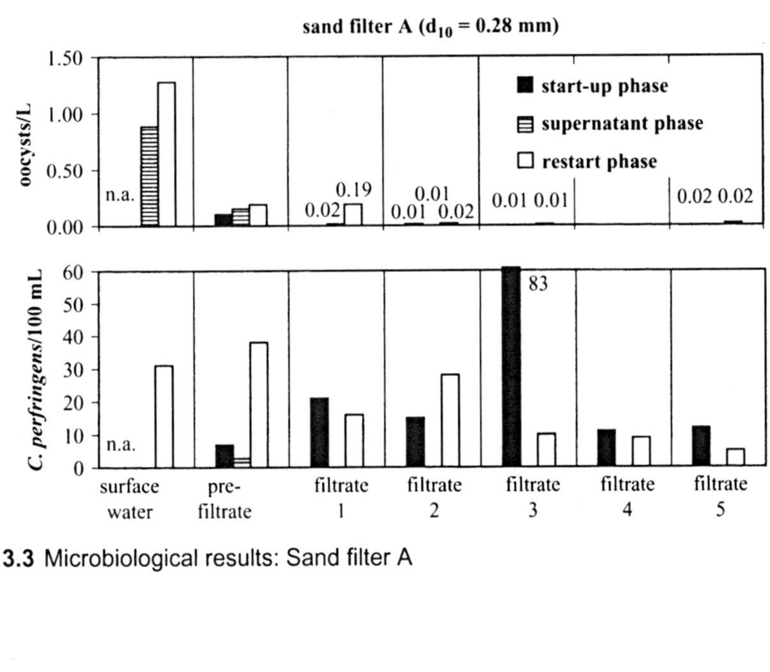

Figure 3.3 Microbiological results: Sand filter A

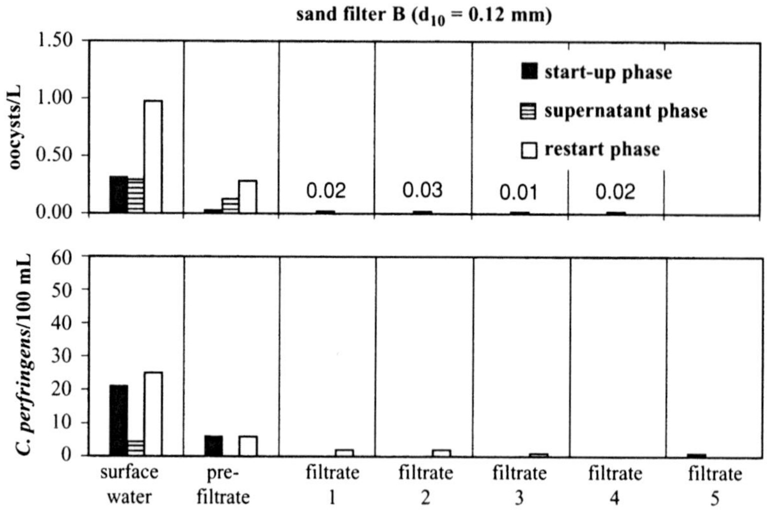

Figure 3.4 Microbiological results: Sand filter B

efficiency of the different filter materials investigated in this study. Also listed in the table are the influents concentrations in the pre-filtrate respectively filtrate. The calculated separation efficiencies confirm already a removal of oocysts in the gravel filter between 62 to 87%.

During the second filtration step the three tested filter sands were not able to retain the *Cryptosporidium* oocysts completely in all operating stages. The filter sands with a smaller effective particle size (d_{10} – value) achieved better reduction efficiency but this

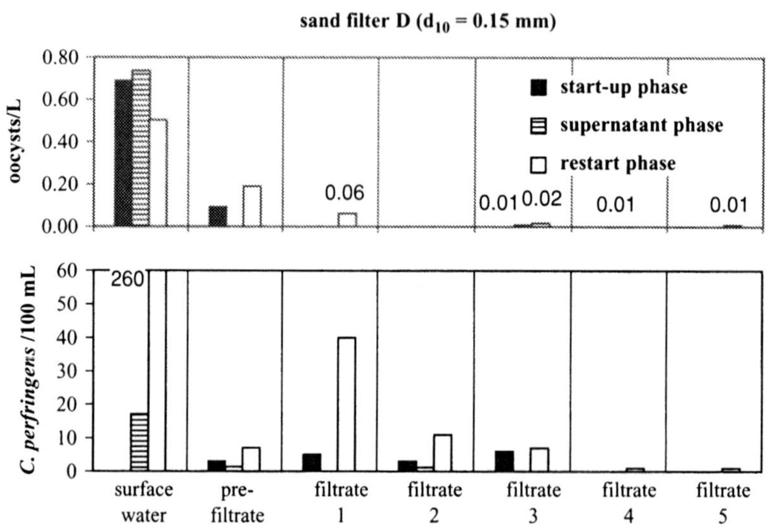

Figure 3.5 Microbiological results: Sand filter D

Table 3.1 *Cryptosporidium* oocysts reduction efficiency (%) of gravel and sand filters in various stages of operation

	gravel filter	**sand filter A**	**gravel filter**	**sand filter B**	**gravel filter**	**sand filter D**
start-up phase	n. a.	97% (0.10**)	n. a.	100% (0,03**)	87% (0.69*)	100% (0.09**)
supernatant phase	82% (0.83*)	93% (0.15**)	74% (0.29*)	83% (0.08**)	n. a.	n. a.
restart phase	86% (1.28*)	80% (0.18**)	71% (0.98*)	100% (0.28**)	62% (0.50*)	93% (0.19**)

* input concentration *Cryptosporidium* oocysts per L surface water,
** input concentration *Cryptosporidium* oocysts per L pre-filtrate
n. a. = not available

is not an absolute safety against a pathogen breakthrough. Increased filtration rates e.g. in restart stages could cause a remobilisation of retained oocysts which leads to increased concentrations in the initial filtrates. Sand filter B is the only filter which achieved an oocysts removal of 100% in the restart stage with a higher filtration rate. The study of Harter *et al.* (2000) indicated that a significant part of the initial oocysts separation is reversible probably because of a change of surface charge.

Conclusions

The river Ruhr is characterized through varying however low concentrations of *Cryptosporidium* oocysts and *Giardia* cysts. The presence of *Clostridium perfringens* was no indicator for the presence of oocysts and vice versa. For an assessment of seasonal effects or load calculations more frequent investigations over a longer period are needed. Up to now this is limited by the time and cost expensive sampling and analysis of parasites.

The results achieved at the water works and the pilot scale plant proved that gravel pre-filtration already leads to a noticeable elimination of parasites. Slow sand filtration using filter sands with a small effective particle size is more suitable for avoiding an oocysts break through. A subsequent underground passage can be estimated as a further purification process.

In all, the permanent further developed and locally adapted multi-barrier-system of the water works at the river Ruhr can be considered as an effective concept for the elimination of parasites. The combination of sedimentation, respectively gravel pre-filtration, slow sand filtration and underground passage showed parasite elimination efficiencies comparable to other treatment processes (e.g. membrane filtration) under real operating conditions. Suitable operating conditions of water works (e.g. multi-stage filtration stages with good filter capacity, sufficient filter depth, steady filter operating with low filtration rates) in combination with a comprehensive protection of the surface water against the entry of pathogens belong to the concept of a multi-barrier-system and guarantee the quality of drinking water.

Acknowledgements

The authors would like to gratefully acknowledge financial support of Dortmunder Energie- und Wasserversorgung GmbH (DEW21) and Wasserwerke Westfalen GmbH (WWW).

References

Bischoff, K. and Feuerpfeil, I. (2001) Belastungen von Trinkwassertalsperren und ihrem Einzugsgebiet mit Parasiten (*Cryptosporidium*-Oocysten und *Giardia*-Cysten) und ausgewählten potentiell pathogenen Bakterien als hygienisches Risiko bei der Trinkwasseraufbereitung. *Umweltmedizinischer Informationsdienst*, 1, 3–11.

Drinking water ordinance (2001) *Bundesgesundheitsblatt* Teil I, Nr. 24, 959–980.

Gornik, V. and Exner, M. (1997) Verhalten von Mikroorganismen und Viren bei der Trinkwasseraufbereitung. *Cryptosporidium sp.. DVGW-Schriftenreihe Wasser*, 91, 173–203.

Harter, T., Wagner, S. and Atwill, E.R. (2000) Colloid Transport and Filtration of *Cryptosporidium parvum* in sandy soils and Aquifer sediments. *Environmental Science and Technology* **34**(1), 62–70.

Hütter, U. and Preuß, G. (2001) Exemplarische Untersuchungen zur Rückhaltung von Cryptosporidien-Dauerstadien in verschiedenen Filtermaterialien. *Ruhrgütebericht*, 131–138.

Hütter, U., Preuß, G., Zullei-Seibert, N. and Schöttler, U. (2005) Entfernung von Parasiten durch künstliche Grundwasseranreicherung-Erfahrungen aus Forschung und Praxis. *gwf-Wasser/ Abwasser*, **146**(9), 665–672.

Moulton-Hancock, C., Rose, J. B., Vasconcelos, G. J., Harris, S. I., Klonicki, P. T. and Sturbaum, G. D. (2000) *Giardia* and *Cryptosporidium* occurrence in groundwater. *JAWWA* **92**(9), 117–123.

United States Environmental Protection Agency (US-EPA), Method 1623 (1999) *Cryptosporidium* and *Giardia* in Water by Filtration / IMS / FA. Office of Water. EPA 821-R-99-006, Washington, URL: http://www.epa.gov/nerlcwww/1623.pdf.

71 Assessing the impact of local boundary conditions on the fate of organic micropollutants during underground passage

C.K. Schmidt*, F.T. Lange*, H.-J. Brauch*

*DVGW-Technologiezentrum Wasser (TZW), Karlsruher Str. 84, 76139 Karlsruhe, Germany

Abstract In Germany and other highly populated countries, several waterworks use riverbank filtration and artificial groundwater recharge in the treatment of river water for water supplies. Passage of water underground provides several benefits for drinking water treatment. Experience demonstrates that during infiltration and underground transport, processes such as filtration, sorption, and biodegradation cause significant improvements in raw water quality. Due to industrial municipal discharges and the influence of agriculture, rivers and lakes are often polluted by a number of organic chemicals. In order to assess the impact of these organic micropollutants on the quality of drinking water it is necessary to clarify their fate during infiltration and underground passage. This contribution will report on the impact of different boundary conditions like redox milieu and retention times on the purification capacity of bank filtration and artificial groundwater recharge with regard to different classes of organic micropollutants.

Keywords Riverbank filtration, underground transport, micropollutants, complexing agents, pharmaceuticals, X-ray contrast media, aromatic sulfonates, MTBE, organohalogen compounds, pesticides, redox milieu, retention time.

Introduction

The draw-down of the groundwater table during extraction of groundwater from aquifers in river valleys via wells with the creation of a pressure head difference between river and well inevitably causes an infiltration of river water into the aquifer. This process is called riverbank filtration. Along the rivers Rhine and Ruhr, riverbank filtration was introduced already in the mid 19th century in the course of industrialisation and claim for an increased drinking water supply that was associated with the beginning population growth in these regions. Furthermore, massive hygienic problems required a renunciation of the direct utilisation of surface water from rivers, ponds and open ditches (Kühn et al. 2006, Schmidt et al. 2003).

Nowadays, waterworks with riverbank filtration can be found in Germany mainly at the lower Rhine in the region between the tributaries Sieg and Ruhr, at the Elbe between

Dresden and Torgau, and in the Berlin area. Outside of Germany, wells extracting riverbank filtrate are mainly operated along Rhine and Meuse in the Netherlands and along the Danube starting at Vienna/Austria and further downstream in the Slovakian Republic, in Hungary, and in Serbia/Montenegro. Recently, riverbank filtration is also utilised by some waterworks in non-European countries such as Brazil, USA, India, and South Korea (Kühn et al. 2006).

Surface waters are environmentally sensitive and are often contaminated by anthropogenic organic micropollutants, in particular in industrialised and urbane areas. The great variety of physico-chemical properties of chemical substances accordingly provokes many substance-specific patterns of behaviour during riverbank filtration. In general, the fate of organic micropollutants in a river water – groundwater infiltration system is mainly determined by hydrological boundary conditions and by adsorption mechanisms and biological transformations in the infiltration zone and the aquifer. The purification process of the underground passage starts in the infiltration zone that can be characterised as a gelatinous, biological highly active biolayer which consists of algae, bacteria, fungi and protozoa as well as of organic and inorganic particles. During infiltration, the river water with its dissolved components meets multifaceted biogenous and abiogenous surface structures that aid in intensifying most of the self-cleaning mechanisms that are in principle also present in the free surface water (Schmidt et al. 2003).

The formation of the hydrochemical milieu during riverbank filtration is closely linked to redox reactions catalysed by microorganisms. In these redox reactions, available organic substances represent electron donators, while reducible substances in water (O_2, NO_3^-, SO_4^{2-}, CO_2) and solid phases (Fe(III)-, Mn(IV)-oxides and -hydroxides) act as electron acceptors. In the underground, microorganisms compete for available resources, utilising out of the pool of potential electron acceptors always those that provide a maximum of energy in the redox process. The energy gain is highest for aerobic respiration, so that during microbial degradation of organic substances in water dissolved oxygen is consumed first. Varying microorganisms, that are adapted to the particular milieu conditions, can cause, by successive depletion of alternative, less efficient electron acceptors (NO_3^-, Mn(IV), Fe(III), SO_4^{2-}), continued degradation of organic compounds in a sequence of reduction steps. At the end of this sequence stands the reduction of carbon dioxide with the corresponding formation of methane. The spatial expansion of this redox zone succession along the infiltration pathway is not well-defined and can, depending on aquifer thickness, organic carbon content of the substrate and the infiltrating water, and hydrochemical, biological and hydraulic boundary conditions, vary between a few centimetres and several hundred metres. Since many microorganisms can perform their microbial activity only at certain redox potentials and since the extent of biodegradation is dependent on sufficient residence times and sometimes proceeds retarded, the degradation of organic micropollutants is linked to both the presence of favourable redox conditions and sufficient residence times in these zones. As a consequence, elimination rates of intrinsically degradable substances may vary considerably depending on local geological and hydrochemical conditions as well as on organic loads of surface waters and infiltration zones. Redox conditions can vary from aquifer to aquifer. At the Rhine, for example, aerobic riverbank filtration sites dominate, whereas aquifers in the Berlin area and at the Elbe are typically characterised by anoxic redox conditions. At an individual riverbank filtration site, however, also temporal changes of surface water quality or temperature may cause redox shifts in the corresponding bank filtrate (Schmidt 1988, Preuß and Schulte-Ebbert 2000, Schmidt et al. 2005).

Methods

So far, data from systematic studies investigating the impact of residence times and redox conditions on the fate of organic micropollutants commonly present in surface waters during riverbank filtration is not available. In the present study extensive measurement campaigns were carried out at four well-characterised aquifers along the rivers Rhine, Ruhr, and Elbe.

Figure 1 shows the particular cross section of each riverbank filtration site studied and table 1 summarises their characteristics. The different characteristics of these field sites allowed for detailed investigations concerning the elimination capacity of riverbank filtration at various redox conditions (aerobic, aerobic-denitrifying (suboxic), denitrifying

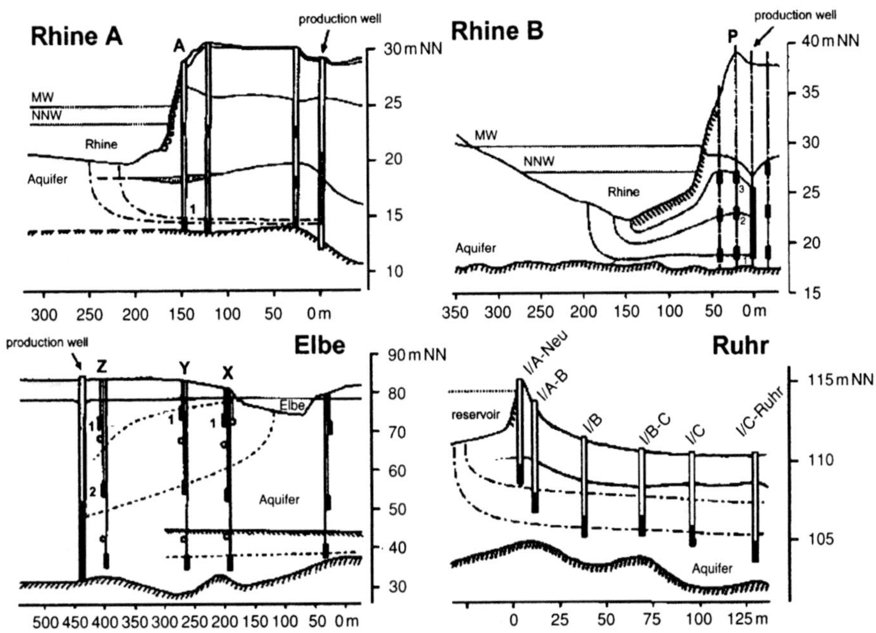

Figure 1 Cross-section profiles of the investigated aquifers.

Table 1 Characteristics of the investigated riverbank filtration sites.

	Rhine A	Rhine B	Elbe	Ruhr
aquifer thickness	12–15m	10–12 m	40–55 m	5 m
material	gravel/sand	gravel/sand	gravel/sand	gravel/sand
residence time	7–20 days	12–60 days	45–300 days	5–15 days
K_f-value	$12 \cdot 10^{-3}$ m/s (deep, A1) $2 \cdot 10^{-3}$ m/s (high)	$3.0 \cdot 10^{-3}$ m/s to $6{,}0 \cdot 10^{-3}$ m/s	$0.6 \cdot 10^{-3}$ m/s (depth 2) $2{,}0 \cdot 10^{-3}$ m/s (depth 1)	10^{-1} m/s to 10^{-3} m/s
redox conditions	aerobic	aerobic to denitrifying	denitrifying	strictly anaerobic

(anoxic), strictly anaerobic) and residence times of the infiltrated water in the underground (5–300 days). As can be seen from figure 1, each field site provided several sampling wells located between river and production wells. These sampling wells are in contrast to the production wells not influenced by natural land-sided groundwater. Typically, samples from the surface water and from several sampling wells along the aquifer were investigated at each field site. At field site Rhine A, however, only one sampling well was sampled. Sampling frequencies varied between 5 and 33 dates depending on field site and parameter. Analysed target compounds comprised several contaminants relevant for the aquatic environment, such as complexing agents, aromatic sulfonates, pharmaceuticals, X-ray contrast media, and MTBE. Furthermore, an intensified literature survey was conducted concerning the fate of selected pesticides and organohalogen compounds during riverbank filtration and artificial groundwater recharge with focus on the impact of varying redox conditions (aerobic or anoxic/anaerobic).

Results and discussion

Field studies

Figure 2 shows a part of the findings for field site Rhine B. Given are the results of all compounds of the pharmaceutical and iodinated X-ray contrast media group that were permanently detectable in the surface water. During the time of sampling in hot summer 2003 redox conditions were at the edge of aerobic to denitrifying (suboxic).

The riverbank filtrate at this site was investigated at sampling point P (see figure 1) that provides three independent sampling wells with three different filter depths. Travel times of the water between the point of infiltration at the river bottom and the passage of the sampling

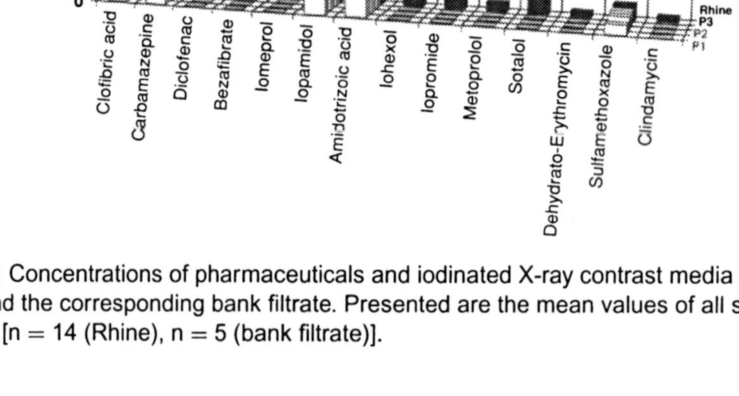

Figure 2 Concentrations of pharmaceuticals and iodinated X-ray contrast media in the Rhine and the corresponding bank filtrate. Presented are the mean values of all spot samples [n = 14 (Rhine), n = 5 (bank filtrate)].

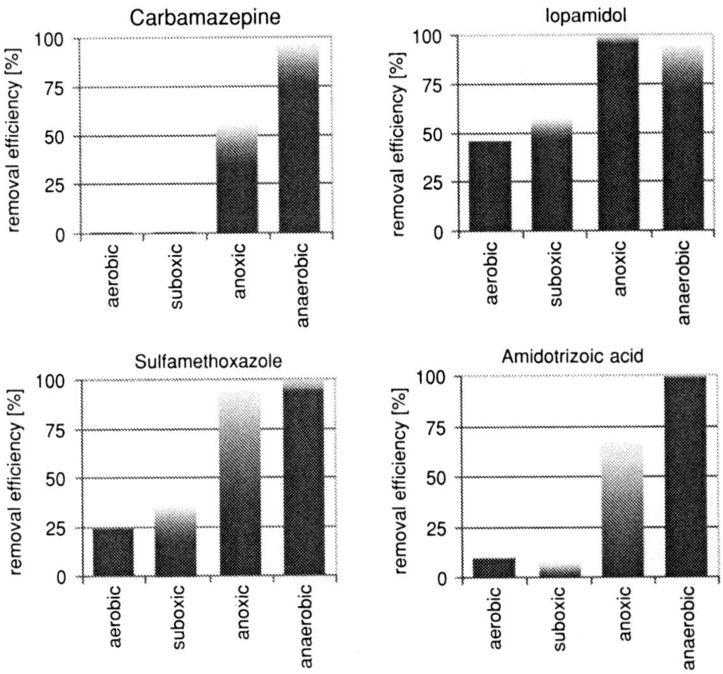

Figure 3 Removal efficiency of riverbank filtration for carbamazepine, sulfamethoxazole, iopamidol and amidotrizoic acid in aquifers with different redox milieu. The black part of the bars represents the extent of elimination that was found for all sampling wells, grey parts of the bars characterise the part of the elimination that was variable between the investigated sampling wells.

well are approximately 60 days for well P1 (lower layer), 30 days for well P2 (middle layer) and 12 days for well P3 (upper layer). The results demonstrate that most compounds of this group are obviously well eliminated during riverbank filtration. Only carbamazepine, sulfamethoxazole, iopamidol and amidotrizoic acid show a rather persistent behaviour. This was also confirmed by the findings from the aerobic field site Rhine A. At anoxic (Elbe) or strictly anaerobic (Ruhr) milieu conditions, however, riverbank filtration turned out to be quite efficient for the removal of carbamazepine, sulfamethoxazole, iopamidol, and amidotrizoic acid (figure 3).

The essential findings of all field investigations are summarised in table 2. This compilation shows that many organic micropollutants detectable in surface waters can be eliminated via riverbank filtration. However, chemically relatively similar compounds sometimes show quite different elimination rates (e.g. NTA and EDTA). One of the most important findings is certainly the substantial impact of the redox milieu on the removal efficiency. On closer examination of different substance classes this influence turned out to be highly non-uniform. Aminopolycarboxylate chelating agents and naphthalene sulfonates, for instance, seem to be better removable at aerobic redox conditions, whereas pharmaceuticals and iodinated X-ray contrast media are preferentially eliminated in the anoxic and anaerobic milieu.

Table 2 Efficiency of riverbank filtration with regard to the elimination of selected organic micropollutants [o (0–25%), + (26–50%), ++ (51–70%), +++ (>70%), ++++ (>80%)].

	aerobic (Rhine A)	**suboxic (Rhine B)**	**anoxic (Elbe)**	**anaerobic (Ruhr)**
Nitrilotriacetic acid (NTA)	++++	++++	++++	++++
Diethylenetriaminepentaacetic acid (DTPA)	+++	+++	+++	++
Ethylenediaminetetraacetic acid (EDTA)	++	++	o	o
Naphthalene-1-sulfonate	+++	+++	++++	+++
Naphthalene-2-sulfonate	++++	++++	++++	++++
Naphthalene-1,5-disulfonate	o	+	o	+
Naphthalene-1,6-disulfonate	++++	++++	+++	o
Naphthalene-1,7-disulfonate	++++	++++	++++	o
Naphthalene-2,6-disulfonate	++++	++++	+++	++
Naphthalene-2,7-disulfonate	++++	++++	++++	+
Naphthalene-1,3,5-trisulfonate	o	+	o	o
Naphthalene-1,3,6-trisulfonate	+	++	+	o
Naphthalene-1,3,7-trisulfonate			++	o
8,8′-Methylenebis-2-naphthalenesulfonate	+	o	+	o
2-Amino-1,5-naphthalenedisulfonate	o	+	o	o
2-Amino-4,8-naphthalenedisulfonate	+	++	++	o
4,4′-Diamino-1,1′-bianthraquinone-3,3′-disulfonate	+++	++++		
2-Hydroxy-4,6-bis(4-sulfanilo)-1,3,5-triazine	o			
cis-4,4′-Dinitrostilbene-2,2′-disulfonate	o			
MTBE	+		+	
Bezafibrate	++++	++++	++++	++++
Carbamazepine	o	o	+	++++
Clofibric acid		+++		++++
Fenofibric acid	+++			
Diclofenac	++++	++++	++++	++++
Ibuprofen	++++		++++	++++
Indomethacin	+++		+++	
Naproxen	++++		+++	+++
Pentoxifylline	++++		++++	
Clarithromycin	+++		+++	+++
Dehydrato-Erythromycin A	++++	+++	+++	+++
Roxithromycin				+++
Sulfamethoxazole	o	o	+++	++++
Clindamycin	++++	+++	+++	+++
Trimethoprim	++++		++++	++++
Atenolol	++++		++++	++++
Metoprolol	++++	++++	++++	++++
Bisoprolol	+++			+++
Sotalol	++++	++++	++++	++++
Iopamidol	+	++	+++	++++
Iopromide	++++	++++	++++	++++
Iomeprol	++++	++++	++++	++++
Ioxhexol	++++	++++	++++	++++
Amidotrizoic acid	o	o	+	++++
Ioxitalamic acid	++++	++++	++++	++++
Iotalamic acid	+			

Literature survey

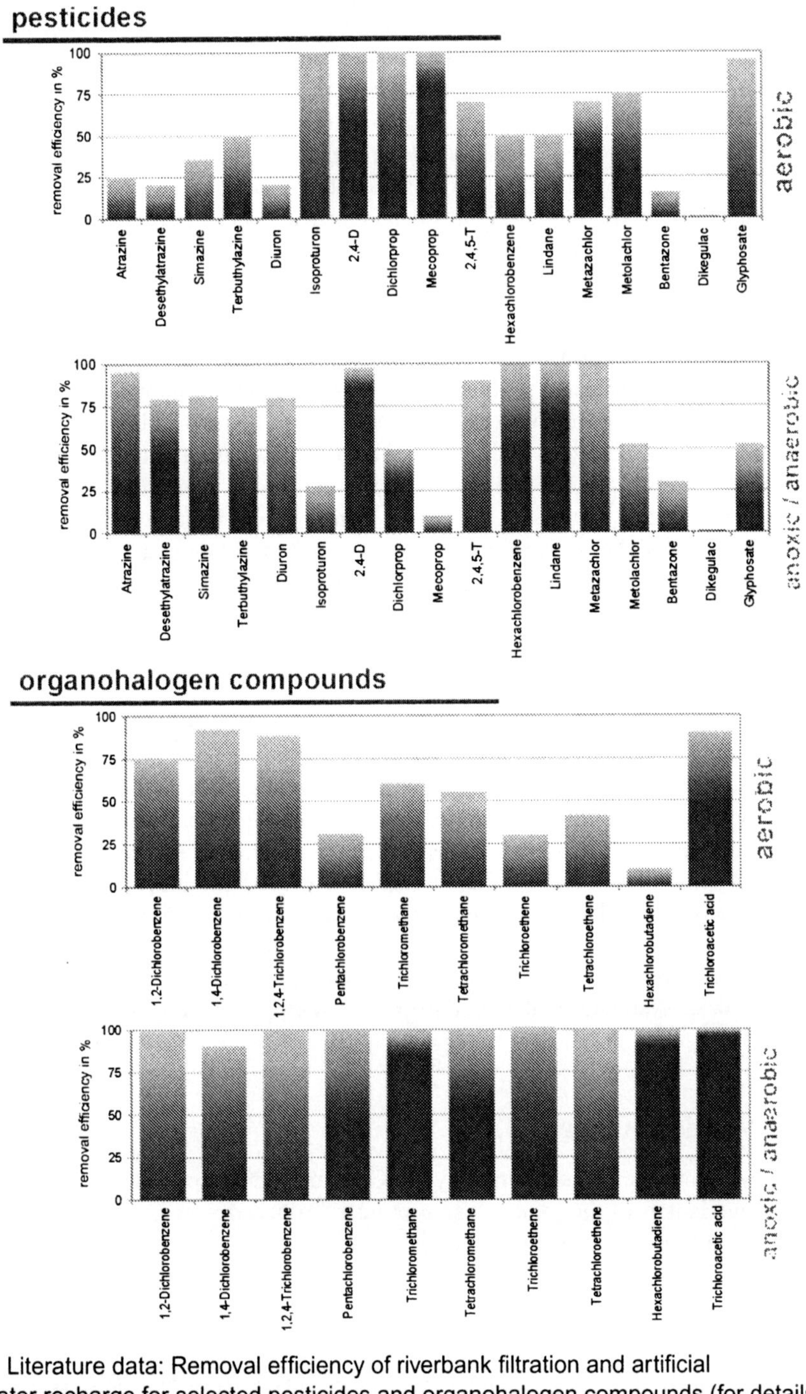

Figure 4 Literature data: Removal efficiency of riverbank filtration and artificial groundwater recharge for selected pesticides and organohalogen compounds (for details see Schmidt 2006). The black part of the bars represents the extent of elimination that was found in all investigations, grey parts cover the range of additional elimination observed only in some of the studies.

It was generally observed that the major part of the elimination proceeds already in the infiltration zone or in the first few metres of the infiltration pathway. This part of the purification effect is practically independent from the distance of the production wells from the river. The elimination extent in the subsequent underground passage is typically lower, but also this part of the aquifer causes still an appreciable purification effect for a number of compounds (e.g. naphthalene-1,3,6-trisulfonate). It became however obvious that the underlying redox conditions have typically a more pronounced effect on the removal effect than residence times of the water in the underground.

Literature survey

A comprehensive illustration of the results from the literature survey concerning the elimination behaviour of selected pesticides and organohalogen compounds during riverbank filtration and artificial groundwater recharge is given in figure 4. It becomes apparent that redox conditions also have a major impact on the elimination pattern of several single compounds from these substance classes. Particularly with regard to chlorinated aliphatics, triazines (atrazine, simazine), hexachlorobenzene, and lindane anoxic and anaerobic aquifers provided higher removal rates than riverbank filtration under aerobic conditions. Isoproturon, dichlorprop, mecoprop, and glyphosate, however, were better removable in aerobic than in anoxic and anaerobic aquifers.

Conclusions

Based upon the results of the present investigation and the literature survey, riverbank filtration is certainly an efficient process for removal of many organic micropollutants frequently found in surface waters. The systematic investigations at aquifers with different geological and hydrochemical boundary conditions revealed that the extent of elimination of some organic micropollutants is decisively dependent on the underlying redox conditions. This influence is non-uniform. Some pollutants are more degradable under anaerobic conditions than under aerobic conditions and *vice versa*. Overall, redox conditions turned out to have a more pronounced effect on the removal effect than residence times of the infiltrated water in the aquifer, implying that the redox environment of an aquifer has to be regarded as the chemical "master-variable" with regard to organic micropollutant removal during riverbank filtration.

Acknowledgements

This study was financially supported by the German Ministry of Education and Research (BMBF) and is part of the joint research project "Adjusted water treatment technologies and water distribution under regional conditions".

References

Kühn, W., Müller, U. et al. (2006) Exportorientierte F&E auf dem Gebiet der Wasserver- und-entsorgung. Teil I: Trinkwasser, Band 2, Leitfaden. DVGW-Technologiezentrum Wasser (TZW) Karlsruhe, Eigenverlag.

Preuß, G. and Schulte-Ebbert, U. (2000) Artificial Groundwater Recharge and Bank Filtration. In *Rehm, H.-J. and Reed, G. (Eds.) Biotechnology Vol. 11c, 2nd Ed.*, Weinheim, Wiley-VCH, 425–444.

Schmidt, C.K., Lange, F.T., Brauch, H.-J. and Kühn, W. (2003) Experiences with riverbank filtration in Germany. In *Proc. of the International Symposium on Artificial Recharge of Groundwater*, Daejon, Korea, 14 November, 115–141.

Schmidt, Kh. (1988) Pufferwirkung der Untergrundpassage – Leistungsfähgikeit und Grenzen der Uferfiltration und künstlichen Grundwasseranreicherung bei der Eliminierung von Schadstoffen. *Gewässerschutz Wasser Abwasser* **131**, 181–194.

Schmidt, C.K., Lange, F.T. and Brauch, H.-J. (2005) Verhalten von pharmazeutischen Wirkstoffen und Röntgenkontrastmitteln bei der Uferfiltration unter standortspezifischen Randbedingungen. In *Proc. of the 71. Jahrestagung der Wasserchemischen Gesellschaft*, Bad Mergentheim, 2–4 May, 118–122.

Schmidt, C.K. (2006) Datenbank zum Verhalten organischer Spurenstoffe bei der Uferfiltration. In *DVGW-Technologiezentrum Wasser (TZW) Karlsruhe (Ed.) Exportorientierte F&E auf dem Gebiet der Wasserver- und –entsorgung. Teil I: Trinkwasser, Band 2, Leitfaden. Eigenverlag.*

72 Deep bed regeneration of infiltration basins

Mathias Schöpel*, Horst Losen**

*RWW Rheinisch-Westfälische Wasserwerksgesellschaft mbH, Am Schloß Broich 1 – 3, D-45479 Mülheim an der Ruhr

**Hydrogeologisches Büro Prof. Dr. H. Losen, Klettenberggürtel 43, D-50939 Köln

Abstract Deep bed regeneration was carried out in slow sand filtration basins in order to remove the colmation of the underlying quaternary gravel. Due to decades of infiltration of surface water incrusted aquifer layers up to 2 m thickness had been located. The colmated gravel was removed, washed, sieved and step by step refilled in graduated stratums. The slurry also was treated in order to minimise the disposal costs. Some aspects are specified as to the evaluation of the degree of the colmation with simple field methods before the regeneration and finally to the examination of the efficiency of the regeneration by hydrogeological field measurements. An increase of the infiltration capacity up to a minimum factor of 3.5 could be shown.

Keywords deep bed regeneration, slow sand infiltration, infiltration capacity, colmation, regenerated gravels, slurry treatment.

Introduction

The water treatment plant Styrum-East, located near the conference centre in Muelheim (Germany), is one of the biggest waterworks of Rheinisch-Westfälische Wasserwerksgesellschaft mbH (RWW). The Muelheim Treatment Process, presented in detail by keynote speaker G. Bundermann, is in operation in this waterworks since 1981.

In the first step of water purification, raw water from the Ruhr river dams is pumped to infiltration basins (slow sand filters) for artificial ground-water recharge and subsequent passage through the natural quarternary aquifer systems of the Ruhr valley. After a travel time of several days or weeks the soil filtrated water is recovered and pumped to the water treatment plant for further processing (ozonation, double-layer filtration, BAC-filtration, deacidification and final disinfection by UV-light irradiation).

Most of the infiltration basins are have been in operation for more than seven decades and a gradual loss of infiltration capacity has being observed during the last years. The causes are colmation (caking up) in the upper layer of the quarternary gravel due to the deposition of filter sand and matter suspended in the river water, compression by the cleaning machines and the precipitation of iron and manganese hydroxides.

Therefore RWW decided to carry out deep bed regeneration (Figure 1) in cooperation with USG Umweltservice GmbH & Co (Velbert, Germany). The upper two meters of the

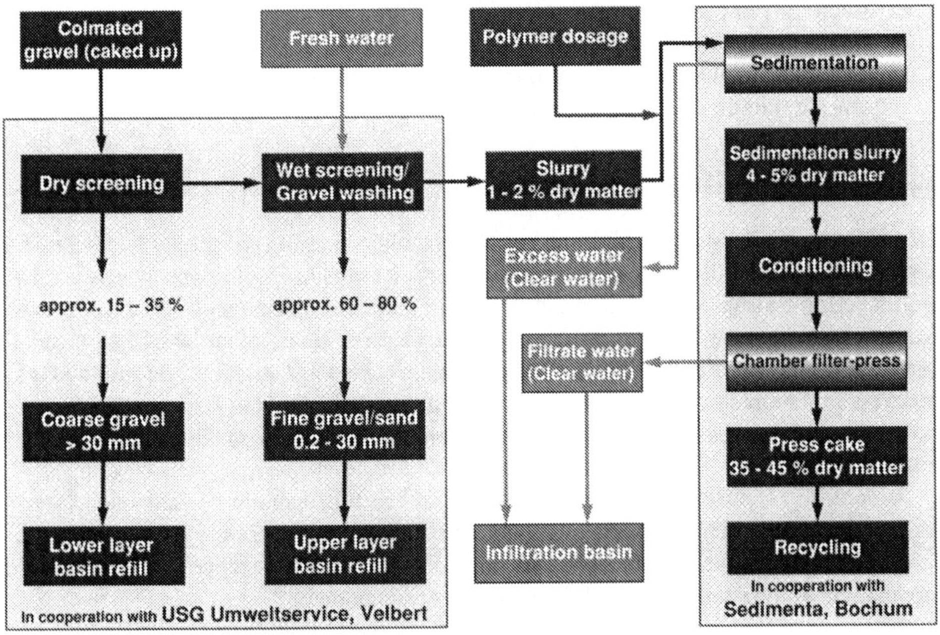

Figure 1 Flow chart of the deep bed regeneration of infiltration basins

colmated gravel layers were removed and cleaned by special washing processes. After this, the cleaned gravel was filled back.

Regeneration procedures

The first step of the cleaning process was carried out in the infiltration basins. The colmated gravel was removed and washed in a mobile dry screening plant. The coarse gravel (above 30 mm) was refilled directly as lower layer of the basins. The fine gravel (0.2–30 mm), consisting of a 65–85 percentage of the total volume, was transported to a mobile wet screening plant outside the infiltration basins. In the second step the fine gravel was washed, then sieved and reused as top layer of the basins. The undersize material (less than 0.2 mm), a 5–6 percentage of the total volume, was flushed out.

The slurry (1–2 % dry matter) of the wet screening plant was treated in a small waste water plant by flocculation and sedimentation processes. The treated slurry (4–5 % dry matter) was drained in a mobile chamber filter press. The filtrate water and the excess water of the sedimentation unit were used for ground water recharge. The slurry treatment plant was operated in cooperation with Sedimenta GmbH (Bochum, Germany).

For managing deep bed regeneration process in slow sand filtration basins the following aspects should be considered:

(1) Evaluation of the degree of colmation in the upper aquifer layers via survey trenches in the sand filtration beds and the upper gravel-aquifer layers. Monitoring of the ground-water flow rate in the surrounding areas of the basins during the infiltration process.

(2) Removal and cleaning of the colmated layers and subsequent refilling and consolidation of the cleaned material.
(3) Examination of the results by infiltration tests within the slow sand filtration basins, and recording of changes in the ground-water flow in the surrounding areas.

Hydrological (re-)examinations

The efficiency of the slow sand filtration basins is mainly depending on the vertical hydraulic conductivity of the filter beds of the aquifer. Survey trenches in the upper levels of the aquifer permit a visual evaluation of the extent of colmation within the Ruhr terraces. The colmated sections show black iron and manganese incrustations. At the site of the Styrum-East waterworks, the incrusted layers in the aquifer are 0.2 m to 2.0 m thick. The variations in thickness are caused by differences in aquifer hydraulic conductivities, variation concerning the depth to the water table, water infiltration configuration and the hydrochemistry of the infiltrated water.

In order to evaluate the efficiency of the deep bed regeneration of slow sand filtration basins, additional hydro-geological field investigations are necessary to determine the hydraulic conductivities before and after deep bed regeneration process. The hydraulic conductivity of the aquifer can be determined by recharge tests.

At the site of the Styrum-East waterworks a mini-excavator was used to dig an open pit into the colmated aquifer gravels. The infiltration test-cylinders were steel pipe-segments with a length of 1.91 m and an interior radius of 0.49 m. Due to the wide diameter of the pipes the main source of error, breaking up of the grain structure at the rim while ramming the tube into the gravel, is minimized leaving the main part of the infiltration layer undisturbed. The infiltration test cylinders were driven into the upper stratum of the Ruhr terraces by a front loader. The depths of penetration of the steel pipes varied from a few centimetres to a maxi-mum of 0.45 m depending on soil resistance. The area between the open pit and the steel cylinders was filled up with sand and gravel considering constant soil consolidation. After filling up the infiltration test pipe with ground-water the drawdown of the water level within the test-cylinder was measured continuously (Figure 2).

The inlet of the water from the filled up tube into the porous aquifer can only occur through the bottom of the cylinder. To carry out the infiltration tests one boundary condition is to have a saturated ground-water flow within the infiltration zone in the former unsaturated aquifer. In order to get a constantly saturated vertical ground-water flow rate several pre-infiltrations have to be carried out before the test can be started.

The hydraulic conductivity of the aquifer (K) is given by:

$$K = [r/(4^{*}dt)]^{*} 2.303^{*} \log[h1/h2] \text{ m/s}$$ (Kollbrunner 1946 & Maag 1941)

where

K = Hydraulic conductivity in [m/s]

r = Interior radius of the infiltration test cylinder [m]

dt = Time-interval t2 - t1 between two water levels h1 and h2 in the cylinder [s]

$h1$ = Water level in test cylinder at time t1 [m]

$h2$ = Water level in test cylinder at the end of the test, at time t2 [m]

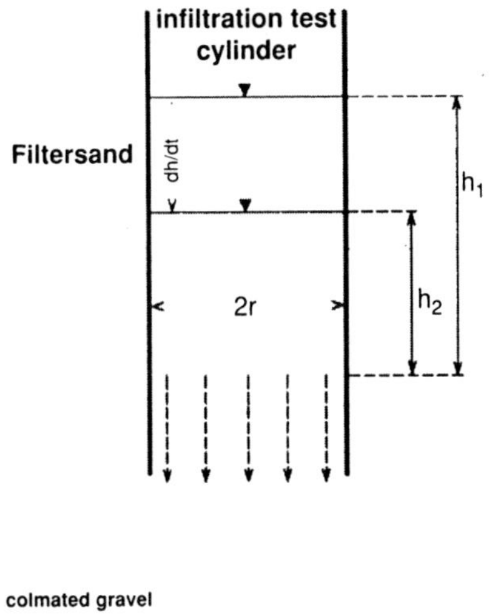

Figure 2 Pattern of the experimental design of the infiltration tests

After testing the colmated aquifer zones additional infiltration tests were carried out in order to quantify the hydraulic conductivity of the cleaned aquifer layers and sand beddings of the slow sand filtration basins.

The results of infiltration tests in the slow sand filtration basins at the site of the Styrum-East waterworks are:

- Variable thickness of the colmated aquifer parts and different degree of colmation cause a wide spread of vertical hydraulic conductivities of non-regenerated aquifer layers. In the upper part of the colmated aquifer, below the slow sand filtration basins, the vertical hydraulic conductivity is varying between 5.67 * 10–6 m/s and 1.06 * 10–4 m/s. The average vertical hydraulic conductivity is about 3.0 * 10–5 m/s.
- Vertical hydraulic conductivity of the filter sands is varying between 3.47 * 10–5 m/s and 4.57 * 10–5 m/s. Vertical hydraulic conductivities of the filter sands are higher than the average vertical hydraulic conductivities in the colmated aquifer parts at the base of the filter sand. Consequently the infiltration capacity of the slow sand filtration basins at the Styrum-East waterworks is depending primarily on the vertical hydraulic conductivity of the colmated aquifer and not on the hydraulic conductivity of the filter sands.
- The regenerated gravels were refilled into the upper part of the aquifer while taking into consideration a constant soil consolidation. By carrying out additional infiltration tests in this part of the aquifer it could be shown that the average vertical hydraulic conductivity could be increased 2- to 40-times. The average hydraulic conductivity of the regenerated beds is 8-times higher than that of the colmated material.

After the completed refilling of the basins the results of initial infiltration tests showed an increase of infiltration capacity by up to a factor of 3.5. By means of a higher discharge at the wells the infiltration capacity could even be raised. The vertical hydraulic conductivity of the regenerated part of the Ruhr terrace aquifer is about 7-times higher than that of the overlying filter sand.

Results

Deep bed regeneration process of colmated upper aquifer parts of long term used slow sand filtration basins is an excellent method to restore the infiltration capacity of these basins. By means of infiltration tests the vertical hydraulic conductivities can be evaluated. Continuous monitoring of the development of the ground water flow rate allows controlling the infiltration capacity. It is recommended that additional trenches be dug to check for incrustations.

References

Kollbrunner, C. F. (1946) Fundation und Konsolidation. Bd. 1, Zürich, Swiss.

Maag, E. (1941) Methode zur feldmäßigen Bestimmung der Wasserdurchlässigkeit - Straße und Verkehr 19, 335–338, Solothurn, Zürich, Swiss.

Author index

Keyword index

Lightning Source UK Ltd.
Milton Keynes UK
UKOW02n2345250813

215838UK00011B/112/A

9 781843 391203